The Dance Music Manual

This book is dedicated to my wife Lindsay and my daughter Neve Trinity

The Dance Music Manual Web Site

Want to know about the history of dance music?

Want to download free audio samples?

Want to download synthesizer patches for a variety of synthesizers?

Want to know the dates of the dance music seminars?

Want to know which sample CDs to buy?

Need more advice on writing dance music?

Need advice on instruments or sequencers?

Have any comments or suggestions for this edition?

Have any suggestions for the second edition?

Point your browser at the web site dedicated to this book:

www.dancemusicproduction.com

The Dance Music Manual
Tools, toys and techniques

Rick Snoman

ELSEVIER

AMSTERDAM • BOSTON • HEIDELBERG • LONDON • NEW YORK • OXFORD
PARIS • SAN DIEGO • SAN FRANCISCO • SINGAPORE • SYDNEY • TOKYO
Focal Press is an imprint of Elsevier

Focal Press

Focal Press
An imprint of Elsevier
Linacre House, Jordan Hill, Oxford OX2 8DP
200 Wheeler Road, Burlington MA 01803

First published 2004

British Library Cataloguing in Publication Data
A catalogue record for this book is available from the British Library

Library of Congress Cataloguing in Publication Data
A catalogue record for this book is available from the Library of Congress

ISBN 0 240 51915 9

For information on all Focal Press publications visit our website at:
www.focalpress.com

Typeset by Newgen Imaging Systems (P) Ltd., Chennai, India
Printed and bound in Great Britain

Contents

Preface   xi

Acknowledgements   xiii

Part 1   Technology and Theory   **1**

    1   Computers, music, MIDI and audio   **3**
       MIDI controllers   6
       Universal System Exclusive Messages   8
       NRPN and RPN controllers   11
       Computer-based sequencers   12
       Resolution   15
       MIDI-based dance studios   16
       Audio sequencing   22

    2   Dance music theory   **28**
       Basic music theory   28
       The major scale   28
       Minor scales   31
       Chords   33
       Musical relationship   38
       Tempo and time signatures   40
       Creating melodies/motifs   45
       Creating progression   47

    3   Basic synthesis   **52**
       Acoustic science   52
       Subtractive synthesis   58
       Other synthesis methods   73

    4   Sampling and digital audio   **76**
       Sample CDs   78
       Practical sampling   80
       Working with loops   81
       Recycling loops   83
       Recording samples   84
       Bit rates   86
       Copyright law   89

    5   Compression, processing and effects   **92**
       Compression theory   92
       Practical compression   97

Creative compression   99
Limiters   105
Noise gates   107
Transient designers   109
Reverb   109
Chorus   112
Phasers and flangers   112
Digital delay   113
EQ   114
Distortion   114

6   Cables, mixing desks and effects busses   **115**
Mixing desk structure   120
Routing effects and processors   123
Automation   130

7   Programming theory   **132**
Timbre effects   133
Theory of programming   135
Customizing sounds   135
Programming basics   138
Programming timbres   140
Programming pads   141
Programming analogue drums   144
Programming bass   153
Programming leads   160
Programming sound effects   168
Programming new timbres   171
General programming tips   174
Creative sampling and effects   175

8   Recording vocals   **179**
Microphone technology   179
Recording preparation   182
Recording techniques   184
Editing vocals   189
Vocal effects   190
Commercial vocals   190
'Telephone' vocals   191
Pitched vocals   192
Vocoders   192

9   Tools of the trade   **195**
Novation BassStation (hardware and software)   195
Novation V-Station (software and hardware (K-Station))   196
Novation SuperNova II (hardware only)   196
Access Virus C (hardware and software)   197
E-Mu Orbit 3 (hardware only)   198
E-Mu Planet Phatt (hardware only)   198
Roland XV5080 (hardware only)   199
Korg Triton (hardware only)   199

AKAI MPC 2000XL (hardware only)   200
Native Instruments FM7 (software only)   201
Native Instruments Pro-53 (software only)   201
Native Instruments B4 (software only)   202
Propellerhead ReBirth RB-338 (software only)   203
Waldorf Attack (hardware and software)   203
Native Instruments Battery (software only)   204
Native Instruments Kontakt (software only)   205
Steinberg HALion (software only)   205
Native Instruments Reaktor (software only)   207
Native Instruments Absynth v2 (software only)   207
Mutronics Mutator (hardware only)   208
Sherman Filterbank 2 (hardware only)   209
Novation ReMote 25 (hardware only)   210
Spectrasonics Atmosphere, Stylus and Trilogy (software only)   210
Waves bundle – effects (software only)   211
TC Native bundle – effects (software only)   211
PSP Audioware effects (software only)   213
GRM Tools Vols 1 and 2 effects (software only)   214

Part 2   Dance Genres   217

10   Trance   219
Musical analysis   220
The rhythm section   220
Rhythm timbres   222
Bass rhythm   224
Bass timbre   225
Trance melodies   228
Melodic timbres   230
Motifs and chord progressions   231
The arrangement   235
Recommended listening   236

11   Hip-hop (rap)   237
Musical analysis   239
The rhythms   240
Rap vocals   245
The bass rhythm   247
Chords and motifs   250
Recommended listening   252

12   Ambient/chill-out   253
Musical analysis   254
The rhythms   255
Rhythm timbres   257
Melodic leads   259
Bass   262
Chords/pads   264
Arrangement   266
Recommended listening   267

13	House	268
	Musical analysis	271
	House rhythms	272
	Bass	277
	Melodies and motifs	281
	Recommended listening	283
14	Techno	285
	Musical analysis	286
	Recommended listening	293
15	Trip-hop	294
	Musical analysis	295
	Recommended listening	306
Part 3	**Mixing and Promotion**	**307**
16	Mixing	309
	Mixing theory	310
	Practical mixing	323
	Volume adjustments	325
	Panning	325
	EQ	326
	Processing and effects	334
	Common mixing problems	335
17	Mastering	340
	Mastering theory	341
	Noise reduction	341
	Equalization	343
	Mastering reverb	347
	Dynamics	348
	Harmonic exciters	354
	Stereo widening	356
	Loudness maximizers	356
	General mastering tips	358
18	Publishing and promotion	361
	Record companies and contracts	362
	Independent release	370
	Distributing via the Internet	377
	Music charts and play lists (UK only)	380
19	Remixing and sample clearance	382
	Session tapes	383
	Bootlegs and Internet remixing	384
	Arranging	392
	Contracts	393
	Clearing samples	395
20	Designing an audio web site	398
	How the web works – the basics	398
	HTML and web browsers	398

Web design software packages 400
Design considerations and techniques 400
Using the tutorial files 406
Tutorial 1: Creating a basic web site 406
Tutorial 2: HTML used for the Dance Music Production web site 415
Adding audio 447
Viewing the finished Dance Music Production web site 450
Setting your site up on the web 451

21 A DJ's perspective 454
Promos and reaction sheets 454
What 'works' on the dance floor? 456
DJ techniques 457
Recommended equipment for those wanting to be a DJ 457
Tips for aspiring DJs 463
How do DJs promote themselves? 464
More advice 465
Useful links 467

Appendix A: Binary and hex 469
Counting in binary 469
Counting in hexadecimal 470
Appendix B: Decimal to hexadecimal conversion table 473
Appendix C: General MIDI instrument patch maps 474
Appendix D: General MIDI CC list 477
Appendix E: Sequencer note divisions 478
Appendix F: Tempo delay time chart 480
Appendix G: Musical note to MIDI and frequencies 481

Glossary 483

References 497

CD content copyright information 499

Index 501

Preface

If a book is worth reading then it's worth buying ...

The purpose of this book is to guide you through the technology and techniques behind creating professional dance music. While there have been numerous publications written about this specialist subject, the majority have been written by authors who have little or no experience of the scene nor the music, but simply rely on 'educated guesswork'. With this book, I hope to change the many misconceptions that abound and offer a real-world insight into the techniques on how professional dance music is written, produced and marketed.

I've been actively involved in the dance music scene since the late 1980s and, to date, I've produced and released numerous white labels and remixed over 30 professional artists solely for the dance floor. I've held seminars across the country on remixing and producing club-based dance music, and authored numerous articles and reviews for leading music technology magazines. This book is a culmination of the knowledge I've attained over the years and I believe it is the first publication of its kind to actively discuss the real-world applications behind producing and remixing dance music for the twenty-first century.

The Dance Music Manual has been organized so as to appeal to professionals and novices alike, and to make it easier to digest it has been subdivided into three parts. The first part discusses the latest technology used in dance music production, from the basics of MIDI, synthesis and sampling to music theory, effects, compression, microphone techniques and the principles behind the all-important sound design. If you're new to the technology and theory behind much of today's dance music, then this is the place to start.

The second part covers the techniques for producing musical styles including, amongst others, trance, trip-hop, rap and house. This not only discusses the general programming principles behind drum loops, basses and leads for the genres, but also the programming and effects used to create the sounds. If you already have a good understanding of sampling rates, bits, MIDI, synthesis programming and music theory, then you can dip into these sections and start producing dance music straight away.

The third part is concerned with the ideology behind mixing, mastering, remixing, pressing and publication of your latest masterpiece. This includes the theory and practical applications behind mixing and mastering, along with a realistic look at how record companies work and behave; how to copyright your material, press your own records and the costs involved. At the end of the book you'll also find two contributed chapters: one in which an international DJ has submitted his view on dance music and DJ-ing in general, and another on how to design and create your own web site to promote your music.

Demonstration tracks of all the genres of music discussed can be found on the CD, including before and after mixing and mastering, and to further help in the production I've also carefully selected some of the most widely recognized software used in the production of dance music. These, amongst others, include the PSP audioware, TC Native and Waves collection of useful processors and effects, the now infamous Novation V-Station and Bass Station synthesizers, Big Ticks Rainbow, Cheeze Machine and Rhino Synth, Speedsoft's V-Sampler and Native Instruments widely celebrated Pro 53, Kontakt, FM-7 and Battery.

Of particular note, while producing the CD I decided against including demos of sequencers since not only are these a very personal choice but many potential readers of this book will probably already own a computer and sequencer. If you have no experience of sequencers or are struggling on which to choose, the web site devoted to this book has an area that discusses the different sequencers available. These have been written by dance musicians who actively use their sequencer on a day-to-day basis and offer arguments on why they use their particular choice of sequencer. Furthermore, if you are struggling to understand the basics of computers, need some new preset banks or audio samples to download, need help in choosing a sample CD for your particular choice of music or have some suggestions for additions to the book then point your internet browser at: http://www.dancemusicproduction.com (please note that to access some of the content you will need to have this book to hand).

Of course, even with the included software and the associated web site, I must stress that this book will not turn you into a superstar overnight and it would be presumptuous to suggest that it will guarantee you the next dance chart number one. Dance music will always evolve from musicians pushing the technology further. Nonetheless, it will give you an insight into how the music is produced and from there, it's up to you to push in a new direction. Creativity can never be encapsulated in words, pictures or software and it's our individual creative instincts and twists on a theme that produce the dance floor hits of tomorrow. Experimentation always pays high dividends.

Finally, I'd also like to take this opportunity to thank you for buying *The Dance Music Manual*. In these times of global Internet piracy, I have little doubt that it will appear on some illegal sites as a poorly scanned, almost unreadable PDF document for those too afraid to part with some money. By purchasing this book, you are rewarding me for all the time and effort I've put into producing it and that deserves some gratitude. I hope that, by the end, you feel it was worth your investment.

Acknowledgements

I would like to personally thank the following for their invaluable help, contributions and/or encouragement in writing this book:

Andrew Le Couteur Bisson
Andy Jones at Future Music
Andy Parker
Antoni Ozynski at PSP Audioware
Beth Howard and the Focal Press Team
Big Tick Software
Darren Gash at the SAE Institute, London
DJ Cristo
Emma Lawrence
Ian Jannaway at Novation Music
Ian Shaw
James Froggatt
John Froggatt
John Mitchell
Mark at Izotope Inc. Software
Matthias 'Maz' Ziegs at Speedsoft Software
Mike Waldren
Mike at the Whipping post
Neil Crane at Square Circle Software
Patrick at Native Instruments Software
RC Waves Software
Richard Hoffmann at Synapse Software
Richard James
Ronan MacDonald at Computer Music
Steve Marcus
The hardworking guys at Sony, EMI and Warner
The TC Native Support Team

And all those DJs who played my records

…You know who you are…

Part 1
Technology and Theory

1 Computers, music, MIDI and audio

Computer games don't affect kids; I mean if Pac-Man affected us as kids, we'd all be running around in darkened rooms, munching magic pills and listening to repetitive electronic music.

Kristian Wilson, Nintendo, 1989

The most obvious place to start on any book concerning dance music is with the basics of sequencing and sequencers. Their earlier incarnations were responsible for the typically metronomic sounding beats that are still evident in today's music, and no dance musician would seriously consider writing a track without the aid of one. Although sequencers have moved on from offering just simple MIDI (*Musical Instrument Digital Interface*) functionality, an understanding of it is imperative to producing a good dance record.

Note: As with most subjects, to understand one aspect requires some understanding of another and this is especially true of music that is generated using computers and sequencers. Consequently, when first starting out, the learning curve may appear quite steep. If something doesn't make sense at this stage, or perhaps seems too obscure to be related to dance music production, stick with it; its use and reasoning will be mentioned again in later chapters.

The most basic requirement for using any sequencer is a thorough understanding of the MIDI protocol. MIDI is a widespread communications protocol that enables any MIDI-equipped instruments to be interconnected and controlled from a sequencer. This means that both the sequencer and any instruments you wish to control must have MIDI IN and OUT ports fitted to them for the two devices to perform two-way communication.

In a simple studio configuration, a MIDI-equipped keyboard is connected to the sequencer, which is usually a computer, via a MIDI cable. With the sequencer then set to record, any notes that are played on the keyboard are recorded as MIDI data to a MIDI track on the sequencer. Once a number of tracks have been recorded, they can be played back to any MIDI-equipped synthesizer to recreate the performance.

In addition to the IN and OUT MIDI ports, some of the more substantial synthesizers also feature a THRU port. Using this, information received at the IN port can be transmitted back out (i.e. repeated) at the device's THRU port. Thus, several MIDI sound modules can be connected

Figure 1.1 A simple MIDI set-up.

Figure 1.2 A more elaborate MIDI set-up.

via the THRU output of one device to the IN connector of the next device downstream in what's called a 'daisy-chain'. Connecting devices in this way creates more elaborate MIDI systems by allowing the sequencer to communicate with a number of synthesizers and samplers at once.

A common misconception is that the information the sequencer records is the sound itself when in fact this isn't the case. Instead, it only records digital information to tell it, for example, which key was pressed and for how long. By way of analogy, MIDI information could be considered as a musical score, where the attached synthesizers are the performers and the MIDI information tells the synthesizers what notes to play. Because most synthesizers today are 16-part multi-timbral (meaning that they can play 16 individual instruments simultaneously), it's possible to play back an entire performance on a single synthesizer. Sixteen-part multi-timbral playback is accomplished by sending 16 individual channels of note information through the MIDI cable to the attached synthesizer, but also means that the synthesizer's channels must be set to play the correct instrument on each channel.

For example, suppose channel 1 of the synth is set to a piano sound and channel 2 to a bass sound. Any MIDI information transmitted to channel 1 of the synth will sound as a piano, even if the MIDI file is programmed with a bass sound in mind. Consequently, a common practice is to insert 'program changes' at the beginning of each channel's sequence so the attached synthesizer knows which instrument (often called a 'patch') should be used on any particular MIDI channel. These patch change messages must be situated at the 'header' (the beginning) of a

MIDI arrangement and occur before any notes are present, to give the attached synth time to load the right sound before the notes begin sounding the synthesizer.

> Note: Patch change messages usually consist of information denoting the bank where the sound is located and the patch number. Consequently, there are no generic settings for sending patch change information, so if you want to send these, refer to your synthesizer manual.

For this to work accurately there must be a standard defining the relationship between patch numbers and sounds within all synthesizers, otherwise a MIDI sequence will select different sounds when played on different synthesizers.

To prevent this, the General MIDI (GM) specification was developed. This is a generalized list of requirements that any synth must adhere to for it to feature the GM symbol. Amongst many things, it determines what sounds are assigned to which patch numbers, how percussion instruments should be mapped and the type of MIDI messages that should be recognized. With this capability, you're assured that any MIDI sequence that has been generated for use on a General MIDI instrument will play correctly on any General MIDI synthesizer.

To make the GM format easier to adhere to, all the instruments in a GM synth are grouped into collections of related sounds: program numbers 1–8 are piano sounds, program numbers 9–16 are chromatic percussive sounds, 17–24 are organ sounds, 25–32 are guitar sounds, and so forth. These are not limited to specific sequencer channels, so channels 1–9 and 11–16 can be used to play any of these sounds. The program change number contained in each respective channel of the sequencer will ensure that the correct patch number is selected for any particular channel.

> Note: The full GM list can be located in Appendix C.

As in the majority of cases drum sounds are not chromatic (i.e. they only have one pitch), channel 10 is set aside for drum kits. The entire drum kit is contained within this one channel, so rather than each different key on the MIDI keyboard corresponding to higher or lower pitch of the same sound, they will relate to a different drum sound. Again, the GM specification plays a part here, detailing which drum sounds should occur on which key to ensure that a MIDI arrangement will play back the percussion track correctly.

More importantly, though, while these MIDI standards specify what instrument or sound corresponds with each patch number, it doesn't specify how these sounds should be produced. So despite the fact that program number 1 will select an acoustic grand piano on any GM-compatible synth, the quality of the sound will differ according to the technique used by the synth to create the sound. This isn't necessarily a problem, as very few dance musicians will actually trouble themselves with the General MIDI sound set, which is often considered as uninspiring. Rather, the standard GM sound set is used to get the general idea of the track down and then exchanged for more suitable sounds later in the arrangement. Indeed, it's the variety of other sounds contained within a synth, along with the MIDI editing features, that makes MIDI so appealing.

MIDI controllers

So far we've only considered the use of MIDI to send program change messages to select an initial sound and to send simple note-on and note-off data that starts and stops the synth playing, but if this was all MIDI was capable of it would be impossible to create realistic performances. Every 'real' musician will perform with expression by consciously or subconsciously using a series of different playing styles. For instance, a pianist may hit some notes harder than others, a guitarist may slide their fingers up their guitar's fret board bending the sound, or may sustain some notes while cutting others short. All of these playing nuances contribute to the human element behind the music and without it music would sound dull and lifeless.

Of course, some may argue that there is no human element behind any electronic dance music, as the whole principle is that the metronomic beats and melodies sound as though a machine has created them. In most cases, this isn't true and under close scrutiny you'll find that even the most metronomic techno is crammed full of tiny sonic nuances to prevent it from becoming tedious. This is the key reason why some songs appear to ooze energy and soul while others seem motionless and drab, and is also why it is imperative that you fully understand how MIDI can be used to its full potential. By knowing the ins and outs (pun intended) of MIDI it can be exploited to give the impression that a track has energy, drive and an indefinable character.

Much of the energy and flow of any music can be attributed to slight or aggressive development of the sounds throughout the length of an arrangement. While the musician playing the piece naturally injects this into a performance, capturing this feel over MIDI requires you to use a series of MIDI controllers. For instance, when you play a note and move the pitch-bend wheel, the wheel sends a continual stream of control messages to the synth's engine, which in turn bends the pitch either upwards or downwards. If this played an important part in the performance, then you would need to perform it in real-time every time you played the music.

Fortunately, this effect can be simulated using a series of MIDI control change (CC) messages transmitted from the sequencer into the synthesizer, thus avoiding the need to continually employ any real-time movements. These pre-programmed forms of expression not only free your hands, but they can also be used to modify certain parameters of a sound if the synth doesn't have the appropriate hardware controller on its fascia.

Because some synth interfaces appear as rack-mountable modules, to reduce the space they use there may be no real-time controllers on them at all, even though the synth could respond to them. If this is the case, the only way to add any musical expression is to use a series of MIDI control messages.

According to the General MIDI protocol, there are 128 possible controllers on any MIDI device, numbered from 0 to 127. Some of these controllers are hard-wired to control particular functions on a MIDI unit, while other controller numbers are free to be randomly interpreted by the connected device. Some of the most commonly used of these are listed in Table 1.1.

Note: A full list of the available CC controllers can be found in Appendix D.

Each of the defined GM controllers has variable parameters associated with it, the nature of which depends on the controller being used. Indeed, every CC controller message must be

Table 1.1 Some of the most commonly used CC messages

CC number	Function	Action
1	Modulation	This message is used for pitch modulation, allowing you to add vibrato (vary the pitch) to a musical note
7	Channel volume	This controller can be used to control the volume of any particular MIDI channel
10	Pan	Using this message, you can pan sounds across the stereo image from fully left to fully right or anywhere in between
11	Expression	Expression is a dynamic controller that is very similar to volume. A typical example of its use would be to dynamically alter the volume of a pad sound while it is playing
74	Brightness	This controller is used to dynamically modify the brightness of notes. A typical example of this is where a sound gradually becomes brighter over its length, producing the archetypal dance music filter sweep

followed by a second variable that informs the receiving device how much that particular CC controller is to be adjusted. Some CC controllers offer up to 128 variables (0–127), while others have only two variables (0 or 1) that are essentially on/off switches.

As a practical example, if you wanted to increase the volume of a MIDI snare roll to contribute to the building sensation, you would need to transmit a series of CC 7 messages to the receiving device and set that controller's variable in each message to gradually increase in value, as shown in Table 1.2.

Table 1.2 Sending CC 7 messages to increase volume

Song position	CC number	Variable value
1 min 32 seconds	7 (Volume)	67
1 min 33 seconds	7 (Volume)	68
1 min 34 seconds	7 (Volume)	69
1 min 35 seconds	7 (Volume)	70
1 min 36 seconds	7 (Volume)	71
1 min 37 seconds	7 (Volume)	72
1 min 38 seconds	7 (Volume)	73
1 min 39 seconds	7 (Volume)	74
1 min 40 seconds	7 (Volume)	75

By programming these commands into the sequencer, each time the CC message is received the volume will increase by one unit.

Note: Some CC messages will have both positive and negative values. A typical example of this is the pan controller, which takes the value 64 to indicate central position, so when panning a sound, any values lower than this will send the sound to the left of the spectrum while values greater than this will send the sound to the right of the stereo spectrum.

Although most synths will respond to the fundamental controllers required in music, such as pitch bend, modulation, expression, pan and volume, it's important to note that the age and quality of the synth will determine whether some or all of the messages listed in the MIDI specification are supported. Additionally, many synths have now outgrown the original specification of MIDI and offer many more user-accessible parameters. Because of this, two major synthesizer manufacturers have developed their own MIDI standards, which act as an extension to the GM standard.

The first of these to appear was Roland's own GS system. Contrary to popular belief, GS does not stand for General Standard but is simply the name of the microprocessor used to power the engine (and only Roland seems to know what it stands for). Nevertheless, this protocol is compatible with the GM standard whilst also adding variations to the original sound set, access to two drum kits simultaneously and a collection of new CC messages.

Yamaha soon followed this technique by introducing the XG system, which is also fully compatible with GM but offers more advanced functions than both the GM and GS formats. Yamaha's XG system allows you to use three drum kits simultaneously, along with three simultaneous effects, and have a minimum of 32-note polyphony (as it can play 32 voices concurrently). What's more, the XG system is also partly compatible with the Roland GS system. This is only partly compatible, since Roland constantly updates the GS system with the release of new instruments.

As if these three formats were not enough, the developers of the General MIDI standard have recently released the General MIDI 2 standard. This is yet again fully compatible with the original GM standard, but also adds a series of extensions to the original, including MIDI Tuning, more controllers, Registered Parameter Numbers (RPN) and Universal System Exclusive Messages (USEM). These latter two forms of messages play a fundamental role when using any MIDI equipment, since even with the advanced GM2, XG and GS structures there will be instances where you need to access parameters that are not covered by these standards.

Most dance music relies heavily on cleverly programmed sounds as well as samples. By accessing and controlling these 'hidden' elements of a synth from a MIDI arrangement using RPN and USEM messages, it is possible to switch the functions of a synth part way through a performance or develop sounds that are not accessible through the CC list over the length of the arrangement.

Both RPN and USEM are renowned as notoriously difficult to use, but ignoring them means that you're not exploiting the full potential of the synth, so the next two sections cover the basics of RPN and USEM synth control.

Universal System Exclusive Messages

System Exclusive (Sysex) was first conceived to provide a way of controlling synth settings without having to waste time pushing a mountain of buttons on the fascia. By sending system exclusive messages it's possible to control any aspect of a synth, including controls listed on the MIDI specification that are not recognized by the synthesizer.

For example, say that part way through an arrangement you want to change the waveform of a low-frequency oscillator (LFO). This is obviously not covered in the MIDI CC list but can be enabled using Sysex, usually listed in the final pages of a synthesizer manual.

Note: On some synthesizers this particular parameter may be accessible through CC messages, and if this is the case it is always worth using CC rather than USEM. This is because CC messages generate less data than USEM.

To program a certain type of behaviour using Sysex messages, you need to know what Sysex message to send to the synth. Sysex messages typically consist of a hexadecimal Address, which pinpoints the function within the synth that you want to adjust, and a hexadecimal Sysex value, telling the function how you want to adjust it. Thus, by sending a Sysex message to the synth reading **00 00 11 04** (derived in this case from the values shown in Table 1.3), the LFO rate would be set to waveform number 4.

Table 1.3 Sysex message structure

Address	Parameter	Sysex value	Adjustable amount
00 00 11	LFO waveform	00	0
		01	1
		02	2
		03	3
		04	4

It's all a little more difficult than this, however, because you can't just send the Address and Sysex information. You also need to provide basic 'identification information' about the synthesizer that you are transmitting the message to, even if there is only one synth in the entire set. If this additional information is not also sent, the message will simply be ignored. Thus, all synths require that the full Sysex message is either eight or nine strings long, combining all the elements described, to complete a message. A typical message is: **F0, 41H, 10H, 42H, 12H, 00 00 11H, 04H, 10H, F7**. To better understand what this means, the message can be divided up into nine distinct parts, as shown in Figure 1.3.

1	2	3	4	5	6	7	8	9
F0	41H	10H	42H	12H	00 00 11H	04H	10H	F7

Figure 1.3 Example Sysex message.

Looking at Figure 1.3, the Sysex message we want to send, comprising the Address and the Sysex value, can be recognized in parts 6 (00 00 11) and 7 (04). The other seven parts of the message provide the required header and synth identification information.

Parts 1, 2 and 9 are defined by the MIDI specification and are required in all Sysex messages. The first part simply informs the synth that a Sysex message is on its way and sends the value F0, while the last part – part 9 – informs the synth that it is the end of the message with the value F7.

The second part of the message is specific to the manufacturer of the synth. Each manufacturer employs a unique number, which is quoted in the back of the synth's manual. In this example,

41H is the identification tag used by Roland synthesizers; thus, only a Roland synth will prepare to receive this particular message and any other manufacturer's synths will ignore it.

The third part of the message is used to identify a particular synth even if they are from the same manufacturer. This can be changed in the parameter pages of the synth in case another device in the set-up shares the same number, so that you can send Sysex messages to the one that you want to control, rather than to all of them.

The fourth part of the message contains the manufacturer's model identification code for that particular synth, to ensure that the message is received and processed only by synths with this particular model identifier (ID).

The fifth part can be one of two variables – 12H or 11H – used to specify whether the message is sending (12H) or requesting (11H) information. If a synth receives a Sysex message it recognizes, it will look at this part to determine whether it needs to change an internal setting or reply with its own Sysex message. An 11H message is usually employed to dump the entire synth's patch settings into a connected sequencer. By sending this, along with the appropriate follow-on messages and pressing record, on a sequencer it's possible to save any user-constructed patches. This is useful if the synth has no user patch memories of its own.

As already observed, part 6 contains the address of the function on which the Sysex message is to act, which in this case is the LFO rate. Most synth manuals provide an address map of the various functions and the corresponding addresses to use.

Part 7 can serve two different functions. If the fifth part contains the value 12H (indicating that the message is being sent), then part 7 will contain the data being sent, as shown in the example we're using (where the Sysex value 04 is sent to set the LFO rate). If the fifth part contains the value 11H, indicating that information is being requested, the value of this part indicates the number of bytes you want the synth to return in its reply message.

Whether part 8 is included will depend on the type of synth you use. Some manufacturers employ this 'extra' byte as a checksum, which is used to validate the message and to ensure that the receiving synth does only what the hexadecimal code asks it to. Although MIDI is a fairly stable platform, there will be the odd occasion when notes stick and messages become corrupted during transmission. If this happens, it's possible that the original message could end up asking the synth to do something entirely different, such as erasing all user presets. Errors like this are avoided with a checksum because if the checksum and message do not match the Sysex message will be ignored. A checksum is calculated using the following formula, taking the Sysex message shown in Figure 1.3 as the basis for the calculation.

> Note: For anyone unfamiliar with the binary and hexadecimal systems and converting numbers between systems, further details are provided in Appendix A.

Example: Convert the Address and Sysex hex values from parts 6 and 7 of the message to decimal values and add them together to give you the value H. Based on the values from Figure 1.3:

  Part 6 = 00 00 11 (hex) and converts to a decimal value of 17.
  Part 7 = 04 (hex) and converts to a decimal value 4.
  Therefore, H = 4 + 17 = 22.

If the value of H is greater than 127 then we need to minus 128 (H − 128) from it to produce the value X. In this case H = 22, which is less than 127, so X is also 22.

H = 22, which is less than 128, therefore X = 22.

Finally, we convert the value X to hexadecimal to derive the checksum value for the message.

The decimal value 22 in hexadecimal is 16. Therefore, the checksum value is 16.

Although creating your own messages can be boring and time-consuming, it does give you complete control over every aspect of a synth, allowing you access to parameters that would otherwise remain dormant. Commonly, a string of Sysex messages is inserted at the beginning of any MIDI arrangement, so that when play commences, the sounds in the synth can be programmed to suit your particular music. This means that you could take any MIDI file of yours to a friend's house and, provided that they have the same synth, the Sysex at the header of the file would not only set the sounds for each channel, but also reprogram the synth to all the sounds you used. Indeed, Sysex is often used by synthesizer manufacturers to update the operating system in the synth, to fix bugs, or provide you with an entirely new sound set. What's more, Sysex can also be used to alter specific parts of a synth during a MIDI sequence, like changing the type of filter used on a sound or, as in our previous example, changing the waveform the LFO uses.

It should, however, be noted that only one Sysex message can be transmitted at any one time and additional messages cannot be received while the synth is using that particular function. Furthermore, any synth will pause for a few microseconds while it receives and processes the command, so any two messages must not be sent too close together so as to give the synth time to process the previous command. If a second message is transmitted whilst the current command is being processed, the synth may 'lock up' and refuse to respond and a 'hardware boot' (switch off then on again) must be performed to reset it. Consequently, the timing between each Sysex message should be considered when transmitting any messages.

These timing problems make Sysex pretty useless if you want to continually increase or decrease a parameter while it is playing back. For instance, if you wanted to emulate the archetypal dance music 'filter sweep' where the sound becomes gradually brighter, you would need to send a continual stream of messages to the filter cut-off to steadily open it. As this is often considered to play an important part of music, it is included in the list of CC messages (CC74 Brightness), but if you wanted to progressively adjust the rate of an LFO during playback this is not covered, so many manufacturers designate these types of non-registered messages to Non-Registered Parameter Numbers (NRPN).

NRPN and RPN controllers

Non-Registered Parameter Numbers, or 'Nerpins', are similar to regular CC messages but can be set to control elements that are not available in the MIDI protocol or that need to be finely controlled. Nerpins are similar to Sysex and are high-resolution controllers that address up to 16 384 separate parameters on a synth. They were introduced into the MIDI format as a means of allowing access to a wider range of parameters that are specific to any one particular synth.

For instance, while the rate of an LFO is not accessible through CC messages, the manufacturer could assign any NRPN number to select this function in the synth, which could then be followed by two other messages (Data Button Increment and Data Button Decrement controllers) to adjust the parameters values.

This negates the need to use any Sysex messages and makes it possible to adjust any parameter on a synth, provided that the manufacturer has previously assigned it. Unfortunately, very few synths actually utilize NRPN controllers, so it is always worth checking the back of a manual to see if they are supported and, if so, what they can control.

It is important to note that if a manufacturer has assigned Nerpins you can only access and adjust one at once. For instance, if you were using them to adjust the LFO rate and then wanted to adjust, say, the amplifier's attack, you would need to redefine the current NRPN so that the Data Button Increment and Data Button Decrement adjust this rather than the LFO rate. There may be over 16 000 NRPNs but there are only three controllers to adjust their properties.

Fortunately, there are some settings that most manufacturers now define as a standard, such as the Bend Range of the pitch wheel and the master tuning. It's for this reason that we also have a pair of controllers for these defined parameter numbers. These are referred to as Registered Parameter Numbers (generally called 'RePins'). These are universal to all MIDI devices but at the moment only six of them are actually used: Pitch Bend, Fine Tuning, Coarse Tuning, Tuning Program Select, Tuning Bank Select and Null. As the MIDI specification continues to evolve this list will undoubtedly continue to grow and should eventually provide universal access to the basic sound editing and effect processing parameters that are currently the non-standard domain of Nerpins.

Computer-based sequencers

Since all these messages are transmitted as binary data it provides any sequencer with unparalleled editing features, allowing them to offer numerous real-time editing facilities. Indeed, you can transpose individual notes or entire sequencers to a different pitch, delete or replace incorrect notes, mute individual parts, and route individual parts to play on different synthesizers or use the same musical phrase more than once by copying and pasting it to new locations. The copy and paste editing tools are particularly important to the dance musician, since most dance records consist of repeated phrases, so you only need to record the phrase once and then copy it to any place in the sequence where you'd like it to appear again. Most of these editing facilities are also non-destructive, so if a mistake is made it can be quickly undone.

To accomplish this, most of today's computer sequencers have adopted more or less the same style of interface and working principles as Steinberg's flagship sequencer Cubase. Even though this wasn't the first commercially available software sequencer to reach the computer market, it was believed to be the most comprehensible and so the style has been adopted by most other packages. Thus, they comprise a main page to handle the recording and arranging duties, and a series of further pages allow for more in-depth editing.

The main window within a sequencer, often referred to as the arrange page, consists of a series of MIDI tracks (and audio tracks if the sequencer is capable) listed down the left-hand side, each of which can be assigned a MIDI channel. During playback or recording a timeline moves from left to right, playing the information contained on each track to the respective channel of the connected synthesizer.

Figure 1.4 A sequencer's arrange page.

Notably, the number and style of any further editing pages depends on the make of sequencer. Some computer sequencers will use a drum editor that allows you to 'draw' in any notes that will then trigger the appropriate drum hits on the attached synthesizer at the correct time. This is the format used throughout this book to describe the programming of drum loops in later chapters.

Most sequencers will also feature a graphical display showing the notes placed along a piano-roll display. This latter editor is the most commonly used editing screen and is where a fair number of dance musicians will spend the most time. Similar to the drum editor, here it's possible to move or delete specific notes, change velocity (how hard the note is hit) or apply any number of real-time MIDI parameters.

Using this piano-roll editing screen, it's possible to 'draw' notes directly onto the grid and then use the mouse pointer to stretch them to the desired note length, after which you can play back your work at any tempo. Alternatively, notes can be recorded in real-time and then moved to more accurate positions, or they can be recorded in 'step-time'. Using this, any notes played on an attached keyboard will be automatically positioned onto specific parts of the grid. In fact,

Figure 1.5 The drum editor in Steinberg's Cubase SX sequencer.

Figure 1.6 A typical piano-roll editing screen.

these grid positions will play an incredibly important part of any dance music written on a sequencer, including drum sequencers, so it is important to have a thorough understanding of why they are present.

Resolution

The main purpose of any MIDI sequencer is to transmit the appropriate message (such as a note-on message) to a sound-generating device at the specific times defined by the user. However, no computer is capable of rendering events at just any time and they are limited by the clock rate of the software or hardware circuit performing the process. Every time this clock pulses, the computer can perform another operation; the maximum number of clock pulses, also called clock 'ticks', that can occur within a given time period is known as the 'resolution'.

All MIDI sequencers will, or should, specify this resolution in terms of Pulses Per Quarter Note (PPQN), and the higher this value the more useful the sequencer will generally be. As an example, if a sequencer has a PPQN of 4, each quarter note would contain four pulses, while each eighth note would contain two pulses and each 16th note would contain one. Consequently, you couldn't render notes that are less than a 16th because there isn't a high enough resolution to yield an integer number of clock pulses. Thus, not only would you be unable to create notes smaller than a 16th, it would also be impossible for any note timing to be adjusted to less than this either. This resolution is typical of some of the early analogue drum machines that were used to create dance rhythms and are responsible for the metronomic nature of the drum patterns.

From this we could determine that you would only really need a PPQN of 16 to write a typical dance drum loop and although this is true in some cases, the elements that sit upon this basic rhythm need a higher PPQN, otherwise the groove of the record will suffer. For instance, say you were to record a three-note chord live from a keyboard into the sequencer. No matter how precise you may think you are, it's highly unlikely that you would actually play all the notes at precisely the same time. The notes may sound as though they've been played simultaneously, but there would be very small timing differences.

If the sequencer being used to record these notes has a lower resolution than the difference in time between the two notes, it has no option but to shift one or both notes to the next pulse. This is obviously going to change the 'feel' you're trying to attain and although many dance rhythms appear robotic in nature, there are underlying elements that do require some human feel; thus, to achieve the sound you want the resolution needs to be higher.

In the early 1990s, Roland came to the conclusion that a PPQN of 96 would be suitable for capturing most human nuances, while a PPQN of 192 would be able to capture the most delicate human 'feel'. Subsequently, most drum sequencers today use this latter resolution, while software sequencers have adopted a much higher PPQN of 480 and above. A sequencer that boasts a PPQN of 480 would work as shown in Table 1.4.

Note: If you have no musical knowledge of bars and musical notation, we'll be looking at these in the next chapter.

Table 1.4 Musical notation and the related ticks

Traditional notation	Length of the note	Number of clock ticks (pulses)
Whole note	One bar	1920
Half note	Half bar	960
Quarter note	Quarter bar	480
Eighth note	Eighth bar	240
Sixteenth note	Sixteenth bar	120
Thirty-second note	Thirty-second bar	60

Each of these clock ticks can be referenced to a grid position within the sequencer, with the number of grids displayed on the screen determined by the quantize value. For instance, if the quantize value were set to 16, the grid would divide into 16ths, as it would be impossible to place notes at smaller integers. The quantize value can obviously be used to affect rhythm, phrasing and embellishments, but its flexibility depends on the options offered by the sequencer and may be limited 1/8th, 1/16th or 1/32nd quantizing. All of these can be applied to previously recorded notes to correct any timing inaccuracies that may appear from recording MIDI on the fly. As it's improbable that you could accurately record a series of notes that sit on exact note divisions live from a keyboard, quantize values can be used to shift all of the notes simultaneously to sit in the correct positions, a process often referred to as strict or 'over-quantizing'.

The strict timing imposed by over-quantizing isn't always a necessity, so another process known as 'percentage quantize' may also be available. This allows you to use a percentage marker so that not all the notes are snapped to a grid but rather are moved by the specified percentage towards it. For instance, by setting a quantize value of, say, 50%, you can move notes to a position that's half way between where you originally played them and the nearest time slot in the quantize grid. This is more useful for tightening up loose playing without losing the feel of the music and is often known as 'iterative quantize' because you can carry out the process as many times as required.

Although both these quantizing features are helpful in correcting sloppy playing, overuse of these techniques can produce anti-musical results. While there are some forms of music that, in theory at least, benefit from strict timing of each note, in reality even the most stringent forms of techno introduce subtle timing differences throughout to add interest and tension. Indeed, the importance of subtle timing differences shouldn't be underestimated when creating music that grooves. We'll look more closely at the elements involved in using more advanced quantizing features to create grooves in a later chapter.

MIDI-based dance studios

Although all dance studios will differ depending on the artist and the type of music that they produce, generally all dance studios will consist of a sequencer (commonly a Mac or PC running sequencing software), a sampler, a number of synthesizers, a drum machine, a host of effects units and a mixing desk. Many of these components will not necessarily be in hardware form, as computers are capable of running effects, samplers and instruments in the software domain, but for now we'll concentrate on a typical hardware-based studio and we'll look more closely at software-based studios in a moment.

Figure 1.7 A simple dance-based studio.

In Figure 1.7, the master keyboard's MIDI OUT is sent to the MIDI IN on the sequencer. This allows notes to be recorded by the sequencer or sent directly through it to the sequencer MIDI OUT port into any of the attached devices, permitting you to play the device live. The MIDI OUT from the sequencer is connected back to the MIDI IN on the keyboard, then the THRU port of the master keyboard is used to connect to the MIDI IN port on the sampler, and so forth throughout the chain. The audio outputs of all these devices are connected to a mixing desk, allowing all the signals to be summed together before being sent to the amplifier and finally the loudspeaker monitors as a stereo signal. Additional devices can be daisy-chained in this way up to the maximum number of inputs available on the mixing desk. When all the devices are connected together there are then a number of settings that need to be made to ensure that everything works together as it should before you start writing music.

For any notes entering the sequencer from an attached keyboard to be sent via the computer sequencer and back out to the attached devices, it must have 'Soft Thru' or 'MIDI Echo' enabled in its set-up pages. This works on the same principle as the THRU connections on a hardware device. The incoming MIDI data from the attached master keyboard is not only received by the sequencer, but it is also sent through it to an attached device, allowing you to record the data in the sequencer while also hearing the notes being played on the daisy-chained devices. This approach, however, can result in problems if the master keyboard has its own in-built synthesizer.

There are two types of keyboards available: controller keyboards and synthesizer keyboards. Controller keyboards are specifically designed to be used within a MIDI system and contain no on-board sounds whatsoever. Instead, they are simply hardware controllers that transmit MIDI information (such as note-on, pitch-bend and modulation) to the sequencer that can then be recorded or transmitted through the chain to the device of your choice. Synthesizer keyboards contain a number of on-board sounds which are triggered through the keyboard while also transmitting the MIDI data, and this can result in problems. Since the keyboard will play the on-board synthesized sounds *and* output to the MIDI OUT port, if this transmitted MIDI data is sent to another device in the chain, not only will it trigger the second device it will also trigger the keyboard's own sound too. To prevent this from occurring, it may be necessary to turn the keyboard's local control to off. By doing so, the keyboard's inbuilt synthesizer is ignored and only MIDI note information is transmitted to the MIDI OUT port.

> Note: It's important to remember that most sequences will default to 'Soft Thru' enabled, but if the local control of the keyboard is left on, any messages that are sent to the sequencer will be sent through to any attached devices; thus, the signal leaving the keyboard will be returned back to it, creating a feedback loop. This re-sounding of the instrument will occur almost instantaneously. Since the repeated note data will not be recorded it may not necessarily be so much of a problem, but it can make the synth difficult to play.

While these procedures can prevent any problems from occurring and also allow you to play back sounds from attached devices, any type of daisy-chained MIDI set-up should only be viewed as a stopgap arrangement and should not be considered the perfect studio set-up, particularly where dance music is concerned.

If a large amount of devices are daisy-chained together the timing of notes can drift severely. In fact, for dance music that uses complex rhythms (such as drum 'n' bass or big beat), daisy-chaining devices is best avoided altogether. This is because MIDI has a limited baud rate, so it takes approximately 1 millisecond (ms) to transmit a typical byte of information, such as a MIDI note on messages, to a synthesizer. While this is small enough to fool the human ear into believing that these events occur at the same time, MIDI is also a serial interface, therefore any consecutive messages can only be transmitted after the previous one has been processed. This can introduce a whole host of problems if a series of notes have to be transmitted simultaneously or have to travel through a number of daisy-chained devices.

For example, if the sequencer has to trigger a three-note chorded pad, a kick drum, snare, hi-hat, a bass note and a series of CC messages simultaneously, the serial nature of MIDI can only transmit these one at a time. Therefore, the first note-on message will occur after 1 millisecond, the second note-on message will occur 2 milliseconds after the first, the third will be transmitted after 3 milliseconds, and so on. Thus, if 16 note-on messages were transmitted down a single MIDI cable, by the time the final note is received by the device it would occur at around 16 milliseconds after the first hit message, resulting in an effect known as MIDI choke or delay. What's more, if you were to transmit a message that has to travel through a number of daisy-chained devices, the message has to pass through the first device, adding 1 millisecond (perhaps longer as the device has to identify that the message is to travel 'thru' it), then through a second, adding another millisecond, then the third, etc. This means that the more devices the message has to travel 'thru', the later in time it is going to occur.

Note: Although daisy-chaining devices can result in a perceivable delay, if a large amount of equipment is daisy-chained together it can result in numerous errors being introduced into the MIDI signals as it passes through the various 'thru' ports!

Though we are still working in milliseconds in both these instances, the human ear is capable of detecting very subtle rhythmic variations in music and even slight timing inconsistencies can severely change the entire feel of the music. Indeed, a timing difference of only 12 milliseconds can destroy the strict rhythm that is required in some dance music and can be especially evident with complex break beat loops. Consequently, many artists prefer to write drum loops using drum samplers that feature their own sequencers to avoid any potential timing problems. This ensures that the timing of each hit is accurate because the clock pulses are controlled by independent hardware, and the link between the samples and the sequencer has a much larger bandwidth. This 'tight' sequencer can then be synchronized with the master computer that's playing the rest of the performance using MIDI Time Code (MTC).

MTC provides a method of synchronizing two sequencers, ensuring that they not only synchronize with one another during playback but that they also start and stop simultaneously. The master device (most usually a computer or a specific MTC generator) controls this by sending a number of MIDI messages to cue any of the attached slave devices. The two sequencers are synchronized during playback by the master sending a series of MIDI pulses at evenly spaced intervals. This also means that the master must send these pulses accurately, otherwise the two will drift apart. Accidental but subtle timing differences can be responsible for creating an entirely different and useful feel, but if they drift too far it'll all sound wrong.

To avoid this, many sequencers use a pre-roll-based system whereby a small buffer is used to store any MIDI data in advance, so that if the computer CPU becomes bogged down for a moment (such as when it needs to poll the keyboard while redrawing the current screen), information in the buffer can continue normally for a short while. Even then, it is prudent to disable any processor-intensive programs running in the background to help ensure that it produces pulses accurately. Also, if the sequencer utilizes a particularly high PPQN, see if it is possible to lower it. For any high resolution to be worthwhile, the sequencer has to be able to maintain the clock rate at any tempo, and in some sequencers PPQN can vary the faster that this is set.

Note: If you are to send MTC to any attached devices, the MIDI specification recommends that these pulses are transmitted on their own discrete MIDI cable, as a missed pulse can result in the two devices drifting apart.

While this approach will prevent any timing discrepancies between sequencers, it doesn't solve the potential issue of MIDI choke when transmitting simultaneous messages down a single cable, so many sequencer manufacturers have adopted numerous methods in an attempt to avoid it. Most sequencers utilize a channel priority system whereby the highest channels in the arrange page are given priority over the MIDI chain. For instance, if all the MIDI channels are being used, the first channel will receive the utmost priority, followed by the next and then the next and so forth. Because of this, with any form of dance music it is prudent to place the drums on the first channel of the sequencer (even if it's being transmitted down MIDI channel 10) and the bass on channel 2. This will help to keep the groove as tight as possible. Also, it's a good working principle to move any other tracks forward by a couple of ticks so that they occur fractionally later than the priority channels, as this helps to prevent any potential MIDI choke.

Any additional CC messages may also choke the interface, so many sequencers will also offer a series of data reduction algorithms. These can usually be accessed through the system pages of a sequencer and allow you to remove any unnecessary data, such as Sysex and aftertouch information. Although both these approaches should help to ensure that any MIDI data occurs pretty much on time, it's still not the ideal situation for a dance studio, so the best approach is to invest in a multi-MIDI interface.

Multi-MIDI interfaces are usually external hardware interfaces (although cards that fit in the computer are available) connected to the computer using the parallel or USB port, and offer a number of separate MIDI IN and OUT ports. Using these, and provided that you have more than one MIDI device, you can use different MIDI outputs to feed each different device rather than having to daisy-chain them together. This reduces the possibility of any MIDI choke by allocating each device to an individual MIDI output from the master sequencer. These, however, are single-buss interfaces and should not be confused with MIDI interfaces that have more than one MIDI OUT but do not use multiple busses.

Single-buss interfaces offer a number of MIDI outputs that are all connect to one MIDI buss, meaning that the 16 channels can be divided between the outputs. These were developed as a substitute for devices that do not feature a MIDI THRU port, allowing them all to feed directly from the interface. Nonetheless, because all the outputs are shared you're still only transmitting 16 channels of MIDI data and this can result in choke if numerous notes occur at the same time. True multi-buss devices will offer 16 MIDI channels *per* MIDI output, so if a multi-buss interface offers four MIDI OUT channels, you'll have 64 MIDI channels at your disposal.

Most multi-buss interfaces feature a number of inputs that should also be on separate busses. This allows you to record MIDI data (such as knob movements on a synth) from a number of attached devices simultaneously without the risk of choking the interface. If only one input buss were offered, attempting to record large amounts of CC data from a few synths could result in severe timing problems.

Most professional sequencers offer support for multiple busses, allowing you to set an individual MIDI track to output any one of the numerous MIDI IN connectors, and two major sequencer manufacturers – Steinberg and Emagic – have developed their own interfaces. These are specifically designed to work with the sequencer in question and use custom buffers to ensure that the MIDI transmitted from the sequencer into the interface, and from the interface to the receiving devices, is accurate to within less than a millisecond. Steinberg's Midex 8 and Midex 3 interfaces employ a process known as Linear Time Base, while Emagic's MT8 and MT4 use Active MIDI Transmission. Despite the branding differences, both use the same principle to keep timing as accurate as possible by attaching time stamp information to each MIDI message. The time stamp information is stored in a buffer within the interface before they are transmitted out to the appropriate device on time.

Note: Although it is possible to use Midex with Emagic's Logic and vice versa, you should look towards purchasing the correct interface for the sequencer you are using, as mixing the two will result in the MIDI information not being 'time stamped'.

A single bus setup

16 MIDI channel transmitted

MIDI out   MIDI in

MIDI in   MIDI out

4 MIDI channels

4 MIDI channels

4 MIDI channels

4 MIDI channels

A multi-bus setup

6×16 MIDI channels transmitted

USB out   USB in

16 MIDI channels

16 MIDI channels

16 MIDI channels

16 MIDI channels

16 MIDI channels

16 MIDI channels

Figure 1.8 Single- and multi-buss MIDI set-ups.

Of course, even if multi-buss MIDI interfaces are used, if 16 channels of MIDI information are all transmitted to the same module, then there is still no avoiding the possible choke unless you transmit MIDI one channel at a time, record the results as audio and then re-insert this back into the sequencer, or alternatively employ VST instruments to avoid choking problems altogether (more on these in a moment).

Audio sequencing

Within the past few years sequencers have moved on from simple MIDI sequencing and now offer a variety of audio sequencing parameters. These allow you to record, import and edit audio in much the same way as MIDI. The audio can be displayed graphically and edited with a mouse, and it is also possible to apply real-time effects to the audio, such as delay, reverb and compression, negating the need to purchase additional hardware units to accomplish the same job. Because all the audio is mixed together using a virtual software mixer, there is little need to purchase a hardware alternative. All the audio can be mixed and equalized inside the computer and the resulting stereo signal can be sent from the soundcard directly into the loudspeakers.

This has the obvious advantage of being much cheaper than having to purchase the hardware equivalents, but has the disadvantage that working with audio uses a lot more power than simply working with MIDI hardware. You're no longer simply sending small bytes of information to attached synthesizers or samplers because you're storing and manipulating audio on the computer itself, and unless the computer is of a high enough spec this may not be possible. While all PCs and Macs available today are capable of acting as a digital audio workstation, there are a number of important factors that determine the number of audio tracks you can run at once and how well they will perform when a number of effects are applied to them.

Although most of today's software sequencers are more than capable of playing back more than 200 audio tracks at a time, whether they can depends entirely on the access speed of the hard drive and the amount of memory installed. All audio used by a sequencer is stored on the hard drive and is read directly from the drive during playback. Thus, the speed at which data can be read (or written while recording) directly affects the number of audio tracks that can be played simultaneously. Most computers will use an EIDE (Enhanced Integrated Drive Electronics) hard drive that has an access time of roughly less than 12 milliseconds, but it must also be able to support Mode 4 buss mastering or at least have a minimum DMA (Direct Memory Access) of 33 to be useful in audio applications.

Similar to transmitting MIDI information, sequencers will use a buffering system when playing back audio files. Audio is streamed from the hard drive into the computer's memory, where it is then read and played by the sequencer. Before any process in a computer can access the memory, it must be referenced with the CPU first. In other words, every time the sequencer accesses the hard drive it first holds a conversation with the CPU to check that there is enough physical memory to buffer the information. This not only makes the access time longer, but also uses some of the CPU power that would be better spent concentrating on applying effects to an audio channel. By using a DMA hard drive, any data can be transferred between the disk and memory without having to go through the processor each time. This not only frees up the CPU, allowing it to spend more time doing other things, but also speeds up the access times, allow-ing you to play more audio tracks simultaneously without suffering from glitches or jumps in the audio files.

You cannot simply connect a DMA 33 hard drive to a computer though, as to work this process must be supported by the motherboard – i.e. the motherboard (often shortened to Mobo) must have direct access from the hard drive to the memory. Furthermore, since digital audio sequencers require a constant stream to be supplied, you will need a large hard disk if you want to record long digital audio tracks. A typical CD-quality stereo file will require 10 megabytes (MB) per minute, so a single 3-minute track will need 30 MB of free hard disk space, and as some music can consist of 10–15 tracks played simultaneously, it may require 300–450 MB of free space.

If the motherboard does not support DMA buss mastering, then SCSI (Special Computer Serial Interface) hard drives can be used in their place. These offer the same or, in some cases, faster transfer speeds than EIDE, but rarely come fitted as standard and unlike EIDE cannot be directly connected to the motherboard unless it has a specific SCSI controller fitted. If this is not the case, you will have to purchase an additional SCSI interface card that can be used to communicate with any attached drives.

SCSI appears in two forms: narrow and wide. These refer to the number of devices that can be daisy-chained together. The narrow SCSI channel supports up to eight hard drives (or any other SCSI devices), while a wide channel supports up to 16. If you have a narrow adaptor, however, you can only attach narrow devices and vice versa for the wide SCSI adaptors. That said, Adaptec, the preferred company to use for SCSI connections, have released a controller card that allows you to connect both narrow and wide connections to the same interface.

Alongside SCSI, another more recent development is Apple's Firewire technology. Sometimes referred to as IEEE 1394, it's a high-speed serial data buss that can not only move 400 megabits per second but is also hot swappable. This means that provided you have a Firewire connection fitted to the computer, you can simply plug in an external Firewire hard drive through a small connection cable while the computer is still switched on. What's more, up to 64 devices can be chained together using this connection, allowing you to attach numerous hard disks to the computer and access them simultaneously.

> Note: Apple has recently introduced a new Firewire protocol labelled as IEEE 1394b. This features all the same capabilities as Firewire IEEE 1394, but allows data transfer speeds of up to 800 megabits per second!

Another important consideration, if a PC is to be used as the sequencer, is the motherboard itself. For any audio applications a board that uses an Intel-based chipset will perform much more reliably than any other. All audio software manufacturers test their equipment using Intel chipsets, but not as many will test them on others and there have been innumerable reports from users of AMD-based systems of incompatibilities between soundcards, SCSI adaptors and USB devices (such as USB-based soundcards and multi-MIDI interfaces). Indeed, there have been some serious issues reported with PCI-to-USB host controller chips that use non-Intel-based chipsets, such as problems that prevent any attached USB devices from working at all.

Though Intel-based chipsets are more expensive, the reliability offered by them is worth it in the long run, especially if you want to make music rather than spend most of the time chasing incompatibility issues. Notably, many sequencers and other audio software packages are specifically programmed to utilize the SSE II instruction sets used on Pentium 4 processors and some companies will not offer product support if there are incompatibilities with their software and non-Pentium-based systems. With the computer as the basis on which the entire studio will be built, money should be no object.

> Note: Apple Mac computers are all constructed to the same specific standards, so any potential incompatibilities are not a problem.

The most important link in the chain, however, is the soundcard, as this defines the ultimate quality of the music you produce. Essentially, a soundcard performs two distinct functions: providing a basic MIDI interface and enabling audio playback and recording. There are a huge variety of soundcards available, some manufactured specifically for audio work and others aimed more towards computer gamers. Of all the factors to bear in mind when looking for any suitable soundcard for your studio, the three most important questions to ask are:

- How good is the audio quality?
- Are the soundcard's drivers multi-client?
- How many inputs and outputs does it offer?

Each of these is explained in more detail below.

How good is the audio quality?

All soundcards offer much the same audio quality provided that the audio remains inside the computer (digitized audio simply consists of numbers), but the quality of the audio being recorded from the soundcard's inputs is entirely dependent on the quality of the card's audio-to-digital converter (ADC). Cheaper cards tend to have a poor signal-to-noise ratio (SNR), resulting in an audible hiss on any recordings taken using them, and many of these cards also remove some of the higher and lower frequencies of a sound. It is also likely that they will use 16-bit rather than 24-bit converters, so they cannot capture the dynamic range as well (don't panic, we'll be looking at all this in Chapter 4). That said, while 24-bit soundcards offer superior recording quality when recording real instruments or vocals, to record this high requires more memory and CPU power and, generally, 16-bit soundcards can still achieve excellent results.

Are the soundcard's drivers multi-client?

As soon as any audio sequencer is opened it captures the audio card's inputs and outputs to use within the program. If the drivers on the card are not multi-client compatible, the sequencer will be the only program that can access the card's connections, and any further audio applications will not be able to access the soundcard while the sequencer is running. Promises that multi-client capabilities will be included in later driver updates shouldn't be taken literally and if the card does not already have them you should look for one that does. Keep in mind that many companies will promise updated drivers but instead release newer cards in their place. Visiting the manufacturer's web site and checking the currently available drivers and their frequency is probably the best way to discover whether they are willing to update them to keep up with moving technology.

How many inputs and outputs (I/O) does it offer?

All soundcards will offer a stereo input and output, but this may not be enough to suit your needs. In some situations you may wish to output each audio channel of the sequencer to a different output so that you can mix down on a hardware mixing desk. Additionally, if you have plans to record more than one instrumentalist at a time then you'll need an input for each performer. Although you may not feel that you need more than one stereo input or output, it is wise to plan for the future.

Provided that you have a capable computer and a good soundcard, possibly the largest benefit from using an audio sequencer is the availability of plug-ins, which are small auxiliary programs that add features to the host program. For digital audio software they can give you access to additional effects such as EQ units, filters, vocoders and reverb algorithms.

Plug-ins for 'home' sequencing platforms are usually in one of three formats: AU (Audio Units), Direct X or VST (Virtual Studio Technology). Audio Units is a Mac-only interface for use on their OSX operating system, Direct X is a Windows-only application and VST is available for both platforms. Nonetheless, AU, Direct X and VST are essentially one and the same, allowing access to real-time effects, but where Direct X and AU can be used in a wider range of sequencing programs, VST is limited to sequencers that support the VST standard. This limited support is advantageous, with the high level of integration providing greater stability, in addition to using less processor power.

More recently, Steinberg developed the Virtual Instrument plug-in interface. This works on similar principles to the VST and Direct X plug-ins, but allows you to use software emulations of real-world instruments within any VST-compatible sequencer. Software samplers have also started to appear, offering all the benefits of sampling but from within the sequencer, negating the need to use any external MIDI devices. Significantly, their audio output can be directed to a number of channels in the audio sequencer's mixer, so the signal can be mixed with the other audio channels contained in the virtual mixer while allowing you to EQ and apply plug-in effects to their output.

To further explain the advantages of a software-based studio, Ronan Macdonald, an accomplished musician and editor of the UK's best selling *Computer Music* magazine,[1] kind-heartedly agreed to be interviewed.

Q: What are the benefits of a studio based entirely on software?
A: The benefits of the software-only studio are many, and so compelling that the music technology hardware industry is genuinely suffering for it (as, indeed, are recording studios the world over), as more and more musicians get their home studios up to professional standards thanks to the power and price of software.

First, then, price. Music software, generally, is much more cost-efficient than hardware. For the price of a decent hard disk recorder, you can now buy a very powerful computer *and* a MIDI/audio sequencer that leaves the hardware equivalent standing.

Second, convenience and mobility. The advantage of having your entire studio running inside your Mac or PC is self-evident in terms of the space it takes up and the lack of need for cables,

---

[1] The *Computer Music* web site can be located at www.computermusic.co.uk.

racks and all the rest of it. And if that Mac or PC happens to be a laptop, the freedom to take your entire studio wherever you want is nothing short of a modern miracle.

Third, total recall. Generally, with a hardware-based studio, it's only possible to work on one mix at a time, since all mixer settings, patch bays and instrument and effects settings have to be cleared for each new project. A totally software-based studio doesn't have this problem, as all settings are saved with the song, meaning you can work on as many projects at once as your brain can handle.

Fourth, sheer power. Today's computers are really, really powerful. It's now possible to run well over 60 tracks of plug-in-processed audio alongside numerous virtual instruments in real-time. Sequencers also make MIDI and audio editing highly user-friendly, very quick and very easy, and in terms of sound quality, since nothing need ever leave the digital domain, low noise and crystal clarity can be taken for granted with a software-only set-up.

Q: What are the advantages of using VST instruments over the hardware alternative?
A: Again, price is a major factor here. Hardware synths can cost hundreds or even thousands of pounds; soft synths very rarely cost more than £400 and usually a lot less. These days it's possible to buy some incredibly powerful virtual instruments on the net for as little as £30 – and that's not to mention the countless free ones out there! On top of that is the fact that you can run multiple instances of most virtual instruments at the same time. As if having a totally realistic software Prophet-5 (a staple instrument in the creation of dance music) wasn't cool enough, how about having eight of them all running at once, and all for the price of just one?

When it comes to samplers in particular, hardware simply can't compete with computer-based systems. Editing a big sample is infinitely easier on a computer monitor than it is on a sampler's LCD screen, and being able to store gigabytes of samples on a computer hard drive makes maintaining and archiving a large sample library a mundane task rather than a constantly fraught one.

Q: But do Virtual Instruments sound as good as hardware instruments?
A: Absolutely. While early VST synths were limited in terms of quality by the speed of computers a few years ago, these days there are no such problems. The latest generation of analogue emulations – such as GMedia's Oddity and Arturia's Moog Modular V – have shown that we now have enough power at our disposal to be able to process exceptionally realistic oscillator algorithms, leading to sound quality that simply can't be distinguished from 'the real thing'.

Q: So you can play these in real-time like any hardware instrument?
A: While there is always a certain amount of latency involved in playing soft synths (the delay between pressing a key and actually hearing the sound come out of the computer), with a high-quality (which doesn't necessarily mean expensive) soundcard, this can be easily brought down to less than 10 ms, which is all but unnoticeable. And once you've actually recorded the performance into the computer, there's no latency at all on playback.

Q: What would you recommend as the basic requirements to those starting out writing dance music on a computer?
A: While, as with everything in computing, the more money you spend, the more power you'll get, any off-the-shelf PC will be capable of running even the most demanding of today's music software. The soundcard is more of an issue than the PC itself, and although perfectly good results can be had using any built-in sound system, it's definitely better to get at least a Sound Blaster Live or Audigy, or – even better – a pro solution such as M-Audio's Audiophile 2496, which offers incredibly low latency and stunning sound quality, and can be bought for around £150.

In terms of software, the budding dance music producer is spoilt for choice. Propellerhead Software's Reason offers a beautifully conceived and totally self-contained virtual studio for around £220, featuring synths, samplers and effects and sounding simply awesome.

Cakewalk's Project5 (£249) takes a similarly dance-orientated approach but has the added benefit of being compatible with Direct X and VST plug-ins. And for what qualifies as unarguably the cheapest way into computer-based music production, *Computer Music* magazine features the CM Studio free on the cover CD of every issue. This comprises a powerful sequencer (Computer Muzys), a virtual analogue synth, a drum synth, a drum sample player, a sampler and a rack of effects.

2 Dance music theory

All you need is a feel for the music. There are people that have been to college to study music and they can't make a simple rhythm track, let alone a hit record.

Farley 'Jackmaster' Funk

Basic music theory

Having discussed the MIDI protocol in Chapter 1, we can look at how this is used in a more traditional musical context. In order to do that, it's handy to have a basic understanding of musical theory. While some dance musicians have released records without any prior knowledge of music theory, familiarity with the basic principles will be helpful, even it only helps you to understand why some different genres of dance music may use different time signatures.

It's important to note right from the start that this is a basic introduction to musical theory aimed at those who have little or no musical knowledge. An in-depth discussion of the finer points of music theory is beyond the scope of this book, so we're only going to cover areas that are relevant to the dance musician. This information will be developed in the discussions of different musical genres in later chapters, so if you feel lost with what follows, stick with it, it will all come together by the end of the book.

We need to begin by first examining the building blocks of music, starting with the musical scale.

The major scale

The major scale consists of a specific pattern of pitches that are named after the first seven letters of the alphabet. These always follow a specific sequence and always begin and end on the same letter to create an octave. For example, the C major scale always begins and ends with a C, and the distance between the two Cs is always one octave. For those who've endured Julie Andrews' warbling in the *Sound of Music* or attended singing lessons, we can relate this to 'Doh – Ray – Me – Fah – So – Lah – Ti – Doh' (the final 'Doh' being one octave higher than the first). What's more, each pitch in the octave has its own name and number, the latter of which is referred to as a 'degree'. Because most aspects of musical theory are based around the distances between notes, it's important to understand these relationships, which are shown in Table 2.1 for the C major scale.

To comprehend the relationship between notes in a scale we need to examine the placement of notes on a typical keyboard and their representation on the musical staff.

Table 2.1 The C major scale

Vocal	'Doh'	'Ray'	'Me'	'Fah'	'So'	'Lah'	'Ti'
Key	C	D	E	F	G	A	B
Degree	1	2	3	4	5	6	7
Name	*Tonic*	*Supertonic*	*Mediant*	*Subdominant*	*Dominant*	*Submediant*	*Subtonic*

One octave

Figure 2.1 Layout of the keyboard.

The keyboard is made up of five black notes and seven white notes, forming one octave, repeated several times along the length of the instrument. Each key on the keyboard relates to the pitch of that particular key and is equal to one semitone. The black keys raise or lower the pitch by one semitone. If they raise the pitch they are called 'sharps' (♯). If they lower the pitch they are called 'flats' (♭), so to the right of C is a raised black note called C♯ and the same note to the left of D is called D♭. Likewise, the black note to the right of D is called D♯ or E♭ and so on (excluding the notes E and B that do not have sharp notes associated with them and the notes F and C that do not have flats associated with them). Because the black keys on the keyboard can be either sharps (♯) or flats (♭), depending on the key of the song or melody, they are sometimes referred to as 'enharmonic equivalents'.

The written musical notation of the octave from C to C on the keyboard is shown in Figure 2.2.

C C♯ D D♯ E F F♯ G G♯ A A♯ B C

Figure 2.2 Notes in the octave from C to C.

Figure 2.2 clearly shows how notes rise in pitch through the octave. Because each note of a particular pitch is equal to one semitone, these can be added together and classed as one tone. Taking the notes C and C♯, the two can be added together to produce one tone (one semitone + one semitone = one tone). Consequently, the notes C, D, F, G and A, which all have associated sharps, are known as whole tones, while the notes E and B, which don't have sharps, are referred to as semitones. Using this logic, the C major scale shown in Figure 2.3 can also be written:

Tone–Tone–Semitone–Tone–Tone–Tone–Semitone

Figure 2.3 The C major scale.

This is the pattern of tones and semitones that defines a major scale. If, rather than starting at C, we start at D ending on the D an octave higher, this arrangement changes to:

Tone–Semitone–Tone–Tone–Tone–Semitone–Tone

Although this may not seem significant on paper, it has an impact on the sound because the key has changed from C to D. That is, D is now our root note. The best way to understand this difference is to import or program a melody into a sequencer and then pitch it up by one tone. Although the melody remains the same, the tonality will change because the notes are now playing at different pitches than before. This is shown in Figure 2.4.

Figure 2.4 C major pitched up one tone to D.

In order to pitch the melody up from C to D without changing the tonality, we need to take the sharps into account. So in pitching the melody from C to D, whilst retaining the tone–tone–semitone pattern associated with the major scale, the relevant sharp notes must be introduced. In the key of D, the major scale has two sharps – F♯ and C♯ – written as shown in Figure 2.5. Looking back at the keyboard layout shown in Figure 2.1, these correspond to the black notes to the right of F and C.

Figure 2.5 The D major scale.

If the melody is then pitched up from C by five semitones to F, F becomes the root note and the pattern of tones changes to:

Tone–Tone–Tone–Semitone–Tone–Tone–Semitone

Again, this looks fine on paper, but the tonality is no longer the same as in the original melody, where C was the root note. In order to correct this, we need to alter the pitch of B, lowering it by one semitone to B♭, as shown in Figure 2.6.

Figure 2.6 The F major scale.

Where the scale is constructed will depend on whether the black keys are sharps or flats, but in defining the key they will only be one or the other.

Minor scales

Along with the major scale, there are three types of minor scales called the 'natural', 'harmonic' and 'melodic' minor scales. These differ from the major scale and from one another because of the order of the tones and semitones. As a result, each of these minor scales has a unique pattern associated with it.

The natural minor scale has semitones at the scale degrees of 2–3 and 5–6. Thus, they all follow the pattern:

Tone–Tone–Semitone–Tone–Tone–Semitone–Tone

Figure 2.7 C minor (natural).

The melodic minor has semitones at the scale degrees of 2–3 and 7–8 when ascending, but reverts back to the natural minor when descending. Thus, they will all follow the pattern of:

Tone–Semitone–Tone–Tone–Tone–Tone–Semitone

Figure 2.8 C minor melodic.

Figure 2.9 C minor harmonic.

The harmonic minor scale has semitones at the scale degrees of 2–3, 5–6 and 7–8. Thus, they all follow the pattern:

Tone–Semitone–Tone–Tone–Semitone–Tone–Semitone

Generally speaking, the harmonic minor scale is used to create basic minor chord structures to harmonize with riffs or melodies that are written using the melodic and natural minor scales. Every major key has a related minor scale and vice versa, but this doesn't mean that the closest relationship to the key of C major is C minor. Instead, the relationship works on the principle that the most closely related major and minor keys are those that have the most notes in common. The closest related minor scale to C major is A minor, because these two scales have the most notes in common (neither contains any sharps or flats). As a guideline, you can define the closest related minor from a major by building it from the sixth degree of the major scale and the closest related major from a minor scale building from the third degree of the minor scale.

The structures of major and minor scales are often referred to as modes. Rather than describing a melody or song by its key, it's usual to use its modal term.

Table 2.2 Modal terms

Key	Name	Mode
C	Tonic	Major (Ionian)
D	Supertonic	Dorian
E	Mediant	Phrygian
F	Subdominant	Lydian
G	Dominant	Mixolydian
A	Submediant	Minor (Aeolian)
B	Subtonic	Locrian

Modes are incredibly important to grasp because they determine the emotion the music conveys. Although there are scientific reasons behind this, principally it's because we subconsciously reference everything we do, hear or see with past events. Indeed, it's impossible for us to listen to a piece of music without subconsciously referencing it against every other piece we've previously heard. This is why we may feel immediately attracted too or feel comfortable with some records but not others. For instance, some dance music is written in the major Ionian mode. Changing this to the minor Aeolian mode would make the mood seem much more serious. This is not only because the Ionian mode sounds less harmonious, but also because we associate the Aeolian pattern of sounds with the serious and sombre sounds of traditional church organ music.

Chords

No matter what key or mode you play in, if only one note is played at a time the music will be rather boring to listen to. Indeed, it's the interaction between notes played simultaneously that provides music with its real interest. When more than one note is played simultaneously, it's referred to as a 'harmony' and the difference in the pitch between the two played notes is known as the 'interval'. Intervals can be anything from one semitone to above an octave apart, and the sonic quality of a sound varies and multiplies with each additional note that's played.

Harmonies that utilize three or more notes are known as chords. The size of these intervals influences the 'feel' of the chord being played and some intervals will create a more pleasing sound than others. Intervals that produce a pleasing sound are called consonant chords, while those less pleasing are called dissonant chords. Although consonant chords are more harmonious than dissonant chords, the combination of the two can be used in music to create tension, so each type is of equal importance when you are writing a track. Indeed, creating a track using nothing but consonant chords results in a track that sounds quite insipid, while creating a track with only dissonant chords will make the music sound unnatural. The two must be mixed together sympathetically to create the right amount of tension and release. To help clarify this, Table 2.3 shows the relationships between consonant and dissonant chords and the interval names over the octave.

Table 2.3 The relationships between consonant and dissonant chords

Interval	Semitones apart	Description
Unison	Same note	Strongly consonant
Minor second	1	Strongly dissonant
Second	2	Mildly dissonant
Minor third	3	Strongly consonant
Major third	4	Strongly consonant
Perfect fourth	5	Dissonant
Tritone	6	Mildly dissonant
Perfect fifth	7	Strongly consonant
Minor sixth	8	Mildly dissonant
Major sixth	9	Consonant
Minor seventh	10	Mildly dissonant
Major seventh	11	Dissonant

Intervals of more than twelve semitones span the octave and have a somewhat different effect.

To better understand how chords are constructed we need to look again at the layout of C major. Using the chart shown in Figure 2.10 we can look at the simplest chord structure – the

Pitch	C	D	E	F	G	A	B
Degree	1	2	3	4	5	6	7

Figure 2.10 Grid of the C major scale.

triad. As the name suggests, this contains three notes and is constructed from the first, third and fifth degree notes of the scale that forms the root note of the chord. Looking at Figure 2.3, the root note of the C major scale is C, so C is the first note of the chord. To create the C major triad, the third and fifth degree notes must be added, giving us the chord C–E–G (or first, third and fifth degree). This major triad is the most basic and common form of chord and is often referred to as 'CMaj'. This chord template can be moved to any note of the scale, so, for example, a chord constructed from the root key of G, taking the first, third and fifth degree notes, the chord consists of the notes G–B–D. Similarly, a triad in the root key of F is made up of the notes F–A–C. This also means that when writing a triad with a root key of D, E, A or B, the third tone will always be a sharp. The major triads of C, D, F and G are shown in Figure 2.11.

There are three variations to the major triad scale, which are the 'minor', 'diminished' and 'augmented' triads. Each of these works on a similar principle to the major scale – using the first, third and fifth degree notes – with the difference that the position of third and/or fifth degree notes are raised or lowered by one semitone.

Figure 2.11 Major triads of C, D, F and G.

For instance, taking the major triad C–E–G from the root key of C and lowering the third tone gives us the minor triad C–E♭–G. For a diminished triad both the third and fifth tones are lowered, resulting in C–E♭–G♭. To create an augmented triad, the fifth note is increased by a semitone, producing C–E–G♯. These and other basic chord groups are shown in Table 2.4.

Figure 2.12 Variations to the major triad.

Table 2.4 Basic chord groups

Chord/root	Major	Minor	Diminished	Augmented
C	C–E–G	C–E♭–G	C–E♭–G♭	C–E–G#
D	D–F#–A	D–F–A	D–F–A♭	D–F#–A#
E	E–G#–B	E–G–B	E–G–B♭	E–G#–C
F	F–A–C	F–A♭–C	F–A♭–B	F–A–C#
G	G–B–D	G–B♭–D	G–B♭–D♭	G–B–D#
A	A–C#–E	A–C–E	A–C–E♭	A–C#–F
B	B–D#–F#	B–D–F#	B–D–F	B–D#–G
C# (enharmonic D♭)	C#–E# (F)–G#	C#–E–G#	C#–E–G	C#–E#(F)–A
D♭	D♭–F–A♭	D♭–E–A♭	D♭–E–G	D♭–F–A
E♭	E♭–G–B♭	E♭–G♭–B♭	E♭–G♭–A	E♭–G–B
F# (enharmonic G♭)	F#–A#–C#	F#–A–C#	F#–A–C	F#–A#–D
G♭	G♭–B♭–D♭	G♭–A–D♭	G♭–A–C	G♭–B♭–D
A♭	A♭–C–E♭	A♭–B–E♭	A♭–B–D	A♭–C–E
B♭	B♭–D–F	B♭–D♭–F	B♭–D♭–E	B♭–D–F#

So, to recap:

• To create a major triad take the first, third and fifth degree notes of the scale.
• To create a minor triad lower the third degree by a semitone.
• To create a diminished triad lower the third and fifth degree notes by a semitone.
• To create an augmented triad raise the fifth degree by a semitone.

In a musical context, the transition between these chords is what gives music a sense of movement and a unique character. To introduce this sense of movement, the centre note of the chord (the third degree) is often inverted or moved as the chords progress. If the third degree is moved up a semitone it creates an interval known as a major third, whereas if it were positioned or moved to three semitones above the root note, it becomes a minor third.

Inversions are created in much the same way as major/minor thirds. A chord is inverted by moving one of the notes by a melodically suitable degree. This is achieved largely by experimentation, but as an example, if a tune is written using the chord of C, the first degree (C) can be transposed up an octave so that the E becomes the root note. This creates a 'first inversion'. Next, the E can be transposed up an octave, leaving the G as the root. This creates a 'second inversion'.

C major   First inversion   Second inversion

Figure 2.13 First and second inversions of C major.

This kind of progression is especially suitable for pads or strings that need to change smoothly as the track progresses without attracting too much attention. That said, movement of any degree within the chord might be used to create chord progressions.

Simple chord triads, such as those shown in Table 2.3, can be extended by adding another note to create a four-note chord. The most commonly used four-note chord sequences extend the major triad by introducing a seventh degree note. This results in a 1–3–5–7 degree pattern, so the four-note chord of C major is made up of the notes C–E–G–B, as shown in Figure 2.14.

C   F

Figure 2.14 C and F major sevenths.

When the root note is in either major C or F, the seventh degree (B or E respectively) is 11 semitones above the root key, creating a major seventh. In any key other than C or F, the seventh degree will be 10 semitones above the root, creating a minor seventh.

D   G

Figure 2.15 D and G minor sevenths.

In dance music it's uncommon for four-note chords to be used. This is because major seventh chords tend to sound like Latino rhythms while minor sevenths are synonymous with jazz music. That said, you should experiment with different chord progressions as the basis for musical harmony. For those who still feel a little unsure about the construction of chords, Table 2.5 provides a list of popular chord sequences.

The chords in Table 2.5 can be used to select a number of chords to program them into the sequencer, and experimenting by adjusting the order of the chord progressions. Although this approach will not work all the time – you may choose totally dissonant chord structures – never underestimate serendipity!

Note: To help you decipher these chord structures, on the CD you'll find the PC software program Kord Kruncher.[1]

---

[1] Kord Kruncher is copyright to Andrew Le Couteur Bisson.

Table 2.5 Popular chord sequences

Keys →	← Chord	Keys →	← Chord
C–E♭–G–A	C MN 6	C–E♭–G–A–D	MN 6/9
C–E♭–G–D	C MN AG9	C–E♭–G–B♭	C MN 7
C–E♭–G–B♭–F	C MN 7 AG11	C–E♭–G–B♭–A	C MN 7 AG13
C–E♭–G–B♭–D	C MN 9	C–E♭–G–B♭–D–F	C MN 11
C–E♭–G–B♭–D–F–A	C MN 13	C–E♭–G–B	C MN MR
C–E♭–G–B–D	C MN 9 MR	C–E♭–F♯–B♭	C MN 7♭5
C–E♭–F♯–B♭–D	C MN 9♭5	C–E♭–F♯–B♭–D–F	C MN 11♭5
C–E♭–F♯	C DM	C–E♭–F♯–A	C DM 7
C–E♭–F♯–A–B	C DM7 AG MR	C–E–F♯–B	C MR♭5
C–E–G♯–B	C MR 5	C–E–G–B–F♯	C MR 11
C–E–G–B	C MR	C–E–G–B–A	C MR
C–E–G♯	C AG	C–F–G–B♭–C♯	C7sus♭9
C–F–G–B♭–C♯–A	C13sus♭9	C–F–F♯–B	Csus♭5
C–F–G	Csus	C–F–G–B♭	C7sus
C–F–G–B♭–D	C9sus	C–F–G–B♭–D–A	C13sus
C–E–G–B–D	Cmj9	C–E–G–B–D–A	Cmj13
C–E–G–B–D–F♯	Cmj9♯11	C–E–G–B–D–F♯–A	Cmj13♯11
C–E–F♯–B♭	C7♭5	C–E–F♯–B♭–D	C9♭5
C–E–G♯–B♭	C7♯5	C–E–G♯–B♭–D	C9♯5
C–E–G–B♭–C	C7♭9	C–E–G–B♭–E♭	C7♯9
C–E–F♯–B♭–C♯	C7♭5♭9	C–E–A♭–B♭–E♭	C7♯5♯9
C–E–G♯–B♭–C♯	C7♯5♭9	C–E–G–B♭–F♯	C7♯11
C–E–G–B♭–D–F♯	C9♯11	C–E–G–B♭–C♯–F♯	C7♭9♯11
C–E–G–B♭–E♭–F♯	C7♯ 9♯11	C–E–F♯–B♭–D–A	C13♭5
C–E–G–B♭–C♯–A	C13♭9	C–E–G–B♭–D–F♯–A	C13♯11
C–E–G–A	C6	C–E–G–D–A	C6/9
C–E–G–D	C AG 9	C–E–G–B♭	C7
C–E–G–B♭–D	C9	C–E–G–B♭–D–A	C13

Key: MN, minor; MR, major seventh; AG, augmented; DM, diminished.

With the basic chords laid down a bass line can be constructed around them. This is not necessarily how all music is formed, but it is common practice to derive the chord structure from the bass or melody or vice versa, and it's useful to prioritize the instruments used in the track in this way.

While there are no absolute rules as to how music is written, or at least there aren't as far as I know, it's best to avoid getting carried away programming complex melodies unless they form an essential part of the music. Although music works on the principle of contrast, too many complicated melodies and rhythms playing together will not necessarily produce a great track, and you'll probably end up with a collection of instruments that are all fighting to be noticed.

For example, typical '*hands in the air*' trance with big, melodic, synthetic leads doesn't require a complex bass melody because the lead is so intricate, so the bass is usually very simple. Indeed, whenever the melodic lead is the focus of the track, it may make sense to work on that first, to avoid putting too much rhythmical energy into the bass. If the bass were developed first, you may make it too melodic and it will make harmonizing the lead to sit with much more difficult and, in our example, may detract from the trance lead rather than enhance it. If, however, you construct the most prominent part of the track first, in this case the melodic synth, it's likely that you'll be more conservative when it comes to the bass. This problem does not rear its ugly head if the chord structure is fashioned first, because chord structures are relatively simple and this simplicity will give you the basic key from which you can derive both the melody and bass to suit the genre of music you are writing.

Indeed, because we subconsciously reference everything we hear, it pays to know the chord progressions that we are all familiar with. Chord progressions that have never been used before tend to alienate listeners. Most clubbers expect themes that they've heard a thousand times before and are not interested in originality: an observation that will inevitably stir up some debate. Nonetheless, it's true to say that each genre – whether techno, trance, drum 'n' bass, chill-out, hip-hop or house – is based around an identifiable theme. For example, if we compare Room 5's *Make Love* with Stardust's *Music Sound Better with You*, DJ Modjo's *Lady* and Supermen Lovers' *Starlight*, we can hear the similarities. Each of these has been a massive hit with clubbers, yet the styles, sounds and arrangements are very similar. This is not to say that you should attempt to copy other artists, but it does raise a number of questions about originality.

The subject of originality is a multifaceted and thorny issue, and often a cause of heated debate among musicians. Although the finer points of musical analysis are beyond the scope of this book, we can reach some useful conclusions by looking briefly at the works of L. Bernstein and A. Marx, two people who have made large discoveries in this area and are often viewed as the founders of musical analysis. They believed that if any musical piece were taken apart, aspects of it would be similar to most other records. Whether this is because musicians are often influenced by somebody else's style is another subject for debate, but while the similarities are not immediately obvious to the casual listener, to the musician these similarities can be used as the basis of musical compositions.

We can summarize their discoveries by examining the musical scale. As we've seen, there are 12 major and 12 minor keys, each consisting of eight notes. What's more, there are three basic major and three basic minor chords in each key. Thus, sitting at a keyboard and attempting to come up with an original structure is unlikely. When you do eventually come up with a progression that sounds 'right', it's likely that it sounds that way because you've subconsciously recognized the structure from another record. This is why chord structures often trigger an emotional response. In many cases, dance hits use a series of familiar underlying chord progressions. While the decoration around these common progressions may be elaborate and differ from other tracks, the underlying harmonic structure is very similar and seldom moves away from a basic three-chord progression. Because of this, there's nothing wrong with forming the bass and melody around a popular chord structure then removing the chords later on. In fact, popular chord structures are often a good starting point if inspiration takes a holiday and you're left staring at a blank sequencer window.

Musical relationship

When writing music of any style it's vital that the chords, bass and melody complement one another. Of these three elements, the most important relationship is between the bass and the melody, so we'll examine this first. This relationship has three musical possibilities, called parallel, oblique or contrary motion.

Parallel motion is when the bass line follows the same direction as the melody. This means that when the melody rises in pitch the bass also rises, but does not follow it at an exact interval. If both the melody and bass rise by a third degree, the resulting sound will be too synchronized and programmed and all of its feeling will be lost. Rather, if the melody rises by a third and the bass by a fifth or alternate thereof, the two instruments will sound like two voices and the overall sound will be enhanced. The bass should also borrow some aspects from the progression of the melody but should not play the same riff.

When the melody and bass work together in oblique motion, the melody or bass move up or down in pitch while the other instruments remain in the same key. In music that is driven by the melody, such as trance, the melody moves while the bass remains constant. In other genres, such as house, where the bass is often quite funky with plenty of movement, the melody remains constant while the bass dances around it.

The relationship known as contrary motion provides the most dynamic movement and occurs when the bass line moves in the opposite direction to the melody: if the melody rises in pitch, the bass falls. Using this technique it's usual that if the melody rises by a third degree, the bass falls by a third. Contrary motion can be a vital component of dance music that works because it draws attention to the harmonic structure of the bass and the melody and makes the whole record gel together.

This leads onto the complex subject of 'harmonization' – how the chord structure interacts with both the bass and melody. Despite the fact that dance tracks don't always use a chord progression or in some cases even a melody, the theory of how chords interlace with other instruments is beneficial on a number of levels. For instance, if you have a bass and melody but the recording still sounds drab, harmonizing chords could be added to see whether they help. Adding these chords can also give you an idea of where the vocals should be pitched.

The easiest way to construct a chord sequence is to take the key of the bass and/or melody and build the chords around it. For instance, if the key of the bass is in E, then the chord's root could be formed around E. If you construct your chords in this way it is, however, unwise to duplicate every change of pitch in the bass or melody as the root for your chord sequence. While the chord sequence will undoubtedly work, the mix of consonant and dissonant sounds will be lost and the resultant sound will be either manic or sterile. To ensure that the chord structure instils emotion, the interactions between the bass, melody and harmony must be more complex, and attention to this will quite often make the difference between great work and an average track.

Often, the best approach is to develop a chord progression based around any of the notes used in the bass. For instance, if the bass is in E, then you shouldn't be afraid of writing the chord in C major, because this contains an E. Similarly, A minor, the minor to C major, would work just as well (come on, keep up at the back ...).

For the chords to work, they must be closely related to the key of the song, so it doesn't necessarily follow that because a chord utilizes E in its structure it will work in harmony with the song as a whole. For a harmony to really work it must be inconspicuous, so it's important that the right chords are used. A prominent harmony will deter from the rest of the track, so harmonies are best kept simple, avoiding frequent or quick changes and changes every time the bass note or melody changes. The best harmony is one that is only noticeable when removed.

While the relationship between the bass and chord structures is fundamental to creating music that gels properly, not every bass note or chord should occur dead on the beat, every beat, or follow the other's progression exactly. This kind of sterile perfection is tedious to listen to and will quickly bore listeners, but it's a trap that is easy to fall into and difficult to get out of, particularly when working with software sequencers. To avoid this, you must experiment by moving notes around so that they occur moments before or after each other. This deliberately inaccurate timing, often referred to as the 'human element', forms a crucial part of most musical styles.

This mistiming is such a natural occurrence that our brains tune into this phenomenon, refusing anything that repeats itself perfectly without any variation. Although you may not recognize exactly what the problem is, the brain instinctively knows that a machine has created the performance. It could be argued that the human element is missing from most forms of dance music, but that assumes that the human element cannot be recreated electronically. By offsetting various notes deliberately or by utilizing quantize functions in the sequencer, the slight inaccuracies we expect to hear can be written into the music, and it is these techniques that define the groove in the most well-known dance records. Before we look at this in more detail, however, we need to understand tempo and time signatures.

Tempo and time signatures

In all popular music the tempo is measured by counting the number of beats that occur per minute (BPM). BPM measurements are also sometimes referred to as 'metronome mark' and musicians, along with some producers, use these along with other musical terms to describe the general tempo of any given genre of music. For instance, the hip-hop genre is characterized by a slow, laid-back beat, but individual tracks will use different tempos that can vary from 80 BPM through to 110 BPM. Thus, it's usual that different musical styles are described using general musical terms. Using musical terms, then, hip-hop can be described as 'andante' or 'moderato', meaning at a walking pace or at a moderate tempo. Similarly, the house or trance genres may be described as 'allegro', meaning quickly, while drum 'n' bass may be described as 'prestissimo', meaning very fast.

Several of the most common musical terms are listed below:

- *Largo* – slowly and broadly.
- *Larghetto* – a little less slow than largo.
- *Adagio* – slowly.
- *Andante* – at a walking pace.
- *Moderato* – at a moderate tempo.
- *Allegretto* – not quite allegro.
- *Allegro* – quickly.
- *Presto* – fast.
- *Prestissimo* – very fast.

It is important to bear in mind that the actual number of beats per minute in a piece of music marked presto, for example, will also depend on the music itself. A track that is constructed of half notes can be played a lot quicker (in terms of BPM) than one that consists almost entirely of 16th notes, but it can still be described using the same word. To make sense of this we need to examine time signatures and different note lengths.

Time signatures determine the rhythmic feel and 'flow' of the music, and so an understanding of them is essential. To understand the principle of time signatures, you first have to learn how to count. Of course, we all learnt how to do this in primary school, but it's not quite the same in music because of the way music is transcribed onto a musical staff.

Figure 2.16 C major scale.

Figure 2.16 shows the notes of the C major scale on the musical staff. As we learned earlier in this chapter, each note's pitch can be seen in relation to the other notes according to its vertical position. While the position of the notes indicates the pitch, the symbol used to represent each note indicates the duration. Table 2.6 shows the timing relationships between notes and their symbolic representation.

Table 2.6 Timing relationships between notes

Symbol	Name	Common name	Value in relation to a whole note	Equivalent in timing to:
o	Whole note	Semibreve	1	♩♩
♩	Half note	Minim	1/2	♩♩♩♩
♩	Quarter note	Crotchet	1/4	♪♪♪♪♪♪♪♪
♪	Eighth note	Quaver	1/8	♪♪♪♪♪♪♪♪♪♪♪♪♪♪♪♪
♪	Sixteenth note	Semiquaver	1/16	♪♪♪♪♪♪♪♪♪♪♪♪♪♪♪♪
♪	Thirty-second note	Demisemiquaver	1/32	–

Each note's duration can be described in terms of its relation to a whole note, or 'semibreve'. If notes were simply dropped onto a musical stave, as they were in Figure 2.3, the musician would play them one after the other but there would be no rhythm to the piece. To avoid this, music is broken down into a series of bars and each bar has an associated rhythm. Generally, this rhythm remains the same throughout every bar of music, but in some instances the rhythm can change during the course of a song.

All music must have rhythm: it's what gets your feet moving on the dance floor. To grasp this we need to be able to count like a musician. The ticking of an ordinary clock (provided it isn't a digital clock, of course!) is a good place to start. When we listen to a clock, every tick corresponds to a number from 1 to 60 seconds. To start counting musically, we count in time with the regular clock ticks but count up to four and then start all over again from one instead of counting all the way up to 60. For instance:

One…two…three…four…one…two…three…four…one…two…three…four…and so forth

This gives us a pattern of four beats in the bar.

Now suppose we emphasize every number one of our count by saying it slightly louder, we now have the pattern:

ONE…two…three…four…**ONE**…two…three…four…**ONE**…two…three…four…and so forth

Next, rather than counting up to four, count up to three while still placing an accent on the number one, changing the pattern to that of three beats in the bar.

ONE…two…three…**ONE**…two…three…**ONE**…two…three…**ONE**…two…three…and so forth

You'll notice there is a distinct change in the rhythm. Counting up to four produces a rather static rhythm (typical of most dance music), while counting up to three produces a rhythm with a little more swing to it (typical of some chill-out, trip-hop and hip-hop tracks). This is the basic concept behind time signatures; they allow the musician to determine what kind of rhythm the music employs.

Time signatures are placed at the beginning of a musical piece to inform the musician of the general rhythm, how many and what kind of notes there are per bar. They are written with one number on top of another, similar to mathematical fractions such as 1/4 or 1/2. In music, the number at the top of the time signature indicates the number of 'beats' per bar (do not confuse this with beats per minute!) and the bottom number indicates the length of these beats.

To explain this, let's consider the time signature of 3/4 (pronounced 'three-four'). The top number determines that there are three beats to the bar, while the bottom number tells us the length of these beats. If we look again at Table 2.6, we see that we have different sizes of notes, from whole notes through to 32nd notes. To determine the size of each beat indicated by a time signature, we take a whole note (1) and divide it by the bottom number (4). This gives us:

1 whole note ÷ 4 = 4 quarter notes (i.e. 4 crotchets)

This gives us the beat size of one crotchet. From this we can determine that each bar of music can accept no more than three crotchets. Of course, this doesn't limit us to using just three crotchets, it simply means that no matter what notes we decide to use we cannot go above the total sum of three crotchets, so in other words we could use any of the examples shown in Table 2.7.

This technique can be applied to other time signatures. If we work in the time signature of 5/2 (five-two) we can tell from the top note that there are five beats to the bar. As before, we work

Table 2.7 Some possible notes in 3/4 time

Note lengths	Musical notation	Equivalent value in crotchets
Four eighth notes + one quarter note	♪♪♪♪ + ♩	= ♩♩♩
One half note + two eighth notes	♩ + ♪♪	= ♩♩♩
One half note + one quarter note	♩ + ♩	= ♩♩♩

out the size of these beats by dividing a whole note by the bottom number in the time signature, in this case two. This gives us five half notes (minims) to the bar. If applied to the time signature 6/8 (six-eight) there are six eighth notes (quavers) in the bar. These different time signatures, while appearing quite abstract and meaningless on paper, produce very different rhythmic results when played.

Although most dance music is written in 4/4 or 3/4, writing in different time signatures can occasionally aid in the creation of grooves. 'Groove' is a vital element in dance music because its purpose is to make you want to move your feet, so having the right kind of vibe is essential. The problem is that, while it is possible to dissect and describe the principles behind the programming of different music genres, defining what actually makes the groove of one record better than another isn't as straightforward. Indeed, creating a killer dance floor groove is something of a Holy Grail that all dance music producers are continually searching for. Although finding it has often been accredited to a mix of skill, creativity and serendipity, there are some general rules of thumb that often apply.

Injecting groove into a performance is something that any good musician does naturally, but it can be roughly measured by two things: timing differences and variations in the dynamics. A drummer, for instance, will constantly differ how hard the kit is hit, resulting in a variation in volume throughout, while also controlling the timing to within microseconds. These variations in volume (referred to as dynamic variation) inject realism and emotion into the performance, while the slight timing differences add to the groove.

By deliberately adjusting the timing and dynamics of each instrument we can inject groove into a recording. If a kick and hi-hat play on the beat or a division thereof, moving the snare's timing forward or backward will make a huge difference to the feel. Similarly, programming parts to play slightly behind the beat creates a bigger, more laid-back feel, while positioning them to play in front of the beat creates a more intense, almost nervous feel. These are the principles on which swing and groove quantize in sequencers operate, both of which can be used to affect rhythm, phrasing and embellishments.

Using swing quantize, the grid is moved away from regular slots to alternating longer and shorter slots. This can be used to add an extra feel to music or, if heavily applied, can change a 4/4 track into a 3/4 track. Groove quantize is a more recent development that allows you to import third-party groove templates and then apply them to the current MIDI file. This differs from all other forms of quantize because it redefines the grid lines over a series of bars rather than just one bar, thereby recreating the feel of a real musical performance. In many instances groove templates also affect note lengths and the overall dynamics of the current file, creating a more realistic performance. In more adept sequencers, groove templates extracted from audio files can be applied to a MIDI file to recreate the feel of a particular performance.

The best results are obtained by using swing and groove quantizing features creatively – for example, by creating complex drum loops from two or three drum loops that each use different quantize settings. This approach works equally well when creating complex bass lines (if the track is bass driven), lead lines and motifs. It is important to note, however, that quantize shouldn't be relied on to instantly add a human feel, so creating a rigid bass pattern in the hope that quantizing will introduce some feel later isn't a good idea. Instead, you should try to program the bass with plenty of feel from the start: by physically moving notes by a few ticks and then experimenting further with the quantize options to see if it adds anything extra.

In addition to groove, another important aspect of dance music is the drive, which gives the effect of pushing the beat forwards as if the song is attempting to run faster than the tempo. This feeling of drive comes from the position of the bass and drums in relation to the rest of the record. For instance, moving the drums and bass forward in time by just a couple of ticks gives the impression that they are pushing the song forward, in effect producing a mix that appears as if it wants to go faster than it actually is. This feeling can be further accentuated if any melodic elements, such as pianos, are programmed to sit astride the beat rather than dead on it and if different velocities are used to accentuate parts of the rhythm.

This brings us to the subject of dynamics and syncopation, both of which play a vital role in capturing the requisite rhythm of dance. All musical forms are based upon patterns of strong and weak phrases working together. In classical music, it is the constant variation in dynamics that produces emotion and rhythm from these contrasts. In dance music these dynamic variations are often accentuated by the rhythmic elements. By convention, the first beat of the bar is the strongest. Armies on the march provide a good example of this, with the 2/4 rhythm, Left–Right–Left–Right–Left–Right. In this rhythm there are two beats (Left and Right). The first beat is the strongest, with the second beat being a little weaker. With the typical dance music 4/4 kick drum the first beat in the bar is usually the strongest (that is, it has the greatest velocity), with any subsequent hits varying in dynamics. An example of dynamic variations is shown in Table 2.8.

Table 2.8 Dynamic variations across four bars

No. of 'kicks' in the bar	Beat pattern	Pattern over four bars
1	Strong	S–S–S–S
2	Strong–weak	SW–SW–SW–SW
3	Strong–medium–weak	SMW–SMW–SMW–SMW
4	Strong–weak–medium–weak	SWMW–SWMW–SWMW–SWMW

Syncopation also plays an important role in the rhythm of a track. This is where the stress of the music falls off the beat rather than on it. All dance music makes use of this and it's often a vital element that helps tie the whole track together. While the four-to-the-floor rhythm pounds out across the dance floor, a series of much more elaborate rhythms interweave between the beats of the main kick drum. For instance, if we count in the same manner as we did for the time signatures:

ONE...two...three...four...**ONE**...two...three...four...**ONE**...two...three...four...and so forth

Each number corresponds to the kick drum, but if we were to add 'and' into the counting we get:

ONE *and* two *and* three *and* four...**ONE** *and* two *and* three *and* four...and so forth

Each '*and*' occurs on the off beat, or, in musical terms, on the quaver of the beat. A good example of this is found in trance music, where the bass notes occur between the kicks sat on the beat.

More emphasis can be added using velocity commands, which simulate or recreate the effect of hitting some notes harder than others, thereby creating different levels of brightness in the sound. These are commonly used with bass notes. Using velocity commands you can keep the first note of the bar at the highest velocity and then lower the velocity for each progressive note in the bar and, at the next bar, repeat this pattern again. It's important to also consider here that the length of any bass notes can also affect the perceived tempo of the track. Shorter, stabbed notes, such as 16ths, will make a record appear slightly more frantic than using notes that are longer.

> Note: If the drum and bass timbres have been programmed, it's worth experimenting by lengthening and/or shortening the duration of notes to observe the difference.

While it is impossible to define a groove precisely, the best way to know whether you've hit upon a good vibe is to try dancing to it. If you can, the groove is working. If not, you'll need to look at the bass and drums again before moving any further. Keep in mind that the groove forms the backbone of all dance records, so it's imperative that the groove works during this early stage. Don't hope that adding additional melodic lines later will help it along: if it sounds great at this point then further melodic elements will only help improve it. If you're stuck for inspiration or are struggling to create a groove, try downloading MIDI files of other great songs with a similar feel and incorporate and manipulate the ideas to create your own groove. This isn't stealing, it's research and *every* musician on the planet started this way!

The only real 'secret' behind creating that killer groove is to never settle for second best, which is why many artists freely admit to spending weeks making sure that a track has a good vibe. This includes giving it a quick mix-down so that it sounds exactly as you would envisage it when the track is completed. If at this point the vibe is still missing, go back and change the elements again until it sounds exactly right. Take into consideration that the mixing desk, as well as being a useful creative tool, only polishes the results of a mix; hence, sounds going into a mix must be precise in the first place.

Creating melodies/motifs

Although melodies and motifs often take second place to the general groove of the record they still have an important role in dance music. Notably, very few dance tracks employ long, elegant, melodic lines (some trance being an exception), since these can deter from the rhythmic elements of the music but, if programmed correctly, they are extremely powerful tools. Consequently, most dance music uses short motifs, as these are simple yet powerful enough to drive the message home. Generally, a good motif is a small, steady, melodious idea that provides listeners with an immediate reference point in the track. For instance, the opening strings in The Verve's *Bitter Sweet Symphony* could be considered a motif, as can the first few piano chords on Britney Spears' *Hit Me Baby One More Time*, because as soon as you hear them you can identify the track.

Knowing what makes a good motif is central to the ability to write one, so what are the principles behind their creation? Well, motifs follow similar rules to the way that we ask and then answer a question. In a motif the first musical phrase 'asks' the question and this is then 'answered' by the second phrase. This is known as a 'binary phrase' (or Da Capo) and some of the best examples can be found in nursery rhymes such as 'Baa baa black sheep' asking the question, answered by 'Yes sir, yes sir, three bags full'.

This is a simple and well-known motif and it's important to observe how the two lines balance each other out. Musically, the first line consists of CCGGABCAG, while the reply consists of FFEEDDC. Also, notice how the first phrase rises but the second phrase falls, whilst maintaining the similarities between the two phrases. This type of balance underlies the most memorable motifs and is key to their creation. Whether a four-to-the-floor remix of Baa baa black sheep would go down well in a club is another question altogether, but as a motif it contains the three key elements:

- Simplicity.
- Rhythmic repetition.
- Some variation.

The use of rhythmic repetition in music is not exclusive to dance music. Most classical music is based entirely around the repetition of small flourishes and motifs – so much so that many dance musicians listen to classical works and derive their ideas from its form. In fact, the principle behind writing a memorable record is to bombard the listener with the same phrases so that they become embedded in the listener's memory, although this must be accomplished in such a way that the music doesn't sound too repetitive. Indeed, one of the biggest mistakes made by musicians just starting out is to instil too many new ideas without taking any notice of what has happened in the previous bars.

When creating a motif, simply repeating the phrases AB–AB–AB–AB can result in too much repetition, so another method is to create a third phrase. This is called 'ternary' repetition and is where you create a third 'answer' and alternate between the three motifs in different ways, such as ABC–BAC.

It's also commonplace for musicians to introduce a fourth motif, resulting in two 'questions' and two 'answers'. This arrangement is called a 'trio minuet' formation and can take on various forms, such as ABA (minuet), CDC (trio), ABA–CDC and so forth. Although not all dance musicians work this way, these are generally accepted methods that have led to two simple rules:

1   If the beginning of the motif rises, the second part will fall.
2   After moving in one direction, there is a step back in the opposite direction.

Having said that, it's not uncommon for the inverse to be true – a relationship that can be used to musical effect. When any motif falls in scale it gives the impression of pulling our emotion downwards, yet if it increases in scale it is more uplifting. This is because there are higher frequencies when a note rises, as is evident when the low-pass filter cut-off of a synthesizer is opened up to allow more high-frequency content through. You can hear this in countless dance records where the music becomes duller, as though it's being played in another room, and then gradually the frequencies increase, creating a sense of expansion as the sound builds. If you listen to most dance records this form of building, using filter cut-off movement, becomes apparent. When the track is building up to a crescendo the filter opens, yet while the track drops away the filter closes. Similarly, the rising key change introduced at the end of many popular songs builds the sound, giving the impression that the song is at full impact, driving the message home.

Note: An excellent example of the effect filter movement has can be heard on Sasha's *Xpander*. The riff used is repetitive and it's the filter movements that carry the track.

Like everything else in music, understanding the theory behind good motifs and actually creating one that captivates attention is an entirely different matter: like great grooves, great motifs are something of a Holy Grail and coming up with them is going to be a mix of skill, creativity and luck. Irrespective of this, there are some ideas that you can try out:

- One approach is to record MIDI information from a synthesizer's arpeggiator and then edit this in a sequencer to produce some basic ideas.
- Another technique that can produce usable results is what's often called the 'two-fingered' approach. This involves tapping out a riff using two keys an octave apart on the keyboard. By alternating between these two keys, keeping the lower key the same and continually moving higher or lower in scale on the second key, the beginnings of interesting rhythmic movements can be developed. If the rhythm is recorded as MIDI it can be edited further by lengthening notes to overlap each other or by adding further notes in between the two octaves to construct motifs.
- The third and most commonly used method is to continually loop the bass and drums and experiment on the keyboard until you come up with something that complements the rhythm section. Once the first bar(s) of the phrase – the 'question' – have been fashioned they can be pasted into the subsequent bar(s) and edited to create the 'answer'.

Above all, it's best to avoid getting carried away with any of these techniques. It's vital that they are kept simple. Many dance tracks use a single pitch of notes played every eighth, 16th or quarter, or are constructed to interweave with the bass to produce a complex groove. Indeed, simple motifs have a much more dramatic effect on the music than complex arpeggios, so it's important that they capture the listener's interest.

Because the sound used for any motif should stand out, it's important to consider the frequencies that are used or are to be used in the mix. In many dance tracks where three of four simple motifs play simultaneously, the combined frequencies can take up a proportionate amount of the available space within the mix. To prevent clogging up the mix, motifs are usually worked with as MIDI rather than transformed to audio, as this allows you to use filter cut-off to reduce the frequencies. This method also allows you to utilize a low-pass filter cut-off and control the sound's sonic content while the other motifs are playing. In working with the optimum frequencies of a motif, its rhythmical pitch must be carefully chosen.

Any motif that uses notes above C4 will contain many higher frequencies and few lower ones, so using a low-pass filter that is almost closed will reduce the frequencies to nothing. Careful use of the filters will create the final result and often you will need to compromise by cutting the frequencies of some motifs while opening them up for others. Manipulating the frequencies in this way creates the impression that the sounds are interweaving, which creates the movement and 'energy' that is characteristic of most club tracks.

Creating progression

Although dance music is heavily based on repetition, it retains interest by building and releasing the groove. This is achieved by adding and removing sonic elements throughout the track and the gratuitous use of filters. To fully understand how a groove is built and released we need to look at sections of the whole track, counting what has gone before, how the next event is prepared, what comes after it and how it ends. For example, when you hear the typical dance snare crescendo building you automatically know that the main part of the groove will follow. The longer this is *reasonably* postponed, the bigger the impact is expected to be. The best way to learn about this is to listen to and dissect popular dance tracks and then create a song map charting where the builds and drops occur, essentially charting the emotion that it creates.

There are various methods of song mapping and each is as valid as the next, but because many psychologists believe that we are naturally disposed to think in pictorial form it may be beneficial to draw a map of the arrangement onto paper. How this is depicted is entirely up to the individual and can consist of anything from simple lines to more complex drawings involving colours and/or symbols to depict each instrument and the parts of the original recording. Because club mixes can be quite long, often averaging around 6–8 minutes, these types of maps can be a blessing. Paul Van Dyk and BT, for instance, are well known for creating a song map before beginning to build any type of track. A song map not only helps you to envisage the final product, but it also helps you plot out the track in terms of crucial builds and drops.

The basic principle behind this strategy is that it gives a visual reference to where the different emotional states occur in the dance track, helping you to envisage where the track will bring listeners up on a 'high' and then drop them down again.

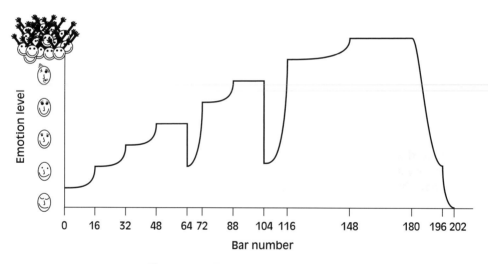

Figure 2.17 Example of a song map.

It's evident from the song map shown in Figure 2.17 that dance music differs from the usual musical structure of verse, chorus, verse, chorus, bridge and outro. Instead, dance music derives its own set of rules based around mathematical repetition taken to extremes, usually broken into sections of four bars. This formal structure may seem overtly technical for what is otherwise a creative art, but try writing a 16-bar loop and introducing an instrument at bar 5, 13 or 15 and it will no doubt sound erroneous.

Looking at Figure 2.17 we can also determine that the typical arrangement consists of an intro, first body, drop, second body, second drop, reprise, main body and finally an outro. Each of these changes relates to the emotional states that, ideally, you need to generate for the listeners. Using a map we can see how this might be transferred and used in an imaginary club track, of which an example is described below.

The intro

Bars 1–16

This section introduces the track and consists of nothing more than a 4/4 drum loop. This is kept simple at the beginning of the track and continues unchanged for the first 16 bars. This standard introduction gives the DJ playing the music a 4/4 beat with no other instrumentation so that he can mix it in with previous track, keeping the beat going from one record to the next to keep everyone up on the floor. 'Dropping a beat' will easily disorientate the clubbers who are dancing: we all need a basis on which to keep time.

Bars 16–32

At this point, a new instrument is introduced. In order not to give the game away too soon a new percussive element comes in. This could consist of another closed or open hi-hat pattern, snares or claps. This completes the drum loop that will play throughout the rest of the mix.

First body

Bars 32–48

After the initial 32 bars, the main groove is introduced. Now both the bass and drums are playing together, laying the foundations of the groove. This plays unaltered to allow the clubbers to become comfortable with the general groove of the record.

Bars 48–64

At bar 48 a motif is introduced. This interweaves with the bass and drums using slow filter cut-off movements. Applying a low-pass filter to the motif and slowly opening it introduces higher frequencies, giving the effect that the song is building up towards something special. Listen to any popular record and you'll notice this effect. As the song progresses more instruments are introduced, usually at the chorus stage, filling out the mix's frequency range, bringing the song towards a climax. These filter movements are typically used to create tension and movement in all forms of dance music and have formed the basis of the most well respected dance records.

The drop

Bars 64–72

At this point all the percussive elements, bar the kick drum, have been dropped from the mix while the motif continually plays along. The filter is closed, giving the impression that the song is dropping emotionally. However, by the middle fourth bar the drum rhythm picks up again, starting with the open hi-hats followed by the snares, signifying that the track is about to come back. As the rhythm picks up, the motifs filter opens up again and is followed by a short snare roll, where the second body of the track is reintroduced.

The second body

Bars 72–88

At the final beat of the preceding bar's drop section, a crash cymbal starts a new passage of music and a new motif is introduced. The filter on both motifs is now contorted, resulting in the motifs' frequencies moving in and out of one another to create a flowing movement that helps the track groove along.

Bars 88–104

Again, a crash cymbal, snare roll, skip or sound effect is employed to denote the next 16 bars of the second body and some chorded strings are introduced. Low-pass filter movements are employed on the strings to prevent them from appearing too static. As before, the filter movements open, giving the impression that the song is building up again, while in the final four bars of this section a snare roll signifies that the song is building further.

The drop

Bars 104–108

A sudden crash cymbal at the end of the preceding bar echoes across the mix while the song drops to just the strings and the motif that was introduced in the second body. Again, the filter cut-off closes to denote a break to the second bar of this drop, whereby it begins to open again, giving the impression that the song is once again building.

The reprise

Bars 108–116

At the end of the drop a snare roll is introduced, slowly gaining in volume as the cut-off frequencies of both the motif and chords are opened further. This snare roll lasts the entire length of the eight bars, with the snares becoming progressively closer together the closer they get to the main body of the track. The principle aim in this section is to build emotion to a maximum, so that when the main body is reached the track is at its apex. Ideally, this is a culmination of all the instruments that have been previously introduced throughout the mix and a new motif.

The main body

Bars 116–148

The main body is used to drive the message of the track home and is signified by all of the previous instruments playing together, creating a crescendo of emotion. The drums, bass, previous motifs and a new motif are introduced, filling up the entire frequency spectrum of the mix. Again, filter cut-offs play a role here to prevent boredom and each instrument interacts with one another by closing the filter of one motif while opening it on the other.

Bars 148–180

At this point, additional snares are introduced to add a skip to the rhythm and a few sweeping effects are introduced that pan across the mix.

The outro

Bars 180–196

One of the motifs is dropped from the mix after a cymbal crash and the filters on the other motifs and chords are beginning to close, bringing the track back down in emotion. Finally, all the motifs and chords are removed, leaving just the drums and bass playing.

Bars 196–212

The bass is dropped from the mix, leaving the full drum rhythm playing solo for the final 16 bars, allowing the DJ to mix in the next record. The track draws to a close.

Of course, this imaginary track is quite simple and in reality building a track is much more complex, but nonetheless this is a good example of typical dance club structure. As with all music this is open to an individual's interpretation, but bear in mind that dance music is an exact science and the audience likes to know what to expect. Building a snare crescendo and then dropping it back again would not be greeted with smiling or gurning faces.

In many cases, the arrangement will differ according to the genre of music being written. While this type of arrangement may work for trance, it won't work too well for hip-hop and trip-hop, as these require a more gentle approach. We'll look more closely at the difference between these genres and their programming in later chapters.

3 Basic synthesis

I know what I'm looking for in a sound as soon as I hear it.

Peter 'Baby' Ford

To be proficient at creating dance music, it's necessary to fully comprehend the technology used in its creation. Because dance music relies as much on the actual production values as on anything else, no matter how musical or well programmed an arrangement is, if it doesn't utilize the 'right' sounds for the genre then it certainly isn't going to light up a dance floor. Even simple things – such as using a typical rock drum kick rather than the archetypal 909 or 808 kicks – can cause a track to lose its dynamic edge. Consequently, to write successful dance music a good understanding of synthesizer programming and the ability to process the results effectively is essential.

Before the various programming techniques can be understood, it's necessary to decipher the purpose of the various knobs and buttons that adorn a typical synth and observe the effects that each has on a sound, and for that we need to start at the very beginning by examining acoustics: the science of sound.

Acoustic science

When any object vibrates, air molecules surrounding it begin to vibrate sympathetically in all directions. These sound waves then cause vibrations in the eardrum that the brain perceives as sound. The movement of sound waves is analogous to the way that waves spread when a stone is thrown into a pool of water. The moment the stone hits the water, the reaction is immediately visible as a series of small waves spread outwards in every direction. This is almost identical to the way in which sound behaves, with each wave of water being similar to the vibrations of air particles.

For instance, when a tuning fork is struck the forks first move toward one another, compressing the air molecules directly in front of them before moving in the opposite direction. In this movement from 'compression' to 'rarefaction' there is a moment where there are less air molecules filling the space between the forks. When this occurs, the surrounding air molecules crowd into this space and are then compressed when the forks return on their next cycle. As the fork continues to vibrate, the previously compressed air molecules are pushed further outwards by the next cycle of the fork, and a series of alternating compressions and rarefactions pass through the air.

The number of rarefactions and compressions, or 'cycles', that are completed every second is referred to as the operating frequency and is measured in hertz (Hz). Any vibrating object that completes, say, 300 cycles per second has a frequency of 300 Hz, while an object that completes 3000 cycles per second has a frequency of 3 kilohertz (kHz).

The frequency of a vibrating object determines its perceived pitch, with faster frequencies producing sounds at a higher pitch than slower frequencies. From this, we can determine that the faster an object vibrates, or 'oscillates', the shorter the cycle between compression and rarefaction. An example of this is shown in Figure 3.1.

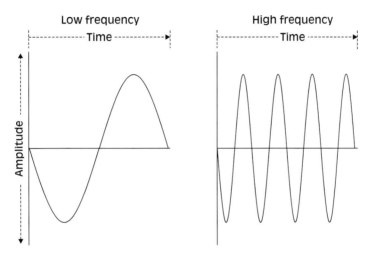

Figure 3.1 Difference between low and high frequencies.

Any object that vibrates must repeatedly pass through the same position as it moves back and forth through its cycle. Any particular point during this movement is referred to as the 'phase' of the cycle and is measured in degrees, similar to the measurement of a geometric circle. As shown in Figure 3.2, each cycle starts at position zero, passes back through this position, known as the 'zero crossing', and returns to zero.

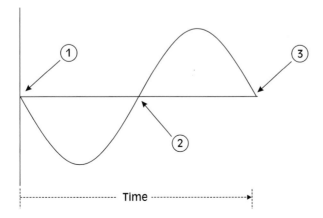

Figure 3.2 The zero crossing in a wave form.

Consequently, if two objects vibrate at different speeds and the resulting waveforms are mixed together, both waveforms will start at the same zero point, but the higher frequency waveform will overtake the phase of the lower frequency. Provided that these waveforms continue to oscillate they will eventually catch up with one another and then repeat the process all over again. This produces an effect known as 'beating'.

The speed at which waveforms 'beat' together depends on the difference in frequency between them. It's important to note that if two waves have the same frequency and are 180 degrees out of phase with one another, i.e. one waveform reaches its peak while the second is at its trough, no sound is produced. This effect, where two waves cancel one another out and no sound is produced, is known as 'phase cancellation' and is shown in Figure 3.3.

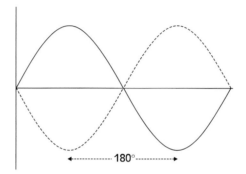

Figure 3.3 Two waves out of phase.

As long as waveforms are not 180 degrees out of phase with one another, the interference between the two can be used to create more complex waveforms than the simple sine wave. In fact, every waveform is made up of a series of sine waves, each slightly out of phase with one another. The more complex the waveform this produces, the more complex the resulting sound. This is because, as an increasing number of waves are combined, a greater number of harmonics are introduced. This can be better understood by examining how an everyday piano produces its sound.

The wires in a piano are adjusted so that each wire oscillates at an exact frequency. When a key is struck a mallet strikes the corresponding string, forcing it to oscillate. This produces the fundamental pitch of the key and, also, if the vibrations from this wire are the same as any of the other wires' natural vibration rates, sets these into motion too. These are called 'sympathetic vibrations' and are important to understand because most musical instruments are based around this principle. The piano is tuned so that the strings that vibrate sympathetically with the originally struck string create a series of waves that are slightly out of phase with one another, producing a complex sound.

Any frequencies that are an integer of the lowest frequency (i.e. the fundamental) will be in harmony with one another, an occurrence that was first realized by Pythagoras, from which he derived the following three rules:

1   If a note's frequency is multiplied or divided by two, the same note is created but in a different octave.

2   If a note's frequency is multiplied or divided by three, the strongest harmonic relation is created. This is the basis of the western musical scale. If we look at the first rule, the ratio 2:3 is known as a perfect fifth and is used as the basis of the scale.
3   If a note's frequency is multiplied or divided by five, this also creates a strong harmonic relation. Again, if we look at the first rule, the ratio 5:4 gives the same harmonic relation but this interval is known as the major third.

The relationships between harmonics and frequency ratios are shown in Table 3.1.

Table 3.1 Relationships between harmonics and frequency ratios

Name	Distance in semitones	Frequency ratios
Unison	0	1:1
Minor second	1	16:15
Major second	2	9:8
Minor third	3	6:5
Major third	4	5:4
Perfect fourth	5	4:3
Augmented fourth	6	45:32
Diminished fifth	6	64:45
Perfect fifth	7	3:2
Major sixth	8	8:5
Minor sixth	9	5:3
Major seventh	10	16:9
Minor seventh	11	15:8
Octave	12	2:1

A single sine wave produces a single tone known as the fundamental frequency, which in effect determines the pitch of the note. When further sine waves that are out of phase from the original are introduced, integer multiples of the fundamental frequency are added to the signal. These integer multiples of the fundamental are known as 'harmonics' and make the sound appear more complex. Harmonics that are not integers of the fundamental are called 'partials', which also contribute to the complexity of the sound. Through the introduction and relationship of these harmonics, an infinite number of sounds can be created.

The harmonic content determines the type of sound we hear, whether it's a piano, an electric guitar or a kick drum, and is commonly referred to as the timbre (French for tonal colour and pronounced 'tamber'). However, while describing a sound is simple enough, representing it on paper is extremely difficult because the waveform produced by any instrument is incredibly complex.

In an attempt to overcome this, Jean Fourier, a French scientist, discovered that no matter how complex any sound is, it could be broken down into its frequency components and, using a given set of harmonics, it was possible to reproduce it in a simple form. To use his words: 'Every periodic wave can be seen as the sum of sine waves with certain lengths and amplitudes, the wavelengths of which have harmonic relations.' This is based around the principle that the content of any sound is determined by the relationship between the level of the fundamental frequency and its harmonics and their evolution over a period of time. From this theory, known as the Fourier Theorem, the waveforms that are common to most synthesizers are derived.

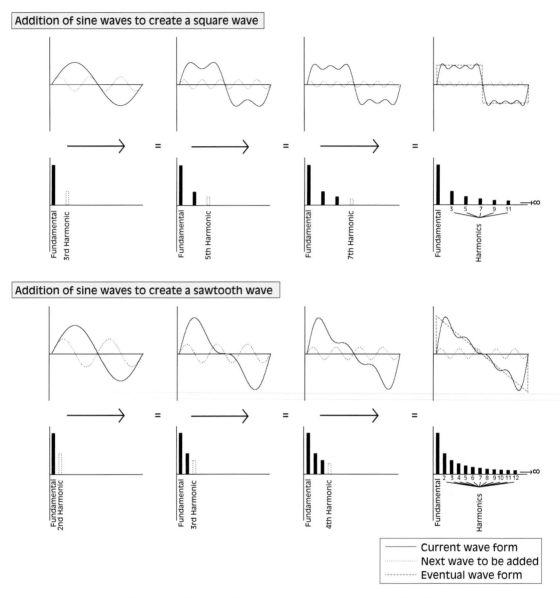

Figure 3.4 How multiple sound waves create harmonics (1).

So far, we've looked at how both the pitch and timbre are determined. The final characteristic to consider is volume. Changes in volume are caused by the amount of air molecules an oscillating object displaces. The more air an object displaces, the louder the perceived sound. This volume, also called 'amplitude', is measured by the degree of motion of the air molecules within the sound waves, corresponding to the extent of rarefaction and compression that accompanies a wave. The problem, however, is that many simple vibrating objects produce a sound that is inaudible to the human ear because so little air is displaced, so for the sound wave to be heard most musical instruments must amplify the sound that's created. To do this, acoustic instruments use the principle of forced vibration that utilizes either a sounding board, as in a piano or similar stringed instrument, or a hollow tube, as in the case of wind instruments.

Addition of sine waves to create a triangle wave

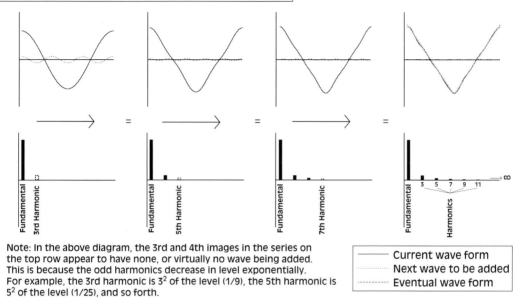

Note: In the above diagram, the 3rd and 4th images in the series on the top row appear to have none, or virtually no wave being added. This is because the odd harmonics decrease in level exponentially. For example, the 3rd harmonic is 3^2 of the level (1/9), the 5th harmonic is 5^2 of the level (1/25), and so forth.

──── Current wave form
············ Next wave to be added
---------- Eventual wave form

Figure 3.5 How multiple sound waves create harmonics (2).

When a piano wire is struck its vibrations not only set other strings in motion, but also vibrate a board located underneath the strings. Because this sounding board does not share the same frequency as the vibrating wires, the reaction is not sympathetic and the board is forced to resonate. This resonance moves a larger number of air particles than the original sound alone, in effect amplifying the sound. Similarly, when a tuning fork is struck, if it's placed on a tabletop, the table's frequency is forced to match that of the tuning fork and the sound is amplified.

Of course, neither of these methods of amplification offers any physical control over the amplitude. If the level of amplification can be adjusted then the ratio between the original and the changed amplitude is called the 'gain'. Gain can either be positive or negative and is usually measured in decibels (dB). This form of measurement, named after Alexander Graham Bell, the inventor of telephony, indicates the relative difference between sound intensities or sound pressure levels. Thus, a decibel represents the relative amplitude of two different sounds, which is in turn related to loudness, but it is *not* a measurement of it.

Note: If we measure the intensity of one particular sound, it is measured by its Sounds Pressure Level and is expressed as SPL dB rather than just dB. More importantly, though, it's vital to comprehend that decibels are always measured on a logarithmic scale in base 10. For instance, if you have two audio signals and there is a 3 dB difference between them, one would be a third louder than the other, although this would be entirely subjective to the listener.

Loudness is difficult to quantify, because it's entirely relative to the listener. Generally, the human ear can detect frequencies from as low as 20 Hz up to 20 kHz; however, this depends on a number of factors. Although most of us are capable of hearing frequencies as low as

20 Hz, the perception of higher frequencies changes with age. Most teenagers are capable of hearing frequencies as high as 18 kHz, while the middle-aged tend not to hear frequencies above 14 kHz. A person's level of hearing may also have been damaged, for example, by over-exposure to loud noise or music. Whether it is possible for us to perceive sounds higher than 18 kHz with the presence of other sounds is a subject of debate that has yet to be proven.

Note: Sounds that are between 3 and 5 kHz appear perceivably louder than frequencies that are outside of this range (more on this in later chapters).

Subtractive synthesis

Having looked into the theory of sound, we next need to understand how this relates to synthesis. Subtractive synthesis is the basis of many forms of synthesis and is usually related to analogue synthesizers. It is achieved by combining a number of sounds or 'oscillators' together to create a timbre that is very rich in harmonics. This rich sound can then be sculpted using a series of 'modifiers'. The number of modifiers available on a synth is entirely dependent on the model, but all synths offer a way of filtering out certain harmonics and of shaping the overall volume of the timbre.

The next part of this chapter looks at how a real analogue synthesizer operates, although any synthesizer that emulates analogue synthesis (i.e. Digital Signal Processing (DSP) analogue) will operate in essentially the same way, with the only difference being that the original analogue synthesizer voltages do not apply to their DSP equivalents.

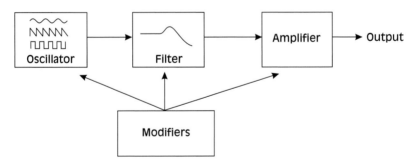

Figure 3.6 Layout of a basic synth.

An analogue synth has three components:

- An oscillator, to make the initial sound.
- A filter, to filter the sound.
- An amplifier, to define the overall level of the sound.

Each of these components and their role in synthesis is discussed in the sections below.

Voltage-controlled oscillator (VCO)

When a key on a keyboard is pressed, a signal is sent to the oscillator to activate it and a specific control voltage (CV) is also sent to determine the pitch. The CV that is sent is unique to the key that is pressed, allowing the oscillator to determine the pitch it should reproduce. For this approach to work correctly, the circuitry in the keyboard and the oscillator must be incredibly precise in order to prevent the tuning from drifting, so the synthesizer must be serviced regularly. In addition, changes in external temperature and fluctuations in the power supply may also cause the oscillator's tuning to drift. This instability gives analogue synths their charm and is the reason why many purists will invest small fortunes in second-hand models rather than use the latest DSP-based analogue emulations. Although, that said, if too much detuning is present it will be immediately evident and could become a major problem!

> Note: The argument over whether it's possible for DSP oscillators to faithfully reproduce analogue-based synths is ongoing, but the argument in favour of DSP synthesizers is that they offer more waveforms and therefore more possibilities for manipulating sound.

In most early subtractive synthesizers the oscillator generated only three types of waveforms: square, sawtooth and triangle waveforms. Today, this number has increased and many synthesizers now offer additional sine, noise, tri-saw, pulse and numerous variable wave shapes as well. Although these additional waveforms each produce different sounds, they are all based around the three basic wave shapes and are often introduced into synthesizers to prevent having to mix numerous basic waveforms together, a task that would reduce the number of oscillators. For example, a tri-saw wave is commonly a sample of three sawtooth waves blended together to produce a sound that is rich in harmonics, with the advantage that the whole sound is contained in one oscillator. Without this waveform it would take three oscillators to recreate this sound, which could be beyond the capabilities of the synth. Even if a synth could utilize three oscillators to produce this one sound, the number of available oscillators would be reduced. Subsequently, while there are numerous oscillator waves available, only knowledge of the six types is required.

1. The sine wave

A sine wave is the simplest wave shape and is based on the mathematical sine function. A sine wave consists of the fundamental frequency alone and does not contain harmonics. This means that they are not suitable for sole use in a subtractive sense, because if the fundamental is

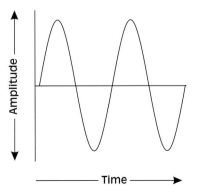

Figure 3.7 A sine wave.

removed no sound is produced (and there are no harmonics upon which the modifiers could act). Consequently, the sine wave is used independently to create sub-basses or whistling timbres or is mixed with other waveforms to add extra body or bottom-end to a sound.

2. The square wave

A square wave is the simplest waveform for an electrical circuit to generate because it only exists in two states: high and low. This wave produces only odd harmonics, resulting in a mellow, hollow sound. This makes it particularly suitable for emulating wind instruments, adding width to strings and pads, or for the creation of deep, wide bass sounds.

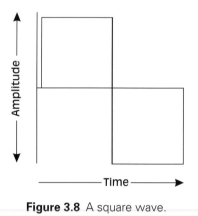

Figure 3.8 A square wave.

3. The pulse wave

Although pulse waves are often confused with square waves, there is a significant difference between the two. Unlike a square wave, a pulse wave allows the width of the high and low states to be adjusted, thereby varying the harmonic content of the sound. Today, it is unusual to see both square and pulse waves featured in a synth. Rather the square wave offers an additional control allowing you to vary the width of the pulses. The benefit of this is that reductions in the width allow you to produce thin reed-like timbres along with the wide hollow sounds created by a square wave.

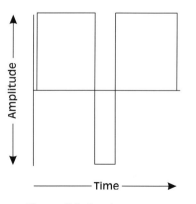

Figure 3.9 A pulse wave.

4. The sawtooth wave

A sawtooth wave produces all the even and odd harmonics in the series and therefore produces a bright sound that is an excellent starting point for brassy, raspy sounds. It's also suitable for creating the gritty, bright sounds needed for leads and raspy basses. Because of its harmonic richness it is often employed in sounds that will be filter swept.

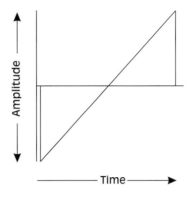

Figure 3.10 A sawtooth wave.

5. The triangle wave

The triangle wave shape features two linear slopes and is not as harmonically rich as a sawtooth wave, since it only contains odd harmonics (partials). Ideally, this type of waveform is mixed with a sine, square or pulse wave to add a sparkling or bright effect to a sound and is often employed on pads to give them a glittery feel.

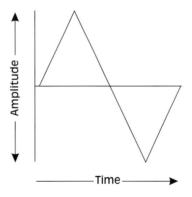

Figure 3.11 A triangle wave.

6. The noise wave

Noise waveforms are unlike the other five waveforms because they create a random mixture of all frequencies rather than actual tones. Noise waveforms can be 'pink' or 'white' depending on the energy of the mixed frequencies they contain. White noise contains the same amount of energy throughout the frequency range and is comparable to radio static, while pink noise contains differing amounts of energy at different frequencies and therefore we perceive it to produce a heavier, deeper hiss. Noise is useful for generating percussive sounds and was commonly used in early drum machines to create snares and handclaps. Although this remains its main use, it can also be useful for simulating wind or sea effects, for producing breath effects in wind instrument timbres, or for producing the typical trance leads.

Figure 3.12 Noise.

Creating complex waveforms

Whether oscillators are created by analogue or DSP circuitry, listening to individual oscillators in isolation can be a mind-numbing experience. To create interesting sounds, several oscillators should be mixed together and used with the available modulation options. This is achieved by first mixing different oscillator waveforms together and detuning them all or just those that share the same waveforms so that they are out of phase from one another, resulting in a beating effect. Detuning is accomplished using the detune parameter on the synthesizer, usually by odd rather than even numbers. This is because detuning by an even number introduces further harmonic content that may mirror the harmonics already provided by the oscillators, causing the already present harmonics to be summed together.

> Note: There is a limit to the level that oscillators can be detuned from one another. As previously discussed, oscillators should be detuned so that they beat, but if the speed of these beats is increased by any more than 20 Hz the oscillators separate, resulting in two noticeably different sounds. This can sometimes be used to good effect if the two oscillators are to be mixed with a timbre from another synth, because the additional timbre can help to fuse the two separate oscillators. As a general rule of thumb, it is unusual to detune an oscillator by more than an octave.

Additional frequencies can also be added into a signal using ring modulation and sync controls. Oscillator sync, usually found within the oscillator section of a synth, allows a number of oscillators' cycles to be synced to one another. Usually, all oscillators are synced to the first oscillator's cycle, so no matter where in the cycle any other oscillators are when the first oscillator starts its cycle again the others are forced to begin again too. For example, if two oscillators are used, with both set to a sawtooth wave and detuned by −5 cents (hundredths of a tone), every time the first oscillator restarts its cycle so too will the second, regardless of the position in its own cycle. This tends to produce a timbre with few harmonics and can be ideal for creating big, bold leads. Furthermore, if the first oscillator is unchanged and pitch bend is applied to the second to speed up or slow its cycle, screaming lead sounds typical of the Chemical Brothers are created as a consequence of the second oscillator fighting against the syncing with the first.

After the signals have left the oscillators they enter the mixer section, where the volume of each oscillator can be adjusted and features such as ring modulation can be applied to introduce further harmonics. (The ring modulation feature can sometimes be found within the oscillator section but is more commonly located in the mixer section, directly after the oscillators.) Ring modulation works by providing a signal that is the sum and difference compound of two signals (while also removing the original tones). Essentially, this means that both signals from a two-oscillator synth enter the ring modulator and come out the other end as one combined signal, with no evidence of the original timbre remaining.

As an example, if one oscillator produces a signal frequency of 440 Hz (A4 on a keyboard) and the second produces a frequency of 660 Hz (E5 on a keyboard) the frequency of the first oscillator is subtracted from the second:

$$660\,Hz - 440\,Hz = 220\,Hz\ (A3)$$

Then, the first oscillator's frequency is added to that of the second:

$$660\,Hz + 440\,Hz = 1100\,Hz\ (C\sharp 6)$$

Based on this example, the difference of 220 Hz provides the fundamental frequency while the sum of the two signals, 1100 Hz, results in a fifth harmonic overtone. When working with synthesis, though, this calculation is rarely performed. This result is commonly achieved by ring modulating the oscillators together at any frequency, then tuning the oscillator. Ring modulation is typically used in the production of metallic-type effect and bell-like sounds. If ring modulation is used to create actual pitched sounds, a large number of in-harmonic overtones are introduced into the signal, creating dissonant, unpitched results.

Note: The option to add noise may also be included in the oscillator's mix section to introduce additional harmonics, making the signal leaving the oscillator/mix section full of frequencies that can then be shaped further using the options available.

Voltage-controlled filters

Following the oscillator's mixer section are the filters for sculpting the previously created signal. In the synthesizer world, if the oscillator's signal is thought of as a piece of wood that is yet to be carved, the filters are the hammer and chisels that are used to shape it. Filters are used to chip away pieces of the original signal until a rough image of the required sound remains. This makes filters the most vital element of any subtractive synth, because if the available filters are of poor quality few sound sculpting options will be available and it will be impossible to create the sound you're after. Indeed, the choice of filters, combined with the oscillators' waveforms, is often the reason why specific synthesizers must be used to recreate certain 'classic' dance timbres.

The most common filter used in basic subtractive synthesizers is a low-pass filter. This is used to remove frequencies above a defined cut-off point. The effect is progressive, meaning that the further the control is reduced, the more frequencies are removed from the sound, starting with the higher harmonics and gradually moving to the lowest. If this filter cut-off point is reduced far enough, all harmonics above the fundamental can be removed, leaving just the fundamental frequency. While it may appear senseless to create a bright sound with oscillators only to remove them later with a filter, there are several reasons why you may wish to do this:

- Using a variable filter on a bright sound allows you to determine the colour of the sound much more precisely than if you tried to create the same effect using oscillators alone.
- This method enables you to employ real-time movement of a sound.

This latter movement is an essential aspect of sound design because we naturally expect dynamic movement of sound throughout the length of the note. For instance, when a piano string is struck the initial sound is very bright, becoming duller as it dies away. This effect can be simulated by opening the filter as the note starts then gradually sweeping the cut-off frequency down to create the effect of the note dying away.

Notably, when using this effect, frequencies that lie above the cut-off point are not attenuated at right angles to the cut-off frequency; therefore, the rate at which they die away will depend on the transition period. This is why different filters that essentially perform the same function can make beautiful sweeps, whilst others can produce quite uneventful results.

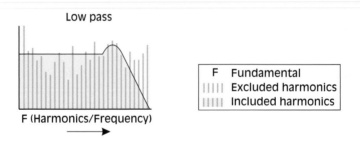

Figure 3.13 Action of the low-pass filter.

When a cut-off point is designated, small quantities of the harmonics that lie above this point are not removed completely and are instead attenuated by a certain degree. The degree of attenuation is dependent on the transition band of the filter being used. The gradient of this transition is important because it defines the sound of any one particular filter. If the slope is steep, the filter is said to be 'sharp' and if the slope is more gradual the filter is said to be 'soft'. To fully understand the action of this transition, some prior knowledge of the electronics involved in analogue synthesis is required.

When the first analogue synths appeared in the 1960s, different voltages were used to control both the oscillators and the filters. Any harmonics produced by the oscillators could be removed gradually by physically manipulating the electrical current. This was achieved using a resistor (to reduce the voltage) and a capacitor (to store a voltage), a system that is often referred to as a resistor–capacitor (RC) circuit. Because a single RC circuit produces a 6 dB transition, the attenuation increases by 6 dB every time a frequency is doubled.

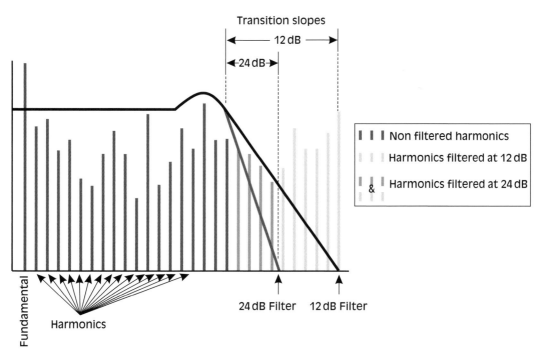

Transition slopes

Non filtered harmonics

Harmonics filtered at 12 dB

Harmonics filtered at 24 dB

24 dB Filter   12 dB Filter

Fundamental

Harmonics

Figure 3.14 The difference between 12 and 24 dB slopes.

One RC element creates a 6 dB per octave one-pole filter that is very similar to the gentle slope created by a mixing desk's equalizer (EQ). Consequently, manufacturers soon implemented additional RC elements into their designs to create two-pole filters, which attenuated 12 dB per octave, and four-pole filters, to provide 24 dB per octave attenuation. Because four-pole filters attenuate 24 dB per octave, making substantial changes to the sound, they tend to sound more synthetic than sounds created by a two-pole filter, so it's important to decide which transition period is best suited to the sound. For example, if a 24 dB filter is used to sweep a pad it will result in strong attenuation throughout the sweep, while a 12 dB filter will create a more natural flowing movement.

If there is more than one filter available, some synthesizers allow them to be connected in series or parallel, which gives more control over the timbre from the oscillators. This means that two 12 dB filters could be summed together to produce a 24 dB transition, or one 24 dB filter could be used in isolation for aggressive tonal adjustments with the following 12 dB filter used to perform a real-time filter sweep.

Although low-pass filters are the most commonly used type, there are numerous variations, including high pass, band pass, and notch and comb. These utilize the same transition periods as the low-pass filter, but each has a wildly different effect on the sound.

A high-pass filter has the opposite effect to a low-pass filter, first removing the low frequencies from the sound, gradually moving toward the highest. This is less useful than the low-pass filter because it effectively removes the fundamental frequency of the sound, leaving only the fizzy harmonic overtones. Because of this, high-pass filters are rarely used in the creation of instruments and are predominantly used to create effervescent sound effects or bright timbres that can be laid over the top of another low-pass sound to increase the harmonic content. The typical

Figure 3.15 Action of a high-pass filter.

euphoric trance leads are good examples of this, as they are often created from a tone with the fundamental overlaid with numerous other tones that have been created using a high-pass filter. This prevents the timbre from becoming too muddy as a consequence of stacking together fundamental frequencies. In both remixing and dance music, it's commonplace to run a high-pass filter over an entire mix to eliminate the lower frequencies, creating an effect similar to a transistor radio or telephone. By reducing the cut-off control, gradually or immediately, the track morphs from a thin sound to a fatter one, which can produce a dramatic effect in the right context.

Note: An example of this can be heard on 'Another Chance' by Roger Sanchez.

If high- and low-pass filters are connected in series, then it's possible to create a band-pass, or band-select, filter. These permit a set of frequencies to pass unaltered through the filter while the frequencies either side of the two filters are attenuated. The frequencies that pass through unaltered are known as the 'bandwidth' or the 'band pass' of the filter, and clearly, if the low-pass is set to attenuate a range of frequencies that are above the current high-pass setting, no frequencies will pass through and no sound is produced.

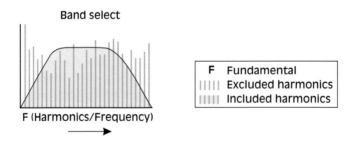

Figure 3.16 Action of the band-pass filter.

Band-pass filters, like high-pass filters, are often used to create timbres consisting of fizzy harmonics. They can also be used to determine the frequency content of a waveform, as by sweeping through the frequencies each individual harmonic can be heard. Because this type of filter frequently removes the fundamental, it is often used as the basis of sound effects or lo-fi and trip-hop timbres, or to create very thin sounds that will form the basis of sound effects.

Although band-pass filters can be used to thin a sound, they should not be confused with band-reject filters, which can be used for a similar purpose. Band-reject filters, often referred to as notch filters, attenuate a selected range of frequencies, effectively creating a notch in the sound – hence the name – and usually leave the fundamental unaffected. This type of filter is handy for scooping out frequencies, thinning out a sound while leaving the fundamental intact, making them useful for creating timbres that contain a discernible pitch, but do not have a high level of harmonic content.

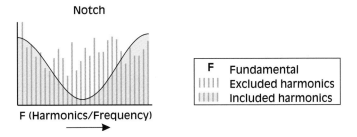

Figure 3.17 Action of the notch filter.

One final form of filter is the comb filter. With these, some of the samples entering the filter are delayed in time and the output is then fed back into the filter to be reprocessed to produce the results, effectively creating a comb appearance, hence the name. Using this method, sounds can be tuned to amplify or reduce specific harmonics based on the length of the delay and the sample rate, making it useful for creating complex sounding timbres that cannot be accomplished any other way. Because of the way that they operate, however, it is rare to find these featured on a synthesizer and they are usually only available as a third-party effect.

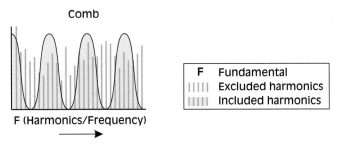

Figure 3.18 Action of the comb filter.

As an example, if a 1 kHz signal is put through the filter with a 1-millisecond delay, the signal will result in phase, because 1 millisecond is coincident with the inputted signal, equalling one. However, if a 500 Hz signal with a 1-millisecond delay were used instead, it would be half of the period length and so would be shifted out of phase by 180 degrees, resulting in a zero. It's this constructive and deconstructive period that creates the continual bump then dip in harmonics, resulting in a comb-like appearance when represented graphically, as in Figure 3.18. This method applies to all frequencies, with integer multiples of 1 kHz producing ones and odd multiples of 500 Hz (1.5, 2.5, 3.5 kHz, etc.) producing zeros. The effect of using this filter can at best be described as highly resonant and its use is limited to extreme sound design rather than the more basic sound sculpting.

One final element of sound manipulation in a synthesizer's filter section is the resonance control. Also referred to as Peak or Q, this refers to the amount of the output of the filter that is fed back directly into the input, emphasizing any frequencies that are situated around the cut-off frequency. This has a similar effect to employing a band-pass filter at the cut-off point, effectively creating a peak. Although this also affects the filter's transition period, it is more noticeable at the actual cut-off frequency than anywhere else. Indeed, as you sweep through the cut-off range the resonance follows the curve, continually peaking at the cut-off point. In terms of the final sound, increasing the resonance makes the filter sound more dramatic and is particularly effective when used in conjunction with low-pass filter sweeps.

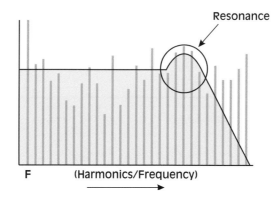

Figure 3.19 The effect of resonance.

On many analogue and DSP analogue-modelled synths, if the resonance is turned up high enough it will feed back on itself. As more and more of the signal is fed back, the signal is exaggerated until the filter breaks into self-oscillation. This produces a sine wave with a frequency equal to that of the set cut-off point and is often a purer sine wave than that produced by the oscillators. Because of this, self-oscillating filters are commonly used to create deep, powerful sub-basses that are particularly suited to the drum 'n' bass and rap genres.

Note: Some filters may also feature a saturation parameter which essentially overdrives the filters. If applied heavily, this can be used to create distortion effects, but more often it's used to thicken out timbres and add even more harmonics and partials to the signal to create rich-sounding leads or basses.

The keyboard's pitch can also be closely related to the action of the filters, using a method known as pitch tracking, keyboard scaling or, more frequently, 'key-follow'. On many synths the depth of this parameter is adjustable, allowing you to determine how much or how little the filter should follow the pitch. When this parameter is set to its neutral state (neither negative nor positive), as a note is played on the keyboard the cut-off frequency tracks the pitch and each note is subjected to the same level of filtering. If this is used on a low-pass filter, for example, the filter setting remains fixed, so as progressively higher notes are played fewer and fewer harmonics will be present in the sound, making the timbre of the higher notes mellower than that of the lower notes. If the key-follow parameter is set to positive, the higher notes will have a higher cut-off frequency and the high notes will remain bright. If, on the other hand, the key-follow parameter is set to negative, the higher notes will lower the cut-off frequency, making the high notes even mellower than when key-follow is set to its neutral state. Key-follow is useful for recreating real instruments such as brass, where the higher notes are often mellower than the lower notes, and is also useful on complex lead lines that jump over an octave, adding further variation to a rhythm.

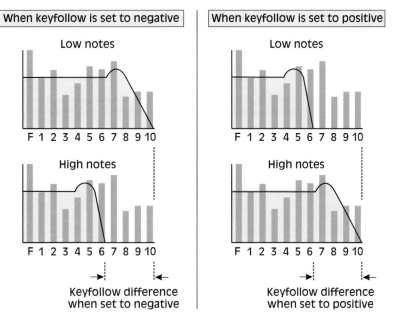

Figure 3.20 The effect of filter key-follow.

Voltage-controlled amplifier (VCA)

Once the filters have sculpted a sound, the signal then moves into the final stage of synthesis: the amplifier. When a key is pressed, rather than the volume rising immediately to its maximum and falling to zero when released, an 'envelope generator' is employed to emulate the nuances of real instruments.

Few, if any, acoustic instruments start and stop immediately. It takes a finite amount of time for the sound to reach its amplitude and then decay away to silence again; thus, the 'envelope generator' – a feature of all synths – can be used to shape the volume with respect to time. This allows you to control whether a sound starts instantly the moment a key is pressed or builds up gradually, and how the sound dies away (quickly or slowly) when the key is released. These controls usually comprise four sections, called Attack, Decay, Sustain and Release (ADSR), each of which determines the shaping that occurs at certain points during the length of a note. An example of this is shown in Figure 3.21.

Attack

The attack control determines how the note starts from the point when the key is pressed and the period of time it takes for the sound to go from silence to full volume. If the period set is quite long, the sound will 'fade in', as if you are slowly turning up a volume knob. If the period set is short, the sound will start the instant a key is pressed. Most instruments utilize a very short attack time.

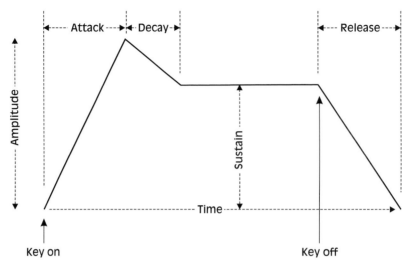

Figure 3.21 The ADSR envelope.

Decay

Immediately after a note has begun it may initially decay in volume. For instance, a piano note starts with a very loud, percussive part, but then drops quickly to a lower volume while the note sustains as the key is held down. The time the note takes to fade from the initial peak at the attack stage to the sustain level is known as the 'decay time'.

Sustain

The sustain period occurs after the initial attack and decay periods, and determines the volume of the note while the key is held down. This means that if the sustain level is set to maximum any decay period will be ineffective, because at the attack stage the volume is at maximum so there is no level to decay down to. Conversely, if the sustain level were set to zero, the sound peaks following the attack period and will fade to nothing even if you continue to hold down the key. In this instance, the decay time determines how quickly the sound decays down to silence.

Release

The release period is the time it takes for the sound to fade from the sustain level to silence after the key has been released. If this is set to zero the sound will stop the instant the key is released, while if a high value is set the note will continue to sound, fading away as the key is released.

Although ADSR envelopes are the most common there are some subtle variations, such as AR, TADSR and ADSTR. Because there are no decay or sustain elements contained in most drum timbres, AR (Attack–Release) envelopes are often used on drum synthesizers. They can also appear on more economical synthesizers simply because the AR parameters are regarded as having the most significant effect on a sound, making them a basic requirement. Both TADSR (Time–Attack–Delay–Sustain–Release) and ADSTR (Attack–Delay–Sustain–Time–Release) envelopes are only usually found on more expensive synthesizers. With the additional period, T (Time), in TADSR, for instance, it is possible to set the amount of time that passes before the Attack stage is reached.

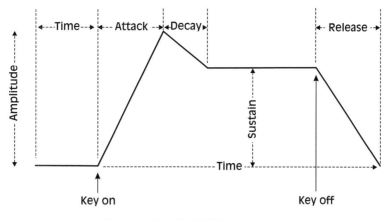

Figure 3.22 The TADSR envelope.

It's also important to note that not all envelopes offer linear transitions, meaning that the attack, decay and release stages will not necessarily consist entirely of the straight line shown in Figure 3.22. On some synths these stages may be concave or convex, while other synths may allow you to state whether the envelope stages should be linear, concave or convex. The differences between the linear and exponential envelopes are shown in Figure 3.23.

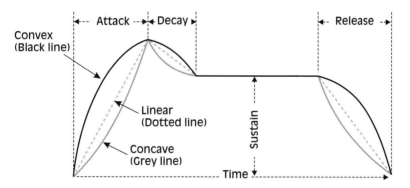

Figure 3.23 Linear and exponential envelopes.

Modifiers

Most synthesizers also offer additional tools for manipulating sound in the form of modulation sources and destinations. Using these tools, the response or movement of one parameter can be used to modify another totally independent parameter, hence the name 'modifiers'. The number of modifiers available, along with the destinations they can affect, is entirely dependent on the synthesizer. Many synths feature a number of envelope generators that allow the action of other parameters alongside the amplifier to be controlled. For example, in many synths an envelope may be used to modify the filter's action and by doing so you can make tonal changes to the note while it plays. A typical example of this is the squelchy bass sound used in most dance music. By having a zero attack, short decay and zero sustain level on the envelope generator, a sound that starts with the filter wide open before quickly sweeping down to fully closed is produced.

Note: This movement is archetypal to most forms of dance music but does not necessarily have to be produced by envelopes. Instead, some synthesizers offer one-shot LFOs which can be used in the envelope's place. For instance, by using a triangle waveform LFO to modulate the amp, it will slowly rise in volume before slowly dropping again.

Low-frequency oscillator (LFO)

LFOs produce output frequencies in much the same way as VCOs. The difference is that a VCO produces an audible frequency (within the 20 Hz to 20 kHz range), while an LFO produces a signal with a relatively low frequency that is inaudible to the human ear (in the range 1–10 Hz). The waveforms an LFO can utilize depend entirely upon the synth in question, but they commonly employ sine, saw, triangle, square, and sample and hold waveforms. The sample and hold waveform is usually constructed with a randomly generated noise waveform that momentarily freezes every few samples before beginning again.

Synths include LFOs because they can be used to modulate other parameters, known as 'destination', to introduce additional movement into a sound. For instance, if an LFO is set to a relatively high frequency, say 5 Hz, to modulate the pitch of a VCO, the pitch of the oscillator will rise and fall according to the speed and shape of the LFO waveform and an effect similar to vibrato is generated. If a sine wave is used for the LFO then it will essentially create an effect similar to a wailing police siren. Alternatively, if this same LFO is used to modulate the filter cutoff, then the filter will open and close at a speed determined by the LFO, while if it were used to modulate an oscillator's volume, it would rise and fall in volume, recreating a tremolo effect.

This means that an LFO must have an amount control (sometimes known as depth) for varying how much the LFO's waveform augments the destination, a rate control, to control the speed of the LFO's waveform cycles, and some may also feature a fade-in control. This is particularly useful for creating pads and string patches that gradually begin to warp and twist since, using this, you can control how quickly the LFO begins to affect the waveform after a key has been pressed. An example of this is shown in Figure 3.24.

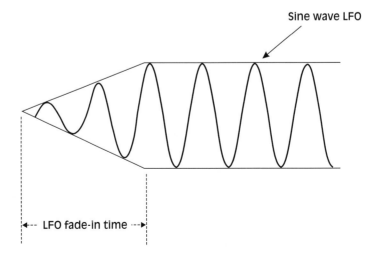

Figure 3.24 LFO fade-in.

The LFO on more capable synths may also have access to its own envelope. This gives control of the LFO's performance over a specified time period, allowing it not only to fade-in after a key has been pressed, but also decay, sustain and fade away gradually. It is worth noting, however, that the destinations an LFO can modulate are entirely dependent on the synth being used. Some synthesizers may only allow LFOs to modulate the oscillator's pitch and the filter, while others may offer multiple destinations and more LFOs. Obviously, the more LFOs and destinations that are available, the more creative options you will have at your disposal.

If required, further modulation can be applied with an attached controller keyboard or the synth itself in the form of two modulation wheels. The first, pitch bend, is hard-wired and provides a convenient method of applying a modulating control voltage to the oscillator(s). By pushing the wheel away from you, you can bend the pitch (i.e. frequency) of the oscillator up. Similarly, you can bend the pitch down by pulling the wheel towards you. This wheel is normally spring loaded to return to the centre position, where no bend is applied, if you let go of it and is commonly used in synth solos to give additional expression. The second wheel, modulation, is freely assignable and offers a convenient method of controlling any on-board parameters, such as the level of the LFO signal sent to the oscillator, filter or VCA or to control the filter cut-off directly. Again, whether this wheel is assignable will depend on the make of synth. On some synths the wheels are hard coded to only allow oscillator modulation (for a vibrato effect), while some others do not have a separate modulation wheel and instead a pitch bend lever can be pushed forward to produce LFO modulation.

Other synthesis methods

Frequency modulation (FM)

Frequency modulation is a form of synthesis developed in the early 1970s by Dr John Chowning of Stanford University, then later developed further by Yamaha, leading to the release of the now legendary DX7 synthesizer: a popular source of bass sounds for numerous dance musicians.

Unlike analogue, FM synthesizers produce sound by using operators; these are very similar to oscillators in an analogue synthesizer but can only produce simple sine waves. Sounds are generated by using the output of the first operator to modulate the pitch of the second, thereby introducing harmonics. Like an analogue synth, this means that each FM voice requires a minimum of two oscillators in order to create a basic sound, but because FM only produces sine waves the timbre produced from just one carrier and modulator isn't very rich in harmonics.

In order to remedy this, FM synths provide many operators that can be configured and connected in any number of ways. Because there are numerous ways these connections can be made, many will not produce musical results, so to simplify matters various algorithms are used. These algorithms are preset as combinations of modulator and carrier routings. For example, one algorithm may consist of a modulator modulating a carrier, which in turn modulates another carrier, before modulating a modulator that modulates a carrier to produce the overall timbre. The resulting sound can then be shaped and modulated further using LFOs, filters and envelopes with the same subtractive methods as in any analogue synth.

This means that it should also be possible to emulate FM synthesis in an analogue synth with two oscillators, where the first oscillator acts as a modulator and the second acts as a carrier.

When the keyboard is played both oscillators produce their respective waveforms, with the frequency dictated by the particular notes that were pressed. If the first oscillator's output is routed into the modulation input of the second oscillator and further notes are played on the keyboard, both oscillators play their respective notes but the pitch of the second oscillator will change in time with the frequency of the first, essentially creating a basic FM synth. Although this is effectively frequency modulation, it is usually called 'cross-modulation' in analogue synthesizers.

Due to the nature of frequency modulation many of the timbres created are quite metallic and digital in character, particularly when compared to the warmth generated by the drifting of analogue oscillators. Also, due to the digital nature of FM synths the fascia generally contains few real-time controllers. Instead, numerous buttons adorn the front panel, forcing you to navigate and adjust any parameters through a small LCD display.

Notably, although both FM and analogue syntheses were originally used to reproduce realistic instruments, neither can fabricate truly realistic timbres. If the goal of the synthesis system is to recreate the sound of an existing instrument, this can generally be accomplished more accurately using digital sample-based techniques.

Sample synthesis

Unlike analogue or frequency modulation, sample synthesis utilizes samples in place of the oscillators. These samples, rather than consisting of whole instrument sounds, are actually samples of the various stages of a real instrument (although they do contain samples of oscillators too). For instance, a typical sample-based synth may contain five different samples of the attack stage of a piano, along with a sample of the decay, sustain and release portions of the sound. This means that it is possible to mix the attack of one sound with the release of another to produce a complex timbre.

Commonly, up to four of these individual 'tones' can be mixed together to produce a timbre and each of these individual tones can have access to numerous modifiers, including LFOs, filters and envelopes. This obviously opens up a whole host of possibilities, not only for emulating real instruments but also for creating complex sounds. This method of synthesis has become the de facto standard for any synth producing realistic instruments. By combining samples of real-world sounds with all the editing features and functionality of analogue synthesis, they can offer a huge scope for creating both realistic and synthetic sounds.

Granular synthesis

One final form of synthesis that has started to make an appearance with the evolution of technology is granular synthesis. It is rare to see granular synthesis employed in hardware synths due to its complexity, but software synths are starting to be developed for the public market that utilize it. Essentially, it works by building up sounds from a series of short segments of sounds called 'grains'. This is best compared to the way that a film projector operates: where series of still images, each slightly different from the last, are played sequentially at a rate of around 25 pictures per second, fooling the eyes and brain into believing there is a smooth continual movement.

Granular synthesis operates in this same manner, with tiny fragments of sound rather than still images. By joining a number of these grains together, an overall tone is produced that develops over a period of time. To do this, each grain must be less than 30 milliseconds (ms) in length, as generally the human ear is unable to determine a single sound if they are less than 30–50 ms apart. This also means that a certain amount of control has to be offered over each grain. In any one sound there can be anything from 200 to 1000 grains, which is the main reason why this form of synthesis appears mostly in the form of software. Typically, a granular synth will offer most, but not necessarily all, of these five parameters:

- *Grain length*. This can be used to alter the length of each individual grain. As previously touched upon, the human ear can differentiate between two grains if they are more than 30–50 ms apart, but many granular synths usually go above this range, covering from 20 to 100 ms. By setting this length to a longer value, it's possible to create a pulsing effect.
- *Density*. This is the percentage of grains that are created by the synth. Generally, it can be said that the more grains that are created, the more complex a sound will be, a factor that is also dependent on the grain shape.
- *Grain shape*. Commonly, this offers a number between 0 and 200 and represents the curve of the envelopes. Grains are normally enveloped so that they start and finish at zero amplitude, helping the individual grains mix together coherently to produce the overall sound. By setting a longer envelope (a higher number), two individual grains will mix together, which can create too many harmonics and often results in the sound exhibiting lots of clicks as it fades from one grain to the other.
- *Grain pan*. This is used to specify the location within the stereo image that each grain is created. This is particularly useful for creating timbres that inhabit both speakers.
- *Spacing*. This is used to alter the period of time between each grain. If the time is set to a negative value, the preceding grain will continue through the next created grain. This means that setting a positive value inserts space between each grain; however, if this space is less than 30 ms in length it will be inaudible.

The sound produced with granular synthesis depends on the synth in question. Usually, each grain consists of single frequencies with specific waveforms or occasionally they are formed from segments of samples or noise that have been filtered with a band-pass filter. Thus, the constant change of grains can produce sounds that are both bright and incredibly complex, resulting in a timbre that's best described as glistening. After creating this sound by combing the grains, the whole sound can be modified using envelopes, filters and LFOs.

4 Sampling and digital audio

I went into the studio with all my favourite records and said sample this, this and this.

Goldie

Sampling is the cornerstone of all dance music. In fact, it's fair to say that without the introduction of samplers and sampling, drum 'n' bass, jungle and trip-hop wouldn't have come about, and neither would many house and trance tracks exist. Sampling plays such a vital role in dance music that, without it, it is incredibly difficult to produce most forms of music. This isn't solely due to the numerous genre-specific sample CDs that are available, although these do often play a role, but because samplers are one of the most creative tools you can have at your disposal. With some creative thought, a recording of a ping-pong ball being thrown at a garage door can become a tom drum, hitting a rolled up newspaper against a wall can make an excellent house drum kick, and the hissing from a compressed air freshener can produce great hi-hats. As a way of making music, sampling can be quick, easy and an immensely creative resource, so much so that if you don't own a sampler, it's about time you did!

At a very basic level, a sampler is the digital equivalent of an analogue tape recorder, but rather than recording the audio signal onto a magnetic tape, it is recorded digitally into random access memory (RAM) or directly onto a hard disk drive. After a sound has been recorded, it can then be manipulated using a series of editing parameters, very similar to those on synthesizers, and also played back at varying pitches from any attached controller keyboard. The variations in pitch are created by the sampler artificially increasing or decreasing the original frequency of the sampled sound according to the key that is pressed. To raise the pitch the frequency is increased, while the pitch is lowered as the frequency is decreased.

This principle is analogous to the way that analogue tape behaves, whereby the speed that the tape is played alters the pitch. However, as with analogue tape, if the sample's speed is increased or decreased by a significant amount it will no longer sound anything like the original source. For instance, if a single key strike from a piano is sampled at C3 and is played back at this same pitch from a controller keyboard, the sample will play back perfectly. If, however, this same sample were played back at C4 the sampler would increase the frequency of the original sound by 12 semitones (from 130.81 to 523.25 Hz) and the result would sound more like two spoons knocking together than the original recording.

These extreme pitch adjustments aren't necessarily a bad thing – particularly for a creative endeavour such as dance music – but to recreate a real instrument, sampling a single note doesn't provide a realistic sounding instrument throughout the octaves. Indeed, most samplers only reproduce an acceptable instrument sound four or five keys from the original root key, so if an original instrument is needed throughout the key range, samples should be taken every few keys of the original source instrument. This is called 'multi-sampling' and is when the source is sampled at every couple of keys and the samples assigned to the same keys in the sampler. For example, with a piano it is prudent to sample the keys at C0, E0, G0 and B0, then C1, E1, G1, B1 and so forth until the entire range has been sampled.

Naturally, recording a sound in this way would equate to over 16 samples and the more samples that are taken, the more memory the sampler must have available. Because most hardware (and some software) samplers hold the sampled sounds in their on-board RAM, the maximum sampling time is limited by the amount of available memory. At full audio bandwidth (20 Hz to 20 kHz) and a 44.1 kHz sampling rate, 1 minute of mono recording will use approximately 5 megabytes (MB) of RAM. In the case of sampling a piano, this would equate to 80 MB of memory, and you can double this if you wanted it in stereo! Consequently, over the years samplers have adopted various techniques to make the most of the available memory, the first of which is to loop the samples.

As the overall length of a sample determines the amount of RAM that is required, reducing the sample's length means more samples will fit into the memory space. Because most sounds have a distinctive attack and decay period but the sustain element remains consistent, the sustain portion can be continually looped for as long as the key is held, moving to the release part after the key is released. This means that only a short burst of the sustain period must be sampled, helping to conserve memory. Well, that's the theory anyway. In practice, sustain looping can prove particularly difficult.

The difficulty arises from the fact that what appears to be a consistent sustain period of a sound is, in most cases, rarely static due to slight yet continual changes in the harmonic structure. If only a small segment of this harmonic movement is looped, the results would sound unnatural. Conversely, if too long a section is looped in an effort to capture the harmonic movements, the decay or some of the release period may also be captured and again, when looped, the final sound will still be unusual. In addition, any looping points must start and end at the same phase and level during the waveform. If not, the difference in phase or volume could result in an audible click as the waveform reaches the end of its looped section and jumps back to the beginning.

Some samplers offer a work-around to this latter problem and automatically locate the nearest zero crossing to the position you choose. While this increases the likelihood that a smoother crossover is achieved, if the waveform's level is different at the two loop points there will still be a glitch. These glitches can, however, be avoided if the sampler has a cross-fading feature.

Cross-fading works by fading out the end of the looped section and overlapping this fade-out with the start of the loop that's fading in. This creates a smooth crossover between the two looping points, reducing the possibility that glitches are introduced. Although this goes some way to resolve the problems associated with looped samples, this is not necessarily the best solution, because if the start and end of the looped points are at different frequencies there will be an apparent change in the overall timbre during the cross-fade. Unfortunately, there is no quick fix for avoiding the pitfalls associated with successfully creating a looped segment, so success can only be accredited to patience and experimentation.

Having said that, some samplers now utilize 'sample streaming', which eliminates the need to squeeze all the samples into available RAM. Often referred to as 'ROMplers', samples are stored on a local hard disk and only the start of the sample is held in the RAM. By using a small buffer, the RAM begins playback after the key is pressed and is constantly updated with the new data directly from the hard disk, allowing samples of any length to be played back. With today's disk drives capable of storing up to and above 200 gigabytes (GB) of data, it's possible to store massive multi-samples. Indeed, it's quite usual for ROMplers to use sample sets made up of the source instrument sampled at every note, at different velocities, and with the foot pedals (if applicable) in different positions.

It'll come as no surprise to learn that this form of key multi-sampling is incredibly time-consuming to undertake yourself and, without access to the right equipment, impossible to accomplish accurately. Even most professional dance artists do not have the experience or equipment required to successfully multi-sample real instruments. With this in mind, if real instruments are required it's much easier to buy a collection of well-recorded sounds on a sample CD than to attempt to build them yourself.

Sample CDs

Sample CDs are produced in several formats to suit different samplers and essentially fall into three categories: Audio CD, Wave CD (WAV) and CD-ROM. Due to the popularity of using PCs and Macs for music, Wave CD formats are the most popular format, as these can be copied digitally from the CD into a compatible software sampler, where they can then be key mapped and edited further. Audio CDs, on the other hand, can be played by any conventional CD player and can be recorded into any sampler hardware or software. Alternatively, if a software sampler is used, the audio can be 'ripped' into the computer and used directly with no loss in audio quality and no need to constantly check the recording levels to ensure that you have the best signal-to-noise ratio.

The disadvantage of both these formats, however, is that after the sounds have been sampled, they must be named, loops created if needed, cross-fades created and key ranges defined. This can be incredibly time-consuming, taking anything from a few hours to over a few days to set up a multi-sampled instrument, depending on the sampler's interface. This is where sample CD-ROMs offer the advantage, though they are considerably more expensive than both Wave and Audio CDs. Because the samples are all stored on a CD in a data format, the samples are already named, key mapped and looped (if necessary) to suit the sampler. This means that it's not necessary to spend time arranging the samples across the keyboard, setting up loops and so on. The data files are simply loaded into the sampler ready for work.

While this has the obvious advantage of easing the sampling process, typically there are also some disadvantages. If the sampler doesn't have a CD-ROM drive attached, you'll have to buy one and, as many external CD-ROM drives use SCSI connections, you may also have to purchase an SCSI card for the sampler too! What's more, the actual data on the CD must be compatible with the operating system used by the sampler. This isn't so much of a problem with the most established sampler manufacturers such as AKAI and E-MU, since these are cross-compatible with each other to a certain extent. For instance, the AKAI S6000 sampler will import CDs that are designed for use with E-MU samplers (and vice versa), but keep in mind that there may be slight discrepancies when using a format-specific CD across different samplers in this way.

Even though the basic operating systems of most samplers are quite similar, each has slightly different parameters on offer. If these have been exploited in the CD-ROM then they may not interpret well on a different sampler. The most common consequence is a difference in the timbre, simply because different filter algorithms and settings are employed by different samplers, but in more severe cases the key mapping may not come out quite as expected, the sustain samples may not loop properly, or velocity modulation may not work correctly.

In an attempt to decipher what will (or should!) work on different samplers, a list of the currently available software and hardware samplers and the formats that they can accept are listed in Tables 4.1 and 4.2. These lists are by no means exhaustive, however, as samplers are constantly being updated and replaced to accept different formats.

Table 4.1 Software sampler compatibility

Sampler	TII	Battery	Reaktor	HALion	S5000/6000	WAV	AIFF	Gig	AKAI	Roland	ReCycle	LM4	LM4 II	EXS24	ISO Disc Image S	SDII	S1000/3000	SampleCell	SF2	Soundfont
Steinberg HALion (v2)				×	×	×	×	×	×	×	×	×	×	×	×	×[a]	×			
Emagic EXS24						×	×	×	×		×			×			×	×	×	
Nemesys/ Tascam Gigastudio					×	×	×	×	×	×								×		×
Speedsoft V-Sampler	×				×	×	×		×									×		
Native Instruments Kontakt[b]		×	×	×	×	×	×	×		×				×			×	×		×

[a] Only accessible on a Mac.
[b] Support for E-MU EIII, ESI and EIV is planned.

Although sample CDs have some quite obvious advantages (and are quite heavily used in all dance music genres), it's important to understand the various clauses involved in their commercial use. When a sample CD is purchased, the cost includes a licence to use the sounds in your own projects. This licence does not include the copyright to the sounds, as these belong to the manufacturer. You are simply granted a licence to use them. It's important to understand the content of this licence because if a mistake is made and a commercial track using them is released, you could be in for a heavy ride for copyright infringement.

Most sample CDs specifically mention the 'musical context' for their use, meaning that they can be used in musical productions without having to notify the original artist or pay royalties, but you cannot use them for library music or other sample CDs. It is vitally important that you read all the terms, though, since some sample CDs have quite unusual terms, ranging from acknowledging the sample's source on the documentation accompanying a commercial

Table 4.2 Hardware sampler compatibility

Sampler	TII	Battery	Reaktor	HALion	S5000/6000	WAV	AIFF	Gig	AKAI	Roland	ReCycle	LM4	LM4 II	EXS24	ISO Disc Image S	SDII	S1000/3000	SampleCell	SF2	E-Mu	Soundfont
AKAI S1000																	×				
AKAI S3000										×							×		×		
AKAI S5000/6000						×	×			×							×		×		
AKAI Z Series[a]					×	×	×	×	×	×[b]							×			×[c]	
Roland S760/SP700										×							d				
Roland S750/770										×											
E-Mu E111X/ESI32																	d				
E-Mu E64/EIV																	d				
Kurzweil K2000/K2500																	×				
Ensoniq ASR10																	d				

[a]Future support may appear for the AKAI MPC 2000XL format.
[b]Roland S700.
[c]E-Mu E3/E4 formats.
[d]S1000 only.

release, asking the copyright owner for permission, or paying them a share of the royalties from every sale! These latter clauses are rare, but always check first.

Practical sampling

Nearly all the sample CDs that are aimed at dance musicians consist of single hits rather than multi-sampled instruments. This is because the bass and melodies in dance music are usually simple, with very little movement throughout the octave. Also, single hit samples of synthesized basses and leads, along with most woodwind and brass instruments, can be successfully stretched over more keys than string instruments such as pianos, violins, harps or guitars.

Nevertheless, before any of these are imported into a sampler, the 'root' key must be set. This is the key that the sampler will use as a reference point for stretching the subsequent notes up or down along the length of the keyboard. For instance, bass samples are typically set up with the root key at C1, meaning that if this key is pressed the sample will play back at its original frequency. This can then be stretched across the keyboard's range as much as required. In most hardware samples, the root key is set in the key range or key zone page, along with the lowest and highest keys available for the sample. Using this information the sampler automatically spreads the root note across the defined range and the bass can be played within the specified range of notes.

Note: To gain the best results from the subsequent pitch adjustments on each key, it is preferable that the sampler's key range is set to six notes above and below the root. This allows the sample to be played over the octave, if required, and will produce much more natural results than if the root note were stretched 12 semitones up or down. Even if you're not planning to play the instrument over an octave, it's still prudent to set the key range to span an octave anyway, as this makes more pitches available if they're needed later.

When setting a key zone for bass sounds, it may be possible to set the range much lower than six semitones, as pitch is determined by the fundamental frequency and the lower this is the more difficult it is to perceive pitch. Thus, for bass samples it may be possible to set the lowest key of the key zone to 12 semitones (an octave) below the root without introducing any unwanted artefacts. In fact, pretty much anything sounds good if it's pitched down low enough.

Most competent samplers will also allow a number of key zones to be set across the keyboard. This makes it possible to have, say, a bass sample occupying F0 to F1 and a lead sample occupying F2 to F3. If set up in this way, both the bass and lead can be played simultaneously from an attached controller keyboard. In taking this approach, it is worth checking that each key zone can access different parameters of the sampler's modulation engine. If this is not possible, settings applied to the bass will also apply to the lead. To avoid this it's prudent to allocate each key zone to an individual MIDI channel. Because the majority of samplers are multi-timbral, it's usually possible to set different synthesis parameters on each MIDI channel.

After the samples are arranged into key zones, you can set how the sampler will behave depending on how hard the key is hit on a controller keyboard. This uses 'velocity modulation', which is useful for recreating the movement within sounds. The immediate use for this is to set the velocity to react with a low-pass filter so that the harder the key is hit, the more the filter opens. This can be used to add expression to any riffs or melodies and prevent the static feel that is often a consequence of using a series of sampled sounds.

'Velocity cross-fading' and 'switching' are also worth experimenting with, if the sampler has these facilities. Switching involves taking two samples and using velocity values to determine which one should play. The two samples are imported into the same key range and hitting the key harder (or softer) changes between the two samples. Velocity cross-fading uses this same principle, but morphs the two samples together creating a (hopefully) seamless blend rather than an immediate switch between one and the other. Although velocity cross-fading produces more convincing results, it also reduces the polyphony of the sampler, as it has to play two samples together while it cross-fades between them. Both these velocity-related parameters are typically used to emulate the nuances of real-world instruments such as a piano, where the harder a note is struck the more the tonal content changes.

Working with loops

Alongside single synthesizer hits, the majority of sample CDs also feature a number of programmed or recorded drum loops to suit the genre of the CD. These can be imported/ recorded into a sampler and subsequently re-triggered continually to create the effect of a constant drum track to play throughout the length of the music. More interestingly, because samplers store the audio in RAM (or on a hard drive), once a loop is recorded the sound will start over from the beginning every time a key is pressed. Thus, if a key on a controller keyboard is tapped continually the loop will start repeatedly, producing a stuttering effect. This technique can be used to create breakdowns in dance music. Alternatively, the sampler can be set to 'one-shot trigger mode', so that a quick tap on a controller keyboard plays the sample in its entirety even if the key is released before the sample has finished playback. This is useful if you want the original sample to play through to its normal conclusion while triggering the same sample again to play over the top. This technique can be used to create complex drum loops and break beats; a few instances of the same loop are playing simultaneously, each out of time and perhaps pitch with one another.

If either of these methods are used to trigger a drum loop or motif to play throughout the length of the track, it is important to ensure that the sample is not looped from within the sampler but is, instead, re-triggered every couple of bars by the sequencer. Bear in mind that whenever a sample is triggered, the only part that will be in sync with the rest of the music is the start of the sample. After this, you're relying entirely on the sampler's internal clock to keep in time with the sequencer. This may not necessarily be an issue if the sampler can be locked to a MIDI clock, but if this is not possible the two could drift apart. For example, if a two-bar drum loop is sampled and programmed to constantly repeat for as long as a MIDI note-on is present, unless it has been looped with sample accuracy the timing will begin to drift as the track continues. While a couple of milliseconds drift near the beginning of the track may be unnoticeable, over the length of a 5- or 6-minute track the timing could drift out by a couple of seconds.

A similar technique to looping is phrase sampling, but this consists of sampling a short musical phrase that is only used a couple of times throughout a song. This method is commonly used in dance music for sampling a short vocal phrase or hook that is then triggered occasionally to fit with the music. This phrase sampling technique can also be developed further to test new musical ideas. By slicing your own song into a series of four-bar sections and assigning each loop to a particular key on the keyboard you can trigger the loops in different orders so you can determine the order that works best. What's more, if four bars of each individual track of the mix (drums, bass, lead, etc.) are sampled, each one can be played back from the keyboard, allowing you to quickly change the timing of each track in relation to the others. Many club records are constructed this way, but with the sampler's one-shot triggering deactivated so that either the entire track or the groove of the record can be 'stuttered' to signify a break or build in the music.

A potential problem with using pre-sampled loops or phrases from a sample CD is that it's unlikely they'll be at the same pitch and/or tempo as your current project. Accordingly, all samplers provide pitch-shifting and time-stretching functions so that the pitch and tempo of the phrase or loop can be adjusted to fit with the music. Both these functions are pretty much self-explanatory: the pitch-shift function changes the pitch of notes or an entire riff without changing the tempo and time-stretching adjusts the tempo without affecting the pitch.

These are, of course, both useful tools to have, but it's important to note that the quality of the results is proportionate to how far they are pushed. While most pitch-shifting algorithms remain musical when moving notes up or down the range by a few semitones, if the pitch is shifted by more than five or six semitones the sound may begin to exhibit an unnatural quality. Similarly, while it should be possible to adjust the tempo by 25 BPM without introducing any digital artefacts, any adjustments above this may introduce noise or crunching that will compromise the audio.

Recycling loops

Due to the limitations of time-stretching, software developers Propellerhead designed the product ReCycle. This is a software tool that slices loops into their constituent parts, and provides tools for editing or applying processing to each section. Much of the initial slicing process is fully automated and, after importing a drum loop for instance, the software scans the file for the transients (the fast attack stage) and places a slice point at each, intelligently breaking the loop. Once accomplished, any severe tempo adjustments will produce more natural results than normal time-stretching methods, because only the spacing between the slices is adjusted. This, however, is only the start, since once a loop has been sliced, a series of editing tools become available. These allow you to adjust the length, attack and decay, or apply compression and EQ to each individual slice. The resulting slices can then be transmitted digitally to a sampler while a MIDI file is also created, denoting the position of each slice in the original drum loop. Thus, the MIDI file is imported into a sequencer and the original loop can be changed by simply moving a series of MIDI notes.

Alternatively, if the sequencer supports ReCycle's own propriety format of REX files, these can be imported directly into the audio track of the sequencer. These contain all the slice information along with the audio, making it possible to grab individual audio slices and move them around, changing the feel of the loop. ReCycle is not the only software that performs these tasks (Square Circle's Wavesurgeon offers more comprehensive functions than ReCycle, and Steinberg's premiere sequencer Cubase SX 1 and 2 has a similar feature built in), but whichever tool is used this form of sample slicing is essential not only for producing dance music, but also remixing.

Figure 4.1 Propellerhead's ReCycle in action.

In addition to using sample CDs, many artists record their own samples from the real world. The Future Sound of London (FSOL) are renowned for creating most of their music using samples created in this way. Indeed, the true art of creative sampling lies with listening out for everyday sounds; all it requires is a creative imagination. Try popping paper bags, hitting things with newspapers, kicking footballs against garage doors, or hitting a spoon against a cup. Each of these, if creatively mangled and manipulated in a sampler, can be used as sound effects or even as instruments.

It can also be advantageous to sample some of your own synthesizer's sounds, although the reason for taking this approach may not be immediately obvious. For example, if you want to play a timbre polyphonically but the synth is monophonic (can only play one sound at any one time) or monotimbral (can only play one type of sound at one time) and you want to use more than one type of sound from the same synth, the sampler will allow these possibilities. Multi-sampling the synthesizer at every key essentially recreates the synthesizer in the sampler, allowing for polyphonic or multi-timbral operation. In some instances the sampler may also even offer more synthesis parameters than are available on the synthesizer, enabling you to, say, synchronize the LFO to the tempo, use numerous LFOs, or access numerous different filter types.

There's more to sampling a synth than simply recording the timbre into a sampler, though, and creating an acceptable multi-sampled (or even single-keyed) instrument requires careful consideration. For starters, it's prudent to expose the filter (by setting the cut-off fully open and resonance fully closed so that they have no effect on the timbre), deactivate any LFO modulation, and set the amplifier envelope to its fastest attack and longest release. By doing so, a relatively static sound will be easier to sustain in a loop, and the envelope, filter and LFO movements can be replicated using the sampler's own synthesis engine.

While this approach is generally the best, there will undoubtedly be occasions where it isn't possible to remove the modulation (such as sampling from previous 'hit' records), in which case you'll have to exercise care. Bear in mind that any timbre that evolves over time will be difficult to map across a keyboard, because as the pitch is adjusted by playing over the keyboard, the modulation rate will increase or decrease. This can be particularly apparent when an LFO has been used to augment the timbre, because the timing may be slightly out, making any sustain looping impossible to achieve. Unfortunately, there is no direct way of overcoming these limitations short of applying a series of effects such as reverb or distortion in an attempt to mask some of the modulation.

Recording samples

While samplers are certainly versatile instruments, the quality of the results they produce will depend entirely on the initial sampling techniques used. While the basic premise behind sampling any source isn't too difficult – plug in an instrument, CD/turntables (or whatever) into the sampler and press record – unless the bit rate, sample rate and recording levels are carefully controlled the resulting sound could contain more noise than original sound. Indeed, when recording anything directly into a sampler, computer soundcard or any digital recording device, high-level recordings combined with good sample and bit rates are of paramount importance for producing the best results.

When using any digital recording system, sound has to be converted from an analogue signal (i.e. the sound you hear) to a digital format that the device can work with. Any digital recording device accomplishes this by measuring the waveform of the incoming signal at specific intervals and converting these measurements into a series of numbers based on the position of the waveform. Each of these numbers is known as an 'individual sample' and the total number of samples that are taken every second is called the 'sample rate'. On this basis, the more samples that are taken every second, the better the overall quality of the recording will be. For instance, if a waveform is sampled at 44 000 times per second, it will produce more accurate results than if it were sampled at 22 000 times per second.

Sample rate

Figure 4.2 Example of sample rate measurements.

As simple as this may seem, any sampling rate must be higher than the frequency of the waveform being recorded in order to produce accurate results. If it isn't, then the analogue-to-digital converter (ADC) could miss anything from half to entire cycles of the waveform, resulting in a side-effect known as 'aliasing'. This is the result of an audio waveform not being 'measured' in the correct places. For instance, a high-frequency waveform could confuse the ADC into believing it's actually of a lower frequency, which would effectively introduce a series of spurious low-frequency spikes into the audio file. Worse still, if the sample rate were set at the exact same frequency as the waveform or at exactly twice the frequency, then the same position in the waveform would be sampled repeatedly, resulting in nothing more than a straight line (i.e. total silence).

To avoid this problem, the sampling rate must be greater than twice the frequency of the waveform – a principle called the Nyquist Theorem. This states that to recreate any waveform accurately in digital form, at least two different points of a waveform's cycle must be sampled. Consequently, as the highest range of the human ear is a frequency of approximately 20 kHz, the sample rate should be just over double this. This is the principle from which the CD quality is derived.

Note: The sample rate of a typical domestic audio CD is 44.1 kHz, which is derived from the calculation:

Human hearing limit = 20 000 Hz

20 000 Hz × 2 = 40 000 Hz + 4100 Hz (to make the rate more than twice the optimum frequency)

Although this frequency response is the de facto standard used for all domestic audio CDs, there are higher sampling frequencies available: 48 000, 88 200, 96 000 and 192 000 Hz. Though these sampling rates are far beyond the frequency response of CDs, it is quite usual to work at these higher rates while processing and editing. Although there has been no solid evidence to support the theory, it is believed that higher sample rates provide better spatial signal resolution. Thus, if the ADC supports higher rates, theoretically there is no harm in working at the highest available rate.

Note: Many engineers argue that, because the signal must be reduced to 44.1 kHz when the mix is put onto CD, the downward conversion may introduce errors, so working at higher sampling rates is pointless. If you do decide to work at a higher sample rate, it may be worthwhile using a rate of 88 200 Hz, as this simplifies the down-sampling process.

Bit rates

In addition to the sample rate, the bit rate also determines the overall quality of the results when recording audio into a digital device. To comprehend the importance of this, we need to examine the binary language used by all computers.

All computers utilize the binary language that consists of two values, 1 and 0. The computer can count up to a specific number depending on how many bits are used. For example, using an 8-bit system it is possible to count to a maximum of 255, while in a 24-bit system the maximum value is 16 777 215.

Relating this to a digital audio recording system, the number of bits determines the number of analogue voltages that are used to measure the volume of a waveform, in effect increasing the overall dynamic range. In technical applications the dynamic range is the ratio between the residual noise (known as the noise floor) created by all audio equipment and the maximum allowable volume before a specific amount of distortion is introduced.[1]

In relation to music, the dynamic range is the difference between the quietest and loudest parts. For instance, classical music utilizes a huge dynamic range by suddenly moving from very low to very high volume, the cannons in the *1812 Overture* being a prime example. Most dance and pop music, on the other hand, has a deliberately limited dynamic range so that it remains permanently up front and 'in your face'. Essentially, this means is that if a low bit rate is used for a recording, only a small dynamic range will be achieved. The inevitable result is that the ratio between the noise floor and the loudest part of the audio will be small, so background noise will be more evident. By increasing the bit rate, each additional bit introduces another analogue voltage, which adds another 6 dB to the dynamic range.

For example, if only 1 bit is used to record a sound, the recorder will produce a square wave at the same frequency as the original signal and at fixed amplitude. This is because only one voltage is used to measure the volume throughout the sampling frequency. However, if an 8-bit system is used, the signal's dynamic range will be represented by 255 different analogue voltages and a more accurate representation of the waveform will result. It's clear, then, that a 24-bit audio signal will have a higher dynamic range than a 16-bit (the bit rate used by CDs) signal.

---

[1] According to the IEC 268 standard, the dynamic range of any professional audio equipment is measured by the difference between the total noise floor and the equivalent sound pressure level where a certain amount of total harmonic distortion (measured in THD) appears.

Table 4.3 Bit rates

8-Bit	Binary	16-Bit	Binary	20-Bit	Binary	24-Bit	Binary
0	1	0	1	0	1	0	1
1	1	1	1	1	1	1	1
2	1	2	1	2	1	2	1
4	1	4	1	4	1	4	1
8	1	8	1	8	1	8	1
16	1	16	1	16	1	16	1
32	1	32	1	32	1	32	1
64	1	64	1	64	1	64	1
128	1	128	1	128	1	128	1
		256	1	256	1	256	1
		512	1	512	1	512	1
		1 024	1	1 024	1	1 024	1
		2 048	1	2 048	1	2 048	1
		4 096	1	4 096	1	4 096	1
		8 192	1	8 192	1	8 192	1
		16 384	1	16 384	1	16 384	1
		32 768	1	32 768	1	32 768	1
				65 536	1	65 536	1
				131 072	1	131 072	1
				262 144	1	262 144	1
				524 288	1	524 288	1
						1 048 576	1
						2 097 152	1
						4 194 304	1
						8 388 607	
Total	255		65 535		1 048 575		16 777 215

Figure 4.3 Example of bit depth.

At the time of writing, although 24-bit is the highest resolution available to samplers' and sound-cards' ADCs, a proportionate amount of audio-capable software utilizes internal 32- or 64-bit processing. The reasoning behind using bit rates this high is that whenever any form of processing is applied to a digitized audio signal, quantization noise is introduced. This is a cumulative effect, meaning that as more digital processing is applied, more quantization noise is introduced and this needs to be kept to a minimum. In more practical terms, this means that while a CD may only accept a 16-bit recording, if a 24-bit process is used throughout digital mixing, editing and processing, when the final sound is dropped to 16-bit resolution for burning to CD the quantization noise will be less apparent.

The process of 'dropping out' bits from a recording to reduce the bit rate is known as 'dithering'. The principle behind how this actually works is not vital to comprehend, but what is important is that the best available dithering algorithms are used. Poor quality algorithms will have a detrimental effect on the music as a whole, resulting in clicks, hiss or noise. As a reference, Apogee are well known and respected for producing excellent dithering algorithms and the Waves Native Gold Bundle[2] on the CD features an excellent dithering algorithm.

It isn't always necessary to work at such a high bit rate and some genres of dance music benefit from using a much lower rate. For instance, 12-bit samples are often used in hip-hop to obtain the typical hip-hop sound. This is because the original artists used old samplers that could only sample at 12-bit resolution; thus, to write music in this genre it's quite usual to limit the maximum bit rate in order to reproduce these timbres. Similarly, with trip-hop and lo-fi, the sample rate is often lowered to 22 or 11 kHz, as this reproduces the gritty timbres that are characteristic of the genre.

Ultimately, no matter what sample or bit rate is used, it's important that the converters on the soundcard or digital recorder are of a good standard and that the amplitude of the signal for recording is as loud as possible (but without clipping the recorder). Although all digital editors allow the gain of a recorded signal to be increased after recording, the signal should ideally be recorded as loud as possible so as to avoid having to artificially increase a signal's gain using software algorithms. This is because all electrical circuitry, no matter how good or expensive, will have some residual noise associated with it due to the random movement of electrons. Of course, the better the overall design the less random movement there will be, but there will always be some noise present, so the ratio between the signal and this noise should be as high as possible. If not, when you artificially increase the gain after it has been recorded it will also increase any residual noise by the same amount.

For instance, if a melodic riff is sampled from a record at a relatively low level then the gain is artificially increased to make the sound audible, the noise floor will increase in direct proportion to the audio signal. This produces a loud signal with loud background noise too. If, however, the signal is recorded as loud as possible, the ratio between the noise floor and the signal is greatly increased. There is, however, a fine line between capturing a recording at optimum gain and actually recording the signal too loud so that it clips the recorder's inputs. This isn't necessarily a problem with analogue recorders as they don't immediately distort when the input level gets a little too high, since they have 'headroom' in case the odd transient hit pushes it too hard, but digital recorders will 'clip' immediately. Unlike analogue distortion, which can often be quite pleasant, digital distortion cuts off the top of the waveform, resulting in an ear-piercing clicking sound. This type of distortion obviously needs to be avoided so you need to set a recording level that is not too loud to cause distortion, yet not too low to introduce too much noise.

All recording software or hardware, including samplers, will display a level meter informing you how loud the input signal is to help to determine the appropriate levels. Before beginning to record, it's vital to set this up correctly and this is accomplished by adjusting the gain of the source so that the signal overload peaks light on only the most powerful parts and then back off slightly so that they are just below the clipping level. This approach is suitable only if there isn't too much volume fluctuation throughout the entire recording, though. If a sample is taken from a record, CD or electronic instrument, there's a good chance that the volume will be constant,

---

[2] The Waves Gold Bundle featured on the CD is copyright to RC Waves.

but if a live source, such as vocals or bass guitars, are recorded there can be a huge difference in the dynamics. It's doubtful that any vocalist, no matter how well trained, will perform at a constant level and in the music itself it's quite common for the vocals to be softer in the verse sections than they are in the chorus sections.

If the recording level is set so that the loudest sections don't clip the recorder, any quieter sections will have a smaller signal-to-noise ratio, while if the recording levels are set so that the quieter sections are captured at a high level, any louder parts will send the meters into the red. To prevent this, it's necessary to use a device that will reduce the dynamic range of the performance by automatically reducing the gain on the loud parts, while leaving the quieter parts unaffected, thus allowing the recording levels to be increased on the quietest parts. This dynamic restriction is accomplished by strapping a compressor between the signal source and the input of the sampler or recording device. By setting the compressor to squash any signal from the source above a certain threshold, any peaks that could cause clipping in the recorder are reduced and the recording can be made at a substantially higher volume. The uses of a compressor will be covered in more detail in the next chapter.

Copyright law

I couldn't possibly end a chapter on sampling without mentioning copyright law. Ever since dance music broke onto the scene, motifs, vocal hooks, drum loops, basses and even entire sections of music from countless records have been sampled, manipulated and otherwise mangled in the name of art. Hits by James Brown, Chicago, Donna Summer, Chic, Chaka Khan, Sylvester, Lolita Holloway, Locksmith and Michael Jackson, along with innumerable others, have come under the dance musicians' sample knife and been remodelled for the dance floor.

Although in the earlier years of dance artists managed to get away with releasing records without clearing the samples, this was because it was a new trend and neither the original artists nor record companies were aware of a way that they could prevent it. This changed in the early 1990s, when the sampling of original records proved it was more that just a passing fad. Today, companies know exactly what to do if they hear that one of their artists' hits has been sampled and in most cases they come down particularly hard on those responsible.

At the time of writing, the most recent and well-published case of illegal sample use was when Björk sampled some of Scanner's work and used it on her *One Little Indian* track. Scanner's representatives noticed the sample and demanded that all the copies of Björk's album were withdrawn from circulation. They were, and another version of the track had to be produced at great cost to the record company involved. The moral of the story is quite clear: before you begin sampling another artist's motif, drum, vocals or entire verse/chorus, consider that by law you cannot sample anything from anybody else and use it commercially without clearance. To help clear up some of the myths that still circulate – about how small initial pressings or samples under a certain length are not covered by law – I spoke to John Mitchell, a music solicitor who has successfully cleared innumerable samples for me and many other artists.

Q: What is copyright?
A: Copyright exists in numerous forms and if I were to explain the entire copyright law it would probably use up most of your publication. To summarize, it belongs to the creator and is protected from the moment of conception, exists for the lifetime of the creator and another 70 years after their death. Once out of copyright, the work becomes 'public domain' and any new versions of that work can be copyrighted again.

Q: So it would be okay to sample Beethoven or another classical composer?
A: In theory yes, but most probably no. The question arises as to where did you sample the recording from, because it certainly wouldn't have been from the original composer. Although copyright belongs to the creator, the performance is also copyrighted. When a company releases a compilation of classical records they will have been re-recorded and the company will own the copyright to that particular performance. If you sample it you're breaching the performance copyright.

Q: What if you've transcribed it and recorded your own performance?
A: This is a legal wrangle that is far too complex to discuss here, as it depends on the original composer and the piece of music. When the original copyright expires and it's re-recorded, the subsequent company may own the copyright to both the performance and the creation. My advice is to check first before you attempt to sample or transcribe anything.

Q: So what can you get away with when sampling copyright music or speech?
A: Nothing – there seems to be a rumour circulating that if the sample is less than 30 seconds in length you don't need any clearance, but this simply isn't true. If the sample is only a second in length and its copyright protected you are breaking the law if you commercially release your music containing it.

Q: What if the sample were heavily disguised with effects and EQ?
A: You do stand a better chance of getting away with it, but that's not to say you will. Many artists spend a surmountable amount of time creating their sounds and melodies and it isn't too difficult to spot them if they're used on other tracks. It really isn't worth the risk!

Q: Does this same copyright law apply to sampling snippets of vocals from TV, radio, DVD, etc?
A: Yes, sampling vocal snippets or music from films breaches a number of copyrights, including the script writer (the creator) and the actor who performed the words (the performer).

Q: Is there an organization that monitors all the releases?
A: Not as far as I'm aware, but I do know that the Mechanical Copyright Protection Society employs a number of DJs who actively listen out for records containing illegal samples. Plus, if the record becomes a big hit more people will hear it and the chances are that someone will recognize it eventually.

Q: So what action can be taken if clearance hasn't been obtained?
A: It all depends on the artist and the record company who own the recording. If the record is selling well then they may approach you and work a deal, but be forewarned that the record company and artist will have the upper hand when negotiating monies. Alternatively, they can take out an injunction to have all copies of the record destroyed. They also have the right to sue the samplist.

Q: Does it make any difference how many pressings are being produced for retail with the uncleared samples on them?
A: This reminds me of another urban myth: if you produce and sell fewer than 500 copies then there is no need to obtain sample clearance. This couldn't be further from the truth. If just one record were released commercially containing an illegal sample the original artist has the legal rights to prevent any more pressings being released.

Q: What if the record were released to DJs only?
A: It makes no difference. The recording is being aired to a public audience.

Q: Can you go to prison for illegal use of samples?
A: It hasn't happened yet but that's not to say that it never will. At the moment, breaching copyright law is viewed as a civil not criminal offence, but times change and I would fully recommend acquiring the clearance before using any samples. You don't want to be the first exception to the rule.

Q: How much does it cost to get sample clearance?
A: It depends on a number of factors. How well known the original artist is, the sampled record's previous chart history, how much of the sample you've used in your track and how many pressings you're planning to make. Some artists are all too willing to give sample clearance because it earns them royalties, while others will want ridiculous sums of money.

Q: How do you clear a sample?
A: Due to the popularity of sampling from other records there are companies appearing every week who will work on your behalf to clear a sample. However, my personal advice would be to talk to the Mechanical Copyright Protection Society. These guys handle the licensing for the recording of musical works and started a Sample Clearance Department in 1994. Although they will not personally handle the clearance itself – for that you need someone like myself – they can certainly put you in touch with the artist or record company involved.

5 Compression, processing and effects

Compression plays a major part of my sound. I have them patched across every output of the desk.

Armand Van Helden

Before looking at the principles behind designing sounds for different music genres, it's first necessary to examine the various processors and effects that are available and understand how they are accessed through the mixing desk. This is because effects are often used as part of the sound design process as well as during the mixing stage, so it pays to understand what they are, how they affect the audio and how a mixing desk can determine the outcome of the effect. This chapter concentrates on the behaviours of the different processors and effects that are widely used in the design and production of dance music, including reverb, chorus, phasers, flangers, delay, EQ, distortion, gates, limiters and compressors.

Of all the effects and processors available, a compressor is the most important tool in any dance musician's arsenal, so a thorough understanding of it is essential. Without compression, drums appear wimpy in comparison to the chest-thudding results heard in professionally produced music, mixes can appear lacking in depth, and basses, vocals and leads can lack any real presence. Despite its importance, the compressor remains the least understood processor of them all, largely because if properly applied the effect should be particularly subtle.

Compression theory

The fundamental application of a compressor is to reduce the dynamic range of a performance and this is vital when working with any form of music. As touched upon in the previous chapter, whenever you record any source you should aim to capture the hottest signal possible so that you can avoid artificially increasing the gain. This is because if you record a source that's too low in gain and then artificially increase it, not only will it increase the volume of the recorded source, it'll also increase the background noise. To prevent this, you need to record a signal as loud as possible, but vocals and 'real' instruments have a huge dynamic range, and the difference between clipping and a nominal level can be as much as 20 dB.

Consequently, it's impossible to set a good average recording level with so much dynamic movement, since if you set the recording level to capture the quiet sections, when it becomes louder the recording will clip. Conversely, setting the recorder so that the loud sections do not clip, any quieter sections will be exposed to more residual noise. Of course, you could sit by the recording fader and increase or decrease the recording levels depending on the section being recorded, but this would mean that you need lightning reflexes. Instead, it's much easier to employ a compressor to control the levels automatically. By routing the source sound through a compressor and then into the recorder, you can set a threshold on the compressor so that any sounds that exceed this are automatically pulled down in gain, thus allowing you to record at a more substantial volume overall.

A compressor can also be used to control the dynamics of a sound while mixing. For example, a dance track that uses a real bass guitar will have a fairly wide dynamic range, even if it was compressed during the recording stage. This will cause problems within a mix because if the gain is adjusted so that the loudest parts fit well within the mix, the quieter parts may disappear behind other instrumentation. Conversely, if the fader is set so that quieter sections can be heard over other instruments, the loud parts could be too prominent. Using compression more heavily on this sound during the mixing stage, the dynamic range can be restricted, further allowing the sound to sit better in a mix.

Although these are the key reasons why compressors were first introduced, a compressor's action also has more creative applications that are especially suited towards dance musicians. Since the signals that exceed the threshold are reduced in gain, the parts that do not exceed the threshold aren't touched, so they remain at the same volume as they were before compression. In other words, the difference in volume between the loudest and quietest parts of the recording are reduced, which means that any uncompressed signals will become louder relative to the compressed parts. This effectively boosts the average signal level, which in turn not only allows you to push the volume up further, but also makes it sound louder.

Figure 5.1 A drum loop waveform before compression.

Figure 5.2 A drum loop waveform after compression (note how the volume difference (dynamics) between instruments has changed).

> Note: After reducing the dynamic range of audio it may be perceived to be louder without actually increasing the gain. This is because we determine the overall volume of music from the average volume (measured in Root Mean Square), not from the transient peaks created by kick drums.

What's more, applying heavy compression to certain elements of a dance mix can change the overall character of the timbre, often resulting in a warmer, smoother and rounder tone – a sound typical of most dance tracks around today.

It is important to note, however, that although there are numerous publications stating the compression settings to use, there are no generic settings for any particular genre of music and its use depends entirely on the timbres that have been used. For instance, a kick drum sample from a record will require a different approach to a kick drum constructed in a synth, while if sampled from a CD, synthesizer or film, different approaches are required again. Therefore, rather than dictate a set list of compression settings, you can achieve much better results by knowing exactly what effect each control will have on a sound and how these are used to acquire the sounds typical of each genre.

Threshold

The first control on a compressor is the threshold, which as touched upon sets the signal level where the compressor will begin squashing the incoming signal. These are commonly calibrated in decibels and will work in direct relationship with a gain reduction meter to inform you of how much the compressor is affecting the incoming signal. In a typical recording situation this control is set so that the average signal level always lies just below the threshold, and if any exuberant parts exceed it, the compressor will jump into action and the gain will be reduced to prevent any clipping.

Ratio

The amount of gain reduction that takes place after a sound exceeds the threshold is set using a ratio control. Expressed in ratios, this control is used to set the dynamic range the compressor affects, indicating the difference between the signals entering the compressor that exceed the threshold to the levels that come out of the other end. For example, if the ratio is 4:1, every time the incoming signal exceeds the threshold by 4 dB, the compressor will squash the signal so that there is only a 1 dB increase at the output of the compressor. Similarly, if the ratio set were 6:1, an increase at the compressor's output of 1 dB will occur when the threshold is exceeded by 6 dB and likewise for ratios of 8:1, 10:1 and so on. Subsequently, the gain reduction ratio always remains constant no matter how much compression takes place. In most compressors these range from 1:1 up to 10:1 and may, in some cases, also offer infinity:1.

From this we can determine that if a sound exceeds a predefined threshold, the compressor will squash the signal by the amount set with the ratio control. The problem with this approach, however, is that we gain a significant amount of information about sounds from their initial attack stage and if the compressor jumps in instantaneously on an exceeded signal it will squash the transient, which reduces its high-frequency content. For instance, if you were to set up a compressor to squash a snare drum, the compressor will clamp down on the attack stage, which in effect diminishes the initial transient, reducing it to a 'thunk'. What's more, this instantaneous action will also appear when the sound drops below the threshold again as the compressor stops processing the audio. This can be especially evident when compressing low-frequency waveforms such as basses, since compressors can apply gain changes during a waveform's period. That is, if a low-frequency waveform, such as a sustained bass note, is squashed the compressor may treat the positive and negative states of the waveform as different signals and continually activate and deactivate. The result of this is an unpleasant distortion of the waveform. To prevent this from occurring, compressors will feature attack and release parameters.

Attack/release

Both these parameters behave in a similar manner to those on a synthesizer, but control how quickly the volume is pulled down and how long it takes to rise back to its nominal level after the signal has fallen below the threshold. In other words, the attack parameter defines how long the compressor takes to reach maximum gain reduction while the release parameter determines how long the compressor will wait *after* the signal has dropped below the threshold before processing stops.

This raises the obvious question that if the attack is set so that it doesn't clamp down on the initial attack of the source sound it could introduce distortion/clipping before the compressor activates. While this is true, in practice very short, sharp signals do not always overload an analogue recorder, since these usually have enough headroom to let small transients through without introducing any unwanted artefacts. This isn't the case with digital recorders, though, and any signals that are beyond the limit of digital can result in clipping, so it's quite usual to follow a compressor with a limiter or, if the compressor features a knee mode, set it to use a soft knee.

Soft/hard knee

All compressors will utilize either soft or hard knee compression, but some will offer the option to switch between the two modes. These are not controllable parameters but dictate the shape of the envelopes curve, and hence the characteristic of how the compressor behaves when a signal approaches the threshold. So far we've considered that when a signal exceeds the threshold the compressor will begin to squash the signal. This immediate action is referred to as hard knee compression. Soft knee, on the other hand, continually measures the incoming signal and when it approaches 3–14 dB (depending on the compressor) towards the current threshold, the compressor starts to apply the gain reduction gradually.

Generally, this will initially start with a ratio of 1:1 and as the signal grows ever closer to the threshold it's gradually increased until the threshold is exceeded, whereby full gain reduction is applied. This allows the compressor's action to be less evident and is particularly suitable for use on acoustic guitars and wind instruments, where you don't necessarily want the action to be evident.

Note: The action of the knee is entirely dependent on the compressor being used and some can be particularly long, starting 12 dB before the threshold, while others may start 3 dB before. As a matter of interest, 6–9 dB soft knees are considered to offer the most natural compression for instruments.

Peak/RMS

Not all compressors feature knees, so short transient peaks can sometimes catch the compressor unaware and 'sneak' past unaffected. This is obviously going to cause problems when recording digitally, so many compressors will implement a switch for peak or RMS modes. Compressors that do not feature these two modes will operate in RMS and this means that the compressor will detect and control signals that stay at an average level rather than the short sharp transient peaks. As a result, no matter how fast the attack may be set, there's a chance that the transients will overshoot the threshold and not be controlled. This is because by the time the compressor has figured out that the sound has exceeded the threshold it's too late – the peak's been and gone again. Therefore, to control short transient sound such as drum loops, it's often prudent to engage the peak mode. With this the compressor becomes sensitive to short sharp peaks and clamps down on them as soon as they come close to the threshold, rather than after they exceed it. By doing so, the peak can be controlled before it overshoots the threshold and creates a problem.

While this can be particularly useful when working with drum and percussion sounds, it can create havoc with most other timbres. Keep in mind that many instruments can exhibit a particularly short, sharp initial attack stage and if the compressor perceives these as possible problems it'll jump down on them before they overshoot. In doing so, the high-frequency elements of the attack will be dulled, which can make the instrument appear less defined, muddled or lost within the mix. Therefore, for all instruments bar drums and percussion, it's advisable to stick with the RMS mode.

Make-up gain

The final control on a compressor is the make-up gain. If you've set the threshold, ratio, attack and release correctly, the compressor should compress effectively and reduce the dynamics in a sound, but this compression will also reduce the overall gain by the amount set by the ratio control. Therefore, whenever compression takes place you can use the make-up gain to bring the signal back up to it's pre-compressed volume level.

Side-chaining

Alongside the physical input and output connections on hardware compressors, many also feature an additional pair of inputs known as side-chains. By inputting an audio signal into these, a sound's envelope can be used to control the action that the compressor has on the signal entering the normal inputs. A good example of this is when a radio DJ begins talking over a record and the volume of the record lowers so their voice becomes audible, then when they stop speaking the record returns to its original volume. This is accomplished by feeding the music through the compressor as normal, but with the microphone connected into the side-chain. This supersedes the compressor's normal operation and uses the signal from the side-chain rather than the threshold as the trigger. Thus, the compressor is triggered when the microphone is spoken into, compressing (in effect lowering the volume of the music) by the amount set with the ratio control. This technique should only be viewed as an example to explain the process, though, and more commonly side-chaining is usually used to make space in a mix for the vocals.

In a typical mix, the lead sound will occupy the same frequencies as the human voice, resulting in a cluttered mid range if the two are to play together. This can be avoided if the lead mix is fed into the main inputs of the compressor while the vocal track is routed into the side-chain. With the ratio set at an appropriate level (depending on the tonal characteristics of the lead and vocals) the lead track will dip when the vocals are present, allowing them to pull through the mix.

Hold control

Most compressors that feature a side-chain are likely to also have an associated 'hold' control on the fascia or employ an automated hold function. This is employed because a side-chain measures the envelope of the incoming signal and if both the release and attack are too fast, the compressor may respond to the cycles of a low-frequency waveform rather than the actual envelope. As touched upon previously, this can result in the peaks and dips of the waveform activating and deactivating the compressor, resulting in distortion. By using a hold the compressor is forced to wait a finite amount of time (usually 40–60 ms on automated hold) before beginning the release phase, which is longer than the period of a low-frequency waveform.

Practical compression

Despite the amount of control offered by the average compressor, they are relatively simple to set up for recording audio. As a generalized starting point, it's advisable to set the ratio at 4:1 and lower the threshold so that the gain reduction meter reads between −8 and −10 dB on the loudest parts of the signal. After this, the attack parameter should be set to the fastest speed possible and the release set to approximately 500 ms. Using these as preliminary settings, they can then be adjusted further to suit any particular sound.

Note: It's advisable that compression is applied sparingly during the recording stage, because once applied it cannot be removed. Any exuberant parts of the performance should be prevented from forcing the recorder's meters into the red while also ensuring that the compressor is as transparent as possible. Solid-state compressors are more transparent than their valve counterparts and so are better suited for this purpose.

As a general rule of thumb, the higher the dynamic range of the instrument being recorded, the higher the ratio and the lower the threshold settings need to be. These settings help to keep the varying dynamics under tighter control and prevent too much fluctuation throughout the performance. Additionally, if the choice between hard or soft knee is available, the structure of the timbre should be taken into account. To retain a sharp, bright attack stage, hard knee compression with an attack setting that allows the initial transient to sneak through unmolested should be used, provided of course that the transient is unlikely to bypass the compression. In these instances, and to capture a more natural sound, soft knee compression should be used.

Finally, the release period should be set as short as possible, but not so short that the effect is noticeable when the compressor stops processing. After setting the release at 500 ms, the time should be continually reduced until the processing is noticeable, then increased slowly until it isn't.

Note: Some compressors feature an auto-mode for the release that uses a fast release on transient hits and a slower time for smaller peaks, making this task easier.

Table 5.1 Compression settings

Compression settings	Ratio	Attack parameter (ms)	Release parameter (ms)	Gain reduction (dB)	Knee
Starting settings	5:1 to 10:1	1–10	40–100	−5 to −15	Hard
Drum loop	5:1 to 10:1	1–10	40–100	−5 to −15	Hard
Bass	4:1 to 12:1	1–10	20 or auto	−6 to −13	Hard
Leads	2:1 to 8:1	3–10	40 or auto	−8 to −10	Hard
Vocals	2:1 to 7:1	1–7	50 or auto	−3 to −10	Soft
Brass instruments	4:1 to 10:1	1–7	30 or auto	−8 to −13	Hard
Electric guitars	8:1 to 10:1	2–7	50 or auto	−5 to −12	Hard
Acoustic guitars	5:1 to 9:1	5–20	40 or auto	−5 to −12	Hard

The settings shown in Table 5.1 are naturally only starting points and too much compression should be avoided during the recording stage, something that can only be accomplished by setting both the ratio and threshold controls carefully. This involves setting the compressor to squash audio but ensuring that it stops processing and that the gain reduction meter drops to 0 dB (i.e. no signal is being compressed) during any silent passages.

As a more practical example, with a simple four-to-the-floor kick running though the compressor and the ratio and threshold controls set so that the gain reduction reads −8 dB on each kick, it's necessary to ensure that the gain reduction meter returns to 0 dB during any silent periods. If it doesn't, then the loop is being over-compressed. If the gain reduction only drops to −2 dB during the silence between kicks, then it makes sense that only 6 dB of gain reduction is actually being applied. This means that every time the compressor activates it has to jump from 0 to 8 dB, when in reality it only needs to jump in by 6 dB. This additional 2 dB of gain will distort the transient that follows the silence, making it necessary for the gain reduction to be adjusted accordingly.

Creative compression

While it is generally worth avoiding any evident compression during the recording stage, deliberately making the compressor's action evident forms a fundamental part of creating the typical sound of dance music. To better describe this, we'll use a drum submix to experiment upon. This is a typical dance drum loop consisting of a kick, snare, closed and open hi-hats.

Figure 5.3 The waveform of the drum pattern.

Note: On the audio CD you can hear the basic drum pattern.

It's clear from listening to the drum loop (or simply looking at its waveform) that the greatest energy – that is, the loudest part of the loop – is derived from the kick drum. With this in mind, if a compressor is inserted across this particular drum loop and the threshold is set just below the peak level of the loudest part, each consecutive kick will activate the compressor.

If the release time is set short, the compressor will activate on the kick then quickly release when the kick drops below the threshold. This results in rapid changes in volume, producing a pumping effect as the compressor activates and deactivates on each kick. This is called 'gain pumping' which, although frowned upon in some areas of music, is deliberately used in dance and popular music to give a track a more dynamic feel. The exact timing of the release will depend entirely on the tempo of the drum loop and must be short enough for the compressor to recover before the next kick. Similarly, the release must be long enough for this effect to sound natural, so it's best to keep the loop repeated over four bars and experiment with the attack and release parameters until the required sound is achieved.

Note: On the audio CD you can hear the effect the release has on a sound.

If we expand on this principle further and add a bass line, pad and chords to the previously compressed loop and compress it again in the same way (i.e. with a short release), the entire mix will pump energetically. As the kick is still controlling the compressor (and provided that the release isn't too short or long), every time a kick occurs the rest of the instruments will drop in volume, which accentuates the overall rhythm of the piece.

Note: On the audio CD you can hear the effect of compression on the bass, pad and chord structure.

Gain pumping is also useful when applied across the whole mix, even though each element in the mix may already have been compressed in isolation. Gain pumping across the whole mix is used to balance areas in the track where instruments are dropped in and out. When fewer instruments play, the gain of the mix will be perceived as lower than when all the instruments play simultaneously. The overall level can be controlled by strapping a compressor across the mixing desk's main stereo buss (more on this in later chapters), to make the mix pump with energy.

Gain pumping across the entire mix ('mix pumping') should be applied with caution, because if the mix is pumped too heavily it will sound strange. Setting a 20–30 ms attack with a 250 ms release and a low threshold and ratio to reduce the range by 2 dB or so should be sufficient to produce a mix that has the 'right' feel. That said, there are many instances, for example on *MTV Dance* or the more recent *Rapture* channel, where gain pumping is so forceful that it's blindingly obvious, so the degree to which you employ these techniques in your own mix is entirely up to you.

Compression can also be used on individual sounds to change the tonal content of a sound. For example, by using heavy compression (a low threshold and high ratio) on an isolated snare drum, the snare's attack avoids compression but the decay is squashed, which brings it up to the attack's gain level. This technique is often employed to create the snare '*thwack*' typical of trance, techno and house music styles. Similarly, if a deeper, duller, speaker-mashing, kick drum 'thud' is required, the compressor's attack should be set as short as possible so that it clamps down hard on the initial attack of the kick. This eradicates much of the initial high-frequency content and as the volume is increased with the compressor's make-up control a deeper and much more substantial 'thud' is produced.

Note: Because hip-hop, rap, house and big beat often have particularly loud bass elements that play consecutively with the kick, the different sounds can conflict, muddying the bottom end of the mix. This can be prevented by feeding the kick drum separately into the side-chain inputs of the compressor, with the ratio set to 3:1, with a fast attack and medium release (depending on the sound). If the bass is fed into the compressor's main inputs, every time the kick occurs the bass will drop in volume, making space for the kick and thereby preventing any conflicts.

It is, however, important to note that the overall quality of compression depends entirely on the type of compressor being used. Compressors are one of two types: solid-state or valve. Solid-state compressors use digital circuitry throughout and will not tend to pump as heavily or sound as good as those that utilize valve technology. Some valve compressors will be solid-state in the most part, using a valve only at the compressor's make-up gain. Solid-state compressors are usually more transparent than their valve-based counterparts and are usually used during the recording stages. Valve compressors are typically used after recording to add warmth to drums, vocals and basses, an effect caused by small amounts of second-order harmonic distortion that are introduced into the final gain circuitry[1] of the compressor. This distortion is a result of the free movement of electrons within the valves and occurs at exactly twice the frequency of the amplified signal. Despite the fact that this distortion only contributes 0.2% to the amplified signal, the human ear (subjectively!) finds it appealing.

More importantly, the characteristic warmth of a valve compressor differs according to the model of valve compressor that is used, as each will exhibit different characteristics. These differences and the variations from compressor to compressor are the reasons why many dance producers will spend a fortune on the right compressor and also why it isn't uncommon for producers to own a number of both valve and solid-state types.

> Note: Most dance producers agree that solid-state circuitry tends to react faster, producing a more defined, less forgiving sound, while valve compressors add warmth that improve the overall timbre.

While it isn't essential to know why these differences exist from model to model, it's worth knowing which compressor is most suited to a particular style of work. Failing that, it also makes for an excellent conversation stopper, so what follows is a quick rundown of the five most popular methods of compression:

- Variable MU.
- Field effect transistor (FET).
- Optical.
- VCA.
- Computer-based digital.

Variable MU

The first compressors ever to appear on the market were called variable MU units. This type of compressor uses valves for the gain control circuitry and does not have an adjustable ratio control. Instead of an adjustable control, the ratio is increased in proportion to the amount of the incoming signal that exceeds the threshold. In other words, the more the level overshoots the threshold, the more the ratio increases. While these compressors do offer attack and release stages, they're not particularly suited towards material with fast transients, even with their fastest attack settings. Due to the valve design, the valves run out of dynamic range relatively quickly, so it's unusual to acquire more than 15–20 dB of gain reduction before the compressor runs out of energy. Nonetheless, variable MU compressors are renowned for their distinctive, phat, lush character, and can work magic on basses and pumping dance mixes. The most notorious variable MU compressors are made by Manley and can cost in excess of £3500.

---

[1] John Ambrose Fleming originally developed the valve in 1904, but it was 2 years later before Lee De Forest constructed the first Triode configuration. Edwin Howard Armstrong then used this to create the first ever valve amplifier in 1912.

Field effect transistor (FET)

FET compressors are often recognized as simulating the action of a typical valve compressor because they use a field effect transitor to vary the gain. They provide incredibly fast attack and release stages, making them an excellent choice for beefing up kick and snare drums, electric guitars, vocals and synthesizer leads. While they suffer from a limited dynamic range, if they're pushed hard they can pump very musically and are perfectly suited for gain pumping a mix. The only major problem is getting your hands on one. Original FETs are as rare as rocking horse manure and consequently second-hand models are incredibly expensive. Reproduction versions of the early FETs, such as the UREI 1176LN Peak Limiter (approx. £1800) and the LA Audio Classic II (approx. £2000), are a worthwhile alternative.

Optical

Optical (or 'opto') compressors rely on a light bulb and phototransistor assembly to automatically adjust the gain of the compressor. The phototransistor monitors the current illumination level of the light bulb and subsequently adjusts the gain to suit. Because the phototransistor must monitor the light-bulb before it takes any action, some latency is created in the compressor's response, so the more heavily the compression is applied, the longer the envelope times tend to be. Consequently, most optical compressors utilize soft knee compression. This creates a more natural attack and release, but also means that the compressor is not quick enough to catch many transients. Despite this, optical compressors are great for compressing vocals, basses, electric guitars and drum loops, providing that a limiter follows the compression (limiters will be explained later).

There are plenty of opto compressors to choose from, including the ADL 1500 (approx. £2500), the UREI LA3 and UREI Teletronix LA-2A (approx. £2900 each), the Joe Meek C2 (approx. £250) and the Joe Meek SC2.2 (approx. £500). Both Joe Meek units sound particularly smooth and warm considering their relatively low prices, and for the typical gain pump synonymous with dance then you could do worse than to pick up the SC2.2.

> Note: Incidentally, all Joe Meek's compressors are green because, after designing his first unit, he decided to spruce it up by colouring it with car aerosol paint and bright green was the only colour he could find in the garage at the time.

VCA

VCA compressors offer the fastest envelope times and highest gain reduction levels of any of the compressors covered so far. These are the compressors most likely to be found in a typical home studio. As with most things, the quality of a VCA compressor varies wildly in relation to its price tag. Many of the models aimed at the more budget conscientious musician reduce the high frequencies when a high gain reduction is used, regardless of whether you're clamping down on the transients or not. When used on a full mix these also rarely produce the pumping energy that is typical of models that are more expensive. Nonetheless, these types of compressor are suitable for use on any sound. The most widely celebrated VCA compressor is the Empirical Labs Stereo Distressor (approx. £2500), which is a digitally controlled analogue compressor with VCA, solid-state and op amps. This allows switching between the different methods of compression to suit the sound. Two versions of the Distressor are available to date: the standard version and the British version. Of the two, the British version produces a much more natural, warm tone (I'm not just being patriotic) and is the preferred choice of many dance musicians.

Computer-based digital

Computer-based digital compressors are possibly the most precise compressors to use on a sound. Because these compressors are based in the software domain, they can analyse the incoming audio before it actually reaches the compressor, allowing them to predict and apply compression without the risk of any transients sneaking past the compressor. This means that they do not need to utilize a peak/RMS operation. These digital compressors can emulate both solid-state, transparent compression and the more obvious, warm, valve compression at the fraction of the price of a hardware unit. In fact, the Waves RComp (part of the Waves Native Bundle featured on the CD) can be switched to emulate an optical compressor. Similarly, the PSP Vintage Warmer (also featured on the CD) can add an incredible amount of valve warmth.

> Note: Another benefit of using plug-in-compressors that emulate valves is that they're generally less noisy and cost considerably less than their hardware counterparts.

The look-ahead functions employed in computer-based compressors can be emulated in hardware with some creative thought, which can be especially useful if the compressor has no peak function. Using a kick drum as an example, make a copy of the kick drum track and then delay it in relation to the original by 50 ms. By then feeding the delayed drum track into the compressor's main inputs and the original drum track into the compressor's side-chain, the original drum track activates the compressor just before the delayed version goes through the main inputs, in effect creating a look-ahead compressor!

Ultimately, it is advisable not to get too carried away when compressing audio, as it can be easy to destroy the sound while still believing that it sounds better. This is because louder sounds are invariably perceived as sounding better than those that are quieter. If the make-up gain on the compressor is set at a higher level than the inputted signal, even if the compressor was set up by your pet cat, it will still sound better than the non-compressed version. The incoming signal must be set at exactly the same level as the output of the compressor, so that when bypassing the compressor to check the results, the difference in volume doesn't persuade you that it sounds better.

Furthermore, while any sounds that are above the threshold will be reduced in gain, those below it will be increased when the make-up gain is turned up. While this has the advantage of boosting the average signal level, a compressor does not differentiate between music and unwanted noise. So 15 dB of gain reduction will reduce the peak level to 15 dB, while the sounds below this remain the same. Using the make-up gain to bring this back up to its nominal level (i.e. 15 dB), any signals that were below the threshold will also be increased by 15 dB and if there is noise present in the recording it may become more noticeable.

Most important of all, dance music relies heavily on the energy of the overall 'punch' produced by the kick drum, which comes from the kick drum physically moving the loudspeaker's cone in and out. The more the cone is physically moved, the greater the punch of the kick. This degree of movement is directly related to the size of the kick's peak in relation to the rest of the music's waveform. If the difference between the peak of the kick and the main body of the music is reduced too much through heavy compression, it may increase the average signal level but the kick will not have as much energy, since the dynamic range is restricted, meaning that all the music will move the cone by the same amount. So, you should be cautious as to how much you compress, otherwise you may lose the excursion, which results in a loud yet flat and unexciting track with no energetic punch from the kick.

Figure 5.4 A mix with excursion.

Figure 5.5 A mix with no excursion (all the contents of the mix are almost at equal volume).

Note: On the CD you can hear a short mix with a large excursion followed by the same mix with a small excursion.

Limiters

While compression is applied, it's common practice to pass the audio through a limiter, just in case any transients are not captured by the compressor. Limiters work along similar principles to compressors, but rather than compress a signal by a ratio, they stop signals from ever exceeding the threshold in the first place. This means that no matter how loud the inputted signal becomes, it will be squashed down so that it never violates the current threshold setting. This is referred to as a 'brick wall' because no sounds can ever exceed the threshold. Some limiters, however, allow a slight increase in level above the threshold in an effort to maintain a more natural sound.

A widespread misconception is that if the compressor offers a ratio above 10:1 and is set to this it will act as a limiter, but this isn't necessarily always the case. As we've seen, a compressor is designed to detect an average signal level (RMS) rather than a peak signal, so even if the attack is set to its fastest response, there's a good chance that signal peaks will catch the compressor unaware. The circuitry within limiters, however, does not employ an attack control and as soon as the signal reaches the threshold it is brought under control instantaneously. Therefore, if recording a signal that contains plenty of peaks, a limiter placed directly after the compressor will clamp down on any signals that creep past the compressor and prevent clipping.

Most limiters are quite simple to use and only feature three controls – an input level, a threshold and an output gain – but some may also feature a release parameter. The input is used to set the overall signal level entering the limiter, while the threshold and output gain, like a compressor, are used to set the level where the limiter begins attenuating the signal and controlling the output level. The release control is not standard on all limiters, but if included it's straightforward and allows the time it takes the limiter to return to its nominal state after controlling a peak. As with compression, however, this must be set cautiously, giving the limiter time to recover before the next signal is received to avoid distorting the subsequent transients.

Since the main purpose of a limiter is to prevent transient signals from ever overshooting the threshold, there is generally no need for an attack control, but some software plug-ins will make use of one. This is because they employ look-ahead algorithms that constantly analyse the incoming signal. This allows the limiter to begin the attack stage just before the peak signal occurs. In most cases, this attack isn't user definable and a soft or hard setting will be provided instead. Similar to the knee setting on a compressor, a hard attack activates the limiter as soon as a peak is close to overshooting. On the other hand, a soft attack has a smoother curve with a 10 or 20 ms timing. This reduces the likelihood that any artefacts are introduced into the processed audio by jumping in on the audio too quickly. These software look-ahead limiters are sometimes referred to as ultramaximizers.

As discussed, the types of signals that require limiting are commonly those with an initial sharp transient peak. As a result, limiters are generally used for removing the 'crack' from snare drums, keeping the kick drum under control, and are often used on a full track to produce a louder mix during the mastering process. Like compressors, though, limiters must be used cautiously because they work on the principle of reducing the dynamic range. That is, the harder a sound is limited, the more dynamically restricted it becomes. Too much limiting can result in a loud but monotonous sounding signal or mix. On average, approximately 3–6 dB is a reasonable amount of limiting, but the exact figure depends entirely on the sound or mix. If the sound has already been quite heavily compressed, it's best to avoid boosting any more than 3 dB at the limiting stage, otherwise any dynamics deliberately left in during the compression stage may be destroyed.

Figure 5.6 Drum loop before limiting.

Figure 5.7 Drum loop after limiting.

Note: On the audio CD you can hear the effect limiting has on drums.

Noise gates

Noise gates can be described as the opposite of compressors. This is because while a compressor attenuates the level of any signal that exceeds the threshold, a gate can attenuate or remove any signals that are below the threshold. The main purpose of this is to remove any low-level noise that may be present during a silent passage. For instance, a typical effect of many commercial dance tracks is to only have a drum kick playing before the main reprise, so that when the track returns fully, the sudden change from almost nothing into everything playing at once creates a massive impact. The problem with this approach, though, is that if there is some low-level noise in the recording it will be evident when the track falls silent (i.e. noise between the kicks), which not only sounds cheap but reduces the impact when the rest of the instruments jump back in. In these instances, by employing a gate it can be set so that whenever sounds fall below its threshold the gate activates and creates absolute silence. While in theory this sounds simple enough, in practice it's all a little more difficult.

Firstly, we need to consider that not all sounds stay at a constant volume throughout their period. Indeed, some sounds can fluctuate wildly in volume, which means that they may constantly jump above and below the threshold of the gate. What's more, if the sound was close to the gate's threshold throughout, if it fluctuates even slightly in volume it'll constantly leap above and below the threshold, resulting in an effect known as chattering. To prevent this, gates will often feature an automated or user-definable hold time. Using this, the gate can be forced to wait for a predetermined amount of time after the signal has fallen below the threshold before it begins its release stage, thus avoiding the problem.

Note: The action of this hold function is sometimes confused with a similar gate process called hysteresis, but the two processes, while accomplishing the same goal, are very different. Whereas the hold function forces the gate to wait for a predefined amount of time before closing, hysteresis adjusts the threshold's tolerance independently for opening and closing the gate. For example, if the threshold was set at, say, $-12\,dB$ the audio signal must breach this before the gate will open, but the signal must fall a few extra decibels below $-12\,dB$ before the gate will close again. Consequently, while both hold and hysteresis accomplish the same goal in preventing any chatter, it is generally accepted that hysteresis sounds much more natural than simply using a hold control.

A second problem develops when we consider that not all sounds will start and stop abruptly. For instance, if you were gating a pad that gradually rose in volume, it would only be allowed through the gate after it exceeds the predefined threshold. If this threshold happened to be set quite high, the pad would suddenly jump in rather than fade in gradually as it was supposed to. Similarly, rather than fade away, it would be abruptly cut off as it fell below the threshold again. Of course, you could always lower the threshold, but that may allow noise to creep in, so gates will also feature attack and release parameters. These are similar in most respects to a compressor's envelope in that they allow you to determine the attack and release times of the gate's action. Using these on our example of a pad, by setting the release quite long, as soon as the pad falls below the threshold the gate will enter the release stage and gradually fade out rather than cut them off abruptly. Likewise, by lengthening the attack on the gate, the strings will fade in rather than jump in unexpectedly.

The third, and final, problem is that we may not always want to silence any sounds that fall below the threshold. Suppose that you've recorded a rapper (or any vocalist for that matter) to drop into the music. He or she will obviously need to breathe between the verses and if they're about to scream something out, they'll need to take a large intake of breath before starting. This sharp intake of breath will make its way onto the vocal recording and, while you don't want it to be too loud, at the same time you don't want it totally removed either, otherwise it'll sound totally unnatural – the audience instinctively knows that vocalists have to breathe!

Consequently, we need a way of lowering the volume of any sounds that fall below the threshold rather than totally attenuating them and so many gates (but not all!) will feature a range control. Fundamentally, this is a volume control that's calibrated in decibels, allowing you to define how much the signal is attenuated when it falls below the threshold. The more that this is increased, the more the signal will be reduced in gain until – set at its maximum setting – the gate will silence the signal altogether. Using this range control on the imaginary rapper, you could set it quite low so that the volume of the breaths is not too loud but not too quiet either. By setting the threshold so that only the vocals breach it and those below are reduced in volume by a small amount, it will sound much more natural. Furthermore, by setting the gate's attack to around 100 ms or so as he/she breathes it will begin at the volume set by the range and then slowly swell in to the vocal, which produces a much more natural effect.

For this application to work properly, the release time of the gate must be set cautiously. If it's set too long the gate may remain open during the silence between the vocals, which doesn't allow a new attack stage to be triggered when they begin to sing again. On the other hand, if it's set too short it can result in the vocals being cut short. Consequently, it's prudent to use the shortest possible decay time possible, yet long enough to provide a smooth sound. Generally, this is usually somewhere between 50 and 200 ms.

Note: Employed creatively, the range control can also be used to modify the attack transients of percussive instruments such as pianos, organs or lead sounds (not on drums, though; these are far too short).

One final aspect of gates is that many will also feature a side-chain connection. In this context they're often referred to as 'key' inputs, but nonetheless they behave in a similar manner to a compressor's side-chain. Fundamentally, they allow you to insert an audio signal in the key input that can then be used to control the action of the gate, which in turn affects the audio travelling through the noise gate's normal inputs. This obviously has numerous creative uses, but the most common use is to program a kick drum rhythm and feed the audio into the key input. Any signals that are then fed through the gate's normal inputs will be gated every time a kick occurs. This action supersedes the gate's threshold setting, but the attack, release, range, hold or hysteresis controls are often still available, allowing you to contour the reaction of the gate on the audio signal.

Another use for this key input is known as 'ducking' and many gates will feature a push button, allowing you to engage it. When this is activated, the gate's process is reversed so that any signals that enter the key input will 'duck' the volume of the signal running through the gate. A typical use for this is to connect a microphone into the key input so that every time you speak, the volume of the original signal travelling through the gate is ducked in volume. Again, this supersedes the threshold control, but all of the other parameters are still available, allowing you to contour the action of the signal being ducked, although it should be noted that the attack time turns into a release control and vice versa. Also, as a side note, some of the more expensive gates will feature a MIDI in port that prevents you from having to insert an audio signal into the key input; instead, MIDI note-on messages can be used to control the action of the gate.

Note: On the audio CD you can hear a gate in action on a drum loop.

Transient designers

Transient designers are quite simple processors that generally only feature two controls: an attack and a sustain parameter, both of which allow you to shape the dynamic envelope of a sound. Fundamentally, this means that you can alter the attack and sustain characteristics of a pre-recorded audio file the same as you would when using a synthesizer. While this may initially not seem to be too impressive, it has a multitude of practical uses.

Since we determine a significant amount of information about a sound through its attack stage, modifying this can change the appearance of any sound. For example, if you have a sampled loop and the drum is too loud, by reducing its attack (and lengthening the sustain so the sound doesn't vanish) it will be moved further back into the mix. In a more creative application, if a groove has been sampled from a record, it allows you to modify the drum sounds into something else.

Similarly, if you've sampled or recorded vocals, pianos, strings or any instrument for that matter, the transient designer can be used to add or remove some of the attack stage to make the sound more or less prominent, while strings and basses could have a longer sustain applied. Similarly, by reducing the sustain parameter on the transient designer you could reduce the length of the notes. Like the previously discussed processors, these can be an invaluable tool to a dance producer and we'll be looking more closely at their uses in the genre chapters.

Note: On the audio CD you can hear the effect of altering the attack and sustain of a drum loop.

Reverb

Reverberation (often shortened to reverb or just verb) is used to describe the natural reflections we've come to expect from listening to sounds in different environments. We already know that when something produces a sound the resulting changes in air pressure emanate out in all directions, but only a proportion of this reaches our ears directly. The rest rebounds off nearby objects and walls before reaching our ears; thus, it makes common sense that these reflected waves would take longer to reach your ears than the direct sound itself.

This creates a series of discrete echoes that are all closely compacted together and from this our brains can decipher a staggering amount of information about the surroundings. This is because each time the sound is reflected from a surface, that surface will absorb some of the sound's energy, therefore reducing the amplitude. However, each surface also has a distinct frequency response and this means that different materials will absorb the sound's energy at different frequencies. For instance, stone walls will rebound high-frequency energy more readily than soft furnishings, which absorb them. If you were in a large hall it would take longer for the reverberations to decay away than it would if you were in a smaller room. In fact, the further away from a sound you are, the more reverberation there would be until, eventually, if the sound was far enough away and the conditions were right, you would hear a series of distinct echoes rather than reverb.

There should be little need to describe all the differing effects of reverb because you'll have experienced them all yourself. If you were blindfolded, you would still be able to determine what type of room you're in from the sonic reflections. In fact, reverb is such a natural occurrence that if it's totally removed (such as in an anechoic chamber) it can be unsettling almost to the point of nausea. Our eyes are informing the brain of the room's dimensions, but the ears are informing it of something completely different.

Ultimately, while compression is the most important processor, reverb is the most important effect because samplers and synthesizers do not generate natural reverberations until the resulting signals are exposed to air. So, in order to create some depth in a mix you often need to add it artificially. For example, the kick may need to be at the front of a mix, but any pads could sit in the background. Simply reducing the volume of the pads may make them disappear into the mix, but by applying a light smear of reverb you could fool the brain into believing that the sound is further away from the drums because of the reverberation that's surrounding it.

However, there's much more to applying reverb than simply increasing the amount that is applied to the sound. As we've seen, reverb behaves very differently depending on the furnishings and wall coverings, so all reverb units will offer many more parameters and using it successfully depends on knowing the effects all these will have on a sound. What follows is a list of the available controls on a reverb unit, but it should be noted that in many cases all of these will not be available – it depends on the quality of the unit itself.

Ratio (sometimes labelled as mix)

The ratio controls the ratio of direct sound to the amount of reverberation applied. If you increase the ratio to near maximum, there will be more reverb than direct sound, while if you decrease it significantly, there will be more direct sound than reverb. Using this, you can make sounds appear further away or closer to you.

Pre-delay time

After a sound occurs, the time separation between the direct sound and the first reflection to reach your ears is referred to as the pre-delay. This parameter on a reverb unit allows you to specify the amount of time between the start of the unaffected sound and the beginning of the first sonic reflection. In a practical sense, by using a long pre-delay setting the attack of the instrument can pull through before the subsequent reflections appear. This can be vital in preventing the reflections from washing over the transient of instruments, forcing them towards the back of a mix or muddying the sound.

Early reflections

Early reflections are used to control the sonic properties of the first few reflections we receive. Since sounds reflect off a multitude of surfaces, it creates subtle differences between each subsequent reflection reaching our ears. Due to the complex nature of these first reflections, only the high-end processors feature this type of control, but if present, it allows you to determine the type of surface the sound has reflected from.

Diffusion

This parameter is associated with the early reflections and is a measure of how far the early reflections are spread across the stereo image. The amount of stereo width associated with the reflections depends on how far away the sound source is. If a sound is far away then much of the stereo width of the reverb will dissipate, but there will be more reverberation than if it was upfront. If the sound source is quite close, however, then the reverberations will tend to be less spread and more monophonic. This is worth keeping in mind, since many artists wash a sound in stereo reverb to push it into the background and then wonder why the stereo image disappears and doesn't sound quite 'right' in context with the rest of the mix.

Density

Directly after the early reflections come the rest of the reflections. On a reverb unit this is referred to as the density. Using this control it's possible to vary the number of reflections and how fast they should repeat. By increasing it, the reflections will become denser, giving the impression that the surface they have reflected from is more complex.

Reverb decay time

This parameter is used to control the amount of time the reverb takes to decay away. In large buildings the reflections will generally take longer to decay into silence than in a smaller room. Thus, by increasing the decay time you can effectively increase the size of the 'room'. This parameter must be used cautiously, however, as if you use a large decay time on a motif the subsequent reflections from previous notes may still be decaying when the next note starts. If the motif is continually repeated, it will be subjected to more and more reflections until it eventually turns into an incoherent mush of frequencies.

Note: The amount of time it takes for a reverb to fade away (after the original sound has stopped) is measured by how long it takes the sound pressure level to decay to one-millionth of its original value. Since one-millionth equates to a 60 dB reduction, reverb decay time is often referred to as RT60 time.

HF and LF damping

The further reflections have to travel, the less high-frequency content they will have, since the surrounding air will absorb them. Additionally, soft furnishings will also absorb higher frequencies, so by reducing the high-frequency content (and reducing the decay time) you can give the impression that the sound is in a small enclosed area or an area with soft furnishings. Alternatively, by increasing the decay time and removing smaller amounts of the high-frequency content you can make the sound source appear further away. Alternatively, by boosting the lower frequency damping it can be used to emulate a large open space. For instance, singing in a large cavernous area there will be a low-end rumble with the reflections but not as much high-frequency energy.

Despite the amount of controls a reverb unit may offer, it is also important to note that units from different manufacturers will sound very different to one another, as each manufacturer will use different algorithms to simulate the effect. Although it is quintessential to use a good reverb unit (such as a Lexicon hardware unit or the TC Native Plug-ins included on the CD), it's not uncommon to use two or three different models of reverb in one mix.

Note: On the audio CD you can hear the effect of reverb.

Chorus

Chorus effects attempt to emulate the sound of two or more of the same instruments playing the same parts simultaneously. Since no two instrumentalists could play exactly in time with one another, the result is a series of phase cancellations. This is analogous to two synthesizer waveforms slightly detuned and playing simultaneously together; there will be a series of phase cancellations as the two frequencies move in and out of phase with one another. A chorus unit achieves this same effect by continually changing the amount of delay and amplitude applied to the incoming signal.

To provide control over this modulation, a typical chorus effect will offer three parameters, all of which can be directly related to the LFO parameters on a typical synthesizer. The first allows you to select a modulation waveform that will be used to modulate the delayed signal, while the second and third parameters allow you to set the modulation rate (referred to as the frequency) and the depth of the chorus effect (often referred to as delay). However, it should be noted that, because the modulation rate stays at a constant depth, rate and waveform, it doesn't produce the 'authentic' results you would experience with real instrumentalists. Nonetheless, it has become a useful effect in its own right and can often be employed to make oscillators and timbre appear thicker, wider and much more substantial.

Note: On the audio CD you can hear the effect of chorus.

Phasers and flangers

Phasers and flangers are very similar effects with subtle differences in how they are created, but work on a comparable principle to the chorus effect. Originally, phasing was produced by using two tape machines that played slightly out of sync with one another. As you can probably imagine, this created an irregularity between the two machines which resulted in the phase relationship of the audio being slightly different, in effect producing a hollow, phase-shifted sound.

This idea was developed further in the 1950s by Les Paul, as he experimented by applying pressure onto the 'flange' (i.e. the metal circle that the tape is wound upon) of the second tape machine. This effectively slowed down the speed of the second machine and produced a more delayed swirling effect due not only to the phase differences, but also the speed. With digital effects units, both these work by mixing the original incoming signal with a delayed version, but also by feeding some of the output back into the input. The only difference between the two is that flangers use a time delay circuit to produce the effect while a phaser uses a phase shift circuit.

Nevertheless, both use an LFO to either modulate the phase shifting of the phaser or the time delay of the flanger. This creates a series of phase cancellations, since the original and delayed signals are out of phase with one another. The resulting effect is that phasers produce a series of notches in the audio file that are harmonically related (since they are related to the phase of the original audio signal) while flangers have a constantly different frequency because they use a time delay circuit. Consequently, both flangers and phasers share the same parameters. They both feature a rate parameter to control the speed of the LFO effect, along with a feedback control to set how deeply the LFO affects the audio. Notably, some phasers will only use a sine wave as a modulation source, but most flangers will allow you to not only change the shape, but also control the number of delays used to process the original signal. Today, both these effects have become a staple in the production of dance, especially house, with the likes of Daft Punk using them on just about every record they've ever produced.

Note: On the audio CD you can hear the effect of flangers and phasers.

Digital delay

To the dance music producer, digital delay (often referred to as DDL – Digital Delay Line) is one of the most important effects to own, as if used creatively it can be one of the most versatile. The simplest units will allow you to delay the incoming audio signal by a predetermined time, which is commonly referred to in milliseconds or sometimes in note values. The numbers of delays produced by the unit are often referred to as the feedback, so by increasing the feedback setting you can produce more than one repeat from a single sound. This works by sending some of the delayed output back into the effects input, so that it's delayed again, and again, and again and so forth. Obviously, this means that if the feedback is set to a very high value, the levels of the repeats end up collecting together rather than gradually dying away, until eventually you'll end up with a horrible howling sound.

While all delay units will work on this basic premise, the more advanced units may permit you to delay the left and right channels individually and pan them to the left and right of the stereo image. They may also allow you to pitch-shift the subsequent delays, employ filters to adjust the harmonic content of the delays, distort or add reverb to the results and apply LFO modulation to the filter. Of all these additional controls (most of which should require no explanation), the modulation is perhaps the most creative application to have on a delay unit, as it allows you to modulate the filter's cut-off or pitch, or both, with an LFO. The number and type of waveforms on offer vary from unit to unit, but fundamentally most will feature at least a sine, square and triangle wave. Similar to a synthesizer's LFO parameters, they will feature rate and depth controls, allowing you to adjust how fast and by how much it should modulate the filter cut-off or pitch parameters.

One of the most common uses for a delay in dance music is not necessarily to add a series of delays to an audio signal, but to create an effect known as granular delay. As we touched upon in Chapter 3, we cannot perceive individual sounds if they are less than 30 milliseconds apart – which is the principle behind granular synthesis. However, if a sound is sent to a delay unit and the delay time is set to less than 30 milliseconds and combined with a low feedback setting, the subsequent delays collect together in a short period of time, which we cannot perceive as a delay. The resulting effect is that the delayed timbre appears much bigger, wider and upfront. This technique is often employed on leads or in some cases basses if the track is based around a powerful driving bass line.

Note: On the audio CD you can hear the effect of DDL.

EQ

At its most basic, EQ is a frequency-specific volume tone control that allows you to intensify or attenuate specific frequencies. For this, three controls are required:

- A frequency control allowing you to home in on the frequency you want to adjust.
- A 'Q' control allowing you to determine how many frequencies either side of the centre frequency you want to adjust.
- A gain control to allow you to attenuate or intensify the selected frequencies.

Notably, not all EQ units will offer this amount of control and some units will have a fixed frequency or a fixed Q, meaning that you can only adjust the volume of the frequencies that are preset by the manufacturer. EQ plays a role in both mixing and sound design, but this has only been a quick introduction for the purpose of the technology chapters and we'll look much more deeply into its effects when we cover mixing and mastering in later chapters.

Distortion

The final effect for this chapter, distortion, is pretty much self-explanatory; it introduces an overdrive effect to any sounds that are fed through it. However, while the basic premise is quite simple, it has many more uses than to simply grunge up a clean audio signal. As touched upon in Chapter 3, a sine wave does not contain any harmonics except for the fundamental frequency, and therefore applying effects such as flangers, phasers or filters will have very little effect. However, if distortion were applied to the sine wave it would introduce a series of harmonics into the signal, giving the aforementioned effects something more substantial to work with.

Note: On the audio CD you can hear the effect of distortion.

6 Cables, mixing desks and effects busses

I don't just use a (*mixing*) desk to mix sounds together I use it as a creative tool.

Juan Atkins

With a fundamental understanding of the different processors and effects available, it's also important to grasp how they are accessed through a typical mixing desk, how the desk treats these signals and the effects it can impart on the audio signals. Before we jump into the theory behind using a mixing desk, however, there's one vital element that's often overlooked and that's the cables used to connect the various samplers, effects, processors and synthesizers to the mixing desk.

We've already touched upon the importance of capturing a signal with a low noise, yet it's surprising how many users purchase expensive equipment with a low signal-to-noise ratio (SNR) and connect them together with poor quality cables. Any studio is only as capable as its weakest link, so if low-quality cables are used to connect devices together, the SNR of the equipment is meaningless since the cables will be susceptible to introducing interference which results in noise. This problem arises because any cables that carry a current, no matter how small, produce their own voltage as the current travels along them. The level of voltage that is produced by the cable will depend on its resistance and the current passing through it, but this nevertheless results in a difference in voltage from one end of the cable to the other.

Because all studio equipment (unless it's all contained inside a Digital Audio Workstation) requires cables to carry the audio signal to and from the mixing desk, the additional voltage introduced by the cables is then transmitted around the instruments from the mixing desk and through to earth. This produces a continual loop, resulting in an electromagnetic field that surrounds all the cables. This field introduces an electrical hum into the signal, an effect known as 'ground hum'. The best way to reduce this is by using professional quality 'balanced' cables, although not all equipment, particularly equipment intended for the home studio, supports this form of cable. Home studio equipment tends to use 'unbalanced' cables and connectors.

The distinction between balanced and unbalanced cable is determined by the termination connectors at each end. Cables terminated with mono jack or phono connectors are unbalanced, while stereo TRS (Tip–Ring–Sleeve) jack connections or XLR connections will be found on balanced cables. Examples of these are shown in Figure 6.1.

All unbalanced cables are made up of two internal wires contained within the outer plastic or rubber core of the wire (known as the earth screen). One of these internal wires carries the

Mono jack

Audio

Ground/Earth

TRS Jack

Positive audio (+ve)

Negative audio (−ve)

Ground/Earth

Male XLR

Positive audio (+ve)

Negative audio (−ve)

Ground/Earth

Head on XLR view

Figure 6.1 Mono, stereo jack and XLR connectors.

audio signal and is connected to the ring of the connector, while the other carries the ground signal and is connected directly to the connector sleeve. The signal is therefore grounded at each end of the cable, helping to prevent any interference from the device itself, but is still susceptible to electromagnetic interference as it is transmitted from one device to another. Because of this, most professional studios use balanced cables with XLR or TRS terminating connections, if the equipment they connect to supports it.

Note: TRS connections do not necessarily mean that the cable is balanced as they can be used to carry a stereo signal (left channel, right channel and earth), but in a studio environment they are more commonly used to transfer a mono signal.

Balanced cables contain three wires within the outer screen. In this configuration a single wire is still used as a ground but the other two wires carry the audio signal, one of which is a phase-inverted version of the original. When this is received by a device, the phase-inverted signal is put back into phase with the original and the two are added together. As a result, any interference introduced is cancelled out when the two signals are summed together. This is similar to the way two oscillators that are in phase cancel each other out, as described in Chapter 3. That's the theory. In practice, although this reduces the problem, phase cancellation rarely removes all of the interference.

Another favourable advantage of using balanced equipment is that they also utilize a more powerful signal level. Commonly referred to as a professional standard, a balanced signal uses a signal level of +4 dBu rather than the semi-professional signal level of −10 dBV. The reasons for this are more the subject matter of electrical engineering than music and although it's not necessarily important to understand this, a short explanation is given in the note below. If you're not interested you can skip this box, because all you really need to know is that +4 dBu signals have a hotter, louder signal than −10 dBV signals and are generally preferred as the signal is over 11 dB hotter, so the chance of capturing a poor signal is reduced.

Note: Before LED and LCD displays appeared on musical gear, audio engineers used volume unit (VU) meters to measure audio signals. These were featured on any hardware that could receive an input. This meant that every VU meter had to give the same reading for the same signal level no matter who manufactured the device. If this were not the case, different equipment within the same studio would have different signal levels. Consequently, engineers decided that if 1 milliwatt (mW) was travelling through the circuitry then the VU meter should read 0 dB. Hence, 0 dB VU was referred to as 0 dBm (with 'm' standing for milliwatt).

Today's audio engineering societies are no longer concerned with using a reference level of milliwatt because the power levels today are much higher, so the level of 0 dBm is now obsolete and we use voltage levels instead. To convert this into an equivalent voltage level, the impedance has to be specified, which in this case is 600 ohms. For those with some prior electrical knowledge it can be calculated with the equation ($P = V_2/R(0.001\ W) = V_0/600\ W(V_2 = 0.001\ W) \times 600\ W(V = sq(0.001\ W \times 600\ W))$). For the layman, the sum of this equation equals 0.775 volts and that's all you need to know.

The value of 0.775 is now used as the reference voltage and is referred to in dBu rather than dBm. Although it was originally referred to as dBv, it was often confused with the reference level of dBV (notice the upper case V), so the suffix u is used in its place. This is only the reference level, though, and all professional equipment will output a level that is +4 dB, which is where we derive the +4 dBu standard. Consequently, on professional equipment, the zero level on the meters actually signifies that it is receiving a +4 dBu signal.

However, some hardware engineers agreed that it would be simpler to use 1 volt as the reference instead, which is where the dBV standard originates. Unlike professional equipment, which uses the +4 dBu output level, unbalanced equipment outputs at 0.316 V, equivalent to −10 dBV. Therefore, on semi-professional equipment, the zero level on the meters signifies that they are receiving a −10 dBV signal. If the professional and semi-professional signals are compared, the professional voltage of 0.775 V is considerably higher than the 0.316 V generated by consumer equipment. When converted to decibels, this results in an 11.8 dB difference between the two.

Despite the difference in the levels of the two signals, in many cases it is still possible to connect a balanced signal into an unbalanced sampler/soundcard/mixer, with the immediate benefit that the signal that is captured is 11.8 dB hotter. Although this usually results in the unbalanced recorder's levels jumping off the scale, whether the signal is actually distorting should be based on whether this distortion is audible. Most recorders employ a safety buffer by setting the clipping meters below the maximum signal level. This is, of course, a one-way connection from a balanced signal to an unbalanced piece of hardware and it isn't a good idea to work with an unbalanced signal connecting to a balanced recorder, because you'll end up with a poor input signal level and any attempts to boost the signal further could introduce noise.

Within the typical home studio environment, it is most likely that the equipment will be of the unbalanced type; therefore, the use of unbalanced connections is unavoidable. If this is the case, it's prudent to take some precautions that will prevent the introduction of unwanted electromagnetic interference, but this can be incredibly difficult.

While the simplest solution would be to disconnect the earth from the power supply, in effect breaking the ground loop, this should be avoided at all costs. Keep in mind that the whole point of an earth system is to pass the current directly to ground rather than you if there's a fault. Human beings make remarkably good earth terminals and as electricity will always take the quickest route it can find to reach earth, it can be an incredibly painful (and sometimes deadly) experience to be present yourself as a short cut.

A less suicidal technique is to remove the earth connection from one end of the audio cable. This breaks the loop, but it has the disadvantage that it can make the cable more susceptible to radio frequency (RF) interference. In other words, the cable would be capable of receiving signals from passing police cars, taxis, mobile phones and any nearby citizen's band radios. While this could be useful if you want to base your music around *Scanners'* previous work, it's something you'll want to avoid.

Although in a majority of cases hum will be caused by the electromagnetic field, it can also be the result of a number of other factors combined. To begin with, it's worthwhile ensuring that the mains and audio cables are wrapped separately from one another and kept as far away from each other as possible. Mains cables create a higher electromagnetic field due to their large current and if they are bound together with cables carrying an audio signal serious hum can be introduced.

Transformers also generate high electromagnetic fields that cause interference and, although you may not think that you have any in the studio, both amplifiers and mixing desks use them. Consequently, amplifiers should be kept a good distance from other equipment, especially sensitive equipment such as microphone pre-amps. If the amplifiers are rack-mounted, a minimum space of four rack units should be between the amplifier and any other devices. This same principle also applies to rack-mounted mixing desks, which should ideally be placed in a rack of their own or kept on a desk. If the rack that is used is constructed from metal and metal screws hold the mixing desk in place, the result is the same as if the instruments were grounding from another source and yet more hum is introduced. Preferably, plastic washers and screw housings should be used, as these isolate the unit from the rack.

If, after all these possible sources have been eliminated, hum is still present, the only viable way of removing or further reducing it is to connect devices together digitally, or invest in a professional mains suppressor. This should be sourced from a professional studio supplier rather than from the local electrical hardware or car superstore, as suppressors sold for use within a studio are specifically manufactured for this purpose, whereas a typical mains suppressor is designed to suppress only the small amounts of hum that are typically associated with normal household equipment.

Note: From a purely personal opinion, the best place to source professional suppressors, suppressed PDUs, quality cables and most other studio-related gear is Studio Spares (www.studiospares.com).

Digital connections can be used as an alternative to analogue cables if the sampler/sound-card/mixer allows it. This has the immediate benefit that no noise will be introduced into the signal, with the additional benefit that this connection can also be used by the beat slicing software to transmit loops to and from an external hardware sampler. Sampler-to-software connectivity is usually accomplished via a direct SCSI interface, so there is little need to be concerned about the digital standards, but on occasion it may be preferable to transmit audio signals through true digital interfaces such as ADAT, TDIF, S/PDIF or AES/EBU.

The problem is that digital interfacing is more complex than analogue interfacing because the transmitted audio data must be decoded properly. This means that the bit rate, sample rate, sample start and end points, and the left and right channels must be coded in such a way that the receiving device can make sense of it all. The problem with this is that digital interfaces appear in various forms, including (amongst many others) Toshiba's TDIF formats, Sony's SDIF, Sony and Philips' S/PDIF, and the AES/EBU standard, none of which are cross-compatible.

In an effort to avoid these interconnection problems, the American Audio Engineering Society (AES) combined with the European Broadcasting Union (EBU) devised a standard connection format imaginatively labelled the AES/EBU standard. This requires a three-pin XLR connection, similar to the balanced analogue equivalent, although the connection is specific to the digital realm.

Like 'balanced' analogue connections AES/EBU is expensive to implement, so Sony and Philips developed a less expensive 'unbalanced' standard known as S/PDIF (Sony/Philips Digital InterFace). This uses either a pair of phono connectors or an optical TOSLINK interface (Toshiba Optical Source). Most recent samplers and soundcards use a TOSLINK or phono connection to transmit digital information to and from other devices.

With compatible interfaces between two devices, both the receiving and transmitting device must be clocked together so that they are fully synchronized. This ensures that they can communicate with one another. If the devices are not synchronized, the receiving device will not know when to expect an incoming signal, producing 'jitter'. The resulting effect this has on audio is difficult to describe, but rather than fill up the book's accompanying CD with the ensuing ear-piercing noises, it's probably best explained as an annoying high-frequency racket mixed with an unstable stereo image. To avoid this, most professional studios use an external clock generator to synchronize multiple digital units together correctly. This is similar in most respects to a typical multi-MIDI interface, bar the fact that it generates and sends out word clock (often abbreviated to WCLK but still pronounced word clock) messages simultaneously to all devices to keep them clocked together.

The WCLK message works by sending a 1-bit signal down the digital cable, resulting in a square wave that is received in all the attached devices. When the signal is decompiled by the receiving device, the peaks and troughs of the square wave denote the left and right channels, while the width between each pulse of the wave determines the clock rate.

These standalone WCLK generators can be expensive, so within a home studio set-up the digital mixing desk or soundcard usually generates the WCLK message, with the devices daisy-chained together to receive the signal. For instance, if the soundcard is used to generate the clock rate, the WCLK message could be sent to an effects device, through this into a microphone pre-amplifier, and so forth before the signal returns back into the soundcard, creating a loop. The principle is similar to the way that numerous MIDI devices are daisy-chained together, as discussed in Chapter 1. As with daisy-chained MIDI devices, though, the signal weakens as it passes through each device; therefore, if the signal is to pass through more than four devices, it's prudent to use a WCLK amplifier to keep the signal powerful enough to prevent any jitter.

Providing that the clock rate is correctly transmitted and received, another important factor to consider is the sample rate. When recording, both the transmitting and receiving devices must be locked to the same sample rate, otherwise the recorder may refuse to go into record mode. Also, unless you are recording the final master, any Serial Copyright Management System (SCMS) features should be disabled.

SCMS was implemented onto all consumer digital connections to reduce the possibility of music piracy. This allows only one digital copy to be made from an original and prevents further copies being made from this one copy. It does this by inserting a 'copyright flag' into the first couple of bytes that are transmitted to the recording device. If the recorder recognizes this flag it will disable the device's record functions. This is obviously going to cause serious problems if you need to edit a second-generation copy because unless the recorder allows you to disable SCMS, you will not be allowed to record the results digitally. Thus, if you plan to transfer and edit data digitally, it is vital to ensure that you can disable the SCMS system.

> Note: If you own a DAT machine that does not allow you to disable the SCMS protection system, then it's possible to get hold of SCMS strippers which remove the first few flags of the signal, in effect disabling the SCMS system.

Mixing desk structure

While some mixing desks appear relatively straightforward, some professional desks – such as the Neve Capricorn – look more ominous. Whatever the level of complexity, all types of hardware- and software-based mixing desks operate according to the same principles. With a basic understanding of the various features and how the channels, busses, subgroups, EQ, and aux send and returns are configured, you can get the best out of your equipment, no matter how large or small the desk is.

The fundamental application of any mixing desk is to take a series of inputs from external instruments and provide an interface that allows you to adjust the tonal characteristics of each perspective instrument. These signals are then culminated together in the desk and fed into the loudspeaker monitors and/or recorder to produce the finished mix. As simple as this premise may be, though, simply looking at a desk reveals that there's actually a lot more going on and to better understand this we need to break the desk down to each individual input channel.

Typically, a mixing desk can offer anywhere from two input channels to over a hundred – depending on its price. Each of these channels (referred to by engineers as a 'strip') is designed to accept either a mono or stereo signal from one musical source and provide some tonal editing features for that one strip. Clearly, this means that if you have five external hardware instruments each with a mono output you would need a desk with an absolute minimum total of five 'strips' so you could input each into a separate channel of the desk. Or at least it would in theory, since for reasons we'll touch upon, you should always aim to have as many channel strips in a mixer as you can afford, no matter how few instruments you may own.

Generally, the physical inputs for each channel are located at the rear of the desk and will consist of a ¼-inch jack or XLR connection, or both. This latter configuration doesn't mean that two inputs can be directed into the same channel strip simultaneously; rather it allows you to choose whether that particular channel accepts a signal from a jack *or* an XLR connector. Most desks will have the capacity to accept a signal of any level, whether its mic level (−60 dBu) or line level (−10 dBV or +4 dBu) and whether the cables are balanced or unbalanced.

> Note: Some of the older mixing desks may describe these inputs as being Low-Z or Hi-Z, but all this is describing is the input impedance of that particular channel. If Hi-Z is used then the impedance is higher to accept unbalanced line-level signals, while if Low-Z is used the impedance is lower to accept balanced.

As all mixing desks operate at line level to keep the signal-to-noise ratio at a minimum, once a signal enters the desk it is first directed to the pre-amplifier stage to bring the incoming signal up to the operating level of the desk. Although this pre-amp stage isn't particularly necessary with line-level signals, as most desks use this as their nominal operating level, it is required to bring the relatively low levels generated by most microphones to a more respectable level for mixing. This pre-amp will have an associated rotary gain control on the fascia of the desk (often called pots – an acronym for potentiometer), labelled 'trim' or 'gain'. As its name would suggest, this allows you to adjust the volume of the signal entering the channel's input by increasing the amount of amplification at the pre-amp.

> Note: To reduce the possibility of the mixer introducing noise into the channel, you should ensure that the signal entering the mixer is as high as possible rather than input a low-level signal and boost it at pre-amp stage of the mixer. Keep in mind that a good signal entering the mixer before the pre-amp is more likely to remain noise free as it passes through the rest of the mixer.

Some mixers may also offer a 'pad' switch at the input stage and this is used to attenuate the incoming signal before it reaches the pre-amplifier. The amount of attenuation applied is commonly fixed at 12 or 24 dB and is used to prevent any hot signals entering the inputs from overdriving the desk. A typical example of this may be where the mixer operates at the 'semi-professional' −10 dBV and one of the instruments connected to it outputs at the 'professional' +4 dBu. As we've previously touched upon, this would mean that the input level at the desk would be 11.8 dB higher than the desk's intended operational level. Obviously this will result in distortion, but by activating the pad switch, the incoming signal could be attenuated by 12 dB and the volume could then be made back up to unity gain with the trim pot on that particular channel.

> Note: Unity gain is a term used to describe where the audio signal's output level is equal to its input level. In other words there is no amplification or attenuation applied to the signal (i.e. faders at 0 dB).

On top of this, some high-end mixers may also offer a phantom power switch, a ground lift switch and/or a phase reverse switch, the functions of which are described below:

- Phantom power is used to supply capacitor microphones with the voltage they require to operate. This does also mean that the mixer has to use its own amplifier to increase its signal level and, generally, the amp circuits in mixers will not be as good as those found in standalone microphone pre-amplifiers.
- A ground lift switch will only be featured on mixers that accept balanced XLR connections, and when this is activated it disables pin number one on the connector from the mixer's ground circuitry. Again, as touched upon, this will eliminate any ground hum or extraneous noises.
- Phase reverse – sometimes marked by a small circle with a diagonal line through it – switches the polarity of the incoming signal and this can have a multitude of uses. Typically, they're included on mixing desks to prevent any recordings taken simultaneously with a number of microphones from interfering with one another, since the sound of one microphone can weaken the sound from a second microphone that's positioned further away. Subsequently, phase reverse switches are only usually found next to XLR microphone inputs, since they're easy to implement on these – the switch simply swaps over the two signal pins.

Following this input section, the signal is commonly routed through a mute switch (allowing you to instantly mute the channel) and into the insert buss.[1] These allow you to insert processors such as compressors or noise gates to clean up or bring a signal under control before it's routed into the EQ. On a mixing desk EQ is the most important circuitry, as it allows you to modify the tonal content of a sound for creative applications or, more commonly, so that the timbre fits into a mix better. As a result, when looking for a mixing desk, it's worthwhile seeing how well equipped the EQ section is, as this will give a good impression of how useful the desk will be to you. Most cheap consumer mixers will only offer a high and low EQ boost, which can be used to increase or reduce the gain at a predetermined frequency. More expensive desks will offer parametric EQs that allow you to select and boost any frequency you choose and offer filters so that you can remove all the frequencies above or below a certain frequency. This section may also offer a pre- or post-EQ button that allows you to re-route the signal path to *bypass* the EQ section (pre-EQ) and move directly to the pre-fader buss, or move *through* the EQ section (post-EQ) and then into the pre-fader buss. Obviously, this allows you to bypass the EQ section once in a while to determine the effect that any EQ adjustments have had on a sound.

Similar to the EQ buss, the pre-fader buss allows you to bypass the channel's volume fader and route the signal at it's nominal volume to the auxiliary (aux) buss or go through the faders and then to the auxiliary buss. Fundamentally, an aux buss is a way of routing the channel's signal to physical outputs usually labelled as 'aux outs' located at the rear of the desk. The purpose behind this is to send the channel's signal out to an external effect and then return the results back into the desk (we'll look more closely at aux and insert effects in a moment). The number

---

[1] Busses refer to the various signal paths within a mixer that the inputted audio can travel through.

of aux busses featured on a desk varies wildly from one to over 20, and not all desks will give the option of pre- or post-fader aux sends and may be hard-wired to the pre-fader. This means that you would have no control over the level that's sent to the aux buss.

After this aux buss section, the signal is passed through onto the volume control for the channel. These can sometimes appear in the form of rotary controllers on cheaper desks, but generally they use faders with a 60 or 100 mm throw.[2] Although these shouldn't particularly require any real explanation, it's astounding how many users believe that they are used to increase the amplification of a particular channel. This isn't the case at all, since if you look at any desk the faders are marked 0 dB close to the top of their throw instead of at the bottom. This means that you're not *amplifying* the incoming signal by increasing the fader's position, rather you're allowing more of the original signal to travel through the fader's circuitry, which increases the gain on that channel.

After the faders, the resulting signal then travels through the panning buss (allowing you to pan the signal left and right) and into a subgroup buss. The number of subgroup busses depends entirely on the price and model of mixer, but essentially these allow you to group a number of fader positions together and control them all with just one fader control. A typical application of this is if the kick, snare, claps, hi-hats and cymbals each have a channel of their own in the mixer. By then setting each individual element to its respective volume and required EQ (in effect mixing down the drum loop so it sounds right), all of these levels can be routed to a single group track, whereby moving just one subgroup fader, the volume of the entire drum submix can be adjusted to suit the rest of the mix.

Finally, these subgroup channels, if used, along with the rest of the 'free' faders, are combined together into the main stereo mix buss, which passes to the main mix fader, allowing you to adjust the overall level of the mix. It's at this stage that things can become more complicated, since the direction and options available to this stereo buss are entirely dependent on the mixer in question.

In the majority of smaller, less expensive mixers the buss will simply pass through a pan control and then out to the mixer's physical outputs. Semi-professional desks may pass through the pan control *and* an EQ before sending the mix out to the main physical outputs. And professional desks may go through panning, EQ and then split the buss into any number of other stereo busses, allowing you to send the mix not only to the speakers, but also headphones, another pair of monitors (situated in the recording room) and a recording device.

Ultimately, the more EQ and routing options a desk has to offer, the more creative you can become. But, at the same time, the more features on offer, the more expensive it will be. Naturally, most of these concerns are circumvented with audio sequencers, as they generally offer everything a professional desk does; the only limitation is the number of physical connections, dictated by the soundcard you have fitted. Nevertheless, this doesn't deter many from relying entirely on sequencers, as any external instruments can always be recorded as audio, placed on their own track and have software effects applied.

Routing effects and processors

Understanding the internal buss structure of a typical mixing desk (hardware or software) is only part of the puzzle, because when it comes to actually processing signals, the insert/aux

---

[2] 'Throw' refers to the two extremes of a fader's movement from minimum to maximum.

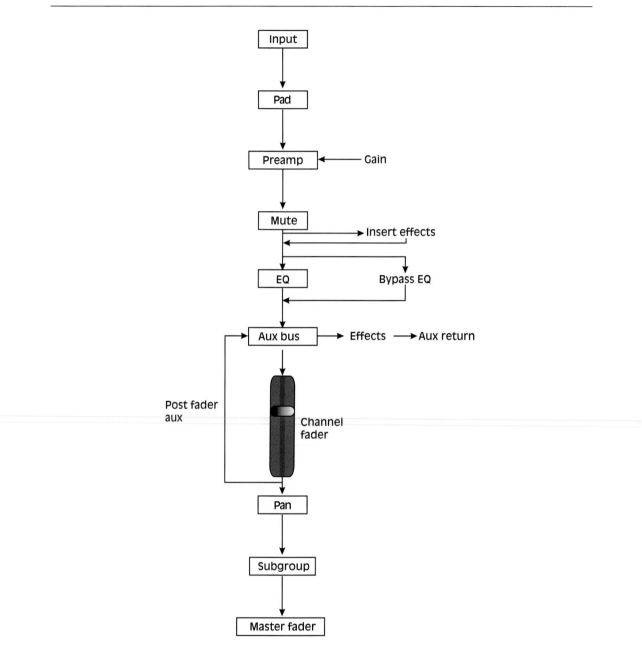

Figure 6.2 The typical structure of a mixing desk.

buss system they are transferred through will often dictate the overall results. Any 'external' signal processing can be divided into two groups: processors and effects. The difference between these two is relatively simple but important to recognize.

All effects will utilize a wet/dry control (wet is the affected signal and dry is the uneffected signal) that allows you to configure how much of the original signal remains uneffected and how much is effected. A typical example of this would be for reverb, whereby you don't necessarily want to run the entire audio through the effect, otherwise it could appear swamped in decays; instead, you would want to effect the signal by a only small amount. For instance, you may keep 75% of the original signal and apply just 25% of the reverb effect. Conversely, all processors, such as compressors, noise gates and limiters, are designed to work with 100% of the signal and thus have no wet/dry parameter. This is simply because in many instances there would be little point trying to control the dynamics of just some of the signal because the rest of it would still retain its dynamic range. Nonetheless, due to the different nature of these two processes, a mixing desk uses two different buss systems to access them: an insert buss for processors and an auxiliary buss for effects. The difference and reasons behind using these two busses will become clear as we look at both.

Auxiliary buss

Nearly all mixers will feature the aux buss after the EQ rather than before, since it's considered that effects are applied at the final stages of a mix to add the final 'polish'. At this point, a percentage of the signal can be routed through the aux buss and to a series of aux outs located at the back of the mixer. Each aux out is connected to an effects unit and the signal is then returned into the desk. With the aux potentiometer on the channel strip set at zero, the audio signal ignores the buss and moves directly onto the fader. However, by gradually increasing the aux pot you can control how much of the signal is split and routed to the aux buss and onto the effect. The more that this is increased the more audio signal will be directed to the aux buss. The aux return (the effected signal) is then returned to a separate channel on the mixing desk, which can then be mixed with the original channel.

Figure 6.3 An aux send configuration.

As you can see from Figure 6.3, two separate audio leads are required. One has to carry the signal from the mixer and into the effects unit while the other has to return the effect back into the desk. Additionally, the effected signal cannot be returned into the original channel, since there will still be some dry audio at the fader; instead, they are usually returned into specific aux returns. These signals are then bussed through the mixer into additional mixer 'subgroup' channels. These are most commonly a group of volume faders or pots (one for each return) that permit you to balance the volume of the effected signal with the original dry channel.

Note: When accessing effects from a mixing desk, the mix control on the effects unit should generally be set to 100% wet, since you can control the amount of wet/dry signal with the mixing desk itself.

While retuning the effected signal to its predefined aux channel return may seem sensible, very few artists will actually bother using the aux returns at all, instead preferring to return the signal to a normal free mixing channel. This opens up a whole new realm of possibilities, since the returning effect has access to the channel's pre-amp, an insert, EQ, pan and volume. For example, a returned reverb effect could be pushed into distortion through increasing the mixer's pre-amp, or the EQ could be used as a low- and high-frequency filter if the effects unit doesn't feature them.

Another benefit of using aux sends is that each channel on the mixer will have access to the same auxiliary busses, meaning that signals from other channels could also be sent down the same auxiliary buss to the effects unit. This approach can be especially useful when using computer-based audio sequencers, since opening multiple instances of the same effect can use up a proportionate amount of the CPU. Instead, you can simply open up one effect and send each channel to the same effect; this also saves time from having to set a number of effects units up, all of which share the same parameters. What's more, in a hardware situation, as only a portion of the signal is being sent to the effect, there is less noise and signal degradation, and as the effect is returned to a separate channel, you have total control over the wet and dry levels through the mixer.

One final aspect of aux busses is that they can sometimes be configured to operate the aux buss either pre- or post-fader. Of the two, pre-fader is possibly the least used, since the signal is split into two before it reaches the fader. This means that if you were to reduce the fader you would not reduce the gain of the signal being bussed to the effect, so with every fader adjustment you would also need to readjust the aux pot to suit. This can have its uses, as it allows you to reduce the volume of the dry signal while leaving the wet at a constant volume, but generally post-fader is much more useful while mixing. Using post-fader, reducing the channel's fader will also systematically reduce the auxiliary send, saving the need to continually adjust the aux buss send level whenever you change the volume.

Insert buss

Unlike aux busses, insert busses are positioned just before the EQ and are designed for use with processors rather than effects. This is because processors require the entire signal and there is no point in applying compression or gating to only part of a signal! The purpose of insert points can be described by examining a typical recording situation where the output of a microphone's pre-amp is connected into a compressor. This signal would then be connected into the mixing desk's channel so that the signal flows out of the mic pre-amp into the compressor and then finally into the desk for mixing. This, however, is a rather convoluted way of working because if you wanted to compress the output of another source with the same compressor you would have to scrabble around at the back of the rack and rewire all the necessary connections. In addition, if the output from the synthesizer was particularly low, there is no way of increasing its gain into the compressor.

This can be avoided if the mixing desk features insert points, which commonly appear directly after the pre-amp stage. Consequently, the mixer's pre-amp stage can be used to increase the signal

Note: On occasion, compressors may be accessed through the aux send buss rather than the insert buss. With this configuration, you can increase the overall level of the signal by mixing the compressed with the uncompressed signal. This can help to preserve the transients of the signal.

Figure 6.4 An insert configuration.

level from the synthesizer before it's passed on to the insert buss. This is then passed into the external compressor, where the signal is squashed before being returned to the same mixing channel.

This is accomplished using only one physical connection – an insert cable – on the mixing desk. Fundamentally, this is an ordinary audio cable with a TRS (Tip–Ring–Sleeve) jack at one end and a TS (Tip–Sleeve) jack at the other. The ring is used to transmit the signal from the desk's channel to the compressor and is returned back into the same mixing channel through the tip connection of the same cable. Most mixers return this processed signal before the EQ and volume buss, allowing shaping of the overall signal after processing.

Tonal shaping after processing won't necessarily create any problems if the processor happens to be a noise gate, but if it's a compressor, it raises the issue that if the EQ were raised *after* compression, the signal would be increased past the level of the previously controlled dynamics. To understand the issue, we first need to consider what would happen if the compressor was returned *after* the EQ section.

To begin with, the EQ section would be used to correct the tone of the incoming signal on that particular channel before it was sent to the compressor. This compressor would then be adjusted so that it controls the dynamics of the timbre *and* the subsequent EQ before the signal is returned to the desk. Due to the nature of compression, however, the tonal content of the sound will be modified, so it would need to be equalized again. This subsequent EQ will re-introduce the peaks that were previously controlled by the compressor, so the compressor must be adjusted again. This would alter the tonal content, so it would need equalizing yet again, and so on. Thus, an ever-increasing circle of EQ–compression–EQ–compression must continue until the signal is over-compressed or pushing beyond unity gain, distorting the desk. By inserting the compressor *before* the EQ section, this continual circle can be avoided.

With the compressor positioned before the EQ, the issue of destroying the previously controlled dynamics when boosting the EQ is still raised, but this shouldn't be the case providing the EQ is correctly used. As touched upon in earlier chapters, it's in our nature to perceive louder sounds to be infinitely better than quieter, even if the louder sound is tonally worse than the quieter one. Thus, when working with EQ it's necessary to reduce the channel's fader while boosting any frequencies, so that they remain at the same volume as the unequalized version. Used in this way, when bypassing the modified EQ to compare it with the unmodified version, a difference in volume cannot cloud your judgement. As a result, the signal level of the modified EQ is the same as the unmodified version, so the dynamics from the compressor remain the same.

This approach maintains a more natural sound, but it is worth experimenting by placing the EQ before the compressor. For example, using this configuration the EQ could be used to create a frequency-selective compressor. In this set-up, the loudest frequencies control the compressor's action. By boosting quieter frequencies so that they instead breach the compressor's threshold, it's possible to change the dynamic action of a drum loop or motif. In fact, it's important to note that the order of any processing or effects, if they're inserted, can have a dramatic influence on the overall sound.

From a theoretical mixing point of view, effects should not be accessed as inserts. This is because, although it is perfectly feasible to route reverb, delay, chorus, flangers and phasers into a mix as inserts, working in this way can introduce a number of problems. If an effect is used as an insert, 100% of the signal will be directed into the external effects unit and because many units introduce low levels of noise while also degrading the signal's overall quality, both the affected signal and the noise will be returned into the desk. What's more, the control over the relationship between dry audio and wet effects would be from the effects interface and in many instances greater control will be required.

Many effects use a single control to adjust the balance between dry and wet, so as the wet level is increased, the dry level decreases proportionally. Equally, increasing the dry level proportionally decreases the wet level. While this may not initially appear problematic, if you decided to thicken out a trance lead with delay or reverb but wanted to keep the same amount of dry level in the mix that you already have, it isn't going to be possible. As soon as the wetness factor is increased the dry will decrease proportionally, which may result in a more wet than dry sound. This can cause the transients of each hit to be washed over by the subsequent delays or reverb tail from the preceding notes, reducing the impact of the sound.

Nevertheless, by using effects as inserts and chaining them together in series, it's possible to create new effects because both dry and wet results from each preceding effect would be transformed by the ones that follow them. This opens up a whole world of experimental and creative opportunity that could easily digest the rest of this book, so rather than list all of the possible combinations (as if I could!), we'll examine the reasoning behind why, theoretically at least, some processors and effects should precede others.

Gate > compressor > EQ > effects

Normally, to maintain a 'natural' sound a gate should always appear first in the line of any processor or effects, since they're used to remove unwanted noise from the signal before it's compressed, equalized or effected. While it is perfectly feasible to place the compressor before the gate, as it would make little difference to the actual sound it's unwise to do so. This is because the compressor reduces the dynamic range of the signal and as a gate works by monitoring the dynamic range and removing artefacts below a certain volume, placing compression first, the gate would be more difficult to set up and may remove some of the signal you wish to keep. For reasons we've already touched upon, the EQ should then appear *after* compression and the effects should follow the EQ section, as they're usually the last aspect in the mixing chain.

Gate > compressor > effects > EQ

Again, the beginning aspects of this arrangement will keep the signal natural, but by placing the EQ *after* the effects it can be used to sculpt the tonal qualities produced by the effect. For example, if the effect following the compression is distortion, the compressor will even out the signal level, making the distortion effect more noticeable on the decays of notes. Additionally, since distortion will introduce more harmonics into the signal, some of which can be unpleasant, it can be carefully sculpted with the EQ unit to produce a more controlled, pleasing result.

Gate > compressor > EQ > effects > effects

The beginning of this signal chain will produce the most natural results, but the order of the effects afterwards will determine the outcome. For instance, if you were to place reverb before distortion, the reverb tails will be treated to distortion, but if it were placed afterwards the effect would not be as strong, since the reverb's tail would not be treated. Similarly, if delay were placed after distortion, the subsequent delays would be of the distorted signal, while if the delay came first, the distortion would be applied to the delays, producing a different sound altogether. If flanger were added to this set-up, things become even more complicated, since this effect is essentially a modulated comb filter. By placing it after distortion the flanger would comb filter the distorted signal, producing a rather spectacular phased effect, yet if it were placed before, the effect would vary the intensity of the distortion.

To go further, if the flanger were placed after distortion but before reverb, the flange effect would contain some distorted frequencies but the reverb's tail would wash over the flanger, diluting the effect but producing a reverb that modulates as if it were controlled with an LFO. The possibilities here are, as they say, endless, and it's worth experimenting by placing the effects in a different order to create new effects.

Gate > effects > EQ > effects > compressor

While the aforementioned method of placing one effect after the other can be useful, the subsequent results can be quite heavy-handed, but by placing an EQ after the first series of effects the tonal content can be modified so that there isn't an uncontrolled mess of frequencies entering the second series of effects, muddying the effect further. Additionally, by placing a compressor at the end of the arrangement, any peaking frequencies introduced by a flanger following distortion or similar arrangement can be brought back under control with the compressor sat at the end of the line.

Compressor > EQ > effects > gate

We've already discussed the effects of placing a compressor before the gate and EQ, but using this configuration the compressor could be used to control the dynamics of sounds before they were equalized and subsequently effected. However, by placing the gate *after* the effects would mean that the effected signals could be treated to a gate effect. Although there is a long list of possible uses for this if you use a little imagination, possibly the most common technique is to apply reverb to a drum kick or trance/techno/house lead and then use the following gate to remove the reverb's tail. This has the effect of thickening out the sound without turning the result into a washed over mush.

Gate > effects > compressor > EQ

Although it is generally accepted that the compressor should come before effects, placing it directly after can have its uses. For instance, if a filter effect has been used to boost some frequencies and this has been followed by chorus or flanger there may be clipping, so the compressor can be used to bring these under control before they're shaped tonally with the EQ. Notably, though, placing compression after distortion will have little effect, since distortion effects tend to reduce the dynamic range anyway.

Above all, though, keep in mind that setting effects in an order that would not theoretically produce great results most probably will in practice. Indeed, it's the artists that are willing to experiment that often produce the most memorable effects on records and, with many of today's audio sequencers offering a multitude of free effects, experimentation is cheap but the results may be priceless.

Automation

A final yet vital aspect of mixing desks appears in the form of mix automation. We've already touched upon the importance of movement within programmed sounds, but it's also important to manipulate a timbre constantly throughout the music. Techno would be nowhere if it were not possible to program a mixer or effects unit to gradually change parameters during the course of the track. With a mixer that features automation, these parameter changes can be recorded as data (usually MIDI) into a sequencer, which when played back to the mixer forces it to either jump or gradually move to the new settings.

Originally, mix automation was carried out by three or four engineers sat by the faders/effects and adjusting the relevant parameters when they received a nod from the producer. This included riding the desk's volume faders throughout the mix to counter any volume inconsistencies. As a result, any parameter changes had to be performed perfectly in one pass, since the outputs of the desk are connected directly into a recording device. If you made a mistake, then you had to do it all again, and again, and again until it was right. In fact, this approach was the only option available when dance music first began to develop and all tracks of that time will have the filters or parameters tweaked live while recording direct to tape. This type of approach is almost impossible today, however, since the development of dance music has embraced the latest forms of mixer automation, so much so that it isn't unusual to have five or six parameters changing at once, or for the mixer and effects units to jump to a whole new range of settings for different parts of the track.

Mix automation is only featured on high-end mixing desks due to the additional circuitry involved, and the more parameters that can be automated the more expensive the desk will generally be. Nonetheless, mix automation appears in two forms: VCA and motorized. Each of these perform the same functions but in a slightly different way. Whereas motorized automation utilizes motors on each fader so that they physically move to the new positions, VCA faders remain in the same position while the relative parameters change. This is accomplished by the faders generating MIDI information rather then audio. This is then transferred to the desk's computer, which adjusts the volume accordingly. While this does mean that you have to look at a computer screen for the 'real' position of the faders, it's often preferred by many professional studios, since motorized faders can be quite noisy when they move.

Alongside automating the volume, most desks will also permit you to automate the muting of each channel. This can be particularly useful when tracks are not currently playing in the arrangement, since muting the channel will remove the possibility of hiss on the mixer's channel. On top of this, some of the considerably expensive desks also allow you to automate the send and return system and store snapshots (or 'scenes') of mixes. Essentially, these are a capture of the current fader, EQ and mute settings which can be recalled at any time by sending the respective data to the desk. Despite the fact that these features are fairly essential to mixing, it is only available on expensive desks and it is not possible to automate any effects parameters. For this type of automation, software sequencers offer much better potential.

Most audio-capable MIDI sequencers will offer mix automation, but unlike hardware this is not limited to just the volume, muting and send/return system. It's usually possible to also automate panning, EQ along with all the parameters on VST instruments and plug-in effects or processors. These can usually be identified by an R (Read automation) and W (Write automation) appearing somewhere on the plug-in's interface. By activating the Write button and commencing playback any movements of the plug-in or mixing desk will be recorded, and can then be played back by activating the Read button. What's more, many sequencers also offer the opportunity to finely edit any recorded automation data with an editor. This can be invaluable to the dance musician, since you can finely control volume or filter sweeps easily.

7 Programming theory

> It's ridiculous to think that your music will sound better if you go out and buy the latest keyboard. People tend to crave more equipment when they don't know what they want from their music.
>
> A Guy Called Gerald

Sound design is one of the most vital elements of creating dance music, since the sounds will more often than not determine the overall genre of music. However, although it would be fair to say that quite a few timbres are gleaned from other records or sample CDs, there are numerous advantages to programming your own sounds. Not only do you have much more control over the parameters when compared to samples, there are no key range/pitch-shifting restrictions and no copyright issues. What's more, you'll get a lot more synth for your money than if you simply stuck with the presets and you'll also open up a whole new host of avenues that you would otherwise have missed.

To many, the general approach to writing music is to get the general arrangement/idea of the piece down in MIDI form and then begin programming the sounds to suit the arrangement. It doesn't necessarily have to be a complete song before you begin programming, but it is helpful if most of the final elements of the mix are present. This is because, when it comes to creating the timbres, it's preferable to have an idea of the various MIDI files that will be playing together so that you can determine each file's overall frequency content. Keep in mind that the sounds of any instrument/programmed timbre will occupy a specific part of the frequency range. If you program each individually without any thought to the other instruments that will be playing alongside it, you can wind up programming a host of complex timbres that sound great on their own, but when placed in the mix they all conflict with one another, creating a cluttered mix. With the MIDI down from the start you can prioritize the instruments depending on the genre of music.

Knowing which frequencies are free after creating the most important parts is down to experience, but to help you on the way it's often worth employing a spectral analyser. These are available in software or hardware form and display the relative volume of each frequency band of every sound that is played through them. For example, if you program the groove and lead element first you can use an analyser to give you an idea of the frequencies left for other instruments.

In some cases, once these fundamental elements are created you may find that there is too much going on in the frequency range, so you can remove the superfluous parts or place them elsewhere in the arrangement. Alternatively, you could EQ the less important parts to thin them out so that they fit in with the main programmed elements. Although they may then sound

Figure 7.1 As the above spectral analysis reveals, the groove and motif take up specific frequencies while leaving others free for new instruments.

peculiar on their own, it doesn't particularly matter so long as they sound right in the context of the mix and they're not played in isolation during the arrangement. If they are, then it's prudent to either use two different versions of the timbre, one for playing in isolation and one for playing with the rest of the mix, or more simply leave the timbre out when the other instruments are playing. Indeed, the main key to producing a good track/mix is to play back sections of the arrangement and program the sounds so that they all fit together agreeably, and it isn't unusual at this stage to modify the arrangement to accomplish this. By doing so, when it comes to mixing you're not continually fighting with the EQ to make everything fit together appropriately. If it sounds right at the end of the programming stage, the mixing will not only be much easier, but the entire mix will sound more professional due to less cluttering. As a result, it shouldn't need explaining that after you've programmed each timbre you should leave it playing back while you work on the next in line, progressively programming the less important instruments until they are all complete.

Timbre effects

One of the main instigators of creating a cluttered mix or sounds that are too powerful is a synthesizer/sampler's effects algorithm. All synthesizers and many sample patches are designed to sound great in isolation to coax you into parting with money, but when these are all combined into the final mix, the effects tails, delays, chorus, etc. will all combine together to produce a muddy result. Much of the impact of dance music comes from noticeable spaces within the mix and a great house motif, for instance, is often created sparingly by keeping the lengths of the samples short so that there are gaps between the hits. This adds a dynamic edge because there are sudden shifts from silence to sound. If, however, there are effects such as delay or reverb applied to the sound from the source, the gaps in between the notes are smeared over, which considerably lessens the overall impact. Consequently, before even beginning to program you should create a user bank of sounds by either copying the presets you like into the user bank or creating your own and turning any and all effects off.

Naturally, some timbres will benefit heavily from effects, but if this is heavily effected to make it wider and more in your face, it's prudent to refrain from using effects on any other instruments. A lead soaked in reverb/delay/chorus, etc. will have a huge impact if the rest of the instruments are dry, whereas if all the instruments are soaked in effects the impact will be significantly lessened.

Note: A good mix/arrangement works in contrast – you can have too much of a good thing and it's better to leave the audience gasping for more than gasping for breath. Also, if many of the instruments are left dry, if when it comes to mixing you feel that they need some effecting, you can always introduce the effects at the mixing desk.

This approach may not always be suitable, since the effects signature may actually contribute heavily towards the timbre. If this is the case, then you'll need to consider your approach carefully. For example, if reverb is contributing to the sound's colour, ask yourself whether the subsequent tails are really necessary, as these will often reduce the impact. If they're not required then it's prudent to run the timbre through a noise gate set to remove the reverb tail. This way, the colour of the sound is not affected but the successive tails are removed, which also prevents it from being moved to the back of the mix. Similarly, if delay is making the timbre sound great with the programmed motif, try emulating the delay by ghosting the notes in MIDI and using velocity. This can often reduce the additional harmonics that are introduced through the delay running over the next note, and as many delay algorithms are in stereo it allows you to keep the effect in mono (more on this in a moment). Naturally, if this overrun is contributing to the sound's overall colour then you will have no option but to leave it in, but you may need to reconsider other sounds that are accompanying the part to prevent the overall mix becoming cluttered.

Another facet of synthesizers and samplers that may result in problems is through the use of stereo. Most tone modules, keyboards, VST instruments and samplers will greatly exaggerate the stereo spread to make the individual sounds appear more impressive. These are created in one of two ways, either through the use of effects or layering two different timbres together that are spread to the left and right speakers. Again, this makes them sound great in isolation, but when placed into a mix they can overpower it quickly, resulting in a wall of incomprehensible sound. To avoid this, you should always look towards programming sounds in mono unless it is a particularly important part of the music and requires the width (such as a lead instrument). Even then, the bass kick drum and snare, while forming an essential part of the music, should be in mono as these will most commonly sit dead centre of the mix so that the energy is shared by both speakers. Naturally, if you're using a sampled drum loop then they will most probably be in stereo, but this shouldn't be the cause of any great concern. Provided that the source of the sample (whether it's a record or sample CD) has been programmed and mixed competently, the instruments will have been positioned efficiently with the kick sat in the centre and perhaps the hi-hats and other percussive instruments spread thoughtfully across the image. This will, however, often dictate the positioning and frequencies of other elements in the mix.

On the subject of using samples, due to dance music mostly being formed around analogue synths, most professional sample CDs will contain samples of them. However, part of the charm of analogue is that they often drift out of tune, so there is no guarantee that the pitch you previously programmed in MIDI will be in the correct key when it's replaced with a sample. Additionally, as touched upon in Chapter 4, most sample CDs will only contain single hits – not multi-samples – so they will have a restricted range. If your choice of sampler has poor algorithms for shifting notes up and down the key range, then this will also limit the previously programmed motif. Since you should aim to program the most important parts first, if the sample plays a part in one of these you may have to change the key of the rest of the track or change the rhythm that the sample plays.

Theory of programming

Fundamentally, there are two ways to program synthesizers. You can use careful analysis to reconstruct the sound you have in mind or alternatively you can simply tweak the parameters of a preset patch to see what you come up with. It's important to keep in mind that *both* these options are viable ways of working, and despite the upturned noses from some musicians at the thought of using presets, a proportion of professional dance musicians do use them. The Chemical Brothers have used unaltered presets from the Roland JP8080 and JV1080; David Morales has used presets from the JV2080, the Korg Z-1, E-Mu Orbit and Planet Phatt; Paul Oakenfold has used presets from the Novation SuperNova, E-Mu Orbit and the Xtreme lead; and Sasha, Andy Gray, Matt Darey and, well, just about every trance musician on the planet has used presets from the Access Virus. At the end of the day, if the preset fits into your music then you should feel free to use it.

When it comes to synth programming, it's generally recommended that those with no or little prior experience begin by using simple editing procedures on the presets so that you can become accustomed to not only the effects each has on a sound, but also the character of the synth you're using. Despite the manufacturer's bumf that their synth is capable of producing any sound, they each exhibit a different character and you have to accept that there will be some sounds that cannot be constructed unless you have the correct synth. Alternatively, those with a little more experience can begin stripping all the modulation away and building on the foundation of the oscillators.

Remember that a proportionate amount of instruments within a dance mix share many similarities. For example, a bass sound is just that – a bass sound. As we'll cover a little later, many specific instruments are programmed in roughly the same manner using the same oscillators, and it's the actual synthesizer used, along with the modulation from the envelope generators (EGs), LFOs and filters, that produces the different tones. Thus, it can often be easier to simply strip all the modulation away from the preset patch so you're left with just the oscillators and then build on this foundation.

Customizing sounds

Unless you've decided to build the entire track around a preset or have a fortunate coincidence, presets will benefit from some editing. From earlier chapters, the effects that each controller will impart on a sound should be quite clear, but for those new to editing presets it can be difficult knowing the best place to start.

First and foremost, the motif/pad/drums, etc. should be playing into the synthesizer or sampler in its entirety. A common mistake is to just keep banging away at middle C on a controller keyboard to audition the sounds. While this may give you an idea of the timbre, remember that velocity and pitch can often change the timbre in weird and wonderful ways, and you could miss a great motif and sound combination if you were constantly hitting just one key. What's more, the controls for the envelope generators, LFOs, filters and effects (if employed) will adjust not only the sound, but also the general mood of the motif. For example, a motif constructed of 1/16th notes will sound very different when the amplifier's release is altered. If it's shortened, the motif will become more 'stabby', while if it's lengthened it will flow together more. Alternatively, if you lengthen the amplifier's attack, the timing of the notes may appear to shift, which could add more drive to the track.

Also consider that the LFOs will restart their cycle as you audition each patch, but if they're not set to restart at key-press they will sound entirely different on a motif than a single key strike. Plus, if filter key-follow is being used, the filter's action will also change depending on the pitch of the note. Perhaps most important of all, though, both your hands are free to experiment with the controls, allowing you to adjust two parameters simultaneously, such as filter and resonance or amp attack and release.

Once you've come across the timbre that shares some similarities with the tonality you require for the mix, most users tend to go for the filter/resonance combination. However, it's generally best to start by tweaking the amplifier envelope to shape the overall volume of the sound before you begin editing the colour of the timbre. This is because the amp EG will have a significant effect on the MIDI motif and the shape of the sound over time, so it's beneficial to get this 'right' before you begin adjusting the timbre. We've already covered the implications the amp EG has on a sound, but here we can look at it in a more practical context.

The initial transient of a note is by far the most important aspect of any sound. The first few moments of the attack stage provide the listener with a huge amount of information and can make the difference between a timbre sounding clear-cut, defined and upfront or more atmospheric and sat in the background. This is because we perceive timbres with an immediate attack to be louder than those with a slower attack and short stabby sounds to be louder than those with a quick attack but long release. Thus, if a timbre seems too sloppy when played back from MIDI, reducing the attack and/or release stage will make it appear much more defined. That said, a common mistake is to shorten the attack and release stage of every instrument to make all the instruments appear distinct, but this should be avoided.

Dance music relies heavily on contrast and not all timbres should start and/or stop immediately. You need to think not only in terms of tone, but also time. By using a fast attack and release on some instruments, employing a fast attack and slow release on others and using a slow attack and fast release on others will create a mix that gels together more than if all the sounds used the same envelope settings.

Naturally, what amp EG settings to use on each instrument will depend entirely on the mix in question, but very roughly speaking, trance, big beat and techno often benefit from the basses, kicks, snares and percussion having a fast attack and release stage, with the rest of the instruments sharing a mix of slow attack/fast release and a fast attack/slow release. This latter setting is particularly important in attaining the 'hands in the air' trance leads.

Conversely, house, drum 'n' bass, lo-fi and chill-out/ambient benefit from longer release settings on the basses, almost to the point that the notes are all connected together. Sounds that sit on top of this then often employ a short attack and release to add contrast. Examples of this behaviour can be heard in the latest stem of house releases that utilize the funky guitar riffs. The bass (and often a phased pad sat in the background using a quick attack and slow release) almost flow together, while the funk guitars and vocals that sit on top are kept short and stabby.

This leaves the decay and sustain of the amp EG. Sustain can be viewed as the body of the sound after the initial pluck of the timbre has ended, while the decay controls the amount of 'pluck'. By decreasing the decay the pluck will become stabby and prominent, and increasing it will create a more drawn out pluck. What's more, by adjusting the shape of the decay stage from the usual linear fashion to a convex or concave structure, the sound can take on a more 'thwacking' sucking feel or a more rounded transient respectively.

Note: A popular sound design technique is to use a lengthy decay (and attack) and reduce the release and sustain parameters to zero. This creates the initial transient of the sound, which is then mixed with a different timbre with only the release and sustain.

Once the overall shape of the riff has been modified to add contrast to the music, the filter envelopes and filter cut-off/resonance can be used to modify the tonal content of the timbre. As the filter's envelope will react to the current filter settings, it's beneficial to adjust these first. Most timbres in synthesizers will utilize a low-pass filter with a 12 dB transition, as these produce the most musically useful results, however, it's worthwhile experimenting with the other filter types on offer, such as band pass and high pass. For instance, by keeping the timbre as is on a low-pass filter but double tracking the MIDI to another synthesizer and using a high pass or band pass to remove the low-frequency elements, you're left with just the fizzy overtones. This can then be mixed with the original timbre to produce a sound that's much richer in harmonic content. If this is then sampled, the sampler's filters can be used to craft the sound further.

Also, keep in mind that the filter's transition will have a large effect on the timbre. As mentioned, most synthesizers will utilize a 12 dB transition, but a 24 dB transition used on some timbres of the mix will help to introduce some contrast as the slope is much sharper. A typical use for this is if two filter sweeps are occurring simultaneously; by setting one to 12 dB and the other to 24 dB the difference in the harmonic movement can produce tonally interesting results and add some contrast to the music. Alternatively, by double tracking a MIDI file and using a 12 and 24 dB on the same timbre, the two transitions will interact, creating a more complex tone that warps and shifts in harmonic content.

With the tone correct, you can move onto the filter's EG. These work on the same principle as the amp's EG but rather than control volume, they control the filter's action over time. This is why it's important to modify the amp's envelope before any other parameters, since if the amp's attack is set at its fastest and the filter's attack is set quite long, the timbre may reach the amp's release portion before the filter has fully been introduced.

While mentioning the attack and decay of the amplifier envelope, we touched upon the pluck being determined by the decay rate, but the filter's attack and decay also plays a part in this. If the attack is set to zero, the filter will act upon the sound on key-press, but if it's set slightly longer than the attack on the amp EG, the filter will sweep into the note, creating a '*quack*' at the beginning of the note. Similarly, as the decay setting determines how quickly the filter falls to its sustaining rate, by shortening this you can introduce a harder/faster plucking character to the note. Alternatively, with longer progressive pad sounds, if the filter's attack is set so that it's the same length as the amp EG's attack and decay rates, the sound will sweep in to the sustain stage, whereby the decay and sustain of the filter begin creating harmonic movement in the sustain.

Note: Keep in mind that, while any form of sonic movement in a sound makes it much more interesting, it's the initial transient that's the most important, as it provides the listener with a huge amount of information about the timbre. Thus, before tweaking the release and sustain parameters, concentrate on getting the transient of the sound right first.

Of course, the filter's envelope doesn't always provide the best results and the synthesizer's modulation matrix can often produce better results. For instance, by using a sawtooth LFO set to modulate the filter's cut-off, the harmonic content will rise sharply but fall slowly and the speed at which all this takes place will be governed by the LFO's rate. In fact, the LFO is one of the most underrated yet important aspects of any synthesizer as it introduces movement within a timbre, which is the real key behind producing great results.

Static sounds will bore the ear very quickly and make even the most complex motifs appear tedious, so it's practical to experiment by changing the LFO's waveforms and its destinations within the patch. Notably, if an LFO is used, remember that its rate is a vital aspect. As the tempo of dance is of paramount importance, it's sensible to sync the timing of any modulation to the tempo. Without this, not only can an arrangement become messy, but in many tracks the LFO speeds up with the tempo and this can only be accomplished by syncing the LFO to the sequencer's clock.

Programming basics

While customizing presets to fit into a mix can be rewarding, it's only useful if the synth happens to have a similar timbre to that you're looking for and, if not, you're going to have to program from the ground up. Before going any further, though, there are a few things that should be made clear.

Firstly, while you may listen to a record and wonder how they programmed that sound or groove, it's important to note that the artist may not have actually programmed them and they may be straight off a sample CD, or in some instances another record. If you want to know how Stardust created the groove on *Music Sounds Better with You*, how David Morales programmed *Needin' You* or Phats and Small managed to produce *Turn Around*, you should first listen to the original music they sampled to produce them. In many cases it will become abundantly clear how they managed to inject such a good groove or sound – they sampled from previous hits or funk records.

Table 7.1 contains a list of the most popular sounds and grooves that have actually been derived from other records. It's by no means exhaustive, as to catalogue them all would require approximately 30 pages, but instead it covers the most popular and well-established grooves.

Table 7.1 Popular sounds and grooves derived from other records

Dance artist	Title of track	Original artist	Original title of track
Stardust	*Music Sounds Better With You*	Chaka Khan	*Fate*
Armand Van Helden	*You Don't Know Me*	Carrie Lucas	*Dance With You*
David Morales	*Needin' You*	Rare Pleasure	*Let Me Down Easy*
		The Chi-lites	*My First Mistake*
Daft Punk	*Digital Love*	George Duke	*I Love You More*
Phats and Small	*Turn Around*	Tony Lee	*Reach Up*
		Change	*The Glow Of Love*
Cassius	*1999*	Donna Summer	*If It Hurts Just A Little*
The Bucketheads	*The Bomb*	Chicago	*Street Player*
DJ Angel	*Funk Music*	Salsoul Orchestra	*Take Some Time Out For Love*

Table 7.1 (*Continued*)

Dance artist	Title of track	Original artist	Original title of track
Blueboy	*Remember Me*	Marlena Shaw	*Woman Of The Ghetto*
Full Intention	*Everybody Loves The Sunshine*	Roy Ayers	*Everybody Loves The Sunshine*
Todd Terry	*Keep on Jumpin'*	Lisa Marie Ex Musique	*Keep on Jumpin'* *Keep on Jumpin'*
Byron Stingly	*Come On Get Up Everybody*	Sylvester	*Dance (Disco Heat)*
Soulsearcher	*I Can't Get Enough*	Gary's Gang	*Let's Lovedance Tonight*
GU	*I Need GU*	Sylvester	*I Need You*
Peppermint Jam Allstars	*Check It Out*	MFSB	*TSOP (The Sound Of Philadelphia)*
Nuyorican Soul	*Runaway*	Salsoul Orchestra	*Runaway*
The Bucketheads	*I Wanna Know*	Atmosfear	*Motivation*
Michael Lange	*Brothers and Sisters*	Bob James	*Westchester Lady*
Basement Jaxx	*Red Alert*	Locksmith	*Far Beyond*
Deaf 'n' Dumb Crew	*Tonite*	Michael Jackson	*Off The Wall*
Moodyman	*I Can't Kick This Feeling When It Hits*	Chic	*I Want Your Love*
Cassius	*Feeling For You*	Gwen McCrae	*All This Love I'm Givin'*
Spiller	*Batucada*	Sergio Mendes	*Batucada*
Eddie Amadour	*House Music*	Exodus	*Together Forever*
DJ Modjo	*Lady, Hear Me Tonight*	Chic	*Soup For One*
Spiller	*Groovejet*	Carol Williams	*Love Is You*
Prodigy	*Out Of Space*	Max Romeo And The Upsetters	*Chase The Devil*
Moby	*Natural Blues*	Vera Hall	*Troubled So Hard*
FatBoy Slim	*Praise You*	Camille Yarbrough	*Take Yo Praise*
Dee-Lite	*Groove Is In The Heart*	Herbie Hancock	*Bring Down The Birds*
Massive Attack	*Be Thankful*	William De-Vaughn	*Be Thankful For What You've Got*
Massive Attack	*Safe From Harm*	Billy Cobham	*Stratus*
Eminem	*My Name Is*	Labbi Siffre	*I Got The*
De La Soul	*3 Is The Magic Number*	Bob Dorough	*The Magic Number*
The Notorious BIG	*Mo Money Mo Problems*	Diana Ross	*I'm Coming Out*
A Tribe Called Quest	*Bonita Applebum*	Carly Simon	*Why*
De La Soul	*Say No Go*	Hall and Oates Neneh Cherry	*I Can't Go For That Buddy X*
Groove Armada	*At The River*	Patti Page	*Old Cape Cod*
Dream Warriors	*My Definition Of A Bombastic Jazz Style*	Quincy Jones	*Soul Bossa Nova*
Gang-Star	*Love Slick*	Young Hot Unlimited	*Ain't There Something Money Can't Buy*

Secondly, there is no 'quick fix' to creating great timbres. It comes from practical experience and plenty of experimentation with not only the synthesizer parameters, but also effects and processors. The world's leading sound designers and dance artists didn't simply read a book and begin designing fantastic sounds a day later, they will have learnt the basics and then spent months and in many cases years learning from practical experience and taking the time to study what each synthesizer, effect, processor and sampler can and cannot accomplish. Thus, if you want to program great timbres, patience and experimentation are the real keys. You need to set time aside from writing music and begin to learn exactly what your chosen synth is capable of and how to manipulate it further using effects or processors.

Finally, and most important of all, there is no such thing as a 'hit' sound. While some genres of music are created from using a particular type of sound and it can be fun emulating the timbres from other popular tracks, using them will not instantly make your own music an instant hit. Instead, it will often make it look like an imitation of a great record and will be subsequently judged alongside it rather than on its own merits. Despite the promise of many books offering secret advice to writing hit sounds or music, there is no secret formula; there are no special synthesizers, no hit making effects and no magic timbres. What's more, copying a timbre exactly from a previous hit dance track isn't going to make your music any better, as dance floor tastes change very quickly.

Note: As a case in point, the 'pizz' timbre was bypassed by every dance musician on the planet as an unworthy sound until Faithless soaked the Roland JD990's 'pizz' in reverb and used it for their massive club hit *Insomnia*. Following this, a host of 'pizz' saturated tracks appeared on the scene and it became so popular that it now appears on just about every dance-based module around. But as tastes have changed, the timbre has now almost vanished into obscurity and is skipped past by most musicians.

Keep in mind that Joe Public, the listening audience, are fickle, insensitive and short of attention span, and while some timbres may be doing the rounds today, next week/month they could be completely different. As a result, the following is not devoted to how to program a precise timbre from a specific track, as it would most likely date the book before I've even completed this chapter. Instead, it will concentrate on building the basic sounds that are synonymous with dance music and as such will create a number of related 'presets' that you can then manipulate further and experiment with. Creativity may be tiring and difficult at times, but the rewards are certainly worth it.

Programming timbres

As previously touched upon, despite the claims from the synthesizer's manufacturer, just one synth is not capable of producing every type of sound suited towards every particular genre. In fact, in many instances you will have to accept that it is only possible to create some timbres on certain instruments no matter how good at programming you may be. Just as you wouldn't expect different models of speakers/monitors to sound exactly alike, the same is true of synthesis.

Although the oscillators, filters and modulation are all based on the same principles, they will all sound very different. This is why some synthesizers are said to have certain character and is the reason why many older analogue keyboards demand such a high price on the second-hand market. Musicians are willing to pay for the particular sound characteristics of a synth. As a result, although you may follow the sound design guidelines in this chapter to the letter, there is no guarantee that they will produce exactly the same sound. Consequently, before you even begin programming your own timbres, the first step to understanding programming is to learn your chosen synthesizers inside out.

Indeed, it's absolutely crucial that you set time aside to experiment to learn the character of the synths' oscillators by mixing a sine with a square, a square with a saw, a sine and a saw, a triangle and a saw, a sine and a triangle, and so forth, and noting the results it has on the overall timbre. This is the *only* way you'll be able to progress towards creating your own sounds, short of manipulating presets.

You *have* to know the characteristics of the synthesizers you use and how to exploit their idiosyncrasies to create sounds. All professional artists have purchased one synth and learnt it inside out before purchasing another. This is obviously more difficult today with the multitude of virtual instruments appearing for ridiculously low prices, but even though it may be tempting to own every synth on the market you need to limit yourself to a few choice favourites. You'll never have the chance to program timbres if you have to learn the characteristics of 50 different virtual instruments. Gerald's advice at the beginning of this chapter is very wise counsel indeed and you have to accept that there are no shortcuts to creating great dance music, short of ripping from sample CDs.

Programming pads

Although pads are not usually the most important timbre in dance music, we'll look at these first. This is simply because an understanding of how they're created will help increase your knowledge of how the various processes of LFO and envelopes can all work together to produce evolving, interesting sounds. Of course, there are no predetermined pads to use within dance music production and therefore there are no definitive methods to create them. But, while it's impossible to suggest ways of creating a pad to suit a particular style of music, there are, as ever, some guidelines that you can follow.

Firstly, pads in dance music are employed to provide one of three things:

- To supply or enhance the atmosphere in the music – especially the case with chill-out/ambient music.
- To fill a 'hole' in the mix between the groove of the music and lead or vocals.
- To be used as a lead itself.

Depending on which of these functions the pad is to provide will also determine how it should be programmed. Although many of the sounds in dance will utilize an immediate attack stage on the amp and filter EG so that the sound starts immediately, it is only really necessary to use them on pads if the sound is providing the lead of the track. As discussed, we determine timbres that start abruptly to be perceivably louder than those that do not, but more interestingly we also tend to perceive sounds with a slow attack stage to be 'less important' to the mix, even though in reality this may not be the case at all. As a consequence, when pad sounds are used as 'backing' instruments they should not start abruptly but filter in or start slowly, while if they're used as leads the attack stage should be quite abrupt, as this not only helps it cut through the mix but also gives the impression that it's an important aspect of the music.

A popular, if somewhat clichéd and overused, technique to demonstrate this is the gated pad. Possibly the best example of this in use (which incidentally doesn't sound clichéd) is from Sasha's club hit *Xpander*, which to many clubbers is viewed as one of the greatest trance tracks of all time. A single evolving, shifting pad is played as one long progressive note throughout the track and a noise gate is employed to rhythmically cut the pad. When the noise gate releases and lets the sound back through, a plucked lead is dropped in to accentuate the return of the pad. The gated pad effect can be constructed in one of three ways:

- Sending a series of MIDI notes to a (MIDI-compatible) gate to determine where the pad should cut off. The length of the 'gate' is determined by the length of the MIDI note.
- A short percussive sound is inserted into the key input of the gate, so at every kick the gate is activated. The length of the gate is determined by the hold and release parameters on the gate.
- A series of CC11 (expression) commands are sent to the synthesizer creating the pad, which successively closes and opens the expression to produce a gated sound. The first CC11 command is set at zero to turn the pad off, while this is followed a few ticks later by another CC11 command set to 127 to switch it back on again. The length of the gate is controlled by the distance between the CC11 off and on commands in the sequence.

Naturally, for this to work in a musical context the pad must evolve throughout, because if this technique were used on a sustaining sound with no movement you may as well just retrigger the timbre for its attack stage whenever required rather than gate it. In fact, it's this movement that's the real secret behind a creating a good pad. If the timbre remains static throughout without any timbral variations then the ear soon becomes bored and switches off. This is the main reason why analogue synths are often recommended for the creation of pads, since the random fluctuations of the oscillators' pitch and timbre provide an extra sense of movement that when augmented with LFOs and/or envelopes produces a sound that has constant movement that you can't help but be attracted to. There are various ways you can employ this movement, ranging from LFOs to using envelopes to gradually increase or decrease the harmonic content while it plays. To better explain this, we'll look at the methodology behind how the envelopes are used to create the beginnings of a good pad.

By setting a fast attack and the decay quite long on an amplifier envelope, we can determine that the sound will take a finite amount of time to reach the sustain portion. Provided that this sustain portion is set just below the amp's decay stage, it will decay slowly to sustain and then continually 'loop' until the MIDI note is released, whereby it'll progress onto the release stage. This creates the basic premise of any pad or string timbre – it continues on and on until the key is released. Assuming that the pad has a rich harmonic structure, movement can be added by gradually increasing a low-pass filter's cut-off while the pad is playing – there will be a gradual rise in the amount of harmonics contained within the pad.

If a positive filter envelope was employed to control the filter's action and the envelope amount was set to fully modulate the filter, by using a long attack, short decay, low sustain and fast release, the filter's action would be introduced slowly before going through a fast decay stage and moving onto the sustain. This would create an effect whereby the filter would slowly open through the course of the amplifier's attack, decay and sustain stage before the filter entered a short decay stage during the 'middle' of the amp's sustain stage. Conversely, by using the same filter envelope settings, but applied negatively, the envelope is inverted, creating an effect of sweeping downwards rather then upwards. Nevertheless, the amp and filter envelopes working in different time scales create a pad that evolves in harmonic content over a period of time. Notably, this is a one-way configuration because if the function of these envelopes were reversed (in that the filter begins immediately but the amplifier's attack was set long) the filter would have little effect, since there would be nothing to filter until the pad is introduced.

This leads onto the subject of producing timbres with a high harmonic content and this is accomplished by mixing a series of oscillators together, along with detuning and modulating. The main principle here is to create a sound that features plenty of harmonics for the filters to sweep; saw, triangle, noise and square waves produce the best results, although on some occasions a sine wave can be used to add some bottom-end presence if required. A good starting point for any pad is to use two saw, triangle or pulse wave oscillators, with one detuned from the other by −3 or −5 cents. This introduces a slight phasing effect between the oscillators, helping to widen the timbre and make it more interesting to the ear. To further emphasize this detuning, a saw, triangle, sine or noise waveform LFO set to gently and slowly modulate the pitch or volume of one of the oscillators will produce a sound with more analogue feel, while also preventing the basic timbre from appearing too static.

If the pad is being used as a lead or has to fill in a large 'hole' in the mix, then it's worthwhile adding a third oscillator and detuning this by +3 or +5 to make the timbre more substantial. The choice of waveform for the third oscillator depends on the waveform of the original two, but in general it should be a different waveform. For instance, if two saws have been used to create the basic patch, adding a third detuned triangle wave will introduce a sparkling effect to the sound, while replacing this triangle with a square wave would in effect make the timbre exhibit a hollow character. These oscillators then, combined with the previously discussed envelopes for both the filter and amp, form the foundation of every pad sound and from here it's up to the designer to change the envelope settings, modulation routings and the waveform used for the LFOs to create a pad sound to suit the track. What follows is a guide to how most of the pads used in dance music are created, but it is by no means a definitive list and it's always worth experimenting to produce different variations.

Rise and fall pad

To construct this pad, use two sawtooth oscillators with one waveform detuned by 3 cents. Apply a fast attack, short decay, medium sustain and a long release for the amp envelope and employ a low-pass filter with the cut-off and resonance set quite low. This should result in a static buzzing timbre. From this, set the filter's envelope to a long attack and decay, but use a short release and no sustain and set the filter envelope to maximum positive modulation. Finally, use the filter's key-follow so that it tracks the pitch of the notes being played. This results in the filter sweeping up through the pad before slowly settling down. If the pad is to continue playing during the sustain portion for a long period of time, then it's also worth modulating the pitch of one of the oscillators with a triangle LFO and modulating the filter's cut-off or resonance with a square wave LFO. Both these should be set to a medium depth and a slow rate.

Resonant pads

Resonant pads can be created by mixing a triangle and square wave together and detuning one of the oscillators from the other by 5 cents. Similar to the previous pad, the amp's attack should be set to zero with a short decay, medium sustain and long release, but set the filter's envelope to a long attack, sustain and release with a short decay. Using a low-pass filter, set the cut-off quite low but set the resonance to around three-quarters, so that the timbre appears quite resonant. Finally, modulate the pitch of the triangle oscillator with a sine wave LFO set to a slow rate and a medium depth and use the filter's key-follow. The LFO modulation creates a pad that exhibits the natural analogue 'character', while the filter tracks the pitch and sweeps in through the attack and decay of the pad, and then sustains itself through the amp's sustain. Again, if the pad's sustain is going to continue for a length of time, it's worthwhile employing a sine, pulse or triangle wave LFO to modulate the filter's cut-off to help maintain interest.

Swirling pads

The swirling pad is typical of some of Daft Punk's work and consists of two sawtooth oscillators detuned from one another by 5 cents. A square LFO is often applied to one of the saws to gently modulate the pitch, while a third oscillator set to a triangle wave is pitched approximately six semitones above the two saws to add a 'glistening' tone to the sound. The amp envelope uses a medium attack, sustain and release with a short decay, while the filter envelope uses a fast attack and release with a long decay and medium sustain. A low-pass filter is used to modify the tonal content and this is set to a low cut with the resonance set to midway. Finally, chorus *and then* phaser or flanger effects are applied to the timbre to produce a swirling effect. It's important that the flanger or phaser is inserted *after* the chorus rather than before, so that it modulates the chorus effect as well.

Thin pads

All of the pads we've covered so far are quite heavy and in some instances you may need a lighter pad to sit in the background. For this, pulse oscillators are possibly the best to use, since they do not contain as many harmonics as saws or triangles and you can use an LFO to modulate the pulse width to add some interest to the timbre. These types of pads simply consist of one pulse oscillator that uses a medium attack, sustain and release with a fast decay on the amp envelope. A low-pass or high-pass filter is used, depending on how deep or bright you want the sound to appear, and there is usually little need for a filter envelope, since gently modulating the pulse width with a sine, sawtooth or noise waveform produces all the movement required. If you decide that it does need more movement, however, a very slow triangle LFO set to modulate the filter cut-off or resonance set to a medium depth will usually produce enough movement to maintain some interest, but try not to get too carried away. The purpose of this pad is to sit in the background and too much movement may push it to the front of the mix, resulting in all the instruments fighting for their own place in the mix.

Note: On the CD you can hear the pads being constructed (with narration).

Programming analogue drums

The majority of drum timbres used in the creation of dance chiefly originated from four machines that are now out of production: the Roland TR909, the Roland TR808, the Simmons SDS-5 and the E-Mu Drumulator. Consequently, while these machines (or, to use their proper term, drum synthesizers) are often seen as requisite for producing most genres of dance music, they're very highly sought after and demand an absurd sum of money on the second-hand market – if you can find them. Because of this, most musicians will use software or hardware emulations and although, to the author's knowledge, there are no alternatives to the Simmons SDS-5 and the E-Mu Drumulator, both the infamous TR machines are available from numerous software and hardware manufacturers. Indeed, due to the importance of using these kits when producing dance, most keyboards and tone modules today will feature the requisite TR808 and TR909 kits, and there are plenty of software plug-ins and standalone sequencers that offer them. The most prominent of these is Propellerhead's ReBirth, which imitates both TR machines and a couple of TB303s (more on these later), but Propellerhead's Reason, Imagine line's Fruity-loops and D-lusion's Drum station all offer samples/synthesis of the original machines too.

Kick drums

In the majority of cases, the Roland TR909 kick drum is the most frequently used kick in dance music, but the way it is created and edited with the synth parameters and effects can sometimes determine the genre of music it suits the most. In the original machines this kick was created using a sine wave with an envelope generator used to control its pitch. To add more of a transient to this sine wave, a pulse and a noise waveform were combined together and filtered to produce an initial click. This was then combined with the sine wave to produce the typical 909 kick sound. Notably, due to the age of these machines, there was not a huge amount of physical control over the timbre they created, but by building your own kick you can play with a larger number of parameters once the basic elements are down. In fact, this applies to all drum sounds, not just the kick, and is much better than sampling them from other records or sample CDs, since you can't alter the really important parameters.

When constructing a kick in a synthesizer, the frequency of the sine wave will determine how deep the kick becomes, and while anywhere between 30 and 100 Hz will produce a good kick, it does depend on how deep you want it to be. A sine wave at 30–60 Hz can be used to create an incredibly deep bowel moving thud typical of hip-hop, a frequency of 50–80 Hz can provide a starting block for lo-fi and a frequency of 70–100 Hz can form the beginning of a typical club kick. An attack/decay envelope generator (EG) can then be used to modulate the pitch of the sine wave. These envelopes are most common on drum machines but are also available on some synthesizers, such as the Roland JP8080 and a number of soft synths. Similarly, this action can be recreated on any synth by dropping both sustain and release parameters to zero.

Naturally, the depth of the pitch modulation needs to be set to maximum so that the effect of the pitch envelope can be heard, and it should be set to a positive depth so that the pitch moves downwards and not upwards (as it would if the pitch envelope were negative). Additionally, the attack parameter should be set as fast as possible so that the pitch modulation begins the instant the key is struck. If your synth doesn't have access to a pitch envelope, then the same effect can be produced by pushing the resonance up to maximum so that the filter begins to self-oscillate. The filter's envelope will then act in a similar manner to a pitch envelope, so the attack, sustain and release will need to be set at zero and the decay can be used to control the kick's decay.

Once this initial timbre is created, it's prudent to synthesize a clicking timbre to place over the transient of the sine wave. This can help to give the kick more presence and help it to pull through a mix. Possibly the best way to accomplish this is to use a square wave pitched down and use a very fast amplifier attack and decay setting to produce a short, sharp click. The amount that this wave is pitched down will depend entirely on the sound you want to produce, so it's sensible to layer it over the top of the sine wave and then pitch it up or down until the transient of the kick sounds right for the mix. This, however, is only to acquire a basic kick sound and it's now open to tweaking with all the parameters you have at your disposal, which incidentally are many more than the humble 909 or 808 have to offer.

Firstly, as we've already touched upon, the transient is where we attain most of the information about a sound, so adjusting the filter cut-off and resonance of the square wave will dramatically change the entire character of the kick. A high resonance setting will produce a kick with a more analogue character, while increasing the amplifier's decay will produce a kick that sounds quite 'boxy'. Increasing the filter's cut-off will result in a more natural sounding kick, while increasing the pulse width will create a more open, hollow timbre. Additionally, the oscillator producing the sine wave can also be affected with the synthesis parameters on offer. Using the pitch and pitch envelope parameters you can adjust how the pitch reacts on the sine wave, but more importantly you can determine how 'boomy' the kick is through the pitch decay. In this context, this will set the time that it takes to drop from the maximum pitch change to the sine wave's normal pitch. Thus, by increasing this, the pitch of the sine wave doesn't fall as quickly, permitting the timbre to continue for longer and creating a 'boomy' feel. Similarly, decreasing it will shorten its length, making it appear snappier. More interestingly, though, if you can adjust the properties of the envelope's decay slope you can use it to produce kicks that are fatter or have a smacking/slapping texture.

If the decay's slope remains linear, the sound will die in a linear fashion, producing the characteristic 909 kick sound. However, if a convex slope is used in its place, as the pitch decays it will 'bow' the pitch at a number of frequencies, which results in a kick that's more 'rounded' and much fatter. On the other hand, if the slope is concave, the pitch will curve 'inwards' during the decay period, producing the sucking/smacking timbre similar to the E-Mu Drumulator. By increasing the length of the pitch decay further these effects can be drawn out, producing kicks that are suited towards all genres of dance. It should be noted here, however, that not all synthesizers allow you to edit the envelope's slope in such a manner, but some software samplers (such as Steinberg's HALion) will allow you to modify the slope of a sample. Thus, if you were to sample a kick with a lengthy decay, you could then modulate the various stages of the envelope.

Alternatively, if the synthesizer is quite substantial it may allow you to modulate not only the sine wave, but also certain aspects of the envelope with itself. Fundamentally, this means that the pitch envelope not only modulates the sine wave, but it also modulates its own parameters. Using this you can set the synth's modulation destination to affect the oscillator's pitch and its own decay parameter by a negative or positive amount, which results in the decay becoming convex or concave respectively. This effect is often referred to as 'recursive' modulation, but as mentioned this is only possible on the more adept synthesizers. Nevertheless, whether recursive modulation is available or not, the key, at this stage, is to experiment with different variations of pitch decay, the filter section on the square wave and both oscillators. For instance, replacing the sine wave with a square produces a 'clunky' sound, while replacing it with a triangle produces a sound similar to the Simmons SDS-5.

While these methods will produce a kick that can be modified to suit all genres, for hip-hop you can sometimes glean better results using a self-oscillating filter and a noise gate. If you push the resonance up until it breaks into self-oscillation it will produce a pure sine wave. This is often purer than the oscillators themselves and you can use this to produce a deep tone that's suitable for use as a kick in the track (usually 40 Hz). If you then program a 4/4 loop in a MIDI sequencer and feed the results into a noise gate it can be used as an envelope generator. While playing the loop, lower the threshold so that only the peaks of the wave are breaching, set the attack to zero and use the release to control the kick's delay. This kick can be modified further by adjusting the hold time, as this will allow more of the peak through before entering the release stage.

Once the basic kick element is down it will most likely benefit from some compression, but the settings to use will depend on the genre of music. Typically, house, techno and trance will benefit from the compressor using a fast attack so that the transient is crushed by the compression. This produces the emblematic club kick, while setting the compressor so that the attack misses the transient but grips the decay stage it can be raised in gain to produce the characteristic hip-hop, big beat or drum 'n' bass timbre.

Snare drums

The snare drum in most dance music is derived (somewhat unsurprisingly) from the TR909, or in the case of house the E-Mu Drumulator or the Simmons SDS-5. All these, however, were synthesized in much the same way through using a triangle oscillator mixed in with pink or white noise that was treated to positive pitch movements. This can, of course, be emulated in any synthesizer by selecting a triangle wave for the first oscillator and using either pink or white noise for the second. The choice between whether to use pink or white noise depends on the overall effect you wish to achieve, but by and large pink noise is used for house, lo-fi and ambient snares, while white is used for drum 'n' bass, techno, garage, trance and big beat. This is simply because pink noise contains more low-frequency harmonics and energy than white and so produces a thicker, wider-sounding timbre.

To produce the initial snare sound much of the low-frequency content will need removing, so it's sensible to employ a high-pass, band-pass or notch filter depending on the type of sound you require. Notching out the middle frequencies will create a clean snare sound that's commonly used in break beat, while a band-pass filter will add crispness to the timbre, making it suitable for techno. Alternatively, using a high-pass filter with a medium resonance setting will create the house 'thunk' timbre. As with the kick drum, snares need to start immediately on key-press and remain fairly short, so the amp's EG will need setting to a zero attack, sustain and release, while the decay can be used to control the length of the snare itself. In some cases, if it's possible in the synthesizer, it's prudent to employ a different amp EG for both the noise and triangle wave. This way, the triangle wave can be kept quite short and swift by using a fast decay, while the noise can be made to ring a little further by increasing its decay parameter. The more that this is increased, the more atmospheric the snare will appear, allowing you to move from the typical techno snare, through big beat and trance, before finally arriving at ambient. What's more, this approach may also allow you to use a convex envelope on the noise to produce a smacking timbre similar to the NIN 'Closer' snare.

If two amp EGs are not available, small amounts of reverb can help to lengthen the timbre and, if this is followed with a noise gate with a fast attack and short hold time, the decay can be used to control the amount of ambience in the loop. Even if artificial ambience isn't required, employing a gate can be particularly important when programming house, drum 'n' bass and hip-hop loops, as the snare is often cut short in these genres to produce a more dynamic loop.

Additionally, for drum 'n' bass the snare can then be pitched further up the keyboard to produce the characteristic bright 'snap'.

This initial snare can be further modified using a pitch envelope to modulate both oscillators and can be applied either positive or negative depending on the genre of music. Most usually, small amounts of positive pitch modulation are applied to force the pitch downwards as it plays, but some house tracks will employ a negative envelope to create a snare that exhibits a 'sucking' nature, resulting in a thwacking sound. If you decide to use this technique, however, it's often worth removing the transient of the snare in a wave editor and replacing it with one that uses positive pitch modulation.

The two combined then produce a sound that has a good solid strike but decays upwards in pitch at the end of the hit. If this isn't possible in the synthesizer/wave editor, then a viable alternative is to sweep the pitch from low to high with a sawtooth or sine LFO (provided that the saw starts low and moves high) set to a fast rate or program a series of CC messages to sweep it from the sequencer. Once this is accomplished, small amounts of compression set so that only the decay is squashed (i.e. slow attack) will help to bring it up in volume so that it doesn't disappear into the rest of the mix.

Hi-hats

Hi-hats can be synthesized in a number of ways depending on the type of sound you require, but in the 'original' drum machines they were created with nothing more than filtered white noise. This can be accomplished in most synthesizers by selecting *white* noise as an oscillator and setting the filter envelope to a fast attack, sustain and release with a medium to short decay. Finally, set the filter to a high pass and use it to roll off any frequencies that are too low to create a hi-hat. The length of the decay parameter will determine whether the hi-hat is open or closed (open hats have a longer decay period).

While this is the typical way to produce a typical analogue hi-hat sound, it can sound rather cheap and nasty on some synthesizers, and even if it produces the timbre it can still appear quite dreary. As a result, a much better approach is to use either ring or frequency modulation, as this produces a hi-hat that sounds sparkling and animated, helping to add some energy to the music. Ring modulation is possibly the easiest solution of the two, simply consisting of modulating a high-pitched triangle wave with a lower pitched triangle. Frequency modulation consists of modulating a square or sine wave with a high-pitched triangle oscillator. The result is a high-frequency noise waveform that can then be modified with a volume envelope set to a zero attack, sustain and release with a short to medium decay. If FM is used and the modulator source is modified with a pitch envelope and the amount of FM modulation is increased or reduced, the resulting waveform can take on more interesting properties, so it's worthwhile experimenting with both these parameters (if available). Once this basic timbre is constructed, shortening the decay creates a closed hi-hat, while lengthening it will produce an open hat. Similarly, it's also worth experimenting by changing the decay slope to convex or concave to produce fatter or thinner sounding hats.

Notably, unlike most percussive instruments, compression should not be used on hi-hats, as all but the best compressors will reduce the higher frequencies even if the attack is set so that it skips by the attack stage. This obviously results in a dull sounding top end of a mix, so any form of dynamic restriction should be avoided on high-frequency sounds, and this includes shakers, cymbals, cowbells, claves and claps.

Shakers

Shakers are constructed in a similar fashion to the hi-hats. That is, they are created from white noise with a short attack, sustain and release with a medium decay on the amplifier and filter envelope. A high-pass filter is then used to remove any low-end artefacts to produce a timbre consisting entirely of higher frequencies. Once you have the basic 'hi-hat' timbre, the decay can be lengthened to produce a longer sound and this is then treated to an LFO modulating the high-pass filter to produce some movement. The waveform, rate and depth of the LFO depends entirely on the overall sound you want to produce, but as a general starting point a sine wave LFO with a fast rate and medium depth produces the typical 'shaker' timbre. Again, once this initial timbre has been constructed, like all drum sounds, it's worth experimenting by changing the envelope's attack and decay slope from linear to convex or concave.

Cymbals

Again, cymbals are created in a similar fashion to hi-hats and shakers as they're constructed from noise, but rather than use the noise from an oscillator, it's commonly generated from ring or frequency modulation. Using two square waves played high on the keyboard, detune them so they're approximately two octaves apart and set the amp's EG to a fast attack with no release or sustain and a medium decay (in fact, similar to hi-hats and shakers). Once the tone is shaped, both need to be fed into a ring or cross-modulator to produce the typical analogue cymbal noise timbre.

The attack of the cymbal is particularly important, so it may be worth synthesizing a transient to drop over the top or using a filter envelope to produce more of an initial crash. The filter envelope is set to a fast attack, sustain and release with a medium to short decay, as this produces an initial 'hit', but you can synthesize an additional transient using an oscillator that produces *pink* noise with the same filter settings as previously mentioned. If ring modulation is not available on the synth, a similar sound can be created using frequency modulation. This consists of modulating a high-pitched square wave with another, lower pitched, square or triangle wave. As with the frequency modulation used to create hi-hats, by modifying the source with pitch modulation or increasing/decreasing the amount of frequency modulation you can create a number of different crash cymbals.

Claps

Claps are perhaps the most difficult 'percussive' element to synthesize because they consist of a large number of 'snaps' all played in a rapid, sometimes pitch-shifting, sequence. Although in the interests of theory we'll look at how they're created, generally you're much better off recording yourself clapping (remember to run the mic through a compressor first, though, as the transient will often clip a recorder) and treating them to a chorus or harmonizing effect or, simpler still, sampling some from a CD or record.

Generally, claps are created from white noise passed through a high-pass filter. The filter and amp EGs (as should be obvious by now) are set to a fast attack with no release or sustain and a decay set to suit the timbre you require – midway is a good starting point. The filter cut-off, however, often benefits from being augmented by a sawtooth LFO set to a very fast rate and maximum depth to produce a 'snapping' type timbre. This produces the basic tone, but you'll also need to use an arpeggiator to create the successive snaps to follow. For this, program a series of staccato notes into a MIDI sequencer and use these to trigger the arpeggiator set to one octave or less so that it constantly repeats the same notes in a fast succession. These can be pitched downwards, if required, using a pitch envelope set to a positive depth on the oscillator, but it must be set so that it doesn't retrigger on each individual note, otherwise the timbre turns into mushy pitch-shifted clutter. Although claps are difficult to synthesize, it is often worth the effort required, however, as adjusting the filter section and/or the decay slopes of the amp and filter EG opens up a whole new realm of clap timbres.

Cowbells

Fundamentally, cowbells are quite easy to synthesize and can be constructed in a number of ways. You can use two square oscillators or a triangle and a square depending on the sound you require. If you require a sound with more body then it's best to use two square waves, but if you prefer a brighter sound then a square mixed with a triangle will produce better results.

For a cowbell with more body, set two oscillators to a square wave and detune them so that the first square plays at C#5 (554 Hz) while the other plays at G#5 (830 Hz). Follow this by setting the amp envelope to a fast attack with no release or sustain and a very short decay. This resulting tone is then fed into a band-pass filter, which can be used to shape the overall colour of the sound. Alternatively, if you want to create a cowbell that exhibits a brighter colour to sit near the top of a mix, it's preferable to use just one square wave mixed with a triangle wave. The frequency of the square should be set around 550 Hz and the triangle should be detuned so that it sits anywhere from half an octave to an octave from the square, depending on the timbre you require. Both these are then ring modulated and the result is fed through a high-pass filter, allowing you to remove the lower frequencies introduced by the ring modulation. Once these basic timbres are created, the amp's EG can be lengthened or shortened to suit the current rhythm.

Congas

Congas are constructed from two oscillators with a dash of frequency modulation to produce a 'clunky' timbre. These can be easily constructed in any synth (that features FM) by setting the first oscillator to a sine wave and the second to any noise waveform. The sine wave amp's EG is set to a very fast attack and decay with no release or sustain to produce a click that is then used as frequency modulation for the noise waveform. The noise waveform's amp EG needs to be set to a fast attack with no release or sustain and the decay set to taste.

There is little need to use a filter on the resulting sound, but if it seems to have too much bottom or top end then use a low-pass or high-pass filter to remove the upper or lower frequencies consecutively. In fact, by employing a high-pass filter and reducing it slowly you can create muted congas, while adjusting the noise amp's decay slope to convex or concave can produce the typical slapped congas that are sometimes used in house music.

Tambourines

Tambourines, like claps, are difficult to synthesize as they essentially consist of a number of hi-hats with a short decay each occurring one after the other in a rapid succession which is passed through a band-pass filter. This means that they can initially be constructed through using white noise, frequency modulation or ring modulation in the same manner as hi-hats. Once this basic timbre is down, the tone is augmented with a sawtooth LFO set to a fast rate and full depth. After this, you'll need to program a series of staccato notes to trigger the synthesizer's arpeggiator to create a series of successive hits. The decay parameter of the amp's EG can then be used to manipulate the tambourine's character, while the results are fed into a band-pass filter which can be used to shape the type of tambourine being used. Typically, wide band-pass settings will recreate a tambourine with a large timpanic membrane and thinner settings will recreate a tambourine with a smaller membrane.

Toms

Toms can be synthesized in one of two ways depending on how deep and wide you want the tom drum to appear. Typically, they're synthesized by using the same methods as producing a kick drum, but utilize a higher pitch with a longer decay on the amp's EG and some white noise mixed in to produce some ambience. The settings for the white noise oscillator generally remain the same as the amp's EG for the sine wave (zero attack, release and sustain with a medium decay). Alternatively, they can be produced by creating a snare timbre by mixing a triangle wave with a noise waveform but modulating the noise waveform with pitch so that it falls while the triangle wave continues unchanged.

Percussion guidelines

Although we've looked at the percussive instruments used throughout all dance genres, if you decide to become more creative, or simply want to experiment further with producing percussive hits, there are some general guidelines that you can follow.

The main oscillator usually always consists of a sine or triangle wave with its pitch modulated by a positive pitch envelope. This creates the initial tone of the timbre, while the second oscillator is used to either create the subsequent resonances of the skin after it's been hit or alternatively used to create the initial transient. For the resonance, white or pink noise is commonly used, while to create the transient a square wave is often used. The amp and filter envelope of the first oscillator is nearly always set to a zero attack, zero release and medium decay. This is so that the sound starts immediately on key-press (the drummer's strike), while the decay controls how ambient the surrounding room is. If the decay is set quite long, the sound will obviously take longer to decay away, producing an effect similar to reverb on most instruments. That said, if the decay is set too long on low-pitched percussive elements such as a kick, it may result in a 'whooping' sound rather than a solid hit. If the second oscillator is being used to create the subsequent resonance to the first oscillator then the amp and filter settings are the same as the first oscillator, whereas if it's being used to create the transient, the same attack, release and sustain settings are used but the decay is generally much shorter.

For more creative applications, it's worthwhile experimenting with the slope of the amp and filter EG decay and occasionally attack. Also, by experimenting with frequency modulation and ring modulation it's possible to create a host of new drum timbres. For instance, if the second oscillator is producing a noise waveform, this can be used to modulate the main oscillator to reduce the overall tone of the sound. What's more, by using the filters the sound can be shaped to fit into the current loop. For example, using a high-pass filter you can remove the 'boom' from a kick drum, which as a consequence produces a tighter and more punchy kick. The key is to experiment.

Creating/sampling professional loops

Even with the drum sounds, synthesized or sampled, creating a loop that sounds 'professional' is not an easy task. While it is much easier to produce a professional loop using synthesized sounds than simply using samples, both instances require a great deal of care, various effects and processors, but most important of all a good ear. However, though it would be unethical to suggest that all drum loops are constructed in any one particular way, there are some tricks that you can employ to produce loops that sound professional.

The big 'secret' behind getting a good, professional sounding loop is to keep the length of the hits short to add a dynamic edge to the rhythm. Keep in mind that long drum sounds cover the gaps between each of the drum hits and this lessens the impact of the rhythm. Jumping from silence to sound then back again to silence will have a much more dramatic impact than sounds occurring directly one after the other. What's more, long samples can also cause problems with the bass line as a deep kick will tend to merge with the bass line, resulting in a muddy sound-ing bottom end, so it's prudent to keep the kick drum snappy as this will introduce space between the hits, resulting in a more dynamic rhythm. Generally, if the bass line is quite heavy, the drums are tighter and more controlled, whereas if the bass is quite bright and breezy the drums are more 'sloppy' with less space between hits (i.e. by increasing the decay).

This can be accomplished by playing back the loop with synthesized sounds and reducing the decay of the less important percussion (such as toms, claves, tambourines, etc.) and then mov-ing onto the more important elements such as the kick, snare and hi-hats. If the loop is made of samples (or even if it isn't) it's also valuable to sample the loop and then run a noise gate across the loop. By setting the threshold so that most of the percussive elements lie above it and experimenting with the hold and decay times, it's possible to introduce character to loops that were initially quite superficial. Heavier beats or rhythms can also be made by apply-ing light reverb over the loop as a whole and then employing a noise gate to remove the reverb's tail. The gate can also be used as a distortion tool for drums provided that it's set up correctly. If you play back a loop into a noise gate and set the attack, hold and release param-eters to zero, the gate will open and close in quick succession, often resulting in the gate fol-lowing the individual cycles of the low-frequency kick. As a result, the waveforms that fall below the threshold become a series of square waves while the peaks remain unmolested, resulting in a distorted loop. The amount of distortion can then be controlled by lowering or rais-ing the threshold.

Compression can also be used to produce a crunchy distortion that is particularly suited towards big beat, hip-hop and techno drum loops. For this, two compressors are used in serial with the drum loop fed into the first compressor set to a high ratio and low threshold mixed with a fast attack and release. If the output gain of this compressor is pushed high enough it results in distortion of the mid range, which adds the typical vintage character of these genres. By feeding this output into a second compressor, the distortion can be controlled to prevent it from clipping the inputs of the recording device (or the outputs of a wave editor). Generally, an opto compressor, such as the Waves Renaissance Compressor, produces the best results, but it's worth experimenting with other vintage style compressors. It's also worth noting that hip-hop, lo-fi and big beat will also often use 'crunchy', dirty timbres, which is best accomplished by lowering the bit rate to 12-bit or the sample rate to 22 kHz prior to compression or distortion. This replicates the 'feel' of these rhythms, as many are commonly sampled from other records.

On that point, although I can't condone lifting samples from previous records because of the legal consequences, it would be incredibly naive to suggest that some house and hip-hop artists, in particular, do not sample drum hits, loops and sometimes entire grooves from other vinyl records. In fact, in many instances this sampling is absolutely paramount in attaining the 'feel' of the music. Although there are plenty of sample CDs dedicated to both these genres, they are generally best avoided, as everyone will have access to the very same CDs. The trick is to find old records that have preferably only been pressed in a small amount due to their obscurity. Indeed, the more obscure the better, as it's unlikely any other producers will have access to the same records and, if required, you'll be able to get copyright permission much more easily. These records can be sourced in the majority of second-hand and charity shops, but try to ensure that the records are over 20 years old as it's the character of the sound that matters the most.

Once you have a good collection (over 100 records is often classed as an average collection) you'll need to listen to each for an exposed segment of the drums to sample. Although you can sample these straight off the record and use them as is, it's much better to be a little more creative and drop the loop into a loop-slicing program such as Propellerhead's ReCycle or better still Wavesurgeon. This latter software is often considered easier to use than ReCycle and instantly cuts up the loop while generating a series of MIDI files associated with each hit. Once this is accomplished the rhythm can be shifted around and the timing can be adjusted to produce new variations.

Of course, recycling loops in this fashion can be incredibly time-consuming and in some instances may not be possible if the loop is quite complex or played live, as it's likely that hi-hats may occur part way through a snare or kick. If you were to cut these and move them the rhythm may lose its cohesion and sound dreadful, so as an alternative it may be worthwhile layering other hits over the top of the original loop or write a new pattern to sit over the original. If you take this latter approach, however, you need to exercise caution, as it's a fine line between a great loop and a poor one. The most common problem experienced is that the loop becomes too complex, and if this is the case it's prudent to place the loop in a wave editor and reduce the volume of the offending parts.

Note: On the CD you can hear various drums being constructed (with narration).

Programming bass

Synth basses are difficult to encapsulate in terms of programming, as we have no real expectations of how they should sound and they always sound different when placed into a mix anyway. As such, there are no definitive ways to construct a bass, as pretty much anything goes provided that it fits with the music. Of course, as always, there are some guidelines that apply to all basses and many genres also tend to use a similar bass timbre, so here we'll concentrate on how to construct these, along with some of the basic guidelines.

Generally, most synth bass sounds are quite simple in design as their main function is to supply some underpinning, as it's the lead/vocals that provide the main focal point. As a result, they are not particularly complex to program and you can make some astounding bass timbres using just one or two oscillators. Indeed, the big secret to producing great basses is not from the oscillators but from the filters and a thoughtful implementation of modulation to create some movement.

Whenever approaching a bass sound it's wise to have the bass riff programmed and playing back in your sequencer along with the kick drum and any precedence instruments. By doing so, it's much easier to hear if the bass is interfering with the kick and other priority instruments while you manipulate the parameters. For example, if the bass sound seems to disappear into the track or has no definite starting point, making it appear 'woolly', then you'll need to work with both the amp and filter envelopes to provide a more prominent attack. Typically, the amplifier's attack should be set to its shortest time so that the note starts immediately on key-press and the decay should be set so that it acts as a release setting (sustain and release are rarely used in bass timbres). This is also common with the filter's envelope. Setting the filter's attack stage too long will result in the filter slowly fading in over the length of the note, which can destroy the attack of the bass.

Notably, bass timbres also tend to have fairly complex attack, so if it needs a more prominent attack it's prudent to sometimes layer an initial pluck over the transient. This pluck can be created by synthesis in the same manner as creating a pluck for a drum's kick, but more commonly percussion sounds such as cowbells and wood blocks pitched down are used to add to the transient. If this latter approach is used then it's advisable to reduce the length of the drum sample to less than 30 milliseconds. This is because, as we touched upon when discussing granular synthesis, we find it difficult to perceive individual sounds if they are less than 30 milliseconds in length. This can usually be accomplished by reducing the amp's decay (remember that drum samples have no sustain or release) with the benefit that you can use amplitude envelopes to fade out the attack transient oscillator as you fade in the main body of the bass, helping to keep the two sounds from getting in each other's way. If this is not possible then simply reducing the volume of the drum timbre until it merges with the bass timbre may provide sufficient results.

Another important aspect of creating a good bass is sonic movement. No matter how energetic the bass riff may be in MIDI, if it's playing a simple tone with absolutely no movement our ears can get bored very quickly and we tend to 'turn off' from the music. In fact, this lack of movement is one of the main reasons why some grooves just don't seem to groove at all. If it's a boring timbre, the groove will appear just as monotonous. This movement can be implemented in a number of ways. Firstly, as we'll touch upon in the genre chapters, programming CC messages or using velocity commands can breathe life into a bass provided, of course, that you program the synth to accept these controllers. Secondly, it's often worthwhile assigning the modulation wheel to control the filter cut-off, resonance, LFO rate or pitch. This way, after programming a timbre you can move the wheel to introduce further sonic movement and record these as CC data into the sequencer. And finally, you can employ LFO modulation to the timbre itself to introduce pitch or filter movement.

If, on the other hand, you decide to use a real bass guitar then unless you're already experienced it isn't recommended attempting to program the timbre in a synth or record one live. Unlike synthetic timbres we know how a real bass guitar should sound. If these are constructed in a synth they often sound too synthetic, while recording a bass guitar reasonably well requires plenty of experience and the right equipment. Thus, if you feel that the track would benefit from a real bass it is much easier to invest in a sample CD or alternatively Spectrasonic's Trilogy – a VST instrument that contains multi-samples of a huge number of basses, real and synthetic.

Deep heavy bass

The heavy bass is typical of drum 'n' bass tracks (by artists such as Photek) and is the simplest bass timbre to produce, as it's essentially a kick drum timbre with the amplifier's decay and release parameters lengthened. This means that you'll need to use a single oscillator set to a sine wave with its pitch positively modulated by an attack decay envelope. On top of this it may also be worth synthesizing a short stab to place over the transient of the sine wave. As with the kick, the best way to accomplish this is to use a square wave pitched down and use a very fast amplifier attack and decay with no release or sustain. Once this initial timbre is laid down, you can increase the sine wave's amp EG decay and sustain until you have the sound you require.

Sub bass

Following on from the large bass, another alternative for drum 'n' bass is the earth shaking, speaker melting sub bass. Essentially, these are formed from a single sine wave with perhaps a small 'clunk' positioned at the transient to help it pull through a mix. The best results come from a self-oscillating filter using any oscillator, but if the filter will not resonate a sine wave any synth should provide good results. Obviously, the amplifier's attack stage should be set at zero so that the sound begins the moment the key is depressed, but the decay setting to use will depend entirely on the sound you require and the current bass motif (a good starting point is to use a fairly short decay, with no sustain or release). If the sine wave has been produced by an oscillator then the filter cut-off will have no effect as there are no harmonics to remove, but if it's been created with a self-oscillating filter, reducing the filter's cut-off will remove the high-end artefacts that may be present. Also, it's prudent to set the filter's key-follow, if available, to positive so that the further up the keyboard you play, the more it will open. This will help to add some movement to the sound. If a click is required at the beginning of the note then, as with the drums, a square wave with a fast amplifier attack and decay and a high cut-off setting can be dropped onto the transient of the sine wave. Alternatively, setting the filter's envelope to a zero attack, sustain and release, along with a very quick decay, and increasing the amount of the filter's EG until you have the transient can also provide great results.

This will produce the basic 'preset' tone typical of a deep sub bass, but is of course open to tweaking the initial click with filter cut-off and resonance, along with modulating the sine wave to add some movement. For example, by modulating the sine wave's pitch by 2 cents with an EG set to a slow attack, medium decay and no sustain or release, the note will bend slightly every time it's played. If there is no EG available, then a sine wave LFO with a slow rate and set to restart at key-press will produce much the same results. That said, you will have to adjust the rate by ear so that the tone pitches properly. As a variation of this sliding bass, rather than modulate the pitch and provided that the sine wave has been created with a self-oscillating filter, it's worthwhile sliding the filter.

This can be accomplished by setting the filter envelope to a fast attack with no sustain or release and a halfway setting on the decay parameter. By then increasing the positive depth of the envelope to filter you can control how much the bass slides during playback.

On top of this, keep in mind that changing the attack, decay and release of the amp or/and filter EG from linear to convex or concave will also create new variations. For example, by setting the decay to a concave slope will create a 'plucking' timbre, while setting it to convex will produce one that's more rounded. Similarly, small amounts of controlled distortion or very light flanging can also add movement. A more creative approach, though, is to use a vocal intonation program with experimental settings. The more extreme these settings are, the more the pitch will be adjusted, but with some experimentation it can introduce interesting fluctuations.

Moog bass

The MiniMoog was one of the proprietary instruments in creating basses for dance music and has been used in its various guises throughout big beat, trance, hip-hop and house. Again, this synthesizer is out of production and although it's unlikely that you'll find one on the second-hand market as they're highly prized possessions, if you do they'll demand an extraordinarily high price. Nevertheless, this type of timbre can be constructed on most analogue style synths (emulated or real) from either using a sine or triangle wave as the initial oscillator depending on whether you want a clean rounded sound (sine) or a more gritty timbre (triangle). On top of this, add a square wave and detune it from the main oscillator by either + or −3 and set the amplifier's envelope for both oscillators to its fastest attack, a medium decay, no sustain and no release. The square wave helps to 'thicken' the sound and give it a more woody character, while the subsequent amplifier setting creates a timbre that starts on key-press. The decay setting acts as the release parameter for the timbre. The filter envelope to use will depend entirely on how you want the sound to appear, but a good starting point is to set the low-pass filter cutoff to medium with a low resonance and use a fast attack with a medium decay. By then lengthening the attack or shortening the decay of the filter's envelope, you can create a timbre that 'plucks' or 'growls'. Depending on the type of sound you require it may also benefit from filter key-follow, so the higher up the keyboard you play the more the filter opens. If you decide not to employ this, however, it is worth modulating the pitch of one, or both, of the oscillators with an LFO set to a sine wave running at slow rate and make sure that this restarts with every key-press. This will create the basic Moog bass patch, but it should be noted that the original Moog synthesizer employed convex slopes on the envelopes, so if you want to emulate it exactly you will have to curve the attack, decay and release parameters.

TB303

1. Acid house bass

The acid house bass was a popular choice during the late 1980s but has been making something of a comeback in house, techno, drum 'n' bass and chill-out/ambient. Fundamentally, it was first created using the Roland TB303 Bass Synthesizer, which like the accompanying TR909 and TR808 is now out of production and as such demands a huge price on the second-hand market. Similar to the 909 and 808, however, there are software emulations available, with the most notable being Propellerhead's ReBirth, which includes two along with the TR808 and 909. As usual, though, this timbre can be recreated in any analogue synthesizer (emulated or real), but it should be noted that due to the parameters offered by the original synthesizer, there are thousands of permutations available. As a result, we'll just concentrate on creating the two most popular sounds and you can adjust the parameters of these basic patches to suit your own music.

The acid house ('donk') bass can be created using either a square or sawtooth oscillator depending on whether you want it to sound 'raspy' (saw) or more 'woody' and rounded (square). As with most bass timbres the sound should start immediately on key-press, so it requires the amp's attack to be set to its fastest position, but the decay can be set to suit the type of sound you require (as a starting point try a medium decay and no sustain or release). Using a low-pass filter set the cut-off quite low and then slowly increase the resonance so that it sits just below self-oscillation. This will create a quite bright ('donky') sound that can be further modelled using the filter envelope.

As with the amp settings, the filter envelope should be set so that it fits your music, but as a general starting point a zero attack, sustain and release with a decay that's slightly longer than the amp's EG decay will produce the typical house bass timbre. Filter key-follow is often employed in these sounds to create movement, but it's also prudent to modulate the filter's cut-off with velocity so that the harder the key is struck the more it opens. Generally, by taking this approach you'll have to tune the subsequent harmonics into the key of the song, but this is not always necessary. In fact, many dance and drum 'n' bass artists have used this timbre but made it deliberately dissonant to the music to make it more interesting.

2. Resonant bass

The resonant bass is similar in some respects to the tone produced by the acid bass, but has a much more resonant character that almost squeals. This, however, is a little more difficult to accomplish on many synths, as it relies heavily on the quality of the filters, and ideally these should be modelled around analogue. Start with a sawtooth oscillator and set the amplifier's EG to zero attack and sustain with a medium release and a fast decay. This sets the timbre to start immediately on key-press and then quickly jump from a short decay into the release portion, in effect producing a bass with a quick pluck. Follow this by setting the filter's envelope to a zero attack and sustain with a short release and a short decay (both shorter than the amp's settings), and set the low-pass filter's cut-off and resonance to halfway. These settings on the filter envelope introduce resonance to the decay's 'pluck' at the beginning of the note. This creates the basic timbre, but it's worth employing positive filter key-follow so that the filter's action follows the pitch, helping to maintain some interest in the timbre. On the subject of pitch, modulating the sawtooth using a positive or negative envelope set over a two-semitone range can help to further enhance the sound. Typically, if you want the timbre to 'bow' and drop in pitch as it plays then it's best to use a positive envelope, while if you want to create a timbre that exhibits a 'sucking' motion with the pitch rising, it's best to use a negative envelope.

Sweeping bass (typical of UK and speed garage)

The sweeping bass is typical of UK garage and speed garage tracks and consists of a tight yet deep bass that sweeps in pitch and/or frequencies. These are created with two oscillators, one set to a sine wave to add depth to the timbre while the other is set to a sawtooth to introduce harmonics that can be swept with a filter. These are commonly detuned from one another but the amount varies depending on the type of timbre required. Hence, it's worth experimenting by first setting them apart by 3 cents and increasing this gradually until the sound becomes as thick as you need for the track.

As with all bass sounds, they should start the moment the key is depressed, so the amp EG's attack is set to zero along with sustain, but the release and decay should initially be set midway. The decay setting provides the 'pluck', while the release can be modified to suit the motif being played from the sequencer. The filter cut-off is set to a low pass as you want to remove the higher harmonics from the signal (opposed to removing the lower frequencies first) and this, along with the resonance, are adjusted so that they both sit approximately halfway between fully exposed and fully closed. Ideally, the filter should be controlled with a filter envelope using the same settings as the amp EG, but to increase the 'pluck' of the sound it's beneficial to adjust the attack and decay so that they're slightly longer than the amplifier's settings. Finally, positive filter key-follow should be employed so that the filter will track the pitch of the notes being played, which helps to add more movement.

These settings will produce the basic timbre but it will benefit from pitch-shifting and/or filter movements. The pitch-shifting is accomplished, somewhat unsurprisingly, by modulating both oscillators with a pitch envelope set to a fast attack and medium decay but, if possible, the pitch bend range should be limited to two semitones to prevent it from going too wild. If you decide to modulate the filter then it's best to use an LFO with a sawtooth that ramps upwards so that the filter opens, rather than decays, as the note plays. The depth of the LFO can be set to maximum so that it's applied fully to the waveform and the rate should be set so that it sweeps the note quickly. What's more, if the notes are being played in succession it's prudent to set the LFO to retrigger on key-press, otherwise it will only sweep properly on the first note and any successive notes will be treated differently, depending on where the LFO is in its current cycle.

Techno 'kick' bass

Although a bass is not always used in techno, if it is, it's usually kept short and sharp so as not to get in the way of the various rhythms. That said, as there are few leads employed, the bass is programmed so that it's quite deep and powerful as (if they're used) they play a major role in the music.

To create this type of bass requires four oscillators stacked together, all using the same waveform. Typically, sawtooth waveforms are used, but square, triangle and sine waves can also work equally well so long as the oscillators used are all the same waveform. One waveform is kept at its original pitch while the other three are detuned from this and each other as far as possible without sounding like individual timbres (i.e. less than 20 Hz). Obviously, the sound needs to begin the moment the key is struck, so the resulting timbre is sent to an amp EG with a zero attack along with a zero sustain and release with a medium decay setting. Typically, a techno bass also exhibits a 'whump' at the decay stage and this can be introduced by modulating the pitch of all the oscillators with an attack/decay envelope. Obviously this uses a fast attack so that pitch begins at the start of the note, but the decay setting should be set just short of the decay used on the amp EG. Usually, the pitch modulation is positive so that the 'whump' is created by moving the pitch downwards, but it's worth experimenting by setting this to negative so it sweeps upwards. Additionally, if the synthesizer offers the option to adjust the slope of the envelopes, a convex decay is used, but experimentation is the real key with this type of bass and in some cases a concave envelope may produce more acceptable results. Filter key-follow is rarely used as the bass tends to remain at one key, but if your motif moves up or down in the range, it's prudent to use a positive key-follow to introduce some movement into the riff.

Trance 'digital' bass

This bass is typical of those used in many trance tracks, and while it doesn't exhibit a particularly powerful bottom end, it does provide enough of a bass element without being too rich in harmonics so that it interferes with the characteristic trance lead. It requires two oscillators, both set to square waves and detuned from each other by 3 cents to produce the basic tone. A low-pass filter is used with the cut-off set so that it's almost closed and the resonance is pushed up so that it sits just below self-oscillation. The sound, as always, needs to start immediately on key-press, so the amp's attack is set to zero along with both the release and sustain. The decay should be set about midway between being fully exposed and fully closed. The filter envelope emulates these amp settings using a zero attack, sustain and release, but the decay should be set so that it's slightly shorter than the amp's decay so it produces a resonant pluck. Finally, the filter key-follow is applied so that the filter follows the pitch across the bass motif. Once constructed, if the bass is too resonant it can be condensed by reducing the filter's decay to make the 'pluck' tighter or alternatively you can lower the resonance and increase the filter's cut-off.

'Pop' bass

The pop bass is commonly used in many popular music tracks, hence the name, but it is also useful for some house and trance mixes where you need some 'bottom-end' presence, but at the same time don't want it to take up too much of the frequencies available in the mix. These are easily created in most synths by using a sawtooth and a triangle oscillator, with the triangle transposed up by an octave from the sawtooth. The amp envelope is set to an on/off status whereby the attack, decay and release are all set to zero with the sustain set just below maximum. This means that the sound almost immediately jumps into the sustain portion, which produces a constant bass tone for as long as the key is depressed (remember that the sustain portion controls the volume level of sustain and not its length!). To add a 'pluck' to the sound, a low-pass filter is usually set very low to begin with, while the resonance is increased as high as possible without forcing the filter into self-oscillation. The filter envelope is then set to a zero attack, release and sustain but the decay is set quite long so that it encompasses the sustain portion of the amp's EG, in effect producing a resonant pluck to the bass. By then increasing the depth of the filter envelope, along with increasing the filter's cut-off and resonance, you can control how resonant the bass becomes. Depending on the motif that this bass plays, it may also be worth employing some filter key-tracking so that the filter follows the pitch of the bass.

Generic bass guidelines

Although here we've looked at the main properties that contribute towards the creation of bass timbres and covered how to construct the most commonly used basses throughout the dance genres, occasionally simply creating these sounds as is will not always produce the *'right'* results. Indeed, much of the time you find that the bass is too heavy without enough upper harmonics or that it's too light without the right amount of depth. In these cases it's useful to use different sonic elements from different synthesizers to construct a patch – a process known as layering. Essentially this means that you construct a bass patch in one synthesizer and then create one in another (and possibly another and so forth) and then layer them all together to produce a single patch.

It's important to understand here, though, that these additional layers should not be sourced from the same synth and the filter settings for each consecutive layer should be different. The reason behind this is due to the fact that all synths sound tonally different from one another, even if they use the same parameter settings. For instance, if an analogue (or analogue-emulated) synth is used to create the initial patch, using an analogue emulation from another manufacturer to create another bass (or using a digital synth) will create a timbre with an entirely different character. If these are layered on top of one another and the respective volumes from each are adjusted you can create a more substantial bass timbre. Ideally, to prevent the subsequent mixed timbres from becoming too overpowering in the mix, it's quite usual to also employ different filter types on each synthesizer. For example, if the first synth is producing the low-frequency energy but it lacks any top end, the second synth should use a high-pass filter. This allows you to remove the low-frequency elements from the second synth so that they're less likely to interfere with the harmonics from the first. This form of layering is often essential in producing a bass with the right amount of character and is one of the reasons why many professional artists and studios will have a number of synthesizers at their disposal.

It's also worth bearing in mind that it's inadvisable to apply any stereo widening effects to bass timbres. This is because the bass should sit in the centre of the mix (for reasons we'll touch upon in the mixing chapter), but it's sometimes worthwhile applying controlled distortion to them to increase the harmonic content. As some basses are constructed from sine waves that contain no harmonics, they can become lost in the mix, but by applying distortion the upper harmonics introduced can help the bass cut through the mix. This distortion can then be accurately controlled using filters or EQ to mould the bass to the timbre you require. What's more, some effects such as flangers or phasers require additional harmonics to make any noticeable difference to the sound, something that can be accomplished by applying distortion before the flanger or phaser. Of course, if these effects are applied it's sensible to ensure that they're applied in mono, not stereo, to prevent the bass becoming spread across the stereo image.

Finally, as with the drum timbres, if you have access to a synth that allows you to adjust the linearity of the envelope's attack and decay stage you should certainly experiment with this. In fact, a convex or concave decay on the filter and/or amp is commonly used to produce bass timbres with a harder pluck, allowing it to pull through a mix better than a linear envelope.

Note: On the CD you can hear various basses being constructed (with narration).

Programming leads

Saying that basses were difficult to encapsulate is only the tip of the sound design iceberg, since trying to define what makes a good lead is impossible. Every track will use a different lead sound, ranging from Daft Punk's distorted or phased leads to the various plucks used by artists such as Paul Van Dyk, through to the hundreds of variations on the euphoric trance leads. Consequently, there are no definitive methods to creating a lead timbre, but it is important to take plenty of time producing one that sounds *right*. The entire track rests on the quality of the lead and it's absolutely vital that this sounds precise. However, while it's impossible to suggest ways of creating new leads to suit any one particular track, there are some rough generalizations that can be applied.

Firstly, most lead instruments will utilize a fast attack on the amp and filter envelope so that it starts immediately on key-press, with the filter introducing the harmonics to help it to pull through a mix. The decay, sustain and release parameters, though, will depend entirely on the type of lead and sound you want to accomplish.

For example, if the sound has a 'pluck' associated with it then the sustain parameter, if used, will have to be set quite low on both amp and filter EGs so that the decay parameter can drop down to it to create the pluck. Additionally, the release parameter of the amp can be used to determine whether the notes of the lead motif flow together, or are more staccato (keep in mind that staccato notes appear louder than those that are drawn out). If the release is set so that the notes flow together it isn't unusual to employ portamento on the synth so that notes rise or fall into the successive ones.

As the lead is the most prominent part of the track, it usually sits in the mid range and is often bursting with harmonics that occupy this area. The best approach to accomplish this is to build a harmonically rich sound by using sawtooth, square waves, triangle and noise waveforms stacked together and make use of the unison feature (if the synth has it available). This is a form of stacking a number of the synthesizer's voices together to produce thicker and wider tones, but also reduces the polyphony available to the synth, so you have to exercise caution as to the polyphony available to the synth. Once a harmonically rich voice is created it can then be thinned, if required, with the filters or EQ and modulated with the envelopes and LFOs. These latter modulation options play a vital role in producing leads, as they need some sonic movement to maintain interest.

Typically, these methods alone will not always provide a lead sound that is rich or deep enough; it's worth employing a number of methods to make it 'bigger', such as layering, doubling, splitting, hocketing or residual synthesis. We've already looked at some of the principles behind layering when looking at basses, but with leads this can be stretched further as there is little need to keep the lead under any real control – its whole purpose is to sit above every other element in the mix! Alongside layering the sounds in other synthesizers, it's often worth using different amp and/or filter envelopes in each synth. For example, one timbre could utilize a fast attack but a slow release or decay parameter, while the second layered sound could utilize a slow attack and a fast decay or release. When the two are layered together, the harmonic interaction between the two sounds produces very complex timbres that can then be mixed together in a desk and equalized or filtered externally to suit.

Doubling is similar to layering, but the two should not be confused as doubling does not use any additional synthesizer but the one you're using to program the sounds. This involves copying the MIDI information of the lead motif to another track in the MIDI sequencer and transposing this up to produce a much richer sound. Usually, transposing this copy up by a fifth of an octave will produce musically harmonious results, but it's worth experimenting by transposing it further and examining the effects it has on the sound. A variation on this theme is to make a copy of the original lead and then only transpose some notes of the copy rather than all of them so that some notes are accented.

Hocketing consists of sending the successive notes from the same musical phrase to different synthesizers or patches within the same module to give the impression of a complex lead. Usually, you determine which synth receives the resulting note through velocity by setting one synth or patch to not accept velocity values below, say, 64 while the second synth or patch will only accept velocity values above 64. If this is not possible, then simply copying the MIDI file onto different tracks and deleting notes that should not be sent to the synth will produce the same results.

Splitting and residual synthesis are the most difficult to implement but often produce the best results. Fundamentally, splitting is similar in some respects to layering, but rather than produce the same timbre on two different synthesizers, a timbre is broken into its individual components, which are then sent to different synthesizers. For example, you may have a sound that's constructed from a sine, sawtooth and triangle wave, but rather than have one synthesizer do this, the sine may come from one synth, the triangle from another and the saw from yet another. These are all modulated in different ways using the respective synthesis engines, but by listening carefully to the overall sound through a mixing desk the sound is constructed and manipulated through the synthesizer's parameters and mixing desk as if it were from one synth. Residual synthesis, on the other hand, involves creating a sound in one synthesizer and then using a band-pass or notch filter to remove some of the central harmonics from the timbre. These are then replaced using a different synthesizer or synthesis engine and recombined at the mixer.

Finally, effects also play a large part in creating a lead timbre. The most typical of these that are used on leads are reverb, delay, phasers, flangers, distortion and chorus, but experimentation with different effects and even the order of the effects can all produce great results.

As always, the key to producing great leads, as with all other sounds, is through experimentation and familiarization with effects and the synthesis engines you use.

Euphoric trance lead

The euphoric trance lead is probably the most elusive lead to program properly, but in many cases this is simply because it cannot be recreated on any synthesizer. To capture the sound properly requires an analogue synth (emulated or real) with oscillators that have the right character for the genre. This means that you should use a synthesizer that employs methods to dynamically alter both the tone and pitch of the oscillators in a slightly different manner every time you hit a key. This method, often referred to as phase initialization, produces the characteristics distinctive of any good analogue synth. Most software or hardware analogue emulations will, or should, employ this, but the amount of initialization depends entirely on the synth and, for trance leads, the higher this is, the better the results will be. As a side note, the most commonly used synth to create the trance lead is the Access Virus (in fact, nearly all professional trance musicians will use the Virus!), but the Novation SuperNova, Novation A station, Novation K station or the Novation V station (the software VST Instrument of the K station) can also produce the requisite timbres with a little extra work.

Alongside the 'unreliable' feature of analogue oscillators, the real secret behind creating any good trance lead is through clever use of effects and noise. Not the bad type of noise, of course, but the noise produced by an oscillator in the synth. As we've already touched upon, noise produces a vast number of harmonics and this is essential in creating the hands in the air vibe.

A basic trance lead timbre can be constructed with four oscillators. Two are set to pulse and detuned to produce a wide hollow sound, while the third is a sawtooth to add further harmonics, and the fourth is set to create noise to add some 'top-end' fizzy harmonics to the sound. The two pulse waves are detuned from one another by detuning one to −5 cents and the other to +5 cents to produce a wide sound, while the saw is detuned from these by a full octave to add some bottom-end harmonics to the timbre. The final oscillator is left as is, but can use either pink or white noise depending on how fizzy you want the harmonics it produces to be. Generally, pink noise is the best choice, since it contains a huge range of harmonics from low to high, while white noise is more like radio static and tends to be a little too light and fizzy for trance leads.

Obviously, the note needs to start as soon as the MIDI note is received, so the amp's attack is set to zero, but to create a small pluck use a medium decay with a small sustain and a fairly short release. This release can be lengthened or shortened further depending on the trance riff you've programmed. The filter envelope can use the same settings as the amplifier, although in some cases it may be worth shortening the decay to produce a better 'plucking' sound. This envelope should be applied from three-quarters to full as positive modulation to the filters and the filter key-follow should be on so that they track the pitch of the notes being played. The filter itself is set to a low pass with a high cut-off but a low resonance to prevent the timbre from squealing and it should use a four-pole filter rather then the usual two-pole. This is so the filter sweeps sharply, helping the lead become more prominent and allowing it to cut through the mix.

To add some energy and interest to the timbre, it's an idea to modulate the pulse width of both pulse oscillators and, if at all possible, you should use two LFOs so that each pulse can be modulated differently. The first LFO modulates the pulse width of the first oscillator with a sine wave set to a slow to medium rate and full depth, while the second modulates the pulse width of the second oscillator. This, however, is set to a triangle wave with a faster rate than the first LFO and at a full depth. The resulting effect is that both pulse waves beat against each other, creating a more interesting timbre. Finally, the timbre will need to be washed in both reverb and delay to provide the required sound. The reverb should be applied quite heavily as a send effect, but with 50 ms of pre-delay so that the transient pulls through unmolested and the tail set quite short to prevent it from washing over the successive notes. It's also prudent to set a noise gate to remove the subsequent reverb tail, as this will produce a heavier timbre which cuts through the mix and prevents it from becoming pushed into the background (an effect known as gate-tailing). Decay should also be applied but again this is best used as a send effect so that only a part of the timbre is sent to the delay unit. The settings to use will depend on the type of sound you require, but the delays should be set to less than 30 ms to produce the granular delay effect to make the timbre appear big in the mix.

Plucked leads

Plucked leads are often used in most genres of music from house through trance, but can be typified in most of Paul Van Dyk's work. This is not to say that they will all sound like Paul Van Dyk, however, as they can be constructed in numerous ways depending on the type of sound you want to accomplish and the genre of music you write. Thus, what follows are two of the most commonly used basic patches for plucked leads in dance which, as always, you should manipulate further to suit your own music.

Plucked lead 1

The first plucked lead consists of three oscillators, two of which are set to sawtooths to add plenty of harmonics and the third set to either a sine or triangle wave to add some bottom-end weight. The choice between whether to use a sine or triangle is up to you, but using a sine will add more bottom end and is useful if the bass used in the track is rather thin, while if the bass is quite heavy, a triangle may be better as this will add less of a bottom end and introduce more harmonics. The sawtooth waves are detuned from each other by 3 cents (or further if you want a richer sound) and the third oscillator is transposed down by an octave to introduce some bottom-end weight into the timbre. Obviously, the amp's attack is set to zero as is the sustain portion, but both decay and release are set to a medium depth depending on the amount of pluck you require and the motif playing the timbre. A low-pass filter is used to remove the higher harmonics of the sound and this should be set initially at midway while the resonance should be set quite low. This is controlled with a filter envelope set to a zero attack and sustain, but the decay and release are set just short of the amplifier's decay and release settings, so if you adjust the amp's decay and release, you'll also need to reduce the filter's decay and release. Finally, it's worthwhile applying some reverb to the timbre to help widen it a little further, but if you take this approach you'll also have to employ a tailing gate to prevent the lead from being pushed too far back into the mix.

Plucked lead 2

The second plucked lead is a little easier to assemble and consists of just two oscillators, a saw and a triangle. The saw wave is pitched down to produce a low end, while the triangle wave is pitched up to produce a slight glistening effect. The amount that these are detuned by depends on the timbre you require, so you'll need to experiment through detuning them by different amounts until the frequencies sit into the mix you have so far. As a general starting point, detune them from one another as far as you can without them sounding as two different timbres and then reduce this tuning until the timbre fits to the music. Due to the extremities of this detuning, sync both oscillators together to prevent the two from beating too much and then apply a positive pitch envelope to the saw wave with a fast attack and medium decay so that it pitches up as it plays. The sound needs to start on key-press, so set the attack to zero with no attack, sustain or release and set the decay to halfway. This envelope is copied to the filter envelope, but set the decay to zero and increase the sustain parameter to about halfway. Finally, using a low-pass filter set the cut-off to around three-quarters open and set the resonance to about a quarter. This will produce the basic pluck timbre, but it may be worth using the synth's unison feature to thicken the sound out further depending on the mix. If this isn't possible granular delay can be used in its place to widen the timbre.

TB303 leads

Although we've already seen the use of a TB303 when looking at programming basses, it is a versatile machine and is also equally at home creating lead sounds by simply pitching the bass frequencies up by a few octaves. The most notable example of this was on Josh Wink's *Higher State of Consciousness* and this same effect can be recreated on analogue (or analogue-emulated) synthesizer. Only one oscillator is required and this is a sawtooth, due to the high number of harmonics required for it to cut through a mix. The sound should start immediately on key-press so it requires the amp's attack to be set to its fastest position, but the decay can be set to suit the type of sound you require (as a starting point try a medium decay and no sustain or release). Using a low-pass filter set the cut-off quite low and the resonance so that it sits just below self-oscillation. The filter envelope is set to a fast attack with no sustain or release, but the decay needs to be set so that it's just short of the amp's decay stage. Filter key-follow is often employed to create additional movement, but it's also prudent to modulate the filter's cut-off with velocity, so that the harder the key is struck, the more it opens.

Finally, the timbre is run through a distortion unit (Steinberg's QuadraFuzz is particularly useful for this) and the results are filtered with an external or plug-in filter to finally mould the sound. In some instances it may also be worth increasing the amplifier's decay so that the notes overlap each other and then switch on portamento so that they slur into each other while the riff is played.

Distorted leads

There are literally hundreds of distorted leads, but one of the most popular is the distorted/phased lead used in many house tracks and in particular by Daft Punk and Tomcraft.

The basic patch is created using a single sawtooth or if you need it slightly thicker, two saws detuned from one another by 5 cents. The amp envelope is set to a zero attack with a full sustain, a short release setting and the decay to around a quarter. As always, the decay will have the most influence over the sound, so it's worth experimenting with to produce the sound you need. The filter cut-off should be set quite low, while the resonance should be pushed up pretty high to produce overtones which can be distorted with effects in a moment. The filter's envelope will need adjusting so that it reacts with the sound, so set the attack, sustain and release to zero and the decay so that it's slightly longer than the amp's decay setting. Ideally, the filter's envelope should fully affect the sound, so as a starting point set this so it fully affects the filter and then experiment by reducing the amount until the basic timbre you want appears.

Distortion and phaser are also applied to the sound, but as previously touched upon the distortion should come first so that the subsequent phaser also works on the distortion. It is important, however, not to apply too much distortion, otherwise it may overpower the mix and become difficult to mix. Preferably, you should aim for a subtle but noticeable effect before finally applying a phaser to the distorted signal. How much to apply will obviously depend on the sound you require, but exercise caution that you do not apply too much, otherwise the solidity of the timbre can be lost. In some cases it may also be worth applying portamento with a slow rate to enable the sounds to slew into one another which, as a result, brings more attention to the phased effect.

Theremins

For those with no idea of what a theremin is, it consists of a vertical metal pole approximately 12–24 inches high that responds to movements of your hands and creates a warbling low-pitched whistle depending on your hands' position. The sound was used heavily in 1950s and 1960s sci-fi movies to provide a 'scary' atmosphere when the aliens appeared, but has made many appearances in lo-fi music. The most notable example of this use in music can be heard on Portishead's *Mysteron*.

These are incredibly simple to create and can be reproduced on any synth through using a sawtooth wave with a low-pass filter's cut-off parameter reduced until it produces a soft constant tone. On some digital synthesizers you may need to raise the resonance to produce the correct tone, but most analogue synths will produce the tone with no resonance at all. This is also an example of one lead where the amplifier's attack is not set at zero; instead, this is set at halfway, along with the release, while the sustain is set to full. Theremins are renowned for their random variations in pitch, so you'll need to emulate this by holding down a key (around C3) and using the pitch wheel to introduce the variations from waving your hands around. If the synth has access to portamento then it's also advisable to use this to recreate the slow shifting from note to note.

Hoovers

The hoover sound originally appeared on the Roland Juno synths and has been constantly used throughout all genres of dance music, including techno, house, acid house, break beat, drum 'n' bass and big beat. In fact, these still remain one of the most popular sounds used in dance music today. Originally in the Juno they were aptly named '*What the…*', but due to their dissonant tonal qualities they were lovingly renamed 'hoover' sounds by dance artists, since they tend to share the same sound as vacuum cleaners … Honestly.

The sound is best constructed in an analogue/DSP synth as the tonal qualities of the oscillators play a large role in creating the *right* sound. Two sawtooth oscillators are used to create the initial patch and these need to be detuned as far from each other as possible, but not so far that they become two individual timbres.

The sound starts immediately, so the amp's attack is set at zero mixed with a short decay and release, and sustain set to just below full so there is a small pluck evident from the decay stage. The filter should be a low pass with a low cut-off and a medium resonance setting, which is modulated slightly with the filter's envelope. This latter envelope is set to a medium attack, decay and release but no sustain, but you need to exercise caution as to how much this envelope modulates the filter, since settings that are too high will result in a timbre that's nothing like a hoover. To add the typical dissonance feel that hoovers often exhibit, the pitch of both oscillators will also need modulating using a pitch envelope set to a fast attack and a short decay. This, however, should be applied negative rather than positive so that the pitch bends upwards into the note rather than downwards. Finally, depending on the synth recreating the timbre, it may need some widening and this can be accomplished by washing the sound in chorus or preferably stacking as many voices as possible using a unison mode.

'House' pianos

The typical house piano was drawn directly from the Yamaha DX range of synthesizers and is usually left unmodified (due to the complexity and painful experience that is programming with frequency modulation). In fact, if you want to use the house piano, the best option is to either purchase an original Yamaha DX7 synthesizer or alternatively invest in Native Instruments' FM-7[1] VST instrument. This is a software emulation of the FM synths produced by Yamaha, which can also import the sounds from the original range of DX synthesizers.[2]

Acquiring the FM piano is difficult on most analogue synthesizers and often impossible of most digital synths (well apart from the DX range of course!), because of the quirks of frequency modulation. Nonetheless, if you want to give this a go, it can be accomplished by using two oscillators set to sine waves with one of the oscillators detuned so that it's at a multiple of the second oscillator. These are then frequency modulated to produce the general tone. The amp envelope is then set to a fast attack, short decay and release, and a medium sustain with the filter key-tracking switched on. To produce the initial transient for the note, a third sine wave pitched high up on the keyboard and modulated by a one-shot LFO (i.e. the LFO acts as an envelope – fast attack, short decay, no sustain or release) will produce the desired timbre. As a side note to this, if you want to produce the infamous bell-like or metallic tones made famous by FM synths, use two sine oscillators with one detuned so that its frequency is at a non-related integer of the second and then use frequency modulation to produce the sound.

Organs

Organs are commonly used in the production of house and hip-hop, with the most frequent choice being the Hammond B-4 drawbar organ (which incidentally is also available as a VST

---

[1] A demo of Native Instruments' DX-7 can be located on the accompanying CD-ROM. © Native Instruments.

[2] A selection of the presets from the original DX synths (including the DX-7 piano) can be downloaded from www.dancemusicproduction.com for use with the FM-7.

instrument from Native Instruments). This general timbre, however, can be emulated in any subtractive synthesizer through using a pulse and a sawtooth oscillator. As the sound starts on key-press the amp uses a zero attack with a full sustain and medium release (note that there is no decay since the sustain parameter is at maximum). A low-pass filter is used to shape the timbre with the cut-off set to zero and the resonance increased to about halfway. A filter envelope is not employed since the sound should remain unmodulated, but if you require a 'click' at the beginning of the note you can turn the filter envelope to maximum, but the attack, release and sustain parameters should remain at zero with a very, very short decay stage. Finally, the filter key-follow should be set so that the filter tracks the current pitch, which will produce the typical organ timbre that can then be modified further.

Generic leads

So far, we've looked at some of the most popular timbres used for dance leads, but as previously touched upon, there are literally thousands of combinations available, so what follows is a brief overview of how to produce the basic patches that can be used as the basic building blocks to further develop upon.

1   As a starting point to trance, techno, big beat and lo-fi leads, mix a sawtooth oscillator with a square wave and detune them by at least 3 to produce a wide timbre. The amp EG is set to a fast attack with no sustain and a medium release with a short decay. The filter's (usually a low pass to keep the body of the sound) cut-off is set low with a medium resonance setting, with the envelope using a fast attack, medium decay, low sustain and no release. The amount the filter envelope modulates the filters will determine much of the sound, so start by using a high value and reduce it as necessary. Employ filter key-follow so that the filter tracks the pitch and then experiment with LFOs modulating the filters and pitch. If the sound still appears thin at this stage, make use of the unison feature or add another square or sawtooth to produce a thicker sound.
2   As a starting point for hip-hop (assuming that you don't want to sample it) and chill-out, a triangle wave mixed with a square or saw wave detuned by 5 or 7 will produce the basic harmonic patch. Using a low-pass filter, set the cut-off and resonance quite low and set the key-tracking to full. For the amplifier envelope, use a fast attack with a medium decay and release, and a sustain set just below the decay parameter. For the filter envelope, it's prudent to use an attack that's slightly longer than the amp's attack stage, along with a medium decay, low sustain and a release just short of the amp. Sync both oscillators together and use an LFO to add some vibrato to the filter and the oscillators.
3   As a starting point for UK garage, try starting with a square, sine and triangle wave each detuned from each other as much as possible without them appearing as distinct individual timbres. From this, use a low-pass filter with the cut-off set quite high and no resonance, and set the key-tracking to full. Use a zero attack, long decay, medium sustain and no release for both the filter and amp envelopes, and modulate the pitch of the triangle wave with either positive or negative values.
4   The basic timbres that the vintage techno and house sounds were based around can be easily created on any analogue/DSP synth by detuning saw, triangle or square waves by a third, fifth or seventh. The choice of which of these oscillators to use obviously depends on the type of sound you require, but once selected they should be detuned by an odd amount. The amplifier EG is generally set at a zero attack, sustain and release, while the decay is used to shape the overall sound. The filter envelope is normally set to match the amp's EG, although if you require a pluck in the sound, the decay should be set slightly shorter than the

amp's decay. The filter is always set at low pass and a good starting point is to use a high resonance with a cut-off set about midway. Key-follow is employed so the filter tracks the keyboard's pitch and it's quite common to employ a pitch envelope on both oscillators so that they pitch either up or down while playing.

Note: On the CD you can hear the various leads being constructed (with narration).

Programming sound effects

Sound effects can play a fundamental role in the production of dance music as they have a multitude of uses from creating drops and adding to builds, to sitting in the background and enhancing the overall mix. Indeed, their importance in dance music production should not be underestimated, since without them a mix can sound dry, characterless or just plain insipid.

For creating sound effects in a synth, the most important parameter on a synth is the LFO, as this can be used to modulate various parameters of any sound. For instance, simply modulating a sine wave's pitch and filter cut-off with an LFO set to S&H (sample and hold) or noise waveform will produce strange burbling noises, while a sine wave LFO modulating the pitch can produce siren-type effects. There are no limits and as such no real advice on how to create them, as it comes from experimenting with different LFO waveforms modulating different parameters and oscillators. Generally, though, the waveform used by the LFO will contribute a great deal to the sound effect you receive. Triangle waves are great for creating bubbly, almost liquid sounds, while saws are suitable for zipping-type noises. Square waveforms are good for short percussive snaps such as gunshots and random waves are particularly useful for creating burbling, shifting textures. All of these used at different modulation depths and rates; modulating different oscillators and parameters will create wildly differing effects.

Additionally, if you're using wave editors you shouldn't discount their uses in creating sound effects. Most wave editors today feature their own synthesis engines that allow you to create and mix raw waveforms that can then be affected and treated to any plug-in effects you may own. For example, WaveLab's test signal generator is very similar to an additive synth, allowing you to create sounds containing up to 64 different layers of waveforms and treat each individually to different frequency, tremolo and vibrato values. There's little need to use this many layers as this will often results in a sound that's far too complex to be used as a sound effect, so you don't have to worry about typing in 64 different values for each.

Normally, most sound effects can be acquired from using just two or three waveforms and some creative thinking. The real solution to creating effects is to experiment with all the options at your disposal and not be afraid of making a mess. Even the most unsuitable noises can be tonally shaped with EQ and filters to be more suitable.

To help you along in your experiments, what follows is a small list of some of the most popular effects and how they're created. Unfortunately, though, there are no specific terms to describe sound effects, so what follows is a rough description of the sound they produce, but from what we've covered so far in this chapter simply reading about how the effect is created should give you a good idea of what they will sound like. And, of course, if you're still not sure, try creating them to see how they sound or listen to the book's accompanying CD.

Siren FX

The siren is possibly the easiest sound effect to recreate. Set one oscillator to produce a sine wave and use an amp envelope with a fast attack, sustain and release, and a medium decay. Finally, use a triangle wave or sine wave LFO to modulate the pitch of the oscillator at full depth. The faster the LFO rate is set, the faster the siren will become.

Whooping FX

To create a 'whooping effect' use one oscillator set to a sine wave with a fast attack, no sustain or release and a longish decay on the amp EG. Modulate the pitch of this sine wave with a fast attack and long decay set to a positive amount and then program a series of staccato notes into a MIDI sequencer. Use these to trigger an arpeggiator set to one octave or less so that it constantly repeats the same notes in a fast succession to create a '*whoop, whoop, whoop*' effect.

Zap FX

To create a zapping effect turn the filter cut-off down to zero and increase the resonance until it breaks into self-oscillation, creating a pure sine wave. Set the amplifier's attack to a fast attack, sustain and release, and a medium decay, and use these same settings on the filter envelope. Set this latter envelope to fully modulate the filter and the use either a triangle or saw wave LFO set at a medium speed and full depth to modulate the filter's cut-off.

Explosion FX

To create explosive-type effects requires two oscillators. One oscillator should be set to a saw wave while the other should be set to a triangle wave. Detune the triangle from the saw by +3, +5 or +7 and set a low-pass filter to a high cut-off and resonance (but not so high that the filter self-oscillates). Set the amp's envelope to a medium attack, sustain and release but with a long decay, and copy these settings to the filter envelope but make the decay a little shorter than the amp's EG. Use a sawtooth LFO set to a negative amount and use this to control the pitch of the oscillators along with the filter's cut-off. Finally, use frequency modulation from the saw onto the triangle and play low down on the keyboard to produce the explosive effects. Notably, the quality of the results from this will depend on the synthesizer being used. Most digital and a few analogue synths don't produce a good effect when frequency modulation is used, so the effect will differ from synth to synth.

Zipping FX

This effect is quite popular in all forms of dance and is created with two oscillators and an LFO to modulate the filter frequency. Start by selecting a saw and triangle as the two oscillator waveforms and detune one from the other by +7. Use a fast attack and release along with a medium to long decay and medium sustain on the amp's EG and, using a low-pass filter, set the cut-off quite low but use a high resonance. Finally, set a sawtooth LFO (using a saw waveform that moves from nothing to maximum) at full depth to slowly modulate the filter's cut-off and, if possible, use an envelope to modulate the LFO's speed so that it gradually speeds up as the note plays. If this is not possible, you'll have to increase the speed of the LFO manually and record the results into a sampler or audio sequencer.

Rising speed/filter FX

Another popular effect is the rising filter, whereby its speed increases as it opens further. For the oscillator you can use a saw, pulse or triangle wave, but a self-oscillating filter provides the best results. Set both the amp and filter EG to a fast decay and release but a long attack and high sustain. Use a triangle or sine LFO set to a positive mild depth and very slow rate (about 1 Hz) to modulate the filter's cut-off. Finally, use the filter's envelope to also modulate the speed of the LFO so that as the filter opens, the LFO also speeds up. If the synthesizer doesn't allow you to use multiple destinations, you can increase the speed of the LFO manually and record the results into a sampler or audio sequencer.

Falling speed/filter FX

This is basically the opposite of the previously described effect, so that rather than the filter rising and simultaneously speeding up, it falls while simultaneously slowing down. Again, set both the amp and filter EG to a fast decay and release with a long attack but don't use any sustain. Use a triangle or sine LFO set to a positive mild depth and fast rate to modulate the filter's cut-off. Finally, use the filter's envelope to also modulate the speed of the LFO so that as the filter closes the LFO also slows down. If the synthesizer doesn't allow you to use multiple destinations, you can decrease the speed of the LFO manually and record the results into a sampler or audio sequencer.

Dalek voice

The Dalek voice isn't just for Dr Who fans as it can be used in place of a vocoder to produce more metallic style voices that are suitable for use in any dance genre. This can be accomplished by recording your voice into a sampler or sequencer and feeding it, along with a low-frequency saw wave, into a ring modulator. You can experiment with the results of this by changing or using cyclic modulation on the pitch of the sine wave entering the ring modulator.

Sweeps

Sweeps are generally best created using sawtooth oscillators, as their high harmonic content gives the filter plenty to work with. Start by using two oscillators both set to sawtooth waves and detune them as far as possible without them becoming individual timbres and feed the results into a ring modulator. On the amp's envelope, use a fast attack, decay and release but set the sustain parameter to maximum, and then use a slow saw or triangle wave LFO to modulate the pitch of one of the oscillators and the filters. For the filters, a band pass will produce the best results set to a medium cut-off but a very high resonance (but not so high that is self-oscillates). Finally, use a second saw, sine or triangle LFO to modulate the filter's cut-off to produce the typical sweeping effect.

Ghostly noises

To create ghostly noises from a synth use two oscillators both set to triangle waves. Using a low-pass filter, set the cut-off quite low but employ a high resonance and set the filter to track the keyboard's pitch (filter key-follow). Adjust the amp's EG to a fast attack with a long decay, high sustain and medium release, and set the filter's envelope to a fast attack, long decay but no sustain or release. Finally, using an LFO sine wave, very slowly (about 1 Hz) modulate the pitch of the oscillators and play chords in the bass register to produce the timbres.

Computer burbles

Computer burbling noises can be created in a number of ways, but by far the most popular method is to use two oscillators both set to triangle waves. Detune one of these from the other by +3 cents and then set the filter envelope to a medium attack with a long decay with a medium release and no sustain. Do the same for the amp's EG but use a high sustain and set the filter's envelope to positive, but mildly modulate a low-pass filter with a low cut-off and a high resonance. Finally, ensure that filter key-tracking is switched on and modulate the pitch of one, or both, of the oscillators with a noise or sample and hold waveform. This should initially be set quite fast with a full depth, but it's worth experimenting with the depth and speed to produce different results.

Swoosh FX

To create the typical swoosh effect (often used behind a snare roll to help in creating a build-up), the synthesizer will need both filter and amp envelopes, but also a third envelope that can be used to modulate the filter as well. To begin with, switch all the oscillators off, or reduce their volume to zero, and increase the resonance so that the filter breaks into self-oscillation. Use a filter envelope with no decay, a medium release and attack and a high sustain, and set it to positively affect the filter by a small amount and then use a second envelope with these same settings to affect the filter again, but this time set it to negatively affect the filter by the same amount as before. Set the amp's EG to no attack or sustain but a small release time and a long decay, and use a saw or triangle wave LFO set to a medium speed and depth to positively modulate the filter's cut-off. This will create the basic 'swoosh' effect, but if possible employ a second LFO set to a different waveform from the previous one to negatively modulate the filter by the same amount.

Note: On the CD you can hear the various sound effects being constructed (with narration).

Programming new timbres

Of course, there will come a time when you want to construct the timbre that's in your head, but for this it's important to note that you need to be experienced in identifying the various sounds produced by oscillators and the effects they have when mixed together. Also, you need to be able to identify the effect positive and negative envelopes can impart on a sound, along with the effects produced by an LFO's augmenting parameters. If you have this knowledge then constructing sounds isn't particularly complicated and just requires some careful consideration.

Firstly, when you're attempting to come up with sounds on your own, it's important to be able to conceptualize exactly what you want the instrument to be. This means that you need to imagine the completed sound (and this is much more difficult than you envisage!) and then take it apart in your mind by asking a series of questions, such as:

- Does the timbre start immediately?
- How does it evolve in volume over time?
- Is the instrument synthetic, plucked, struck, blown or bowed?
- What happens to its pitch when it's sounded?
- Are there any pitches besides the fundamental that stand out enough to be important to the timbre?
- Does it continue to ring after the notes have been sounded?
- How bright is the sound?
- How much bass presence does the sound have?
- Does it sound hollow, rounded, gritty or bright and sparkly?
- What happens to this brightness over time?
- Is there any modulation present?
- What does the modulation do to the sound?

The answer to all these should be written down on paper, otherwise you can easily lose track and wind up going in another direction altogether. As a more practical example of how this form of sound design is accomplished, we'll tear a sound apart and reconstruct it. As my mental powers have not reached the stage where I can project the image of a sound into your mind, we'll have to use a more conventional approach and include an 'imagined' sound on the CD.

Note: On the CD you can hear the deep bass.

We need to begin by examining the sound on the CD and then breaking it down into its subsequent parts. For this we can use the information in Table 7.2 as a general guideline.

A useful step when first starting out the process of recreating timbres is to try and emulate the sound's properties with your voice and mouth and record the results. As madcap as this probably sounds (and to those around you while you try it), not only will you have a physical record of the sound, but the movements of the mouth and tongue can often help you determine much about the sound you're trying to construct. The expansion and contraction of the mouth's muscles can often be related to the filter, while movement of the tongue can often be related to the LFO augmenting a parameter. In this instance, the sound is similar to saying 'wwOOOwww' (listen to the timbre and replicate it with your voice and mouth and this will make a little more sense).

From this we can determine that there is some filter augmentation in the timbre since it opens and closes and that, when it does, it allows higher harmonics through as it sweeps. This means that it's using a low-pass filter, but as there is no evidence of the characteristics of using an LFO waveform to modulate the timbre (for instance, a saw waveform would be evident by opening slowly and closing suddenly, etc.) it must be the filter envelope that's modulating the filter's cutoff. What's more, it's quite easy to determine that the filter envelope is being applied positively rather than negatively, otherwise it would sound entirely different.

Table 7.2 Guidelines for breaking down a sound

General sound	Technical term	Synthesizer's parameter
Type of sound	Harmonic content	Oscillator's waveforms
Brightness	Amplitude of harmonics	Filter cut-off and resonance
Timbre changes over time	Dynamic filtering	Filter's envelope
Volume changes over time	Dynamic amplitude	Amplifier envelope
Pitch	Frequency	Oscillator's pitch
Sound has a cyclic variation	LFO modulation	LFO waveform, depth and rate
Tremolo (cyclic variation in volume)	Amplitude modulation	LFO augments the amplifier
Vibrato (cyclic variation in pitch)	Pitch modulation	LFO augments the pitch
Sound is percussive	Transient	Fast attack and decay on the amplifier
Sound starts immediately or fades in	Attack time	Lengthen or shorten the attack and decay stages
Sound stops immediately or fades out	Release time	Lengthen or shorten the release on the amplifier
Sound gradually grows 'richer' in harmonics	Filter automation	Programmed CC messages or a slow rate LFO augmenting the filter cut-off

Therefore, we've determined so far that:

- The filter is augmented with an envelope and it's positive.
- The filter is a low pass.

Next, listening to the sound's overall timbre, it's obviously quite rich in harmonics, all of which change with the movements of the filter, so there is no sine wave present. Additionally, the timbre doesn't exhibit a particularly hollow character or the bright 'fizzy' harmonics related with noise, so it's safe to assume that there is no square or noise waveform present either. This (generally) leaves two waveform options: a saw or a triangle. Finally, as we can hear the filter moving in and out at key-press, and that it most definitely has a pluck to it, we can also determine that the amp envelope is using a fast attack with a long decay, little sustain, if any, and a short release.

Thus, we have:

- The timbre consists of either a saw or triangle wave, or both.
- The amplifier envelope utilizes a fast attack, a long decay, perhaps a small sustain and a short release.
- The filter is augmented with an envelope and it's positive.
- The filter is a low pass.

At this point, it's worth inputting these parameters into a synthesizer and experimenting with the oscillators and envelopes to see how close to the timbre you can get with the parameters you've theorized thus far. If you're not pitch perfect it may first be worth setting the synth to a single sine oscillator with the aforementioned envelope settings and finding the key that the riff is in first, as this will help things along no end (as a shortcut it's at C2). Further experimentation should also reveal that the sound consists of both a saw and a triangle wave that are detuned from each other to make the timbre appear as wide as it does (the saw is transposed down an octave from the triangle).

Now we have the basic timbre and amplifier settings, we can listen back to the sound again. The sound isn't particularly resonant, so the resonance must be set quite low. Also, the filter is sweeping the harmonic content quite rapidly, so the filter must be 24 rather than 12 dB. What's more, listening to the way that the filter's envelope augments the filter's cut-off we can determine that the envelope is using a long attack and decay with a medium sustain and release.

The final result is that to recreate the timbre two oscillators are used, a sine and a triangle wave, with the saw wave transposed down by an octave. The amp envelope uses a fast attack with a long decay, no sustain and a fast release, while the filter envelope has a longish attack and decay with a medium sustain and release.

General programming tips

- Never bang away at a single key on the keyboard while programming, it will not give you the full impression of the patch. Always play the motif to the synth before any programming.
- Ears become accustomed to sounds very quickly and an unchanging sound can quickly become tedious and tiresome. Consequently, it's prudent to introduce sonic variation into long timbres through the use of envelopes or LFOs augmenting the pitch or filters.
- Generally, the simpler the motif, the more movement the sound should exhibit. So for basses and motifs that are quite simple assign the velocity to the filter cut-off, so the harder the key is hit the brighter the sound becomes.
- Don't underestimate the uses of keyboard tracking. When activated this can breathe new life into motifs that move up or down the range, as the filter's cut-off will change accordingly.
- Although all synthesizers share the same parameters, they do not all sound the same. Simply copying the patch from one synth to another can produce totally different results.
- Although the noise oscillator can seem useless in a synth when compared to the saw, sine, triangle and square waves, it happens to be one of the most important oscillators when producing timbres for dance and is used for everything from trance leads to hi-hats.
- To learn more about the character of the synthesizers you use, dial up a patch you don't like, strip it down to the oscillators and then rebuild it using different modulation options.
- If you have no idea of what sound you require, set every synth parameter to maximum and then begin lowering each parameter to sculpt the sound into something you like.
- Many bass sounds may become lost in the mix due to the lack of a sharp transient. In this instance, synthesize a click or use a woodblock timbre or similar to enhance the transient.
- The best sounds are created from just one or occasionally two oscillators modulated with no more than three envelopes: the filter, the amplifier and the pitch. Try not to over-complicate matters by using all the oscillators and modulation options at your disposal.
- Try layering two sounds together but use different amplifier and filter settings on both. For example, using a slow attack but quick decay and release on one timbre and a fast attack and slow decay and release on another will '*hocket*' the sounds together to produce interesting textures.
- Never underestimate the uses of an LFO. A triangle wave set to modulate the filter cut-off on a sound can breathe new life into a dreary timbre.

- Remember that we determine a huge amount of information about a sound from the initial transient. Thus, if you replace the attack of a timbre with the attack portion of another it can create interesting timbres. For instance, try replacing the attack stage of a pad or string with a guitar's pluck.
- For more interesting transients, layer two oscillators together and at the beginning of the note use the pitch envelope to pitch the first oscillator up and the second one down (i.e. using positive and negative pitch modulation).
- For sounds that have a long release stage, set the filter's envelope attack longer than the attack and decay stage, but set the release slightly shorter than the amp's release stage. After this, send the envelope fully to the filter so that as the sound dies away the filter begins to open.

In the end programming good timbres comes from a mix of experience, experimentation and serendipity. What's more, all of the preceding examples are just that – examples to start you on the path towards sound design – and you should be willing to push the envelope (pun intended) much further. Try adding an extra oscillator, changing the filters to high pass or band select, adjust the amp and/or filter envelopes and try changing the LFO's modulation source, destination and/or waveform. The more time you set aside to experiment, the more you'll begin to understand not only the characteristics of your synthesizer, but also the effects each controller can impart on a sound. This will ultimately be time well spent, as professionally programmed sounds will make the difference between a great dance track and an average one.

Creative sampling and effects

While many sounds used in dance production are acquired through synthesis, the use of samples, samplers and effects shouldn't be undervalued either, and it's often a mix of creative synthesis, sampling and effects that creates a good mix. For example, while hip-hop, big beat and lo-fi may make use of some of the timbres used in the creation of trance, techno and house, etc., they do not sound as clean and it's common practice to make them sound as though they've been ripped off another record, or have been through some of the worst effects processors available.

As with much of music, how you accomplish any of this is entirely down to your own creativity and it's impossible to offer advice on all the options that are available through the use of samplers and effects. Consequently, what follows are only some general ideas to get you started, but in the end it's worth experimenting and coming up with your own creative slant, as this is what will pick you out from the crowd.

Creative sampling

Although cheap microphones are worth avoiding if you're after recording good vocals, they do have other creative uses. Simply plugging one into a sampler and striking the top of the mix with a newspaper in your hand can be used as the starting point of kick drums. Additionally, scratching the top of the microphone can be used as the starting point to Guiros. You should also listen out for sounds in the real world to sample and contort – FSOL have made most of their music using only samples of the real world. Todd Terry acquired snare samples by bouncing a golf ball off a wall and Mark Moore used an aerosol as an open hi-hat sound. Hitting a plastic bin with a wet newspaper can be used as a thick slurring kick drum and scraping a key down the strings of a piano or guitar can be used as the basis for whirring string effects (it worked for

Dr Who's TARDIS anyway). Once sampled, these sounds can be pitched up or down, or effected as you see fit. Even subtle pitch changes can produce effective results. For example, pitching up an analogue snare by just a few semitones results in the snare used on drum 'n' bass, while pitching it down gives you the snare typical of lo-fi.

Sample reversing

This is probably the most immediate effect to try, but while it's incredibly simple to implement it can produce great results. The simplest use of this is to sample a cymbal hit and then reverse it in the sampler (or wave editor) to produce a reverse cymbal that can be used to signify the introduction of a new instrument. More creative options appear when you consider that reversing any sound with a fast attack but long decay or release will create a sound with long attack and an immediate release.

For example, a guitar pluck can be reversed and mixed with a lead that isn't. The two attack stages will meet if they're placed together and these can be cross-faded together. Alternatively, the attack of the guitar could be removed so that the timbre begins with a reverse guitar, which then moves into the sharp attack phase of a lead sound or pad. On the other hand, if a timbre is recorded in stereo, the left channel could be reversed while the right channel stays as it is to produce a mix of the two timbres. You could then sum these to mono and apply EQ, filters or effects to shape the sound further.

Pitch-shifting

A popular method used by dance producers is to pitch-shift a vocal phrase, pad or lead sound by a fifth, as this can often create impressive harmonies. If pads or leads are shifted further than this and mixed in with the original it can introduce a pleasant phasing effect. Despite the advice from some, however, this will not work well on drum loops. Keep in mind that these form a crucial part of the track and should be as dynamic as possible.

Time-stretching

Time-stretching adjusts the length of a sample while also leaving the pitch unchanged. It does this by cutting or adding samples at various intervals during the course of the sample so that it reaches the desired length, while, to a certain extent, smoothing out the side-effects of this process on the quality and timbre of the sound. This is a complex, processor-intensive process and is not usually suitable for extreme stretching. For instance, stretching a 67 BPM loop into 150 BPM introduces unpleasant digital noise into sounds, but this isn't something that you should always want to avoid. Continually stretching and shortening loops, vocals, motifs or even single notes will introduce more and more digital noise, which is great for dirtying up sounds for use in hip-hop, lo-fi and big beat. If a timbre is programmed on a synthesizer, sampled and stretched numerous times, the resulting noisy timbre can be sampled and used to play the motif.

Perceptual encoding

Similarly, any perceptual encoding devices (such as MiniDisc) can be used to compress and mangle loops further. Fundamentally, these work by analysing the incoming data and removing anything that the device deems irrelevant. In other words, data representing sounds that are considered to be inaudible in the presence of the other elements are removed and this can sometimes be beneficial on loops.

Physical flanging

Most flanger effects are designed to have a low noise floor, which isn't much use if you need a dirty flanging effect for use in some genres of dance. This can, however, be created if you own an old analogue cassette recorder. If you record the sound to cassette and apply a small amount of pressure on the drive spool, the sound will begin to flange in a dirty, uncontrollable manner.

Gritty sounds

Most dance musicians will not only use the latest samplers, but they'll also own a couple of old samplers. Due to the low bit and sample rate quality of these samples, any sounds that are recorded using them take on a gritty, dirty nature. This same effect can be accomplished in most of today's samplers and wave editors by reducing the bit and sample rates of the audio. This type of effect is especially used in hip-hop and lo-fi to create the gritty drum loops.

Transient slicing

A popular technique for house tracks is to design a pad with a slow resonant filter sweep and then sample the results. Once in a sampler, the slow attack phase of the pad is cut off so that the sound begins suddenly and sharply. As the transient is the most important part of the timbre, this creates an interesting side-effect that can be particularly striking when placed in a mix.

Adding grunge effects

Most dance artists will also own a number of guitar effects pedals, ranging from phasers to flangers to distortion and compressors. As these are designed for live use they have a particularly high noise floor and are perfect for dirtying up loops, riffs and vocals.

Reverb

Although reverb is used to provide a sense of space around a timbre, it also has more creative uses. From the previous chapter, we know that the amount of reverb associated with a signal determines how far away it is from us, but the further these reverberations have to travel through air the more the high-frequency content is reduced. Thus, by patching a graphic EQ into the reverb's output and reducing the higher frequencies, we can make a sound appear even further away. In a more creative context, if this EQ was automated so that it reduces and then accentuates the high frequencies in time with the music, the sound can appear to move forwards and backwards in the mix, in time with the tempo.

Slurring gates

We've already seen one use of the noise gate in a more creative context, but another use is to set the threshold very low but use a very long attack setting; any sounds entering the gate that are above the threshold will be introduced slowly. If you time the attack correctly, it's possible to have the timbre fade in so that it only reaches its maximum level towards the end of its note. This can be used to bring attention to the end of the note rather than the beginning and can be particularly useful for creating sounds that 'slur' in and then end abruptly.

Delay

Lo-fi, techno and hip-hop often make use of delay due to the sparse elements in the mix, but rather than use a delay effects unit to accomplish this, it's often better to design your own delays. This can be accomplished in any audio-capable sequencer by copying the original audio track to a number of other audio tracks. From here the delay tracks can be moved a quarter note (or however periodic you need the delay to be) to the right of the timeline so that they appear later in time. Each 'delay' can then be edited, shortened, equalized or effected as you see fit. A good example of this is to use a mixer's automation to continually lower the volume of the delayed track so that they fade away or alternatively so that they increase in volume. By automating EQ, each consecutive delay can become brighter or darker, or by automating effects such as distortion, each successive delay can become more distorted. What's more, using this technique each delay could be panned around the stereo image.

8 Recording vocals

I recorded one girl down the phone from America straight to my Walkman and we used it on the album. If we'd taken her into the studio it would have sounded too clean.

Mark Moore

Before we move onto creating the various genres of dance, we need to look at one final, yet sometimes vital, element in the chain – the vocals. Although originally vocals played a particularly small role in the dance music scene, over the years the vocal content has grown significantly to the point that, unless you can gain permission to use another artist's vocal performance, you need to be proficient at recording them.

This is much more difficult than it first appears and there's certainly much more to it than simply plugging in a mic, pressing record and hoping for the best. In fact, if you take this approach hope is about the closest you'll get because you certainly won't record a good performance. We've been exposed to the human voice since the day we were born, so we instinctively know how it's supposed to sound. As a result, any poor recording equipment or techniques will be immediately noticeable and, as the vocals can form the centrepiece of a track, if they're wrong, the rest of the track, no matter how well produced, will appear just as bad.

To record vocals proficiently requires not only a good vocalist, but also a good understanding of the effects the surrounding environment can have on a recording and, more importantly, the technology used to record them. This latter knowledge can be particularly significant, since while it would be easy to simply say which microphone you should use and where to place it, with a good understanding of the equipment involved you can make an informed decision yourself as to which microphone and positioning is best suited for the current project.

Microphone technology

Starting with the basics, a microphone simply converts sound into an electrical current that is then transformed into an audio signal at the end of the chain. As straightforward as this appears, though, there are different ways of accomplishing it and the way sound is captured determines the quality of the overall results. Of course, all quality microphones will invariably produce a good sound, but the tonal quality between each is very different so it is important to choose the right microphone for the vocalist and genre of music you produce.

Despite the number of different microphones available, ultimately for vocals, there are only two real choices: electromagnetic or electrostatic. Both of these use the principle of a moving diaphragm to capture sound, but they use different methods for converting it into an electrical signal and so provide entirely different tonality from one another.

Possibly the most instantly recognizable of these is the electromagnetic, which is also referred to as a dynamic microphone. They're always used in live situations and music TV programmes, and are best described as a small stick with a gauze ball sat on top.

Figure 8.1 A dynamic microphone.

Dynamic microphones work under the principle of a moving coil, which is very similar to how a typical loudspeaker operates (although loudspeakers operate in reverse). The microphones consist of a very thin plastic film, known as the diaphragm, which is coupled to a coil of wire that's suspended in a magnetic field. When any sound strikes this diaphragm it begins to vibrate sympathetically, which causes the coil to vibrate too. As this coil vibrates in the magnetic field it creates an alternating current, which the mic then outputs and is subsequently converted into sound.

Generally, this type of assembly is particularly hard wearing and is the reason why it's so often used in live performances. If the roadies drop the mic, there's a good chance that it will still work. On the downside, though, since the vibrations have to move a relatively heavy coil assembly to produce the sound, they're not very sensitive to changes in air pressure and so are relatively slow to respond to transients. Consequently, the higher frequencies are not captured reliably so sounds recorded using them will often exhibit a 'nasal' quality. While this generally makes them unsuitable for recording the delicate vocals that are required for trip-hop and trance, they are perfectly suited towards rap as the 'nasal' sound provides more mid-range frequencies, helping the sound to remain upfront and 'in your face', an effect that can sometimes be vital for the genre.[1]

---

[1] Dynamic microphones are sometimes referred to as cardioid microphones, because if you plot the frequency response on paper it looks very similar to the shape of a heart.

It's because of this nasal quality that many artists prefer to use electrostatic (otherwise known as capacitor) microphones. These can be seen in use in all professional studios and are best described as 'flat looking' with angled gauze at either side. These are much more sensitive than electromagnetic microphones as the design doesn't rely on generating a signal by moving a coil in a magnetic field, but is based on capturing sound using a varying capacitance. This means that the diaphragm can be much lighter and so captures sound much more accurately.

Figure 8.2 A condenser microphone.

Typically, in these microphones the diaphragm consists of a gold-plated light Mylar plastic, which is separated from a back plate by a gap of a few microns. A small electrical charge is imposed onto the back plate or the diaphragm (depending on the model), which creates a capacitance charge between the two. When sound strikes the diaphragm the distance between this and the back plate varies, which in turn creates a fluctuation in the capacitance, resulting in small changes in the electrical current. These changes in the current produce the audio signal at the end of the chain. While this means that less air pressure is required to create the signal (and so produces a more accurate response), it is important to note that different capacitor microphones will use differently sized diaphragms and this also has a direct effect on the frequency response.

A larger diaphragm will obviously have a heavier mass and therefore will react relatively slowly to changes in air pressure when compared to that of a smaller diaphragm. Indeed, due to the smaller overall mass of these, they respond faster to changes in air pressure, which results in a sharper and more defined sound. From a theoretical point of view, this would mean that to capture a perfect performance you would be better using a small-diaphragm microphone since it would produce more accurate results, but in practice this isn't always required. To better understand this, we need to examine the relationship between the wavelength of a sound and the size of the diaphragm.

Typically, large-diaphragm microphones will have a diameter of approximately 25 mm and if this is exposed to a wavelength of equal size (*10 kHz has a wavelength of 25 mm*) there is a culmination of frequencies at this point. Effectively, this results in more directional sensitivity at higher frequencies. If we compare this reaction to a smaller diaphragm mic that has a 12 mm diameter, this build-up would occur much higher in the frequency range (*20 kHz has a wavelength of 12.5 mm*) and is therefore not noticeable to the human ear.

While this means that a larger diaphragm mic doesn't have a frequency response as accurate or as flat as a smaller diaphragm, the subsequent coloration results in a rounder smoother timbre that pulls through a mix much more readily. What's more, a large-diaphragm microphone will also have a lower noise floor than a smaller diaphragm because it has a larger surface area to conduct so the electrons can distribute more readily.

> Note: If you decide to use a condenser microphone rather than electromagnetic, an electrical charge is required to provide the capacitance between the diaphragm and back plate. This can be provided by batteries, phantom power or it may even already have a charge retained in the diaphragm or back plate from the manufacturer. In most instances, though, this charge is 48 V phantom power, which is received through the microphone cable itself. The term phantom power is used since you can't 'physically' see where it receives its power from and is commonly supplied from a mixing desk or a microphone pre-amplifier.

Whichever microphone you choose, since the signal produced is incredibly small (−60 dBu), it needs to be amplified to an appropriate volume for recording using a pre-amp. These are relatively simple in design and will perform just two functions: supplying the microphone with power, if required, and amplify the incoming signal. Unsurprisingly, there is a huge range of pre-amps to choose from; they can range in price from £100 to £6000 and above, depending on the design and parameters they have on offer. If at all possible, though, you should aim for a pre-amp that uses valves in its design, as these add warmth to the signal that's archetypal of all genres of dance, but if this isn't feasible (due to the higher price tag) then you can record using a solid-state design and feed the results into a valve compressor or valve emulation plug-in.

> Note: In an ideal situation you should use a dedicated microphone pre-amp rather than one supplied in a mixing desk, since the circuitry in a pre-amp is specifically manufactured for the purpose rather than an 'added extra' as it is in most mixing desks (which as a consequence are prone to a much higher noise floor).

Recording preparation

While it is important to have a good mic and pre-amp combination, the most common problem doesn't actually come from the equipment used but from a poor vocal technique. Indeed, as long as the equipment isn't truly 'bargain basement' and the microphone is placed thoughtfully, it's perfectly possible to attain respectable results. The only real secret to capturing a great performance is to ensure you that have nothing but a great performance going into the microphone. If you settle for anything less at this stage, no amount of effects or editing you apply later are going to make it sound any better.

With this in mind, perhaps the most obvious start is to ensure that the vocalist has attended at least six to eight singing lessons. Anyone can sing, but not everyone can sing well! While very few vocalists will admit that they need any form of voice training – their gift is natural of course – every professional vocalist on the circuit has attended lessons with a voice trainer, singing teacher, or both. Which lessons they should attend will depend on the vocalist, but if they have trouble singing in the correct key, breathing properly or holding notes then they should see a voice trainer. If, however, they can sing fairly well, then they may benefit from attending lessons with a singing teacher, as they'll teach them how to perform a song with depth, emotion and feeling.

It's doubtful that you'll be able to find a teacher who specializes in house or rap or any other dance-related genres (in fact, you may have difficulty finding one who has actually ever heard of them!), as most will specialize in classical performances or musicals, but the genre is unimportant. Simply attending singing lessons will help the vocalist learn the techniques every professional vocalist will use and this can then be applied to any form of music.

Provided that the vocalist is capable of delivering a good performance (which is much more difficult than it sounds), the location can also have a significant effect on the quality of the end result. No one is going to perform well in a clinical environment with bright fluorescent lighting. Apart from the fact that fluorescent lighting is prone to introducing hum into a recording, it's difficult for a performer to concentrate in such a 'scientific' environment. Ask yourself if you could perform your best if you were sat in a dentist's chair. Instead, vocalists produce a much more energetic and passionate performance if the room has a welcoming feel and this means using bulbs that give off a warm glow.

Note: The temperature of the room is also important and ideally you should look towards maintaining a temperature of around 22°C (approximately 72°F). Any colder and the performers vocal chords may tense up.

If the environment is suitable, the next consideration is the foldback mix. Essentially, this is the *complete* mix plus the vocalist's microphone signal returned to a pair of headphones for the performer to perform along to. It is vital that this foldback mix is complete and not simply a 'rough draft', though, since if you play an unfinished or badly mixed track back for them to perform to, they may find it difficult to determine the pitch, the vocal tone they should use or move closer/ further away from the microphone so that they can balance their voice with the current mix. If, on the other hand, the track has been well mixed and engineered, the performer will not only find it much easier to pitch, but they will also want to make an extra effort to match the quality of what's already down.

Usually, the foldback mix is fed from tape, MiniDisc, DAT or CD to headphones with the computer or dedicated hard disk recording the resulting vocals, but whatever the configuration, the balance between the instruments and the performer's voice is critical. More often than not, a mix that is either too loud or quiet in respect to the microphone's output is responsible for poor timing, dynamics and intonation, so you should spend plenty of time ensuring that both mix and vocals are at the appropriate volume.

Note: The headphones used to play the foldback mix should offer plenty of isolation to prevent any spill encroaching onto the vocal recording. Consequently, open headphones are not suitable as they leak sound, but equally closed headphones aren't suitable either. Even though these will not leak any sound, the closed design surrounds the ears and creates a rise in bass frequencies. Because of this, headphones with a semi-open design are the best, but you should also ensure that they are tight enough on the head so that they don't fall off!

Some professional vocalists will not want to listen to the full mix and instead prefer to listen to just the drum track with a single pitched instrument sat on top, as this can help them to keep better timing and intonation. Always take the effort to check exactly what the vocalist wants and no matter how difficult some of their requests or suggestions may be, try to comply. If they are happy, they're more likely to produce a great performance, but if the vocalist doesn't ask for reverb on the foldback vocals, apply it anyway.

Reverb can play a key role in helping a singer to pitch correctly, plus it also invariably makes them sound much better, which in turn helps them to feel better about their own voice. This only needs to be lightly applied and, typically, a square room setting with a short tail of 0.50 ms is usually enough. Alongside this reverb, some engineers also choose to compress the vocal track for the performer's foldback, although whether this is a good idea or not is under constant debate. From a personal point of view, I find that an uncompressed foldback mix tends to persuade the vocalist to control their own dynamics, which invariably sounds much more natural than having a compressor do it.

Recording techniques

Typically, for most genres of dance, with the possible exception of rap, the preferred choice of microphone is a condenser with a large diaphragm such as the AKG C3000. This is generally accepted as giving the best overall warm response that's characteristic of most genres, but any large diaphragm will suffice provided that the frequency response is not too flat. That said, some large-diaphragm models will deliberately employ peaks in the mid range to make vocals appear clearer, but this can sometimes accentuate sibilance with some female vocalists.

Note: Sibilance is the result of overemphasized 'sh' sounds, which can create unwanted hissing in the vocal track.

Additionally, some mics may also roll off any frequencies below 170 Hz to prevent the microphone from picking up any rumble that may work it's way up the stand, but rolling off this high can sometimes result in the vocals appearing too weak or 'thin', which can be difficult to widen later. It's therefore preferable to use a microphone that doesn't roll off above 120–150 Hz and use it in conjunction with a quality pre-amp. Usually, these employ a filter in the signal path, allowing you to manually define the frequency to roll off.

The positioning of the microphone depends entirely on the style of performance and the performer, but generally it should be positioned away from any walls but not in the centre of a room. Any condenser microphone picks up both the direct sound from the vocalist and the surrounding ambience of the room. Thus, if the microphone is positioned in the direct centre of the room or too close to a wall the reflections can create phase cancellations, making the vocals appear thin, phased or too heavy. It should also be mounted on a fixed stand and situated so that the bottom of the housing is level with the performer's nose (with their back straight!). This will ensure a good posture while also helping to avoid heavy plosive popping or breath noises.

Plosive popping is perhaps the most common problem when recording vocals and is a result of the high-velocity bursts of air that are created when we say or sing words containing the letters 'p', 'b' or 't'. These create short, sharp bursts of air that force the microphone's diaphragm to its extremes of movement, which results in a popping sound. These can be incredibly difficult to remove at a later stage, so you need to remove the possibility of them appearing during the recording by using a 'pop shield' positioned between the vocalist and mic. These consist of a circular frame approximately 6–8 inches in diameter covered with a fine nylon mesh and are available as free-standing units or with a short gooseneck for attachment to the microphone stand. By placing these approximately 5 centimetres (about 2 inches) from the microphone, any sudden blasts of air are diffused by the pop shield, preventing them from reaching the diaphragm.

> Note: Though pop shields are available commercially, many artists make their own from metal coat hangers or embroiderers' hoops and their girlfriends/wives' nylon stockings. Provided that the frame is large enough to act as a pop shield, the nylon can be stretched over the ring, which can then be held in position next to the microphone with some stiff wire.

If you are experiencing constant plosive problems, it is better to attend a series of lessons with a professional singing teacher. This will not only help in learning how to reduce the emphasis on plosive sounds, but also contribute to preventing any sibilance. This latter effect can sometimes be removed by placing a thin EQ cut somewhere between 5 and 7 kHz or by using a professional de-esser. These are frequency-dependent compressors that operate at frequencies between 2 and 16 kHz. Using these, it's possible to boost the EQ on vocals to make them appear brighter and the de-esser will jump in to prevent them from becoming too harsh.

As for the distance between mouth and microphone, ideally it should be placed approximately 20–30 centimetres (8–12 inches) away from the mouth of the vocalist. At this distance, the voice tends to sound more natural, plus any small movements on the vocalist's behalf will not create any serious changes in the recording level. Naturally, keeping this microphone-to-mouth distance can be difficult to achieve, especially if the vocalist is used to performing in live conditions or is a little overexcited.

In these situations, vocalists tend to edge ever closer to the microphone as they continue the performance and if they get too close it will not only boost the lower frequencies, but also create deposits of condensation on the diaphragm, which reduces its sensitivity. Since microphones are high-impedance devices, using them in a cold room or having the vocalist stand too close to them can introduce condensation onto the diaphragm. If this happens, the sensitivity of the mic will fall sharply and crackles or pops can start to appear in the recording. Subsequently, it's prudent to ensure that any capacitor mics are at room temperature before they're used, and it's also worthwhile making sure that the room isn't too cold. If condensation does form on the capsule, the only solution is to put them somewhere warm for a few hours.

Obviously, prevention is always better than cure, so to prevent the vocalist from getting too close to the mic it's sensible to position a pop shield 10 centimetres (approximately 4 inches) away from the microphone. This is the preferred approach to placing the microphone above the singer's mouth and angling the mic downwards. While some engineers believe that this approach ensures a 'fixed' distance from the mouth to mic while also helping vocalists straighten their back to improve breathing and projection, if vocalists are forced to raise their chins, the throat becomes tense, which can severely affect the performance.

If, however, the genre demands a more upfront vocal sound that's typical of commercial pop (also trance and house), then it's worthwhile reducing the pop shield-to-mic distance to just 2 inches and have performers almost pressing their lips to the shield while singing. You may have to reinforce the pop shield with this approach by stretching another stocking over it, but this close positioning will take advantage of the proximity effect. This creates an artificial rise in the bass frequencies that can make the vocals appear warmer, more upfront and in your face. Occasionally, this approach can also increase the chance of plosives, but if this happens, turning the microphone slightly off-axis from the performer can help to prevent them from occurring.

Once the mic has been set up and positioned correctly, it's prudent to insert a compressor between the pre-amp and recording device. Some pre-amps have a built-in compressor dedicated for this function, but if not you will need to use an additional hardware unit to control the dynamics. The human voice has an incredibly wide range of dynamics and these will need to be put under some control to prevent any prominent parts clipping the recording.

Notably, some professional vocalists will use their own form of compression by backing away from the microphone on louder sections and moving closer in quieter parts but, nonetheless, even in these circumstances it is still prudent to use a compressor just in case. A single uncontrolled clip can easily destroy an otherwise great performance! This must be set carefully, however, as once recorded you cannot undo the results later, so you should set the compressor to squash only the very loudest parts of the signal.

Ideally, the compressor should be analogue and employ valves, as the second-order harmonic distortion they introduce is typical of the vocal sound of the genres, but if this is not possible, then you can use a solid-state compressor and compress them after recording with a valve emulation plug-in. A good starting point for recording compression is to set a threshold of −12 dB with a 3:1 ratio and a fast attack and moderately fast release. Then, once the vocalist has started to practise, reduce the threshold so that the reduction meters are only lit on the strongest part of the performance.

These compression settings can also be applied to rap music, but generally this genre benefits more from using a dynamic microphone such as the Shure SM58 to capture the performance. Also, rather than mount the mic on a stand, better results can occur if the vocalist holds the mic. This not only captures the slight nasal quality synonymous to most rap music, but also allows the rapper to move freely. Much of rap music relies on the different vocal tones that are produced by rappers changing their position from upright to bowing down, in effect 'compressing' their abdomen to produce lower tones. If this technique is used, though, ensure that vocalists keep their hands clear of the top of the microphone, since obstructing this will severely change the tonal character on the vocals. Also, try to ensure that the microphone is not held too far away from the mouth. Most dynamic microphones will roll off all frequencies below 150 Hz to prevent the proximity effect, but if the vocalist has it too far away then this can result in a severe reduction in bass frequencies. In general, a dynamic microphone should be approximately 10–20 centimetres (approximately 4–8 inches) away from the mouth.

This is the usual distance that most vocalists will use naturally, but if you have a loud foldback mix with a bass cut they'll tend to hold it closer to their mouths to compensate for the lack in mix presence, so it's vital that the mix is bass accurate before adding vocals. It's also worth noting that, due to the design, these mics have a diminished frequency response and can be quite easy to overload if placed too close to the mouth. This can result in light distortion in the higher and middle frequencies, so to prevent this it may be prudent to connect an EQ unit into the side chain of the vocal compressor to create a frequency selective compressor which can control the problem frequencies.

With the equipment set up to record, the vocalist/rapper will, or should, need an hour to warm up their voice to peak performance. Most professional vocalists will have their own routine to warm up and many prefer to do this in private, so it may be necessary to arrange some private space for them. If they are inexperienced they may insist that they do not need to warm up and in this instance you should charm them into it by asking them to sing a number of popular songs that they enjoy while you 'set up the recording levels …'.

During this time, you should listen out for any signs that the performer has poor intonation or is straining and, if so, ask them to take a break and relax. If they need a drink, fruit juices are better than water, tea or coffee as these can dry the throat, but you should avoid any citrus drinks. From personal experience, two teaspoons of honey with one teaspoon of glycerine mixed with 100 millilitres of warm water provides excellent lubrication and helps to keep the throat moist, but persuading the vocalist to drink it can be another matter altogether. Alternatively, Sprite (the soft drink) seems to help, but avoid letting them drink too much as the sugar can settle at the back of the throat. Failing that, sucking lozenges such as Lockets with honey before the performance can help with the lubrication.

Once the vocalist has warmed up, their voice should be at peak performance and they should be able to achieve depth and fullness of sound. It's at this point that you want them to belt out the track with as much emotion as possible, from beginning to end. As difficult as it may be, you should avoid stopping the vocalist at any mistakes apart from the most severe intonation problems and ask them to continue the song despite any mistakes, as you'll be recording more takes later.

Note: Unequivocally, there is no music today that is completed in just one pass and even the most professional artists will make three of four passes, so keep things light and friendly and do not put any pressure on the performer to get the take right in one.

It's also worthwhile ensuring that the vocalist knows the lines off hand without having to read them off a sheet of paper. Encourage them to learn the lines and, if they forget them, ask them to have fun and improvise. The occasional 'lah' or 'ooh' can help to enhance the performance and may actually sound better than the original line, plus if they know the words they are likely to sound more energetic and lively than a droll reading. More importantly, though, lyric sheets can have an adverse effect on the recording as the vocals will reflect from the paper, which can create a slight phase in the recording. This has often been responsible for numerous scratched heads as the engineer tries to figure out why the vocals sound so strange.

Note: In some situations the performer may insist they have lyric sheets to hand. If this is the case, position the lyric sheets behind the microphone and slightly off-axis, so that the possibility of phase problems is reduced.

Once this first performance has been recorded, listen back to the recording and try to identify any problems, such as very poor intonation, the inability to hold high notes, uneasiness and tenseness, phasing effects or too much bass presence. If the microphone has been placed correctly, there shouldn't be any phasing or bass problems, but if there are, then you'll need to reconsider the microphone's placement. If there is too much bass presence, ask the performer to move further away from the microphone, while if there are phasing problems, move the microphone closer, or further away from the wall.

Severe intonation problems, the inability to hold high notes, uneasiness or tenseness, on the other hand, are a little more difficult to resolve, as they reside in the psychological state of the performer and no amount of equipment, no matter how good, will turn a poor or uninspiring performance into a great one. Instead, you'll need to look at the monitoring environment and your own communication skills.

Assuming that the vocalist can actually sing in tune, the most common reason for poor intonation comes from the performer's monitoring environment or a poor overall mix. As previously touched upon, they will most probably monitor the mix through headphones and this can result in an increase in bass frequencies or difficulty in comprehending the psychological acoustic coupling. If the headphones are tight fitting on the performer's head there will be an increase in bass frequencies and it is difficult to perceive pitch in low-frequency waveforms, which can force the performer to lose intonation. This problem can usually be identified by the vocalist 'involuntarily' moving further away from the microphone and singing louder, as this reduces the bass presence in their voice on the feedback mix. To prevent this, it's worth placing a wide cut of a few decibels at 100–300 Hz (where the bass usually resides) or increasing the volume of a higher pitched instrument that the vocalist can pitch to.

Psychological acoustic coupling is less of a problem with experienced performers, but is a result of the singer's vocal chords vibrating their ear drums along with all the bones in the ear cavity – the reason why you sound different when you listen to a recording of yourself. When you place a pair of headphones on the vocalist, their 'natural' environment is upset, which can make it difficult for them to pitch correctly. In this instance, it's worth muting the vocals returned in the foldback mix to see if this makes a difference, but if not it may be worth muting one side of the headphone feed so that they can hear themselves naturally in one ear. Some of the higher priced mixing desks and headphone distribution amps offer a kill switch to terminate one of the headphones' stereo channels for this purpose, but if this option isn't available you'll need to purchase a single-channel headphone, manufacture a mono converter or ask the vocalist to move one of the phones to uncover their ear.

Of course, if the monitoring volume is quite high this latter solution can result in the foldback mix spilling out of the unused earpiece and working its way onto the recording. However, as long as this isn't ridiculously loud then a little overspill isn't too much of a problem, as it will be masked by the other instruments. Indeed, even professional artists such as Moby, Britney Spears, Madonna and Michael Jackson have overspill present on some of their individual vocal tracks, and it's something you'll undoubtedly have to deal with if you have the chance to remix a professional artist.

> Note: Some vocalists simply can't perform wearing headphones, so you need a way of playing the mix back without recording any spill. The best way to accomplish this is to place both monitors 90 centimetres apart and 90 centimetres from the rear of the microphone. Following this, reverse the wires (i.e. the polarity) of the left speaker and then feed both monitors with a mono signal. This will allow the performer to hear the mix, but the phase of the speakers will match at the microphone, in effect cancelling out the sound.

If the problem is an inability to hold high notes or a tense or uneasy voice, then this is the result of a nervous performer not breathing deeply enough, and can only be resolved through tact, plenty of encouragement and excellent communication skills. A vocalist feels more exposed than anyone else in a studio, so they constantly require plenty of positive support and plenty of compliments after they've completed a take. This also means that eye contact throughout is absolutely essential, but refrain from staring at the vocalist with a blank expression – try to keep an encouraging smile on your face. This can give a vocalist a massive confidence boost that will pull through into the recording. More importantly, though, do not swamp the vocalist with techno babble, or try to impress them by giving them a technical run through of what you plan to do. If the vocalist has too many technical comments to think about it can ruin the performance.

Editing vocals

Once you have the first 'tonally' correct, emotional charged take recorded, there will undoubtedly be a few mistakes throughout that will need repairing or replacing. At this stage, before you send the vocalist home, you should ask them to perform the 'dodgy' parts again. It isn't advisable to ask them to perform the whole take again, as it may exhaust their energy before they reach the part that needs to be replaced, so it's a better idea to punch in and out on the parts that need it.

Essentially this means that you need to make it clear to the performer which parts you want to re-record and these should be entire lines, not single words. Also, the vocalist should perform the lines before and after the faulty part to prevent a cold start and ensure that the same breathing pattern as the original take is more or less maintained.

Above all, remember that very, very few vocalists today can perform a perfect take in one go and it certainly isn't unusual for a producer to record as many as eight takes of different parts to later 'comp' (engineer speak for compiling not compressing) various takes together to produce the final, perfect, vocal track. This is where you need to be willing to spend as much time as required to ensure that the vocal track is the best it can be. In some cases, you may even have to recall the vocalist to re-perform some lines and this can often result in a different tone of voice, but it is better to have a slightly different vocal tone than have a poorly sung line. Although ideally you should have ensured that you had enough to work with originally, re-recording some vocal parts at a different time isn't unusual and even professional recording studios do this from time to time.

If problems do occur later and it isn't possible to recall the vocalist, then you'll have no choice but to use a series of studio processors. This approach should *always* be viewed as the very last option, but can range from cross-fading individual words together, using pitch correction software such as Antares Auto-Tune, harmonizing, double tracking, applying chorus, flange or reverb to thicken out thin vocals, time-stretching some parts while time-compressing others, to reconstructing the entire vocals word for word in an audio sequencer or editor. This may all seem a little excessive, but it is quite normal for a producer to spend days and even weeks editing a vocal track to achieve perfection.

If you are experiencing severe problems with intonation you should look at using a different vocalist and try not to rely too much on pitch correction software. While these are adept at correcting the odd out-of-tune words, they do not work as well when applied to an entire vocal line, phrase or chorus section. Although many like to believe that 'manufactured' bands are incapable of holding a single note and rely entirely on pitch correction software, disappointingly they can actually sing quite well (if chosen for looks alone they'll certainly have attended a series of singing lessons!). Any 'good' singer will naturally bend and modulate notes with vibrato, and if a vocal line, phrase or chorus section is particularly slow during the bend, the intonation unit can become confused and attempt to correct it. This can result in it bending the note when it doesn't require it, bending it in the wrong direction altogether or quickly jumping to the next note rather than keeping the bend. All of these side-effects can produce a synthetic unnatural sound that you would usually want to avoid, although of late 'fake' pitching has come to the fore in music since it was used by Cher's producer on *Believe*.[2]

Vocal effects

When it comes to dance music, it isn't extraordinary to apply various processing to the vocals to make them different. As previously touched upon, we've all heard the human voice naturally since the day we were born, so any effects that make them different will inevitably attract attention – so long as the effect isn't clichéd. In fact, with dance music it's quite usual to treat a vocal like any other instrument and heavily EQ or effect them to make them prominent and hopefully grab attention. Naturally, there is no definitive method to effecting vocals and it all depends on the overall effect you want to achieve. As Cher's producer has proved, experimentation is the real key to effecting the human voice and you certainly shouldn't be afraid of trying out new ideas to see what you can come up with. Having said that, there are some vocal effects that are still in wide use throughout the genres and knowing how to achieve these may help aid you to experiment further.

Commercial vocals

Possibly the first effect to achieve is the archetypal contemporary upfront sound that's typical of most commercial pop songs and, of late, trance. Though much of this is acquired at the recording stage by positioning the vocalist closer to the microphone, effects can be used to help widen it across the image and add some depth if it's required. Stereo widening effects are not used for this as you're working with a mono source to begin with and recording the vocals in stereo is not only difficult, requiring two microphones, but it will rarely produce the right results. Instead, a much better way is to employ a very subtle chorus or flange as a send, not insert, effect. By then setting the effect rate as slow as possible, you can send the vocals to the effect by the required amount, thus keeping the stability of the vocals while also applying

---

[2] Cher's *Believe* track did not use an intonation program but the Digitech Talk-Box to acquire the effect.

enough of an effect to spread it across the stereo image. There are no universal settings for this as it depends entirely on the vocals that have been recorded, but a good starting point is 80% uneffected with a 20% effect.

An alternative method to this is to double-track the vocals by copying them onto a second track before applying a pitch-shift effect to both. This, obviously, should be applied as an insert effect, otherwise you will experience a strange phased effect, but by setting one channel to pitch up by 2–5 cents and the other down by 2–5 cents it can produce a wide, thick, upfront sound. Additionally, if you move the second vocal track forward by a few ticks so that it occurs slightly later than the original, this can help to thicken the overall sound.

Many engineers will also recommend applying reverb to a vocal track to help thicken them in a mix, but if it is not applied cautiously and sparingly it will push them into the background, producing a less defined, muddied sound. Nonetheless, if you do feel the need to apply reverb to vocals, use it as a send effect rather than insert, as this will allow you to keep a majority of the vocal track dry. Also, if reverb is being used on any other instruments within the mix, use a reverb with a different tonal character for the vocals, as this will help to draw a listener's attention to them.

The trick behind using any reverb is to set the effect so that it's only noticeable when it's removed, so keep the decay time short for a fast tempo sing and lengthen it for those of a slower tempo. As a very general starting point, try setting the reverb's pre-delay to 1 ms, with a square room and 0.70 ms decay. By then increasing the pre-delay you can separate the effect from the dry vocal to enable the voice to stand out more.

The most widespread processing used to obtain the upfront sound is compression. If you really want that big, in your face, upfront vocal then compression is the real key to acquiring it, but for this it is imperative that you use compressors with the best sound possible. As every dance genre will use valve compressors, this means you will need to use them too, and whether hardware or software emulations they should ideally exhibit 0.2% THD or more for a good sound. Notably, most musicians use more than one compressor for this and will usually employ two or three. The first is commonly a solid-state compressor that's used to smooth out the overall level with a ratio of 4:1, a fast attack and release, and threshold set so that the gain reduction meter reads 4 dB. The results of this are then fed into a valve compressor and this is generally set with a ratio of 7:1, a fast attack and release, along with a threshold set so that the gain reduction meter reads 8 dB. Naturally, these are general settings and should be adjusted depending on the vocals and the microphone used to capture them. As we've touched upon in earlier chapters, applying compression will reduce the higher frequency as it squashes down on the transients of the attack, but if you find that the high-end detail becomes lost it can often be brought back up again using a sonic exciter or enhancer.

'Telephone' vocals

The second effect that's often used in the dance genre is the 'telephone' vocal, so called because it sounds as though it's being played through a restricted bandwidth (i.e. a telephone or transistor radio). This effect is quite easy to achieve by inserting (not sending) the vocals into a band-pass filter so that frequencies above 5 kHz and below 1 kHz are removed from the signal. If you don't have access to a band-pass filter, the same effect can be acquired by first inserting the vocals into a low-pass filter set to remove all frequencies above 5 kHz with the results of this inserted into a high-pass filter set to remove all frequencies below 1 kHz. If you can adjust the poles on these filters then, generally speaking, a two-pole (12 dB) slope will produce the best results, but depending on the mix that the vocals are sitting in, it may be worth using a four-pole (24 dB) or, if possible, a single-pole (6 dB) slope. Alternatively, an EQ unit can be used to achieve much the same results by cutting all frequencies below 1 kHz and above 5 kHz. Once you have this general effect, it can be expanded upon by automating the filters (or EQ) to gradually sweep through the range to add movement and interest.

Pitched vocals

Another, more recent, effect to apply to vocals is the unnatural pitch-shifting that's acquired through using an intonation unit to alter the pitch. How this is accomplished depends entirely on the effects unit itself, but fundamentally they all work on the principle that you can analyse the incoming signal and then select the key signature that it should be in. By using this as an insert effect so that it affects the entire vocal line and setting it to a key that's at least an octave out of the range of the original input, the processor will attempt to correct the vocals to the selected scale and as a side-effect it will produce a strange moving pitch very similar to the effect used on Cher's *Believe*. On top of this, it is also worth experimenting by running the vocals through the intonation unit a number of times, each time using a different key setting before 'comping' the different results together.

Vocoders

One final effect that's particularly useful if the vocalist is incapable of singing in key is the vocoder. Of all the vocal effects, these are not only the most instantly recognizable, but are also the most susceptible to changes in fashion. The robotic voices and talking synth effects they generate can be incredibly clichéd unless they're used both carefully and creatively, but the way in which they operate opens up a whole host of creative opportunities.

Fundamentally, vocoders are simple in design and allow you to use one sound – usually your voice (known as the modulator) – to control the tonal characteristics of a second sound (known as the carrier), which is usually a synthesizer's sustained timbre. However, as simple as this may initially appear, actually producing musically usable results is a little more difficult, since simply dialling up a synth preset and talking, or singing, over it will more often than not produce unusable results. Indeed, to use a vocoder in a musically useful way, it's important to have a good understanding of exactly how they work and to do this we need to begin by examining human speech.

A vocoder works on the principle that we can divide the human voice into a number of distinct frequency bands. For instance, plosive sounds such as 'p' or 'b' consist mostly of low frequencies, 's' or 't' sounds consist mostly of high frequencies, vowels consist mostly of mid-range frequencies and so forth. When a vocal signal enters the vocoder, a spectral analyser measures the signal's properties and subsequently uses a number of filters to divide the signal into a number of different frequency bands. Once divided, each frequency band is sent to an envelope follower, which produces a series of control voltages[3] based on the frequency content and volume of the

---

[3] Control voltages (often shortened to CV) measure the signal at predetermined points and convert each of these measurements into differing voltages to represent the volume and frequency of the waveform.

vocal part. This exact same principle is also used on the carrier signal and these are tuned to the same frequency bands as the modulator's input. However, rather than generate a series of control voltages, they are connected to a series of voltage-controlled amplifiers. Thus, as you speak into the microphone the subsequent frequencies and volume act upon the carrier's voltage-controlled amplifiers, which either attenuates or amplifies the carrier signal, in effect superimposing your voice onto the instrument's timbre. Consequently, since the vocoder analyses the spectral content and not the pitch of the modulator, it isn't necessary to sing in tune as it wouldn't make any difference.

From this, we can also determine that the more filters that are contained in the vocoder's bank, the more accurately it will be able to analyse and divide the modulating signal, and if this happens to be a voice, it will be much more comprehensible. Typically, a vocoder should have a minimum of six frequency bands to make speech understandable, but it's important to note that the number of bands available isn't the only factor when using a vocoder on vocals.

The intelligibility of natural speech is centred between 2.5 and 5 kHz; higher or lower than this and we find it difficult to determine what's being said. This means that when using a vocoder, the carrier signal must be rich in harmonics around these frequencies, since if it's any higher or lower then some frequencies of speech may be missed altogether. To prevent this, it's prudent to use a couple of shelving filters to remove all frequencies below 2 kHz and above 5 kHz before feeding them into the vocoder. Similarly, for best results the carrier signal's sustain portion should remain fairly constant to help maintain some intelligibility. For instance, if the sustain portion is subject to an LFO modulating the pitch or filter, the frequency content will be subject to a cyclic change that may push it in and out of the boundaries of speech, resulting in some words being comprehensible while others become unintelligible. Plus, it should also go without saying that if you plan on using your voice to act as a modulator it's essential that what you have to say, or sing, is intelligible in the first place. This means you should ensure that all the words are pronounced coherently and clearly.

More importantly, vocal tracks will unquestionably change in amplitude throughout the phrases and this will create huge differences in the control voltages generated by the vocoder. This results in the VCA levels that are imposed onto the carrier signal to follow this change in level producing an uneven vocoded effect, which can distort the results. Subsequently, it's an idea to compress the vocals before they enter the vocoder and if the carrier wave uses an LFO to modulate the volume compress this too. The settings to use will depend entirely on the vocals themselves and the impact you want them to have in the mix (bear in mind that dynamics can affect the emotional impact), but as a very general starting point set the ratio on both carrier and modulator to 3:1 with a fast attack and release, and then reduce the threshold so that the quietest parts only just register on the gain reduction meter. Additionally, remember that it isn't just vocals that will trigger the vocoder and breath noises, rumble from the microphone stand and any extraneous background noises will also trigger it. Thus, along with a compressor you should also consider employing a noise gate to remove the possibility of any superfluous noises being introduced.

With both carrier and modulator under control there's a much better chance of producing a musically useful effect and the first stop for any vocoder is to recreate the robotic voice. To produce this effect, the vocoder needs to be used as an insert effect, not send, as all of the vocal line should go through the vocoder. Once this modulator is entering the vocoder you'll need to program a suitable carrier wave. Obviously, it's the tone of this carrier wave that will produce the overall effect, and two sawtooth waves detuned from each other by + and −4 with a short attack, decay and release but a very long sustain should provide the required timbre. If, however, this makes the vocals appear a little too bright, sharp, thin or 'edgy' it may be worthwhile replacing one of the sawtooth waves with a square or sine wave to add some bottom-end weight.

Though this effect is undoubtedly great fun for the first couple of minutes, after the typical *'Luke, I am your father'* it can wear thin and if used as is in a dance track it will probably sound a little too clichéd, so it's worthwhile experimenting further. Unsurprisingly, much of the experimentation with a vocoder comes from modulating the carrier wave in one way or another and the simplest place to start is by adjusting the pitch in time with the vocals. This can be accomplished easily in any audio/MIDI sequencer by programming a series of MIDI notes to play out to the carrier synth, in effect creating a vocal melody. Similarly, an arpeggio sequence used as a carrier wave can create a strange gated, pitch-shifting effect, while an LFO modulating the pitch can create an unusual cyclic pitch-shifted vocal effect. Filter cut-off and resonance can also impart an interesting effect on vocals and in many sequencers this can be automated so that it slowly opens during the verses, creating a build-up to a chorus section. Also, note that the carrier does not necessarily have to be created with saw waves, and a sine wave played around C3 or C4 can be used to recreate a more tonally natural vocal melody that will have some peculiarity surrounding it.

Note: Vocoders do not always have to be used on vocals and you can produce great results by using them to impose one instrument onto another. For instance, using a drum loop as a modulator and a pad as the carrier, the pad will create a gating effect between the kicks of the loop. Alternatively, using the pad as a modulator and the drums as the carrier wave, the drum loops will turn into a loop created by a pad!

Ultimately these have only been simple suggestions to point you in a more creative direction and you should be willing to try out any effects you can lay your hands on to hear the effect it can have on a vocal. Bear in mind that due to the very nature of dance music it's always open to experimentation and it's much better to initiate a new trend than simply follow one set by another artist.

9 Tools of the trade

If I can't create a sound I like then it's difficult to create a song. Using the right sound is the most important thing but it isn't something you can produce on every synthesizer.

Kevin Saunderson

One of the most fundamental aspects of writing dance music is access to the right type of synthesizers for the genre in question. In fact, if you want to produce dance music then it's important to have the right tools to hand. A joiner/carpenter couldn't exactly complete his job if he were handed a bag of spanners and, similarly, while there is the risk of sounding like Blue Peter, to write any form of dance you do need to use the right instruments. This is not to dispute the fact that music is an entirely creative art and you should feel free to use whatever you wish; after all, the Roland TB303 was originally designed to supply the bass for gigging/pub musicians, but unless you're looking to push boundaries, much of contemporary dance music does rely on employing certain styles of timbres that can only be produced by certain synthesizers.

As an example, if you were to replace the typical euphoric 'hands in the air' trance lead with a General MIDI honky-tonk piano, it would no doubt still remain musical but it wouldn't be particularly euphoric. Additionally, it would be obtuse to suggest that you could produce all the timbres used in dance with nothing more than a GM sound module and a MIDI pattern sequencer. As touched upon in Chapter 7, any one synthesizer, no matter how well touted, is not capable of producing every sound reliably and today, more than ever, synthesizers are created with specific musicians in mind. Considering this, what follows is a short description of *some* of the most popular instruments and effects that are used to produce today's dance music.

Novation BassStation (hardware and software)

The BassStation is a mono-timbral, analogue-modelled synthesizer that's based around a twin oscillator subtractive synthesizer with LFOs and envelopes for both the amp and filter. It offers real-time control over all the vital parameters along with a pitch bend and mod wheel. Although it is perfectly adept at creating lead lines, it's particularly suited towards creating deep analogue basses (hence the name) and can reproduce the sounds of the TB303, Prophet five and the MiniMoog with relative ease. All of the parameters send and receive MIDI controller data. The software model holds space for 100 presets and can import presets from the hardware version and vice versa. It's been used by Nine Inch Nails, Daft Punk, Apollo 440, Chemical Brothers, Dust Brothers, Joey Negro and Massive Attack.

Figure 9.1 The Novation BassStation. (Courtesy of Novation Music Ltd. Reproduced with kind permission)

Note: A software demo is included on the CD.

Novation V-Station (software and hardware (K-Station))

The V-Station is a mono-timbral (although you can open as many instances as the processor can handle), analogue-modelled, three-oscillator synth. With eight-voice polyphony, control over every parameter is offered through an intuitive interface including control over the multi-mode filters, two LFOs, three oscillators, a modulation envelope and a filter envelope. Up to seven effects can be applied to any one sound and an arpeggiator with programmable speed, synchronization and sweep range is also available. Each controller also sends and receives continuous MIDI controller data, allowing you to record and playback any parameter tweaks made on the interface.

Note: A software demo is included on the CD.

Novation SuperNova II (hardware only)

Referred to as the 'definitive synthesizer' by many dance artists, the SuperNova II is a 20-voice, analogue-modelled, three-oscillator synth featuring multi-mode filters, two LFOs, three envelope generators, a vocoder, FM capability and 56 effects, permitting seven effects to be applied to each instrument. Although it is only eight-part multi-timbral, like the effects, arpeggiators can be applied individually to each part and there is a comprehensive modulation matrix allowing over 95 modulation routings in the oscillator section alone. Additional upgrade cards can also be

Figure 9.2 The Novation V-Station. (Courtesy of Novation Music Ltd. Reproduced with kind permission)

installed to up the polyphony to 32 or 48 notes. The SuperNova II is a particularly popular synthesizer amongst many dance musicians and is used heavily by the likes of Orbital, A Guy Called Gerald, William Orbit, Massive Attack, BT and ATB.

Figure 9.3 The Novation SuperNova II. (Courtesy of Novation Music Ltd. Reproduced with kind permission)

Access Virus C (hardware and software)

This synthesizer is considered as absolutely essential by most professional dance artists, especially those who write trance, house and techno. Complete with hands-on control over every synthesis parameter, the Virus C is an analogue modelling synth that offers three oscillators, a sub-oscillator, two multi-mode filters, three LFOs, two envelopes, a vocoder, FM synthesis and a large range of effects. It also features 16 arpeggiators, is 32-note polyphonic and 16-part multi-timbral. Amongst countless others, it's used by Andy Gray (the *Big Brother* theme was written almost entirely on a Virus), Paul Oakenfold, Sasha, Faithless, Sister Bliss, DJ Sammy, William Orbit, David Morales, Roger Sanchez, Air, Matt Darey, Delerium and Thrillseekers.

Figure 9.4 The Access Virus C. (Courtesy of Access Music Ltd. Reproduced with kind permission)

E-Mu Orbit 3 (hardware only)

The Orbit 3 is a sample and synthesis dance module packed with 64 MB of ROM samples that can be manipulated and edited via 12 real-time assignable controls on its fascia. This editing includes control over envelopes, filters, LFOs and various modulation options. It offers 128 voices, is 16-part multi-timbral and its soundest can be expanded further with room for up to two expansion cards. Many of these cards contain the entire soundest from E-Mu's previous modules, including the Vintage Keys and the Planet Phatt. On top of this, the Orbit also features a super BEATS mode and a rhythmic pattern generator permitting you to quickly create drum rhythms and synth patterns. Users of the Orbit 3 include Orbital, Leftfield, David Morales and Andy Weatherall.

Figure 9.5 The E-Mu Orbit 3.

E-Mu Planet Phatt (hardware only)

The Planet Phatt is a sample and synthesis module designed for acid jazz, R'n'B and trip-hop musicians. Sixteen-part multi-timbral and offering 64-voice polyphony, it has no real-time fascia controls, but through various editing screens allows you to manipulate the envelopes, filters, LFOs and various modulation options. Similar to the Orbit 3 it offers a BEATS mode allowing you to knock together drum rhythms quickly, and is used heavily by ATB, Roni Size, Moloko, Radio Head, David Morales and The Orb.

Figure 9.6 The Planet Phatt.

Roland XV5080 (hardware only)

Roland's range of JV and XV synthesizers are the industry standard 'bread and butter' modules. In other words, they're commonly used to provide the 'real' instruments in a record. This includes very convincing acoustic guitars, basses, strings and wind instruments. It offers a 128-note polyphonic engine which is also 16-part multi-timbral and features a vast range of effects. More importantly, though, it allows you to install up eight expansion cards from their current range which, amongst others, includes Orchestral, 1960s and 1970s sounds, Dance and Hip-Hop. However, while these latter two cards are impressive, due to the XV being based on digital S&S and appearing in a rack-mountable form, there is very little real-time control on it's fascia, so many musicians use other hands-on synths for generating the synthetic timbres. It's been used by Trevor Horn, Steve Levine, William Orbit, Scanner, Björk, David Arnold and Nile Rodgers.

Figure 9.7 The Roland XV5080. (Copyright Roland Music Corporation Ltd. Reproduced with kind permission)

Korg Triton (hardware only)

The Triton is Korg's answer to Roland's XV5080 and is seen by many as producing better, more professional sounding timbres. Although it only features 62-voice polyphony, it's 16-part multi-timbral, has a built-in sequencer, 16 MB sampler, numerous arpeggiators and a large touch-screen LCD for editing. There's also a plethora of on-board effects and numerous real-time, user-assignable controls including two switches, four knobs, a pitch/modulation joystick and a PC interface allowing you to edit the sounds in a connected PC. Similar to the XV range it can also be expanded with PCM expansion boards to increase the available timbres. To many artists it also produces a warmer tone than the XV modules, and is a favourite of Orbital, BT, ATB, William Orbit, The Orb and Apollo 440.

Figure 9.8 The Korg Triton. (Courtesy of Korg. Reproduced with kind permission)

AKAI MPC 2000XL (hardware only)

The AKAI MPC is the choice of instrument for hip-hop and most house musicians due to its incredibly solid timing. Fundamentally, it's a sampler, drum machine and multi-track MIDI sequencer in one unit, allowing it to act as the centre of a studio in place of a software sequencer (or MTC alongside one). Twelve velocity-sensitive pads can be used to tap out drum rhythms live (using a shift key you can access a total of 64 pads), and it features all the functionality of a typical sampler and sequencer. It can sample at 16-bit 44.1 kHz, there are dedicated buttons for the most commonly used features, it offers 32-note polyphony and is 32-part multi-timbral (it utilizes two MIDI ports to accomplish this). Up to eight analogue outputs are available with the addition of an optional card and S/PDIF connectivity is fitted as standard.

Figure 9.9 The AKAI MPC 2000XL. (Copyright AKAI UK Ltd. Reproduced with kind permission)

Native Instruments FM7 (software only)

The FM7 is based on the Yamaha DX range of synthesizers, the tone of which still appears in many dance records. A frequency modulation synthesizer, it can import the patches from any of the Yamaha DX range, including the DX7, -21, -100 and the recent DX200 'groovebox', while also offering more sound creation functions than any of the original DX range. Whereas the DX series of synths only used six operators, the FM7 has eight, which mimic the original operators while adding resonant filters and distortion. It also offers 32 different waveforms to use as modulators or carriers, a fully programmable algorithm for connecting the operators together, and independent control over the panning, gain, envelope rates, frequency range and amplitude modulation of each operator. For easier operation, there's also an easy edit page permitting you to control numerous parameters such as the filter, LFO and ADSR with faders.

Figure 9.10 The Native Instruments FM7.

Note: A software demo is included on the CD.

Native Instruments Pro-53 (software only)

The Pro-53 is an emulation of the revered Sequential Circuits Prophet Five, which is one of the most influential instruments in dance music. Basically formed around a simple twin-oscillator subtractive synthesizer, the Pro-53 offers three waveforms, a comprehensive filters section, filter and amp EGs, an LFO, mono- or polyphonic operation, and a delay effects section. And of course, it sounds just like the original Prophet Five!

Figure 9.11 The Native Instruments Pro-53.

Note: A software demo is included on the CD.

Native Instruments B4 (software only)

The B4 is a VST instrument that is based on the Hammond B3 organ. For those with no knowledge of the Hammond, it was a tone-wheel organ that featured two keyboards (known as manuals) and numerous draw-bars allowing you to alter the tone of the sound. Its sound often plays a fundamental role in trip-hop and has been used by Portishead, Herbie Hancock, Air and Red Snapper.

Figure 9.12 The Native Instruments B4.

Note: A software demo is included on the CD.

Propellerhead ReBirth RB-338 (software only)

The Roland TR808, TR909 drum synths and TB303 bass synth are the most infamous dance tools around, but having been out of production for nearly 20 years they've become a highly sought-after commodity demanding ridiculous prices on the second-hand market. Propellerhead's ReBirth is an accurate software emulation of these machines and includes two TB303s, a TR808 and a TR909 all contained within a standalone or sequencer-integrated synth. Each synth is modelled precisely around the original machines, even down to the interface and the method of programming the step sequencers. Additionally, each instrument can store 32 patterns of 16-step sequencers and any parameter tweaks or pattern changes can be stored as part of a song within its own sequencer. All of these instruments also have simultaneous access to a compressor, distortion, delay and a filter.

Figure 9.13 The Propellerhead ReBirth RB-338.

Waldorf Attack (hardware and software)

The Waldorf Attack is a percussion synthesizer permitting you to not only create any drum timbres but also manufacture basses and lead sounds. Offering two oscillators along with samples for the creation of hi-hats and cymbals, up to 24 percussive instruments can be assigned to the keyboard. Alternatively, 12 drum sounds can be used along with one melodic instrument spread over an octave. Each oscillator offers nine waveforms, which can be augmented with two envelopes for creating FM sounds and ring modulation, while six different resonant filter types can be used to shape the overall results. The Waldorf Attack is hailed as one of the best drum synthesizers available and has been used by countless dance artists, including Leftfield, Radiohead, The Orb, Moloko, Jah Wobble, Beck, Björk, Joey Negro, Nine Inch Nails, Andy Weatherall, Daft Punk and The Rapino Brothers.

Figure 9.14 The Waldorf Attack.

Native Instruments Battery (software only)

Battery is a software sampler that has been designed to work solely with drums. As such, rather than utilize a keyboard it consists of 54 'cells', each of which can hold an individual drum sample (note that these must be imported and cannot be recorded directly into Battery). Each cell has access to a maximum of 128 velocity layers along with numerous sound sculpting options, such as an AHDSR EG (Attack–Hold–Decay–Sustain–Release), a pitch EG, bit reduction, sample start adjustment, a transient designer, sample tuning and six modulation options. Although no artists have mentioned using this software, it's an absolute must for producing drum loops from numerous single samples.

Note: A software demo is included on the CD.

Figure 9.15 The Native Instruments Battery.

Native Instruments Kontakt (software only)

Kontakt is a software sampler that can either stand alone or integrate into most audio sequencing applications, although calling it a sampler is probably the incorrect term. While it allows you to import up to 16 different instruments, each of which have access to all the editing features of any sampler, it isn't actually capable of recording audio. Instead, you have to import pre-recorded samples to edit and manipulate. Nonetheless, this type of operation is typical of all soft samplers and there are many benefits to using them. In the case of Kontakt, samples can be assigned to both key and velocity by simply clicking and dragging; you can specify eight different loop points within a single sample, there are a host of filters and envelopes available, automatic start and end loop finding, and a multitude of modulation and destination options. What's more, a Tone- and a Time-machine allow you to perform real-time time-stretching and the option to mangle sounds using granular synthesis. Most dance artists are still warming to the idea of soft sampling, preferring hardware units, but BT, Aphex Twin and Orbital are known to use this sampler.

Note: A software demo is included on the CD.

Steinberg HALion (software only)

Similar to Native Instruments Kontakt, HALion is a software sampler that, while incapable of recording, allows you to drag and drop samples into its interface for instant velocity and key

Figure 9.16 The Native Instruments Kontakt.

mapping. Alongside all the typical options available in a sampler (bar time-stretching), up to 16 instruments can be played at once through MIDI; there are two eight-stage envelopes, a multi-mode filter, 12 modulation sources and four assignable MIDI controllers. On top of this, all the loaded samples are stored on the hard drive rather than RAM, so the length of a sample is not limited by the amount of memory you have installed. There have been complaints that the filters were a little thin and digital sounding, but the recent release of HALion 2 has employed filters designed by Waldorf.

Figure 9.17 The Steinberg HALion.

Native Instruments Reaktor (software only)

The Reaktor is a modular real-time synthesizer, sampler and effects processor. That is, you can create absolutely any synthesizer, sampler of effects unit you can dream of by actually building it from scratch using a series of modules and interconnecting them together with virtual patch cables. Since the modules are the bare bones of any instrument, consisting of single oscillators and ADSR envelopes, etc., you even need to create and assign knobs, buttons and faders to control the parameters available on each module. However, while it is possible to construct any instrument, from subtractive to FM to granular, it is not for the faint-hearted. While there are a large number of preset instruments supplied that can be used standalone or within a sequencer, constructing a new instrument from scratch requires a lot of work and a good in-depth knowledge of synthesis and signal routing. Nevertheless, building your own synth from nothing can be a rewarding experience, as it allows you to create synthesizers, samplers or effects that no one has ever heard of before. Reaktor is commonly used by Coldcut, Squarepusher, Aphex Twin, Autechre, LFO, Edwyn Collins, Andy Weatherall and David Arnold.

Figure 9.18 The Native Instruments Reaktor.

Note: A software demo is included on the CD.

Native Instruments Absynth v2 (software only)

Absynth 2 is similar in some respects to Reaktor in that it uses a semi-modular structure. This means that, while it is not possible to build a synthesizer from scratch, it does offer a bewildering array of functions. It offers a number of synthesis methods, including subtractive, FM and AM, each of which can be layered on top of one another to produce all types of sounds. Sounds can be constructed by layering six oscillators together, which can then be modified with four filters, delay, three ring modulators and a waveshaper. In addition, a waveshape editor can be used to draw in any envelope shape you wish that can then be used to modulate the oscillators or LFOs. Further envelopes offer 68 breakpoints so that you can create multiple attack, decay, sustain, hold and release stages. The latest version of Absynth has also added a sampler to the synthesizer, allowing you to edit samples in the same way as the oscillators.

Figure 9.19 The Native Instruments Absynth 2.

Note: A software demo is included on the CD.

Mutronics Mutator (hardware only)

The Mutator is one of the most respected filters for creating dance music. Featuring two *analogue* resonant filters with plenty of hands-on control, it can be used to modify a stereo or two mono audio signals. It also features an LFO with four waveforms and an envelope follower, but more inventively, it responds to MIDI allowing you to use the pitch bend wheel and keys of an attached controller keyboard to control the action of the filters. Since a good external filter is one of the most important aspects when producing dance, just about every professional artist owns one of these, including the Chemical Brothers, Portishead, Prodigy, Fatboy Slim, Air, Daft Punk, Sasha, Paul Oakenfold, Andy Gray, Leftfield, Andy Weatherall, David Arnold, Dust Brothers, Nine Inch Nails, Beck, Joey Negro and the Future Sound of London.

Figure 9.20 The Mutronics Mutator. (Courtesy of Mutronics. Reproduced with kind permission)

Sherman Filterbank 2 (hardware only)

Like the Mutator, the Filterbank 2 is a resonant *analogue* filter, but due to a larger number of controls and connection options is a much more experimental device. An audio signal can be fed through just one or both of the multi-mode filters and these can be modulated from a number of sources, including positive or negative ADSR, FM, LFO and amplitude modulation. Furthermore, three rotary controllers allow you to cross-fade between parallel and serial configurations, high-, band- and low-pass filters, and between bypass and effect. It can be quite difficult creating a simple resonant filter sweep with the Filterbank, but for more experimental timbres it's very highly regarded, and used by Leftfield, William Orbit, Chemical Brothers, Apollo 440 and Ritchie Hawtin.

Figure 9.21 The Sherman Filterbank 2. (Copyright Sherman Productions B.V.B.A. Reproduced with kind permission)

Novation ReMote 25 (hardware only)

It's not exactly a synthesizer, but a good controller keyboard is absolutely essential, more so if you're using VST instruments. Since the mouse can only control one on-screen parameter at a time, it's useful to have a keyboard with an interface crammed with parameters that can be used or programmed to control various aspects of the synthesizers. The ReMote 25 features a total of 17 rotary controllers, eight faders, transport controls, a touch-sensitive X/Y pad, a pitch and modulation yoke (a multi-directional stick that can modulate both pitch and modulation simultaneously or separately) and a semi weighted two-octave keyboard with aftertouch. Each of these real-time controllers can be configured to control any parameter on any VST synth that accepts CC messages, and comes supplied with a patch memory that can be used to store the settings for 64 synths (a number of which are already pre-configured to control the most popular synthesizers).

Figure 9.22 The Novation ReMote 25. (Courtesy of Novation Music Ltd. Reproduced with kind permission)

Spectrasonics Atmosphere, Stylus and Trilogy (software only)

These virtual instruments can be roughly described as sample and synthesis modules. Developed by Spectrasonics, a well known sample company, they each employ over 3 gigabytes worth of samples that are stored on the hard disk which are subsequently accessed through the user interface. Each interface is different, depending on the instrument, permitting you to modify the tonal content of the samples, which includes control over the filters, the pitch, the pitch envelope, LFOs and the ADSR of the amplifier and filter. To date, there are three instruments available, consisting of:

• *Atmosphere*. A synth geared towards lush sweeping pads and strings. It contains over 3 gigabytes of multi-sampled pads and strings from a variety of analogue equipment.

- *Stylus.* A synth geared towards drum samples and numerous loops that are individually triggered through MIDI, allowing you to change the speed of the loops without experiencing any unwanted time-stretching artefacts. It contains over 3 gigabytes of drum samples and loops.
- *Trilogy.* A synth geared towards creating basses, real and synthetic. It contains over 3 gigabytes of multi-sampled real and synthetic basses from analogue equipment.

Figure 9.23 The Spectrasonics Atmosphere.

Waves bundle – effects (software only)

Waves are synonymous with all professional producers for the unsurpassed audio quality of their numerous effects bundles. The most widely celebrated is the Native Gold, which consists of no fewer than 19 plug-in effects. These consist of the C4 Multiband Parametric Processor, Renaissance Reverberator, Renaissance Compressor, Renaissance Equalizer, L1 Ultramaximizer, MaxxBass, Q10 Paragraphic, S1 Stereo Imager, C1 Parametric Compander, Enigma, Supertap, MondoMod, Doppler, UltraPitch, MetaFlanger, TrueVerb Room Emulator, DeEsser, AudioTrack and the PAZ Psychoacoustic Analyzer. While most of these effects are aimed at general processing duties during recording and mixing, more recently they released the Transform plug-in pack. This consists of four effects that are more suited towards sound design and consist of Sound Shifter (a complex pitching effect), Doubler (a complex harmonizing effect), Morphoder (a fully featured vocoder) and TransX (a fully featured transient designer).

Note: A software demo is included on the CD.

TC Native bundle – effects (software only)

Alongside the Waves bundle, TC electronic also has an excellent professional plug-in bundle. While some of the effects are similar to the aforementioned Waves, they do have a significantly

Figure 9.24 The Waves Morphoder.

different tonal character as they all utilize a saturation algorithm to emulate the warm analogue sound. This bundle offers:

- *Filtrator* (Figure 9.25). This is a multi-mode filter featuring 12 and 24 dB slopes, filter distortion, an LFO that can be synced to tempo and an envelope follower.
- *Native Reverb Plus.* An excellent sounding reverb unit with a clear interface offering control over the shape and size of the 'room', diffusion, colour, pre-delay, decay, and the high and low factor.
- *Limiter.* This is a single-band limiter with an auto make-up gain and an adjustable hold time. It also features a histogram that allows you to monitor the difference in gain between the input and output.
- *Compressor/DeEsser.* A compressor that features ratio, threshold, attack, release, hold time and soft knee controls. Like the limiter, it also uses an automatic make-up gain that automatically brings all the peak levels up to 0 dB. Attached onto this compressor is a DeEsser, allowing you to home in on the problematic frequency and subsequently adjust the gain.
- *Graphic EQ.* A 28-band graphic EQ that allows you to click and drag on the interface window to increase or decrease the volume of each band.
- *Parametric EQ.* A 10-band parametric EQ that can operate as a parametric, notch filter or a hi- or lo-shelf. Seven of the bands are fully configurable while the final three are controlled with a joystick, permitting you to quickly adjust the treble and volume of the currently created EQ curve.

Note: A software demo is included on the CD.

Figure 9.25 The TC Native Filtrator.

PSP Audioware effects (software only)

PSP Audioware produce a number of effects that are highly respected and used amongst the dance community. These consist of:

- *VintageWarmer* (Figure 9.26). A simulation of an analogue style single and multi-band compressor that produces the typical saturation you would expect from analogue cassette recorders.
- *PSP 84*. A highly specified delay unit that can emulate the effects of the old tape delay units. This includes employing LFOs and filters, and allows you to control the left and right channels independently of one another.
- *PSP MixTreble*. A processor that allows you to process the treble frequencies of a mix. This includes hiss removal, transient designing and high-frequency enhancement.
- *PSP MixPressor*. An emulation of an analogue and optical compressor, allowing you to attain the typical sound of old (or incredibly expensive) analogue compressors.
- *PSP MixSaturator*. A two-band saturator that can be used to apply differing amounts of analogue warmth to the bass and high frequencies.
- *PSP MixBass*. A bass maximizer which can add harmonics to the bass frequencies.
- *PSP StereoPack*. A pack of four effects that allow you to create pseudo-stereo files from mono files, enhance the stereo image of a recording, eliminate any errors in a stereo recording (such as signal phasing) and view the current audio through an oscilloscope.

Note: A software demo is included on the CD.

Figure 9.26 The PSP VintageWarmer.

Note: A software demo is included on the CD.

GRM Tools Vols 1 and 2 effects (software only)

GRM Tools Volumes 1 and 2 both contain four experimental effects that are particularly suited towards manipulating standard audio to create strange textures.

Vol. 1:

- *Pitch Accumulator* (Figure 9.27). This effect analyses the incoming audio signal and allows you to transpose the pitch in two different directions, set a delay period between the two, adjust the gain and use random modulation.
- *Band Pass*. This uses high- and low-pass filters to provide a variable band-pass or band-reject filter. The cut-off frequency of these can also be independently configured.
- *Shuffling*. This slices the audio into very small user-definable segments and then applies a shuffling algorithm, allowing you to bounce some, or all, of these segments around the stereo image.

Figure 9.27 GRM Tools Vol. 1 Pitch Accumulator.

- *Comb Filters*. This consists of a bank of resonant comb filters that can all self-oscillate. Using these, the incoming signal can be amplified at user-definable frequencies. This effect is then also applied to integer multiples of the selected frequencies.

Vol. 2:

- *Freeze*. Using this you can select a 3-second sample from any incoming audio and then adjust how many loops occur in this sample, along with defining the timing between them, their speed, duration and synchronization.
- *Reson*. Using this, the incoming audio signal can be treated to a maximum of 128 resonant filters, consisting of a mix of low-, high- and band-pass, each of which act upon a single frequency of the audio file.
- *Doppler*. This simulates the Doppler effects. Fundamentally, this effect occurs naturally in the real world – for example, when a vehicle drives past you with its sirens on. As the sound comes closer, its pitch and volume rises, until it passes by, whereby the pitch and volume drop again.
- *Delays*. This utilizes 128 variable delays, each offering control over the amplitude and timing.

Part 2

Dance Genres

10 Trance

The '*soul*' of the machines has always been a part of our music. Trance belongs to repetition and everybody is looking for trance in life … in sex, in the emotional, in pleasure … So, the machines produce an absolutely perfect trance.

Ralf Hütter (Kraftwerk)

Trance is possibly one of the most ambiguous genres of dance music because it appears in so many different forms and few can actually agree on exactly what makes the music '*trance*'. However, it's fairly safe to say that it can be roughly generalized as the only form of dance music that's constructed around glamorous melodies which are either incredibly vigorous, laid-back or pretty much anywhere in between. In fact, it's this 'feel' of the music that often determines whether it's '*progressive*', '*Goa*', '*psychedelic*', '*acid*' or '*euphoric*'. Even then, some will place euphoric trance in with progressive and others believe that acid trance is just another name for Goa. To make matters worse, some forms of trance are given numerous titles by both DJs and clubbers. For example, euphoric trance may be subdivided into '*commercial*' or '*underground*', each of which are different yet again. This obviously presents a dilemma when trying to encapsulate it for the purpose of writing an example track, since it becomes a Hobson's choice.

Nonetheless, much of how the music is written can be determined by looking at its history and one thing that is certain is that trance, in whichever form, has its roots embedded in Germany. During the 1990s, a joint project between DJ Dag and Jam El Mar resulted in Dance2Trance and their first track, labelled *We Came in Peace*, is considered by many to be the first ever 'club' trance music. Although by today's standards it was particularly raw, consisting solely of repetitive patterns (as techno is today), it laid the basic foundations for the genre with the sole purpose of putting clubbers into a trance-like state.

The ideas behind this were nothing new; tribal shamans had been doing the same thing for many years, using natural hallucinogenic herbs and rhythms pounded out on log drums to induce the tribe's people into trance-like states. The only difference with Dance2Trance was that the pharmaceuticals were man-made and the pounding rhythms were produced by machines rather than skin stretched across a log. Indeed, to many trance aficionados (if there is such a thing), the purpose of placing clubbers into a trance state is believed to have formed the basis of Goa and psychedelic trance.

Although both these genres are still produced and played in clubs to this day, the increased popularity of 3,4-methylenedioxy-*N*-methylamphetamine (MDMA or 'E') amongst clubbers inevitably resulted in new forms of trance being developed. Since this pharmaceutical stimulates serotonin levels in the brain, it's difficult, if not impossible, to place clubbers into states of trance with tribal rhythms, and instead the melodies became more and more exotic, slowly taking precedence over every other element in the mix. The supposed purpose was no longer to induce a trance-like state, but emulate or stimulate the highs and lows of MDMA. The basic principle is still the same today in that the chemicals, arguably, still play a role in inducing a state, albeit a euphoric one, and the name euphoric trance was given to these tracks.

This form of music, employing long breakdowns and huge melodic reprises, became the popularized style that still dominates many commercial clubs and the music charts today. As a result, when the term trance is used to describe this type of music, most view it as the typical anthemic 'hands in the air' music that's filled with big breakdowns and huge melodic leads. As this remains one of the biggest and most popular genres of dance music to date, this form of trance will be the focus of the chapter.

Musical analysis

Possibly the best way to begin writing any dance music is to find the most popular tracks around at the moment and break them down into their basic elements. Once this is accomplished we can begin to examine the similarities between each track and determine exactly what it is that differentiates it from other musical styles. In fact, all music that can be placed into a genre-specific category will share similarities in terms of the arrangement, groove and sonic elements.

In the case of euphoric trance, it is viewed as being an anthemic form of music which essentially means that is has an up-tempo, uplifting feel that's very accessible to most clubbers. As a result, it can best be illustrated as consisting of a relatively melodic synth and/or vocal hook line laid over a comparatively unsophisticated drum pattern and bass line. The drums usually feature long snare rolls to signify the build to the reprise and breakdowns, alongside small motifs and/or chord progressions that work around the main melody.

Consequently, all euphoric trance music will utilize a four-to-the-floor time signature to help clubbers keep time while 'stimulated' and be produced either allegretto or allegro. In terms of physical tempo, this can range from a rather contained 125 BPM and move towards the brain-mashing upper limits of 150 BPM, but most tend to stay around the 137–145 BPM mark. This, of course, isn't a demanding rule, but it is important not to go too fast as it needs to remain anthemic and the majority of the public can't dance to an incredibly fast beat.

The rhythm section

The best place to start when programming any dance music is with the rhythm section as this will define the overall pace of the record. With most trance this is invariably kept quite simple and will almost always rely on a four-to-the-floor pattern. This means a kick drum is placed on every beat of the bar, along with snares or claps laid on the second and fourth beats to add some expression to these two beats. To complement this basic pattern, closed hi-hats are commonly placed on every 16th division, or variation of 16ths, while to introduce some syncopation open hi-hats are often employed and placed on every eighth division of the bar. Notably, any closed hi-hats that occur in this position should be removed to prevent any frequency clashes between the closed and open hi-hat.

Figure 10.1 Typical trance drum loop.

This, of course, is only a general guideline to the drum patterns used and is open to your interpretation. For instance, occasionally a set of triplet closed hats may sit at the end of every few bars to add some extra variation or alternatively the snare/clap may be placed a 16th before the kick occurs on the last beat of the fourth bar. Additionally, the kicks are sometimes used to add variation and a double kick at the end of the fourth bar can be used to signify a change – for example, when introducing a new instrument into the mix.

Figure 10.2 A variation on the trance loop.

Although these techniques produce what is essentially an incredibly basic loop, this is exactly what much of trance relies on. Indeed, bongos, congas and most other percussive instruments are not commonly employed, since a more complex drum loop will not only reduce the room left for the lead melody, but it may also take the attention away from it. On this same theme, despite the discussion of 'dynamics' in Chapter 2, both the kick and snare in trance will often remain at the highest velocity rather than following the *strong–weak–medium–weak* syncopation. Again, this is simply because the drums should pull very little attention away from the main melodic lead. The closed hi-hats, however, often employ different velocities throughout the bar to add some interest to the patterns. As a general rule of thumb, the main emphasis is commonly on the first and fourth (remember that the open hat sits on the third!) 16th divisions of the bar, as illustrated below:

1	2	3	4	5	6	7	8	9	10	11	12	13	14	15	16
S	M	W	S	S	M	W	S	S	M	W	S	S	M	W	S
Kick				Kick				Kick				Kick			

Of course, this is simply convention and convention shouldn't always play a part in your music, so you should feel free to experiment by placing the accents at different divisions on the beat. By doing so, the rhythm of the piece can change quite severely so it is worth experimenting.

Rhythm timbres

Simply playing this rhythm through a GM module you can expect the results to sound absolutely nothing like the pounding beats you hear in the clubs, so it's vital to use the right instruments to produce the sounds. Predictably, all of these are most usually derived from the Roland TR909, but the Waldorf Attack, Novation Drum Station or a drum sampler and a series of 909 samples will perform the job just as well. Alternatively, they can be programmed in any capable synthesizer.

The trance kicks tend to be kept quite tight rather than 'boomy', since this keeps the entire rhythm section sounding fairly strong to produce the four-to-the-floor solidity. To reproduce this, try using a 90 Hz sine wave and modulate it with a positive pitch envelope with no attack and a medium release. If you have access to the pitch envelope's shape, then generally a concave release is more typical than convex, but that's not to say convex doesn't produce the results. Finally, experiment with the decay setting, as the faster the decay is set, the tighter the rhythm will become, but there is a limit as to how far you should go. If the decay is set too short then the kick will become a 'blip', so you should look towards making a kick that has enough body to pull out of the mix, but not so much that it begins to boom. Generally, as with most instruments, this is best left as MIDI triggering the drum synth rather than bounced down to audio as soon as it sounds appropriate, as it allows you to further contort the kick when the rest of the instrumentation is in place.

As discussed in Chapter 5, compression can also play a large role in getting the kick to sound 'right' and compression should be applied to the 4/4 kick loop, rather than when the rest of the loop elements are in place. Although many engineers apply compression after the loop is complete (and feel free to experiment), keep in mind that the principle here is to 'pump' the drums and if the loop only has high-frequency elements sat off-beat, the compressor will pump them. This can work musically with the more expensive compressors, but most budget units will destroy the high frequencies of the hats, resulting in a duller sounding loop. What's more, as the kick and snare may occur on the same beat, a compressor with the required fast attack setting will capture the higher frequencies of the snare, dulling them too.

This can be avoided by applying compression just to the 4/4 kick loop. Although there are no underlying elements for the compressor to pump, the attack is set as fast as possible so that the transient is compressed and some of the high-frequency content is removed to produce a more substantial 'thud'. The lower the threshold and higher the ratio, the more this will be evident, but by increasing the release parameter so that when the compressor is nearing the end of its release the next kick starts, a good compressor will distort in a pleasing manner. Generally, most valve-based compressors will pump musically, but the most commonly used compressors for this are the *Joe Meek SC 2.2*, *UREI LA 3* or the *UREI 1176 LN* due to the amount of second-order harmonic distortion they introduce. If you don't have access to a good valve compressor, then after compressing it with a standard unit it's certainly worth throwing the loop through the *PSP VintageWarmer* to recreate the requisite sound. In fact, even if it is put through a valve compressor it's often worth placing it through the VintageWarmer anyway.

> Note: On the CD you can hear a kick loop compressed and uncompressed (with background noise to make the effect more evident).

Once the kick is sounding 'right' the snares can be added and further modified to suit. Notably, these are not always used in the production of trance and claps are occasionally used in their place – it depends entirely on what your creative instincts tell you. Nevertheless, whichever is used in the track they will, more often than not, be sourced from the TR909 or a sample CD (or more commonly another record). They can, of course, also be created in a synthesizer, but along with the triangle wave it's worth employing a pink noise waveform rather than white, as it produces a thicker timbre. The pitch envelope is generally set to positive modulation with a fast attack and a long decay, but it's also worthwhile creating a slurred effect by using negative modulation. This latter snare can be used as part of a build-up to new instruments.

If a sample has been used and the decay cannot be made any longer, sending them to a reverb unit set to a short pre-delay and tail will increase the decay. This is also worth experimenting with even if the decay is long enough, as it can help 'thicken' the timbre. If this latter approach is taken, though, it's prudent to employ a noise gate to remove some of the reverb's tail-off. After this, much of the low-frequency content will need removing to provide the basic snare timbre, so it's sensible to employ a high-pass filter to remove some of the low end and then experiment with the concave and convex decay parameter (of the amp envelope). Convex decays tend to be more popular in most trance tracks, but concave may produce the results to suit what you have in mind.

As previously mentioned, compressors are often used on the snares to help create the timbre, but rather than compress the transient it's prudent to set the attack so that it just misses and captures the decay stage instead. This will prevent the high frequencies of the snare from being compromised, but by increasing the make-up gain, the decay stage is increased in volume, helping to produce the often used and distinctive '*thwack*' style timbre. Similar to the kick, the lower the threshold and higher the ratio, the more this will be evident, so experiment.

Finally, the closed and open hi-hat patterns will need some modifications to suit the kick and snare. Yet again, these tend to be sourced from the TR909, a sample CD or another record. However, it is sometimes prudent to program your own, as these will play a large part in the rhythm of the record.

Ring modulation is possibly the easiest method for creating hi-hats and is easily accomplished by ring modulating a high-pitched triangle wave with a lower-pitched triangle. The result is a high-frequency noise type waveform that can then be modified with an amplifier envelope set to a zero attack, sustain and release, with a short to medium decay. If there isn't enough noise present, it can be augmented with a white noise waveform using the same envelope. Once this basic timbre is constructed, shortening the decay creates a closed hi-hat, while lengthening it will produce an open hat. Similarly, it's also worth experimenting by changing the decay slope to convex or concave to produce fatter or thinner sounding hats.

As both these timbres depend on the high-frequency content sitting at the top of the mix, compression should be avoided, but adjusting the amplifier decay of synthesized hi-hats can help to change the perceived speed of the track. If the hi-hats are sampled then this can be accomplished by importing them into a 'drum sampler' such as the *Native Instruments Battery* or alternatively you can use a transient designer such as *SPL's Transient Designer*, or the software alternative available in the *Waves Transform Package*. Using these to keep the decay quite short they will decay rapidly, which will make the rhythm appear faster. Conversely, by increasing the decay, you can make the rhythm appear slower. Most trance music tends to keep the closed hi-hats quite short through using a short decay, but the open hats often benefit from lengthening the decay. This can often help the higher frequencies of the loop gel together more appropriately for the genre.

With the loop complete, it's worthwhile looping it over a number of bars and experimenting with the synthesis parameters or a transient designer to each instrument in the loop to create a loop that gels together. This includes lengthening or shortening the decay parameters, pitching the instruments up or down and applying effects such as reverb (*subtly!*). Although at this stage it's inadvisable to use compression to warm up the loop, noise gates can be particularly useful for creating loops with a different rhythmic flow. By feeding the entire loop into the gate and reducing the threshold so that the transients of all the instruments pull though, shortening the release parameter on the gate will cut some instruments short.

Note: On the CD you can hear the complete loop from the example track and view the MIDI file.

Bass rhythm

With the drum groove laid down, most artists then move onto the bass rhythm. This, in many trance records, commonly consists of simple eighth notes with very little movement in pitch or in the timbre itself. The reason for this is to leave more 'room' for other elements, such as the inter-melodic lines and the chorded reprise after the middle eight of the track. For this technique to work the bass most usually sits *off* the beat rather than on it, since they're kept relatively short and play a very simple pattern to help keep the focus on the melodic lead. If they occurred *on* the beat the groove would sound similar to a march, because there would be no LFE (low-frequency energy) in between the kicks. As a result, the rhythm would have a series of 'holes' in between each kick/bass hit.

| Kick | .... | Kick | .... | Kick | .... | Kick | ..... |
| Bass | .... | Bass | .... | Bass | .... | Bass | ..... |

To avoid this, the bass commonly sits an eighth *after* the kick, resulting in:

| Kick | | Kick | | Kick | | Kick | |
| | ...Bass | | ...Bass | | ...Bass | | ...Bass |

What's more, by adopting this technique there is a more substantial groove to the music, since the two LFE signals are in effect working as a very basic Da Capo (binary phrase) throughout the track. That is, the first low-frequency signal (the kick) is asking a question which is answered by the second signal (the bass).

Of course, this, like everything, is open to artistic interpretation and while it is generally recommended that the bass occurs after the kick, it isn't necessary to place it an eighth after. Indeed, by experimenting through moving it subtly around the off-beat, the groove can become more intense or relaxed. Also, it isn't compulsory to use notes that are an eighth in length. While these are commonly used to allow the timbre's characteristics to be evident, smaller notes can be used to great effect. Similarly, if the track is quite simple, even with the main melody, increasing the length of the notes to a quarter allows more of the bass's character to be evident, which can be especially useful if the bass features lengthy timbre augmentation with LFOs or envelopes.

Generally, the bass, as with all the melodic elements of trance, works on a looped eight-bar basis – that is, the melodies, chords and so forth all loop around every eight bars (usually with a cymbal crash or short snare roll at the end of the eighth bar to denote the end of the musical passage). This helps to keep the repetition that forms the basis of the music. Thus, the bass can be programmed over two bars as a simple series of one-pitch notes that can then be pasted to create the other six bars. Working this way, you can select and pitch each consecutive two *bars* of notes up and down to create some movement in the rhythm.

The technique of pitching an entire bar rather than any individual notes within bars is very common, since pitch-shifting constituent notes creates a much more noticeable bass rhythm. This not only dissuades the focus from the main melody, but also lays down a series of musical ground rules as to how the melody can be written.

Keep in mind that if the bass features plenty of movement in a bar, for the track to remain musical, the melody should harmonize with this progression. However, by keeping the bass at one pitch throughout a bar and only moving an entire bar up or down, the only ground rules to the melody are to harmonize with the movements of the entire bar, allowing for more experimentation on the lead. Of course, this again is merely the conventional approach used by most trance artists and the example track, and it's certainly open to artistic licence. For example, some tracks will keep the bass off-beat but not maintain even spaces between each hit, thus creating a rhythmic drive to the bass. Tracks of this nature, however, will usually employ less melodic leads.

Bass timbre

For the bass timbre, analogue synthesizers (or DSP equivalents) are invariably chosen over digital since the bass is exposed from the kick and the 'distortion' introduced by analogue produces a warmer tone to complement the kick.

Trance basses are sometimes sourced from sample CDs, but many artists have commented that they prefer to create their own in a synth. Generally, any analogue synth will produce a good analogue bass, but many trance musicians often rely on the *Novation Bass Station* (VSTi), the *Novation SuperNova*, *Novation V-Station* (VSTi) or the *Access Virus* to produce the timbres. Using any of these, possibly the easiest solution for this timbre is to dial up a preset bass in a

Figure 10.3 A typical trance bass rhythm.

synth and then contort the results to suit the tune. If, however, you wish to program your own, a good starting point is to employ a pulse and sine (or triangle) oscillator and detune them + or −5 to produce the basic tone. In general, a heavy transient pluck is not entirely necessary since it isn't competing with the kick. Thus, the amp and filter's attack and decay parameters can be quite long and in many instances the decays actually act as the release parameters due to the relatively short duration of the notes.

When using the decay as a release parameter, the amplifier's release can then be used as a 'fine-tune' parameter to help the bass sit comfortably in with the drum loop. Additionally, by shortening the attack and decay parameters of the filter envelope, set to modulate at a positive depth, more of a pluck can be introduced into the sound so that the tone also complements the drum rhythm. In fact, it's essential to accomplish this rhythmic and tonal interaction between the drum loop and the bass by playing around with both the amp and filter's A/D EG before moving on. This is the underpinning of the entire track and, if it's weak, any instrumentation dropped on top will not rescue it.

By and large, most trance music doesn't employ any velocity changes on the bass rhythm, as this is usually associated to control the filter cut-off, so the harder it's struck, the more the filter opens and the more of a pluck it exhibits. This type of timbral variation is often avoided in order that the bass doesn't demand too much attention and rather simply 'pulses' along to lay down a solid foundation for the lead and counter-melodies. It is, of course, always prudent to experiment and it's also worth editing the synth so that velocity controls other parameters.

Note: Keep in mind that these are only very general guidelines and so long as the drum loop and bass sit together, then you should feel free to experiment.

On the subject of experimentation, delay, distortion and compression are often used on basses. While most effects should be avoided since they tend to spread the sound across the image (which destroys the stereo perspective), small amounts of controlled distortion can help to pull the bass out of the mix or give it a much stronger presence. Similarly, a conservatively applied delay effect can be used to create more complex sounding rhythms that will not place any musical 'restrictions' on the lead. Any effects should be applied cautiously, though, as the purpose of the bass is not to draw attention to itself, but simply underpin the more important elements.

Also, while it's true to say that the sounds from any synthesizer are already heavily compressed at the source, the principle behind a compressor here is to control the results from any preceding effects and to be used as an 'effect' itself. Using the exact same methods as compressing the kick, lengthening or shortening the release parameter can introduce a different character to the bass. More importantly, though, with both the drums and bass laid down, if they're both fed into a compressor (with the hi-hats muted), it can be set to activate on each kick, which results in it pumping the bass. Generally, the best compressors to use for this should be valve due to the second-order harmonic distortion they introduce, as this helps the mix to pump more musically. Again, this means using the *Joe Meek SC 2.2*, *UREI LA 3* or the *UREI 1176 LN*, but if you don't have access to these, the Waves *C1*, *C4* or *Renaissance* compressors will do the trick or the *PSP VintageWarmer* if you need more warmth.

Naturally, the amount of compression applied will depend upon the timbres used, but as a general guideline start by setting the ratio to 9:1, along with an attack of 5 ms and a medium release of approximately 200 ms. Set the threshold control to 0 dB and then slowly decrease it until every kick registers on the gain reduction meter by at least 3 dB. To avoid the volume anomaly (i.e. louder invariably sounds better!), set the make-up gain so that the loop is at the same volume as when the compressor is bypassed and then start experimenting with the release settings. By shortening the release the kicks will begin to pump the bass, which becomes progressively heavier the more that the release is shortened. Unfortunately, the only guidelines for how short this should be set are to use your ears and judgement, but try not to get too excited. The idea here is to help the drums and bass gel together into a cohesive whole and produce a rhythm that punches along energetically. On the same note, it should not be compressed so heavily that you lose the excursion of the kick altogether!

Once these two elements are working together, it's prudent to export the groove as the four separate two-bar loops of audio and drop them into a sampler. This allows you to trigger the groove from numerous points along the arrangement and also permits you to experiment with different progressions by simply hitting the appropriate key on the sampler. It's also sensible to export the drum track alone and keep a note of all compression or effects settings, along with the original MIDI files to come back to later. Most studios will create these 'track' sheets and it's wise to do so too, since many remixers will expect a track sheet if they're to remix the music. What's more, having the groove as a cohesive whole and sat in a sampler can often speed up the production process, since you already have a groove you're happy with and therefore are less likely to be tempted to make more 'tweaks' when they aren't required. In other words, the rest of the instruments have to be programmed to fit the groove, rather than constantly tweaking all instruments to make them all fit together. This not only prevents you from spending months constantly making unnecessary tweaks, but also most tracks will start with just the groove!

Note: On the CD you can hear the bass through GM and programmed synthesis and view the MIDI file.

Trance melodies

With the basic groove laid down, a good approach is to program the melody before any further elements are programmed, since this will often dictate the direction of any other instrumental counter-melodies.

Creating a melodic lead line for trance is possibly the most difficult part, as a good lead is derived not only from the melody but also the timbre, and both of these have to be 'accurate', so close scrutiny of the current scene and market leaders is absolutely essential in acquiring the right feel for the dance floor. Unfortunately, in terms of MIDI programming, trance leads follow very few 'rules', so how to program one is entirely up to your own creative instincts. That said, as ever there are some general guidelines that can be applied.

Firstly, in many cases the lead is constructed using a 'chorded' structure so that the notes alternate between two notes. This creates the results of jumping from the 'key' of the song to a higher note before returning to the main key again.

Figure 10.4 A typical trance melody.

Also, as Figure 10.4 shows, this lead continues over eight bars before being repeated again and each consecutive two bars of music will tend to rise in pitch rather than fall. Indeed, this building up the scale plays a key role in creating trance leads for the dance floor. Progressively moving up the scale throughout the bars and then setting the few final notes of the last bar higher in scale than any of the preceding notes will result in an uplifting feel. Conversely, by setting the final notes lower than that of the start of a bar the feeling is reversed, resulting in a track that pulls the listener down.

To keep the relative speed and energy of trance, it's also an idea to keep the notes short and stabby. This means using a mix of 32nd, 16th and/or eighth notes, and then using the decay and release envelope of the amplifier to lengthen and contort the timbre if required. This gives much more control over the groove, allowing you to immediately adjust the 'feel' of the music by simply adjusting the amplifier's release envelope rather than having to 'kick' MIDI notes around.

With these guidelines in mind, a lead can be constructed in any manner of ways, but if inspiration has left you there are some techniques that can sometimes produce good results. The first and possibly easiest method is to begin by fashioning a chord structure from the previously programmed bass line. This can follow the progression exactly, but generally it's worthwhile experimenting with a mix of parallel, oblique or contrary motions. For instance, if the bass is playing over eight bars and consists of:

Bar 1	Bar 2	Bar 3	Bar 4	Bar5	Bar 6	Bar 7	Bar 8
G	G	F	F	D♯	D♯	F	F

The key of the lead could follow this progression exactly in a parallel motion that would produce musically 'acceptable' results, even though it would be a little boring to the ear. However, by following the progression in a parallel motion for only the first, second and fourth bars and using a contrary motion on the third, the interaction between the lead and bass exhibits a much better result:

Bar no.:	1	2	3	4	5	6	7	8
Bass	G	G	F	F	D♯	D♯	F	F
Lead	G and D	G and D	F and A	F and A	G and D	G and D	F and A	F and A

Of course, it isn't imperative that the third bar becomes contrary and any or all bars could be contrary, or parallel, or oblique, provided that it sounds right to you.

Once this basic chord is constructed to complement the bass, using a noise gate or MIDI CC messages you can cut up each bar in a rhythmic manner. If using CC messages, though, a much better approach is to write a complex rhythmic hi-hat pattern and then use the sequencer to convert the hi-hat's notes to CC11 messages rather than program all the CCs in by hand. This can often lead to a robotic nature, whereas using a hi-hat pattern you're much more likely to produce a rhythmic flowing gate. This gated effect, once suitable, can be applied physically to the notes (i.e. cut them up to produce the effect) so that not only does each note retrigger the synth, but also it permits you to offset the top notes of a chord against the lower ones to produce the alternating notes pattern that's often used for the leads.

Alternatively, another frequently used method to create a trance melody is to begin with a synthesizer's arpeggiator, as this can help to get the general idea down. With some trial and error it's possible to find both arpeggiator pattern and chord combination to produce some inspiration for a lead melody. Once recorded as MIDI, this can be further attuned in a sequencer's key editor.

Another technique that's often referred to as the 'two-fingered' approach can produce usable results too. This involves tapping out a riff using just two keys an octave apart on the keyboard. Alternating between these two keys, keeping the lower key the same and continually moving higher in scale on the second key while playing the audio through a delay unit can produce interesting rhythmic movements. If this is recorded as MIDI information it can be edited further by lengthening notes so that they overlap each other slightly, or by adding further notes in between the two octaves to construct a melody that builds.

Melodic timbres

Just as the melody for the lead is important, so is the timbre and it's absolutely vital that time is taken to program a good one – the entire track will stand or fall on its quality. As discussed in the chapter on sound design, the lead is the most prominent part of a track and so is commonly the main instrument that sits in the mid range/upper mid range. Consequently, it should be rich in harmonics that occupy this area and should also exhibit some fluctuations in the character to remain interesting. This can sometimes be accomplished through modulating the pitch with an LFO on a digital synth, but the phase initialization from analogue synths, more often than not, will produce much better, richer results. Consequently, the synths preferred by many artists are the *Access Virus*, the *Novation SuperNova*, the *Novation A-Station* or the *Novation V-Station (VSTi)*.

We've already covered the basic principles behind creating a harmonically rich lead, so rather than go into too much detail again here, what follows is a quick overview:

Oscillators
- 2 × pulse detuned by −5 and +4 cents.
- 1 × sawtooth (try pitching this up or down an octave).
- 1 × pink noise.

Amp and filter envelope
- Zero attack.
- Medium decay.
- Small sustain.
- Short release.

Filter
- Low-pass 24 dB.

Modulation
- Sine wave LFO positively modulates pulse width of oscillator 1.
- Triangle wave LFO positively modulates pulse width of oscillator 2.
- Filter key-follow to maximum.
- Filter cut-off modulated with velocity commands.

This is naturally open to artistic licence and once the basics of the lead are down it's prudent to experiment by replacing oscillators, the LFO waveforms, depths and envelope settings to create variations. If this technique does not produce a lead that is rich enough, then it's worth employing a number of methods to make it 'bigger', such as layering, doubling, splitting, hocketing or residual synthesis, as discussed in the sound design chapter.

Finally, the timbre will also benefit heavily from applying both reverb and delay. The reverb is often applied quite heavily as a send effect, but with 50 ms of pre-delay so that the transient pulls through unmolested and the tail set quite short to prevent it from washing over the successive notes.

For the delay, this is best used as a send effect, but the settings will depend on the type of sound you require. Generally, the delays should be set to less than 30 ms to produce the granular delay effect to make the timbre appear big in the mix, but as always experimentation with longer settings may produce results you prefer.

If vocals are employed in the track then there may be little need for any effects as the lead should sit under them. However, if you want to keep a wide, harmonically rich sound and vocals, it's wise to employ a compressor on the lead timbre and feed the vocals into a sidechain so that the lead drops when the vocals are present.

> Note: On the CD you can hear the lead melody through GM and programmed synthesis and view the MIDI file.

Motifs and chord progressions

With both the main melody and the all-essential groove down, the final stage is to add motifs and any chord progressions. The latter progressions should require little explanation as, if used, it's simply a case of producing a chord structure that harmonizes with the lead and the bass. Motifs, on the other hand, are a little more complex and require more thought. These counter-melodies are the small ad-lib riffs best referred to as the icing used to finally decorate the musical cake and play an essential part of any dance music, adding much needed variation to what otherwise would be a rather repetitive track. These are often derived from the main melodic riff as not only do they often play a role in the beginning part of the track before the main melodic lead is introduced, but they are also reintroduced after the reprise.

There are various techniques employed to create motifs, but one of the quickest ways is to make a copy of the MIDI lead and 'simplify' it by removing notes to create a much simpler pattern. Once created, this pattern can be offset from the main lead to make the motif more noticeable, or alternatively it can occur at the same time as the lead but the attack and release of the timbre are lengthened. In fact, this latter approach will often produce the best results since, as discussed, dance music works on contrast and so far all the instruments have had a definite attack stage. By lengthening the attack and release, the motif will take on a whole new rhythmic character, which can quickly be changed by listening back to the arrangement so far and adjusting the amplifier envelope's parameters.

Above all, using any of these techniques it's sensible to avoid getting too carried away and it's vital that they are kept simple. Many dance tracks simply use a single pitch of notes, playing every eighth, 16th or quarter, or are constructed to interweave with the bass rather than the lead. This is because not only do simple motifs have a much more dramatic effect on the music than complex arpeggios, but you don't want to detract too much from the main melodic element.

It should also go without saying that the timbre used for any motif should be different from the lead melody, and it's at this point that you'll need to consider the frequencies that are used in the mix thus far.

Figure 10.5 A typical trance motif.

As touched upon, in many trance tracks the lead is harmonically rich and this reduces the available frequencies for a motif, so it's quite common to utilize a low-pass filter cut-off to reduce the harmonic content while the lead is playing, yet set this filter to open wider while there is no lead playing. Additionally, as with all dance music, this filter is tweaked 'live' to create additional movement and interest throughout the arrangement. This movement obviously has to be restrictive while the lead is playing, otherwise the mix can quickly become swamped in frequencies and difficult to mix properly, but during the beginning of the song the filter can be opened wider to allow more frequencies through to fill up any gaps in the mix.

More importantly, the rhythmical pitch of the motif will have to be carefully chosen. A motif that is mostly written using notes above C4 will naturally contain higher frequencies and few lower ones, so using a low-pass filter that is almost closed will reduce the frequencies to nothing. Therefore, the pitch of this motif and the timbre must occupy the low to mid range to fit with the bass and the lead melody, assuming of course that the lead melody contains higher frequencies.

Indeed, with trance, it's careful use of the filter that creates the final results, and often you will need to compromise carefully by cutting some of the frequencies of the lead to leave room for the motif and vice versa. Having said that, if the bass is quite simple, an often used technique is to program a motif that contains an equal amount of low frequencies as mid, and then use a filter to cut higher frequencies to mix the motif in with the frequencies of the bass, helping to enhance the low-end groove, before proceeding to remove this low-end interaction by cutting the lower frequencies and leaving the higher ones in as the track progresses. This creates the impression that all the sounds are interweaving with one another, helping to create more movement and the typical 'energy' that appears in most club tracks.

Note: On the CD you can hear the motif through GM and programmed synthesis and view the MIDI file.

One final, yet vital, element of trance before we come to the arrangement is the programming of rolls. Although these may not initially appear to be particularly significant, dance music relies on tension and drama, and this is best implemented with well-programmed snare rolls. The quick activist snare rolls that interrupt a groove, along with the huge four- or eight-bar snare roll that lead up to the main reprise, creates a build-up of tension followed by the ecstatic release when the track returns, so it's vital that these are well designed. Typically, the snares are the same as those used in the drum loop and are best programmed in step time, as this allows you to program them precisely while also allowing you to edit each note in terms of time, size and velocity. The most common snare rolls used in this genre are programmed by dropping in MIDI notes every 16th followed by 32nd and then drawing in velocity that continually rises. As a result, the snares become progressively louder as the roll reaches its apex.

Figure 10.6 A typical trance snare roll.

Even though many trance tracks rely on this strictly programmed 16th/32nd snare pattern, these rolls form such an integral part of the emotion of the music that it's worth experimenting further to see what emotion it can convey. For instance, rather than programming a strictly quantized pattern of notes, mixing 16th and 32nd notes together with velocity that does not always climb upwards can produce interesting variations.

Along with the velocity, pitch bend can also help to add some extra anticipation by pitching the snares up as they reach the apex, or alternatively a low-pass filter across the snares that opens as the roll progresses can also help create a building sensation (as used in the example track).

Snare rolls should not just be restricted to the main reprise either, as shorter rolls can be used to denote a new passage of music or the introduction of a new instrument. This commonly consists of placing two snares near the end of the bar a 16th apart. The timbre of these two snares is usually different to indicate a form of signing off the previous bar and introducing a new one. This is best accomplished by pitching the second snare a semitone up from the first while also placing the kick drum to play at the same time, in effect creating a small kick/snare roll.

Figure 10.7 A more intense snare roll.

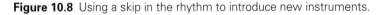

Figure 10.8 Using a skip in the rhythm to introduce new instruments.

Ultimately, these are only suggestions and you should be willing to spend as much time on the snare rolls as every other instrument. The real key is to try out variations at building and applying effects to the snares to create as much tension as possible within the music.

Note: On the CD you can hear the various snare rolls through GM and programmed synthesis and view the associated MIDI files.

The arrangement

With all the basic parts laid down you can begin to look at the arrangement. Unlike most other genres of dance music, this doesn't commonly rely on repeating phrases, but generally follows a more musical structure. Possibly the best way to accomplish this is to mute all the tracks bar the drums, begin playback, close your eyes and *listen*. If you've listened to trance before (and if not it is *strongly* recommended that you do!), you'll instinctively know when a new instrument should be introduced. However, if you're still stuck for where to go, listen to other trance tracks and physically count the bars, making note of where new instruments are introduced, and then implement this same structure in your own music. While this may seem like stealing, it isn't. Many trance tracks follow the very same arrangement principles (and remember that arrangements cannot be copyrighted!).

Indeed, although it would seem unreasonable to say that all euro-trance tracks follow a set pattern and rather that you should use your own integrity to produce the arrangement, trance – like most forms of dance music – is an exact science and it is important that you don't stray too far from what is considered the standard. Staying with this principle, it tends to follow a particular blueprint that normally consists of a combination of bass, drums and inter-melodic lines that culminate into the main uplifting climax. More often than not, this final climax is a culmination of the first part of the track mixed with the main melodic reprise. Occasionally, this main reprise is different thematically from the beginning section, but the timbre used will have been present beforehand to retain some continuity throughout.

With this in mind it's generally best to first plan the track using a song map. Trance, like all dance music, deals mostly in emotional waves, consisting of building and dropping the arrangement to generate an emotional state in the audience. Mathematics play a part in this, created by collecting and introducing sounds every four, eight, 16 or 32 bars, depending on the individual elements used within the track and the required length. This may sound overly mechanical, but it's a natural progression that we have all come to expect from music. As an example, try introducing a new element into an arrangement at the seventh bar rather than the eighth and it'll sound off beam.

Generally, most trance tracks will begin with just a basic drum loop playing over the first 16 bars to allow the DJ to mix the record in. At the end of bar 16, a new instrument is often introduced, which could be signified by a cymbal crash or short snare roll. This instrument is often another percussive element to complement the drum loop or the bass itself to generate the groove. This is often left playing as is for the next 16 or 32 bars to allow the clubbers to become more comfortable with the groove of the record. After these bars, the first motif tends to be introduced with another crash or snare roll and this motif continues for the next 16 or 32 bars. Notably, if this is to play for a more prolonged period of time, it's prudent to employ filter movements to prevent the track from becoming tedious.

After this, the first drop of the record commonly appears. This often consists of dropping the percussive elements bar the kick drum and motif, and this may continue for four or sometimes even eight bars. At the end of the final bar of this drop, a crash or short snare roll is used again to signify the groove of the record returning along with another new element. In many instances, this new element is the lead motif, which has been sliced up and simplified. This is to give the audience a 'taster' of what is to come later in the record. How long this plays for depends on the track itself, but generally it can range from 16 to 32 bars, with a cymbal crash or small snare roll/double kick placed at the end of every fourth bar to keep the interest. Occasionally, the lead riff may also be filtered with a low-pass filter that is gradually opened at the final four bars of this section, often complemented with a four-bar snare roll and perhaps a pad or string section laid underneath.

A crash cymbal placed at the end of the final bar, treated to reverb and delay so that it echoes across the track, is often implemented as the rest of the mix drops, leaving just the filtered down lead or the previously introduced pads. These pads/leads continue for four to eight bars, whereby a four- or eight-bar snare roll building behind the instruments creates an emotional rise towards the reprise of the track. This breakdown and build-up forms an essential part of part of trance and has to be executed carefully, as the whole idea is to link one fast section to another without losing the speed or feel of the track.

At the end of the snare roll, the same crash that was used to signify the start of the breakdown is commonly used to echo across the introduction of the full track and all of the previous instruments are playing together to create a crescendo of emotion. At this point, additional snares are often introduced to add a skip to the rhythm and a few sweeping effects or an additional motif is intro-duced. This often plays over 32 bars but sometimes 64, with a cymbal crash placed at the end of every fourth bar. Finally, a two-bar snare roll often signifies the removal of instruments such as the lead and melody, as the track is slowly broken down again to end with 16 bars of the drum loop.

Notably, trance relies heavily on drive and if at this stage it seems to meander along, it's prudent to inject some additional drive into the patterns. Typically, this can be accomplished through changing the position of some elements in relation to the rest of the rhythm. For instance, by keeping the kicks sat firmly on the first and third beats but moving the kicks (and snares) that occur on the second and fourth beats forward by a tick or two (so that they occur earlier in time), it will add more drive to the piece. Alternatively, by moving the bass a tick or two forward in the arrangement can add a feeling of drive.

Recommended listening

Ultimately, the purpose of this chapter has been to give some insight into the production of trance and there is no one definitive way to produce the genre. Indeed, the best way to learn new techniques and production ethics is to actively listen to the current market leaders and be creative with processing, effects and synthesis. Phaser, flangers, chorus, delay, reverb, distor-tion, compression and noise gates are the most common processor and effects used, so exper-iment by placing these in different orders to create the 'sound' you want. While the arrangement and general premise of every trance track is very similar, it's this production that differentiates it between all the others. With this in mind, what follows is a short list of artists that, at the time of writing, are considered the most influential in this area:

- Ferry Corsten.
- Redd Square.
- Binary Finery.
- Sasha.
- DJ Sammy.

Note: On the CD you can hear an excerpt from the trance track constructed through this chapter, along with the MIDI file to accompany it.

11 Hip-hop (rap)

Rap is like the polio vaccine. At first no one believed in it. Then, once they knew it worked, everyone wanted it.

Grandmaster Flash

Over recent years hip-hop has become a global phenomenon, so much so that what was once frowned upon for seemingly glorifying drugs, guns and general delinquency has now emerged as a multi-billion pound industry. However, it should be noted that despite the record industry's stance of pigeonholing absolutely everything that features a rapper as hip-hop, this isn't the case.

Hip-hop is a culture, not a form of music, and although it does encompass rap it also embraces dancing, language and fashion. Consequently, if you want to produce rap music it has very little to do with programming some MIDI patterns and rapping over the top. To better understand why this is, it's vital to know a little about the history and culture behind it all.

Hip-hop, as a culture, can be defined as consisting of four distinct elements: DJ'ing, breaking, graffiti and MC'ing (emceeing). The roots of the DJ'ing element can be traced back to 1950s Jamaica, where the 'DJs' began to experiment with groove elements of records, resulting in the creation of reggae, ska and the rock steady beat. In 1968, this became even more experimental when King Tubby created the first ever 'dub' record by dropping out all the vocals from the acetate discs he was to press (often called 'dub' plates).

At this same time many Jamaicans were emigrating to the USA, taking these new ideas with them to the ghettos of New York. One particular immigrant, Kool Herc, began to DJ at parties throughout the ghettos and used to chant rhymes over the top of the instrumental breaks of the records he played. As many of these breaks were only short, in 1974 he decided to play two copies of the same record on two decks and then use a mixer to switch between them, in effect creating a longer break beat to rhyme over. Almost simultaneously, in another ghetto Afrika Bambaata founded Zulu Nation, consisting of a group of DJs, break dancers, MCs and graffiti artists, and offered an alternative to the current street gang culture, allowing the youth culture to express themselves in various ways.

Inspired by these new DJ'ing tactics and culture, DJ Grandmaster Flash adopted the style and contorted it into a continuous stream of break beats. This allowed MCs to rhyme over the top of the beats to warm up the crowds, permitting the DJ to concentrate on developing new techniques, such as 'beat juggling', 'scratching', 'cutting' and 'breakdown'. Not being a DJ myself, I can't comment on what some of these techniques involve or who originally developed them but, arguably, it was Grandmaster Flash who introduced this new complex form of DJ'ing to the mass market with the release of *The Adventures of Grandmaster Flash on the Wheels of Steel* in 1981.

These continual break beats also gave rise to a new dance style known as breaking (a.k.a. B-boying), which consisted of a combination of fancy, complex footwork, spins and balancing on hands, heads or shoulders. Many of these moves were enthused through the relentless release of Kung Fu movies in the 1970s, but the inspiration could be drawn from anywhere, including their rivals during 'battles'. This is where 'crews' would compete against one another to see who was the better breaker, and in many instances you had to face off against a crew to be accepted into their clan.

This form of dancing was all encompassed and renamed by the media as 'break dancing', but not all forms of dancing at this time were breaking. Two other styles had developed, including 'locking' and 'popping', but these involved waving arms and sharp robotic movements, which were not classed as part of the hip-hop scene.

Alongside this new music grew another part of hip-hop culture, graffiti. To many, the explosion of graffiti is accredited to TAKI 183 and the publicity he received in the *New York Times* after 'tagging' numerous trains in the subway.

The term tag is used since graffiti refers to any unwelcome defacing of property with spray paint or markers, and the true hip-hop movement was much more artistic. A tag is essentially the writer's signature expressed in an artful and creative way that consisted of three areas: the 'tag', the 'throw-up' and the 'piece'. The tag is classed as the simplest form of graffiti and consisted of a signature in just one colour written using a marker. As time moved on, spray cans were introduced and the tag moved up a step to the throw-up, which was made up of two colours and resulted in more complex styles. The final style, a piece, is the most complex form of graffiti, for which people like Lee Quinones have made a living from selling to art galleries. On the streets, however, to be referred to as a piece it must consist of at least three colours and preferably be 'end-to-end' art. As most graffiti was sprayed onto the subway trains, this latter term should be self-explanatory. This type of graffiti is an incredibly complex form of art, which although viewed by many as destructive, takes a lot of planning, people and an artistic flair (*although for obvious reasons I cannot condone the defacing of any property*).

The last element of hip-hop, and the main purpose of this chapter, is derived from MC'ing. Although the media considers rap to be the same as MC'ing, rap is only one element of it. Indeed, MC'ing encapsulates everything from simply talking over the beats, rapping, or using your voice as an instrument (Human Beat Box). As touched upon, originally MC'ing was used to entertain the crowds by accompanying the break beats rather than taking the focus away from them. The original form was known as a call and response, whereby the MC would typically shout out 'Everybody in the house say yeah', to which the audience would respond with a resounding 'Yeah'.

Although it would be easy to say that rap developed from this basic form of MC'ing, to many it actually existed long before Kool Herc, Afrika Bambaata or Grandmaster Flash began to rhyme over the breaks. Indeed, it's believed to have originated in Jamaica, where stories were told in rhymes, otherwise known as 'toasts'.

In 1974 these were developed into the very first forms of rapping, where the youth would put together boastful rhymes to sit over the top of break beats in an effort to upstage the previous rapper. The first commercial pressing of rap music was by the Fatback Band in 1979 with the title *King Tim III*, but it took the Sugar Hill Gang's *Rappers Delight*, released later that same year, before rapping came to the attention of larger record labels as a viable and acceptable (in other words lucrative) form of music.

Over the following years, rap acts such as NWA, Ice T and Public Enemy brought rap to the forefront of music and demanded a bigger audience through their often hotly debated rhymes that were seen as glamorizing violence, prostitution and guns.

Musical analysis

Although technology has moved on since the early pioneers began to produce breaks from various DJ techniques, the fundamental creation of rap still relies heavily on the sound quality generated using these techniques.

To many the turntable techniques have been replaced with more recent developments such as samplers, hardware workstations and wave editors, but the 'feel' of the early techniques is still of paramount importance when producing rap. As a result, the break beats are *not* programmed through MIDI but are sampled from previous obscure records, and then contorted and manipulated in sample slicing programs such as Wavesurgeon to create new beats. This sampling is fundamental in keeping with the original roots of hip-hop by introducing the vintage 'vinyl' feel to the music that is not possible any other way. Similarly, many of the instrumental riffs and motifs are also sampled from early records but it is, however, possible to program these through MIDI provided that you use the right instruments.

Nonetheless, the preferred option *is* to sample from original vinyl and in many cases these will also produce all the sounds used in the genre, including basses and any 'lead' timbres. Indeed, the general consensus between most professional hip-hop musicians is that if you have a good record collection, a good record deck and a sampling workstation (such as the *AKAI MPC*) you can create rap.

Generally, sample CDs are avoided because everyone has access to them and no matter how they are contorted they can often still be recognized, so obscure vinyl is the preferred choice. Of course, for legal reasons I can't condone stealing riffs, loops or entire segments from other records, but many hip-hop artists agree that so long as the samples chosen are ambiguous or manipulated enough then there is no need to clear them.

Note: However, having said that, Dr Dre (Andre Young), who's considered by many as rap's current godfather, is well known to the American courts for his countless copyright infringements, as he's been sued over eight times in the last 3 years. The most publicized cases are when he asked for permission to use a sample from George Lucas' THX sonic boom and, when permission was refused, used it anyway. Consequently, he was sued for $10 million. London-based Minder Music also successfully sued him for $1.5 million after he sampled their bass line for his 2001 track *Let's Get High* from one of their releases.

More recently, a French jazz musician has started legal proceedings to sue him and Eminem for $10 million because (he claims) they used his music on Slim Shady's *Kill You*. As a result, Dr Dre has now apparently hired a musicologist to advise him on whether he can sample music riffs without infringing copyright. The purpose of this is not to disrespect or accuse Dr Dre or any other rap artists of serial stealing, but to simply act as food for thought before you consider that samples may not need to be cleared at all. While the best approach is to create a track by whatever means you feel necessary, if the track feels right, you should always attempt to clear the samples afterwards. This is a much easier approach than asking permission first, being refused and losing the chance to write what potentially could have been a hit.

Generally, rap can be described as using a slow laid-back beat that can vary from 80 through to 110 BPM. As discussed, most of the break beats used are sampled from other records to create the right vibe, while any further melodic instrumentation takes a back seat to the rapper. Consequently, these instruments play very simple melodies so as not to detract too much attention away from the rap. This means that drum rhythm is usually laid down first, which is followed by the rap vocals and then finally any further melodic instruments are added to 'sit' around the rap.

To accomplish this, you will need to have a good record (sample) collection, most of which have preferably only been small releases pressed in small amounts. Not only can these be picked up for relatively small outlay at any charity/junk shop, but you're more likely to find records that no one has ever considered using before. No matter what the genre of music (unless it's choir or similar), they will all feature a break that can be sampled and manipulated into a hip-hop beat, even unlikely records such as *Don Reeve's Non-Stop Double Hammond Goes Around The World* features some good hip-hop basses if you listen closely enough. The trick is to use your instincts and develop an ear for the right sounds, and this can only be accomplished from being actively involved in the hip-hop scene and gaining experience at listening out for the right sounds.

The rhythms

Primarily, there are two processes used to produce hip-hop rhythms. The first is to use sampled hits from various records and then reconstruct a rhythm live by assigning each pad of a workstation to a specific sample or, more commonly, simply sample a loop or number of loops from records and edit them using software such as Wavesurgeon. For the purpose of this chapter, we'll look at both methods, starting with sampling single hits.

Sampling individual hits is only recommended if you have access to a hardware drum sampler with pads, as creating the hip-hop vibe requires a 'live' feel. Indeed, very few professional rappers will rely on a software sequencer to produce rap, but instead use hardware sampler and sequencers since the timing must be incredibly accurate. Although most of these samplers will default at 16-bit 44.1 kHz, for rap it's recommended to sample the hits at 22 kHz and 12-bit. This reduces the range of the samples, giving them a grittier feel, which is often essential for capturing the feel. Once sampled, each of the hits is assigned its own pad in the sampler, allowing you to tap out a live rhythm, piece by piece.

Rap rhythms are generally kept quite simple and often consist of nothing more than a kick drum, snare, closed and open hi-hats, along with an occasional cymbal crash employed every four or so bars to mark the end of a musical segment. The time signature can also play a fundamental role and rap may use either 4/4 or 3/4 depending on the amount of swing you need the drums to exhibit. What is important, however, is that very few, or in some cases none, of the hits are quantized to the grid. The idea is to create the feel of a real drummer and very few drummers play exactly to the beat. Additionally, the patterns very rarely follow the typical *kick/snare/kick/snare* configuration used in most genres of dance and often rely more on rhythmic interplay between a few kicks followed by a snare, which is then followed by another few kicks and a snare. On top of this, a closed hi-hat pattern often follows a standard 16th pattern ticking away in the background to keep some rhythmic drive and syncopation to the loop.

In the example in Figure 11.1, the closed hi-hats are providing some offbeat syncopation, while both the kick and the snare are occurring ahead of time to increase the feel of a laid-back groove.

Figure 11.1 A typical 'laid-back' hip-hop rhythm.

In the example in Figure 11.2, more emphasis has been placed on rhythmic interaction between the snares and the drum kick. The closed hi-hat pattern has also been reduced and does not occur on the same beat as the open hi-hat to prevent any frequency clashes between the two. More importantly, though, note how few very of the elements actually occur on the beat.

Figure 11.2 A more complex rhythm.

The real key to producing a good rhythm in this manner is to experiment with the interplay between the kicks and snares, and ensure that they are offset from the beat to introduce the laid-back nature of the music. After this has been accomplished, the hi-hats are added to introduce some syncopation to the final groove. Also, as the above examples reveal, rap tends to be based around the constant repetition of two bars of music, allowing for a binary phrase to be applied in the bass or chorded lead if used/required. Following this, it's sensible to create a drum rhythm that continues over two bars before repeating again at the end.

Typically, in most music of this genre there are no variations in the rhythm as this tends to distract from the most important element of the track, but velocity plays a large role in the creation of the rhythm to keep the realism. The most usual method to implement is the *strong–weak–medium–weak* syncopation on the kick drums. The snares too are often subjected to different velocities, with the first and final snares using the highest values and those in between using a mixture of different values.

As touched upon, the common approach to producing this 'live' effect is to play the sampler workstation's pads live and record the subsequent MIDI data for editing later. This is one of the reasons why many rap artists will use the AKAI MPC sampler workstations, as the pads can respond to 16 different velocity values depending on how hard they are hit. There is no need to create the entire loop in one run and typically the drum's kick is recorded first, which is then played back from the unit while the snare rhythms are tapped in over the top. This is then followed with the open hi-hat patterns, followed by the closed patterns and any additional auxiliary percussion. This approach is preferred to 'stamping' in notes in a MIDI sequencer, since the timing becomes far too accurate and the MIDI delay between sequencer and sound source can result in the groove being lost.

Of course, even with the rhythm programmed in the sampler, it's highly probable that they will need some slight quantizing so that any further elements will lock to it, but any strict quantizing should be avoided altogether. The principle is to create a rhythm that has a loose feel but at the same time not so loose that the instruments do not lock to its rhythm. Once this has the right flow, the MPC's various synthesis parameters are used to manipulate each drum timbre to produce a rhythm that flows along. This can involve anything from pitch-shifting individual snares to create more complex rhythms, to tweaking the decay parameters to either lengthen or reduce them to create sharper, more distinct patterns or create slow flowing patterns. Ultimately, experimentation is the true nature here and close listening to the most popular current hip-hop tracks will let you know whether you're on the right track.

> Note: On the CD you can hear a typical hip-hop rhythm tapped in on the AKAI MPC 2000XL, followed by the edited version.

The second process to produce a rap rhythm is to sample entire breaks from records and then manipulate these so that they no longer sound like the original and are more suited towards rap. This is often the preferred approach by many artists, since the principle is much closer to the music produced by its original forefathers. This must be applied cautiously, however, since many record companies do not look lightly on sampling their artists' material. Regardless, this doesn't seem to have discouraged many rap artists and most will sample a loop and then manipulate it with various synthesis parameters to make it totally unrecognizable from the original.

This is accomplished through sampling the loop and importing it into a sample slicing program such as Wavesurgeon. This is often the preferred program to ReCycle by many, since it allows them to rearrange the loop within the program rather than export it into an REX-compatible sequencer before exporting it back to the sampler. Additionally, it offers a number of creative options, such as the ability to randomize the loop, apply any third-party effects such as distortion or apply bit quantizing to 'rough up' the sounds. Plus the option is available to export the samples and associated MIDI file into a sequencer for any MIDI editing if required.

Note: A demo of Wavesurgeon is included on the CD.

Many programmers do not settle for simply moving parts around, and in most cases they will adapt the sounds further by using the mixing desk as a sound design tool before sampling the results. Typical applications of this are to EQ the snares heavily to produce hi-hats or bright kicks. Similarly, transient designers such as *SPL's Transient Designer* or the software plug-in included in the *Waves Transform* bundle, can be used to edit the attack and sustain of audio envelopes. For instance, by setting the attack parameters on these quite long, the initial attack of the hit can be removed to produce a different timbre, while shortening the sustain parameter can make snares appear much tighter, snappy and controlled.

Figure 11.3 Cutting up and rearranging a rhythm.

Note: On the CD you can hear the 'original' loop and the subsequent results from using Wavesurgeon.

As an alternative to simply using and splicing one loop, it is often worth sampling three or four loops so that they can each be lined up in an audio sequencer and time-stretched so that each is playing at the correct tempo. Taking this latter approach, all the loops can be played simultaneously and elements from each loop can be cut up and removed to produce a rhythm composed of constituent parts from each loop.

This can help to create complex polyrhythms, but it's important to note that for this to work each drum loop should be quite simple, since the more complex each is, the more the combined

Figure 11.4 Lining up a series of loops and removing elements to create a basic rhythm.

rhythm will sound muddy and indistinct. This can make it particularly difficult to create a good rhythm through slicing and removing samples. Also, it isn't absolutely necessary to ensure that all the sampled loops occur at the same time, and in many instances by moving one of the loops forward by a few ticks it can help to create a more laid-back feel, since the snares will occur off the quantize values. Indeed, this technique is applied by many rap artists during sampling. Rather than sample the beginning of the drum track (i.e. from the first kick), they often sample the rhythm halfway through and then loop this to create the complete drum loop.

With this technique, the kicks are obviously positioned in a different place than the original record, meaning that the sample may start on a snare or even a hi-hat. To overcome this, it's common practice to remove the original kicks using sample slicing software and then reprogram a new kick loop to sit over the original sample. Typically, this new kick is derived from the Roland TR808 drum synthesizer, which is sampled at 12-bit 22 kHz to produce a gritty timbre. If you don't have access to this synth, then any synthesizer can be used to produce the requisite sounds using a 60 Hz sine wave positively modulated with a pitch envelope set to a fast attack and short decay. Although you can synthesize an additional transient to sit over this through using a square wave, generally the rap kick remains quite 'boomy' without a prominent attack so this isn't particularly necessary, but it is worth experimenting with.

Once the basic tone is present, lengthening the pitch decay will produce a timbre with a heavier boom, but this can only be lengthened so far before it begins to produce a sound that 'whoops'. If this is the case, then changing the pitch decay's envelope from linear to exponential (convex) will help to make it boom more. While these methods will produce a deep 808 kick, it may be worth using a self-oscillating filter and a noise gate in its place. This particular technique has already been discussed in sound design, but to recap quickly begin by increasing the resonance until it breaks into self-oscillation and produces a pure sine wave. After this, program the kick's pattern to suit the current loop and feed the resulting audio into a noise gate with the threshold set so that only the peaks of the wave are breaching it. Finally, set the attack to zero and then use the release to control the kick's delay and the hold time parameter to produce a heavier sound.

Roughly speaking, the drums are not compressed nor do they have any effects applied to the results, as the purpose behind the genre is to keep the sound raw. Besides, if the drums have been sampled from a record they will have already been compressed and effects applied at the original recording, mixing and mastering stages. Naturally, this is only a generalization and if the loop appears quite weak then compression may help to pull it up a little. Typically, the compressors used for this are as transparent as possible, so solid state are often used, such as the Alesis 3630 or the Waves C1.

As a starting point, try a ratio of 8:1 with a fast attack, short release and a threshold set so that the kick registers on the reduction meter by 2 dB. To avoid the volume anomaly, set the make-up gain so that the loop is at the same volume as when the compressor is bypassed and then start experimenting with the release settings. By shortening the release the kicks will begin to pump the rest of the drum timbres more heavily, but you need to exercise caution as to how much this pumps. Keep in mind that the hi-hats and snares are occurring in between the kicks, meaning that the snares and hi-hats will be pumped against the kick, which will deter from the 'raw' flavour of the timbres used! If after compression, the beats seem to exhibit a 'digital' nature, then it may be worth compressing them with a valve compressor such as the *UREI 1176LN*, or alternatively the *PSP VintageWarmer* can be used after the solid-state compression to give the sound warmth.

Above all, these are only suggestions and it's up to your own creativity to produce new variations of old loops through experimentation. One thing that is certain is that, although explaining the principles of creating hip-hop loops is simple, accomplishing it proficiently is an art form. You have to listen to loops with some awareness and be willing to sacrifice some elements to make the loop work properly. Complex loops are not part of hip-hop, but it does take plenty of time and effort to produce a loop that works properly. You have to have a thorough understanding of the genre and how the loops work in context to other instruments and this can only be accomplished by digesting both the culture and the music.

Note: On the CD you can hear the final rap rhythm used on the example track.

Rap vocals

With the drums laid down and presenting a groove, it's usual to record the rapper next, since any other instrumentation will sit around their rhymes. As these form the centrepiece of the record it's of paramount importance that you have a good rapper (anyone can rap, but only a select few can actually rap well!). Equally, it's also essential that the rapper has something to say that can be associated with the genre as a whole; simply rapping about how your car broke down last week will rarely work well (unless you happen to be Eminem) and this is another reason why it's vital that the rapper is actively involved in the whole culture.

Customarily, the lyrics are drawn from the lyricist rhyming about his (or her) skill, the skill of the crew they're associated with or simply 'dissing' (slang for disrespecting) their rivals. This is a form of battle similar to breaking and is essentially a way of competing with other rappers on the same circuit for prominence and respect. In fact, this form of competition is deeply embedded in the history of rap and is still evident today, playing a fundamental role in hip-hop.

Alternatively, the message conveyed may have a more personal political stance, rhyming about the current state of the ghetto, the nation, guns or violence and drug usage. These are usually not to invoke shock value but rather a true story from the rapper's own experience, and is the one of the key reasons why the vocalist must be actively involved in the hip-hop scene. If not, it's very doubtful that the words will have any real meaning to hip-hop aficionados and you could find yourself laughed off the circuit (*Vanilla Ice*).

Whatever the subject, rapping has become an incredibly complex lyrical delivery that basis itself around sophisticated rhythms that syncopate with the drums. Although there are many different styles, ranging from almost being on the verge of singing to more poetic 'talking', the delivery and tone are everything. Any good rapper will be able to produce complex rhythms and word play that complement the instrumentation while also being able to deliver emphasis on the most important words.

Another important aspect is quick thinking and the right delivery. Most professional rappers today are able to come up with rhymes on the spot, a talent that spawned from the earlier years, whereby they may be challenged on the street and had to come up with a rhyme quickly. Of course, this latter talent isn't an absolute necessity and while rappers such as Run DMC can come up with a rhyme off the top of their heads in an instant, they certainly don't walk on stage and make it up on the spot. Indeed, these and other hip-hop artists often take weeks to come up with an entire song's worth of rapping, since the sound and word play must have the right style for the music.

Usually, the rap is recorded before the instrumentation is completely laid down, so that any additional instruments can be laid down to emphasize the rap itself. Of course, this is purely the conventional approach and convention doesn't always play a part in music, so it's up to your own artistic interpretation.

The most typical microphone to record the vocals is the Shure SM58, as this produces the nasal tone typical of the genre. This should be hand held to allow the rapper to move freely, as this helps to keep the authenticity – no rapper stands still while they rhyme to a beat – and should be held approximately 10–20 centimetres (4–8 inches) away from the mouth. This does, however, depend on the vocalist, and you need to listen carefully to the sound produced and either increase or decrease this distance to capture the right sound. The vocal tone should be full and present, and if it seems deep and boomy ask the rapper to move the microphone further away from their mouth, while if it's too bright ask them to hold it closer.

Compression is often required while recording the vocals to prevent any clipping, but this should be applied lightly. If heavy compression is used the dynamics will be too heavily restricted, resulting in a rap that lacks any emotional stature. Generally, a valve compressor, such as the *UREI LA3* or *LN1176*, is the preferred option, but the *Joe Meek SC2.2* can also be used on vocals to add some warmth. Alternatively, a solid-state compressor followed by the *PSP VintageWarmer* can often add the requisite warmth after recording.

Whichever compressor is used, a good starting point is to set a threshold of −12 dB with a 3:1 ratio and a fast attack and moderately fast release. Then, once the vocalist has started to practise, reduce the threshold so that the reduction meters are only lit on the strongest part of the performance. Although the resulting rap is best recorded directly into an audio editor or sampler, some artists will record the vocals to tape cassette first before transferring this to the digital editor afterwards. This technique helps to capture a 'rough' yet warmer tone that is typical of many rap records.

Once recorded, the vocals are very rarely treated to any processing or effects, as the genre tends to move towards a raw sound rather than the 'professional' polish of most popular music. This means that, unlike most vocals, after recording compression is not applied unless it's absolutely necessary and effects such as reverb are applied very lightly, if at all. Additional compression may be needed if the rapper has been slightly off-axis from the mic while performing and in this instance very light settings are used to keep the vocals at a similar volume to allow them to pull through the mix. Unfortunately, there are no generic settings for this as it depends entirely on the vocals, but as a very rough starting point, use a threshold setting so that you have −5 dB on the reduction meter, along with a ratio of 4:1 and a fast attack and moderately fast release (approx. 200 ms). Unless you're after the very commercial form of rap, the ratio should not be set any higher than this and even then you should avoid going over 6:1, otherwise the vocals will lose all of their emotion and 'raw' flavour. If they sound a little thin, then small amounts of reverb may help, but similar to compression this should be kept to a minimum to retain the feel of the music. In fact, reverb should be applied so lightly that its application is only noticeable when it's removed.

Many tracks will also make use of double tracking the vocals to accentuate particular words or phrases. As these are always recorded in mono and sit dead centre of the mix, by double tracking the important phrases the original vocal and the double-tracked phrase can be panned left or right to fill the stereo image and bring attention to particular parts. Many rap tracks will offset one side of the vocals by a few ticks so that they occur moments later than the original, helping to thicken out the vocals and bring more attention to the phrase. These usually lie at the end of a number of bars (similar to how much of dance uses a cymbal crash), but it's up to your own artistic interpretation as to where these accentuated phrases or words should sit. The best advice that can be offered is to *listen* to the rapper and double track him (or her) wherever they place their own accents with the music.

Note: Occasionally, when a rapper's voice needs to be thickened out in the mix, the vocals may be double tracked and one of the tracks de-tuned by −5 semitones to give the impression of two rappers simultaneously vocalizing.

The bass rhythm

Although most old skool hip-hop tracks consisted of nothing more than a break beat and a rapper, today's producers often drop in bass lines to add some bottom-end weight and help the track groove along. In the majority of cases these are real bass guitars that have been sourced from another record and are not normally cut up or edited in any way. In fact, quite a few of today's tracks have used bass lines (and in some cases the entire track they originally belonged too) that are instantly recognizable. Generally, it is more prudent to write your own though, since there's no chance of being caught for copyright infringement and it's much easier to write a bass around a rapper than have a rapper try to rhyme around the bass.

On the whole, most basses are kept relatively simple, as they merely act as the underpinning of the track and do not fundamentally form a major melodic element in themselves. This is to prevent the movements and groove of the bass from drawing attention away from the rapper's rhymes. As a result, a bass line will often consist of nothing more than a few eighth or quarter notes per bar that will tend to stay at one pitch or move very subtly in pitch by no more than three or five semi-tones. As the rhyme at this stage is already down, the key of the bass will most probably already be determined and it's simply a case of repeating the first bar of the rhyme and moving down the keyboard making notes of the pitches that harmonize with the rapper. Once you have these, it's just a case of producing a groove that flows alongside and complements the rapper.

Notably, the bass groove very rarely occurs on the beat of each bar and it's quite usual for it to begin just a few ticks *before* or *after* the bar to offset it from the rest of the record. This is to emulate the nuances of a real bassist, since they very rarely play dead on the beat. What's more, by offsetting it by eight or 10 ticks, the drums will appear to pull or push the record forward, helping to create a more flowing groove. Creating this 'vibe' in between the bass and drums forms an essential part of rap, but whether to sit it before or after the beat is entirely open to artistic licence and after programming it's worth experimenting by moving it before or after the beat to see which produces the best results.

Figure 11.5 A typical bass rhythm (as used in the example track).

For the bass timbre, real bass guitars are customarily preferred over synthetic instruments to not only keep with the original feel of the music, but also because most real basses are recorded by miking up the bass cab and so tend to be particularly noisy. Although some engineers will struggle to get these sounding clean, within hip-hop this noise plays a fundamental role. While it may sound awful soloed, in the context of a mix it often helps it to pull through and add the grit required. Recording live instruments is beyond the scope of this book, as it requires a whole new set of production ethics that would require another book to explain, so for those who do not want to sample from another record but want the 'real' bass sound in the music, here we'll look at how to program a 'realistic' one through MIDI.

The key to programming any real instrument is to take note of how they're played and then emulate this action with MIDI and a series of CC commands. In this case, most bass guitars use the first four strings of a normal guitar (E–A–D–G), but these are tuned an octave lower, resulting in the E being close to three octaves below middle C. Also, they are monophonic, not polyphonic, so the only time notes will actually overlap is when the resonance of the previous string is still dying away as the next note is plucked. This effect can be emulated by leaving the preceding note playing for a few ticks while the next note in the sequence has started.

The strings can also either be plucked or struck and the two techniques produce different results. If the string is plucked, the sound is much brighter and has a longer resonance than if it were simply struck. To copy this, the velocity will need to be mapped to the filter cut-off of the bass module so that higher values open the filter more. Not all notes will be struck at the same velocity, though, and if the bassist is playing a fast rhythm the consecutive notes will commonly have less velocity, since they have to move their hand and pluck the next string quickly. Naturally, this is only a guideline and you should edit each velocity value until it produces a realistic feel.

Depending on the 'bassist' they may also use a technique known as 'hammer on', whereby they play a string and then hit a different pitch on the fret. This results in the pitch changing without actually being accompanied with another pluck of the string. To emulate this, you'll need to make use of pitch bend, so this will first need setting to a maximum bend limit of two semitones, since guitars don't 'bend' any further than this.

Begin by programming two notes, for instance an E0 followed by an A0, and leave the E0 playing underneath the successive A0 for around 100 ticks. At the very beginning of the bass track, drop in a pitch bend message to ensure that it's set midway (i.e. no pitch bend) and just before where the second note occurs drop in another pitch bend message to bend the tone up to A0. If this is programmed correctly, on playback you'll notice that as the E0 ends, the pitch will bend upwards to A0, simulating the effect. Although this could be left as is, it's sensible to drop in a CC11 message (expression) directly after the pitch bend, as this will reduce the overall volume of the second note so that it doesn't sound like it has been plucked.

In addition to this, it's also worthwhile employing some fret noise and finger slides. Most good tone modules will include fret noise that can be dropped in between the notes to emulate the bassist's fingers sliding along the fret board. The pitch bending is best emulated by first programming the notes to overlap slightly and then recording movements of the pitch bend wheel live and editing them in the sequencer.

> Note: On the CD you can hear a programmed 'realistic' bass and view the MIDI file.

For those who break out in a sweat at the mere mention of in-depth MIDI editing, it isn't always necessary to use a real bass and some producers do use synthetic instruments, provided that they're deep enough and have a good 'body'. Although the type of timbre obviously differs from producer to producer, the general tone can be made in any synth (*but Synapse Junglist is recommended*) by using both triangle and pulse waves, with the latter detuned from the triangle by −3 cents.

Set the amplifier and filter's envelope to a fast attack, medium decay, with a short sustain and no release, and use a two-pole low-pass filter, with a low resonance setting. Finally, modulate the pulse width of the pulse with a sine, triangle or sawtooth LFO set to a slow rate and medium depth. This will produce a basic timbre typical of the genre, but it's worth experimenting with the filter's decay, the LFO's rate and depth, and the shape of the decay envelopes on both the amp and filter to mould the timbre to suit your music.

This will not suit all music, though, and some artists ignore synths and construct their own basses from culminating samples together or pitching a sample down the key range, because just about anything sounds great when it's pitched down a couple of octaves into the bass register. The key is to experiment in creating a low tone that has the energy to add some bottom-end weight to the music, but at the same time does not demand too much attention.

On the subject of experimentation, distortion and compression are often used on hip-hop basses. While most effects should be avoided since they tend to spread the sound across the image (which destroys the stereo perspective), small amounts of controlled distortion can help to pull the bass out of the mix and often give it the much needed rawness. Similarly, compression after the distortion can be used to even out the volume, keep the effect under control and bring the overall levels up. As a general guideline, start by setting the ratio to 4:1, along with an attack of 5 ms and a medium release of approximately 150 ms. Set the threshold control to 0 dB and then slowly decrease it until the bass registers on the gain reduction meter by at couple of decibels. Finally, set the make-up gain so that the loop is at the same volume as when the compressor is bypassed, and then start experimenting with the attack and release settings until it produces the level and 'sound' you require. Unlike most other genres of dance music, rap is one of the few that does not use a compressor to 'pump' the low frequencies, so any compression applied should be applied lightly and you should generally try to avoid pumping the low end.

Note: On the CD you can hear the final bass through GM and programmed synthesis.

Chords and motifs

Similar to the bass, old skool rap did not employ any chords or leads, but nonetheless most tracks today will employ some sort of lead sound to sit behind the rapper. Somewhat unsurprisingly, these are usually sampled from other records too, but they can also be programmed in MIDI and then sampled and subsequently reduced in bit and sample rate to add the necessary 'grunge'.

As always, there are no definitive rules as to what instruments should be used as a lead or, indeed, how they should be programmed but, as always, there are some guidelines that will at least help keep you on the right path. Fundamentally, hip-hop tends to stay with real instruments rather than go for synthetic and a proportionate amount of tracks will use Rhodes pianos, old 'electric' pianos, bells, orchestral strings or orchestral 'pizzicatos'. As with the bass these are kept quite simple so as not to take the focus away from the rapper. What's more, they're not quantized and are commonly recorded live to help keep the instrumentation sounding real rather than as if it were generated by a machine. As a very rough generalization these instruments, but particularly the piano, tend to follow the da capo sequence, with the first bar asking the question and the second bar providing the answer. Typically, this 'answer' plays the same melody as the question, but rises in pitch by a few semitones (try semitone shifts of three, five or seven).

Figure 11.6 The piano rhythm using the da capo (from the example track).

If more than one 'lead' instrument is used in the track, only one of these follows the da capo sequence and any other instruments tend to remain the same throughout the bars.

Figure 11.7 The pizz rhythm (from the example track).

Of particular note, even if the elements of the record are *not* sampled, they should nevertheless sound as though they are. This means gratuitous use of bit reduction and sample reduction, along with sampling the vinyl crackle and hiss from a record. This is the preferred approach to using a vinyl plug-in, as these rely on generating random noise or cycling a sample every few bars of music, which often doesn't sound particularly realistic. However, by sampling vinyl hiss from a record you can place it into a wave editor and begin cutting up, looping and creating a good six bars of hiss to sit in the background. If an audio sequencer is being used, this crackle can be laid on an audio track and each six-bar segment can be overlapped at different points throughout the arrangement to create more realistic crackle.

> Note: On the CD you can hear both the pizz and piano, and view the associated MIDI files.

The arrangement of rap differs from most other genres in this book in that it doesn't follow the dance builds or the verse and chorus structure. Instead, the music generally repeats the same bars over and over throughout the track and relies on the rappers to provide the movement of interest. In fact, if you were to strip the rappers away from most rap tracks, the backing would quickly become tedious to listen to, since there are so few changes implemented. On occasion, some tracks will 'drop' the beat close to the end of a bar while the rapper rhymes. This not only accentuates the rapper, but also creates a rush of emotion as the track returns on the next bar. As ever, if you're unsure about how to produce a typical rap arrangement, listen to other similar tracks and physically count the bars, making note of where new instruments are introduced, and then implement this same structure in your own music. This isn't stealing, it's research and every musician on the planet does it. In fact, you'll find that many rap tracks follow the very same arrangement principles (and remember that arrangements cannot be copyrighted!).

Recommended listening

Ultimately, the purpose of this chapter is to give some insight into the production of hip-hop and there is no one definitive way to produce the genre. Indeed, the best way to learn new techniques and production ethics is to actively listen to the current market leaders and be creative with processing, effects and synthesis. Chorus, delay, reverb, distortion, compression and noise gates are the most common processor and effects used within rap music, so experiment by placing these in different orders to create the 'sound' you want. While the arrangement and general premise of each track are similar, it's this production that differentiates it between all the others. With this in mind, what follows is a short list of some of the artists that, at the time of writing, are considered the most influential in this area:

- Africa Bambaata (often seen as the original godfather of hip-hop).
- Eminem.
- Dr Dre.
- Public Enemy.
- Run DMC.
- El-P.
- LL Cool J.
- Cyprus Hill.
- Ice T.

> Note: On the CD you can hear an excerpt from the rap track constructed through this chapter, along with the MIDI file to accompany it.

12 Ambient/chill-out

Ambient is music that envelops the listener without drawing any attention to itself.

Brian Eno

Ambient music has enjoyed a long, if somewhat diverse, history and its subsequent offshoots have formed an important role in dance music since 1989. However, it has only recently re-established itself to many as part of the dance music scene when beats were again dropped over the atmospheric content and it was relabelled by record companies and the media as 'chill-out' music.

The roots of ambient music are nothing if not unusual. It's believed that it first came about in the mid 1970s when Brian Eno was run over by a taxi. While he was recovering in hospital a friend gave him a tape machine with a few dodgy tapes of harp music. The result was that the music didn't remain at a constant volume and on occasion dropped considerably in gain, whereby it mixed with the rain hitting the hospital windows. This second accident formed the beginnings of ambient as Eno began to experiment by mixing in real-world sounds such as whale song and wind chimes with constantly changing synthesized textures. Described by Eno himself as music that didn't draw attention to itself (go figure), it enjoyed some success but was soon relabelled as 'muzak' and subsequently poor imitations began to appear as background music for shopping centres and elevators, and to many the genre was soon written off as 'music suitable only for hippies'.

When the rave generation emerged in the late 1980s and early 1990s, a DJ named Alex Patterson began to experiment with Eno's previous works, playing it back to clubbers in small side rooms who needed a rest from the fast, hard-hitting beats of the main room. These side rooms were soon labelled by clubbers as 'chill-out' rooms, a place where you could go and take a break from the hectic four-to-the-floor beats. As these 'chill-out' rooms began to gain more popularity with clubbers, Patterson teamed up with fellow musician Jimmy Cauty to form The Orb and jointly released what many believe to be the first ever ambient house music for clubbers, somewhat strangely named *A Huge Ever Growing Pulsating Brain That Rules from the Centre of the Ultraworld*. Soon after the release of the album, Patterson and Cauty went their separate ways and while Cauty teamed up with Bill Drummond to form KLF, Patterson continued to write under the moniker of The Orb and continued to DJ in the chill-out rooms.

This form of ambient house began to grow into its own genre and chill-out rooms became a fundamental part of the rave scene. Some DJs became VJs (video jockeys), mixing not only real-world sounds with slow, drawn out drum loops, but also projecting and mixing images onto large screens to accompany the music. This was soon followed by a series of ambient compilations hitting the public market, and artists such as William Orbit and Aphex Twin soon released their own ambient house albums.

In 1992, the genre was in full flow and, as different artists adopted the scene, each putting their own twist on the music, it began to diversify into subgenres such as ambient dub (ambient music with a bass), conventional (ambient with a 4/4 backbeat), beatless (no backbeat but following the same repetition as dance music) and soundscape (essentially pop music with a slow laid-back beat). By 1995 ambient music was everywhere, as the larger record companies took the sound on board and saturated the market with countless ambient compilations (although thankfully there was never any *Now That's What I Call Ambient Music Volume 390*). Even artists who had ignored the genre before began to hop on board in the hopes of making a quick buck out of the new fad.

As with most music that is encapsulated and bastardized in this way, eventually ambient house became a victim of its own success. The general public became tired of the sound and, to the joy of many a clubber, it was no longer the new fashion and it returned to only being played where it had originated from – the chill-out rooms.

Fast forward to the year 2000 and a small Balearic island in the middle of the Mediterranean began to revive the public's and record companies' interest in ambient house. DJs in Ibiza's Café Del Mar began to tailor music to suit the beautiful sunsets by mixing jazz, classical, Hispanic and New Age together to produce laid-back beats for clubbers to once again chill out to. Now repackaged and relabelled 'chill-out music', it's enjoying renewed interest and has become a genre in its own right. However, while chill-out certainly has it roots deeply embedded in ambient music, they have over time become two very different genres. Thus, as the purpose of this book is to cover dance music, for this chapter we'll concentrate on chill-out rather than ambient.

Musical analysis

As always, possibly the best way to begin writing any dance music is to find the most popular tracks around at the moment and break them down to their basic elements. Once this is accomplished we can begin to examine the similarities between each track and determine exactly what it is that differentiates it from other musical styles. In fact, all music that can be placed into a genre-specific category will share similarities in terms of the arrangement and/or sonic elements.

Generally, chill-out is music that incorporates elements from a number of different styles, such as electronica, New Age, classical, Hispanic and jazz. However, it's this very mix of different styles that makes an *exact* definition impossible. Indeed, it could be said that as long as the tempo remains below 120 BPM and it employs a laid-back groove it could be classed as chill-out. In fact, the only way to analyse the music is to take Brian Eno's advice in that it shouldn't draw too much attention and ideally be the type of inoffensive music that most people could easily sit back and relax to.

Defining exactly what this means is difficult, but we can settle for saying that it commonly has a slow rhythmic or trance-like nature, combining both electronic and real instruments, which are often backed up with drop-out beats and occasionally smooth haunting vocals. Generally, many of these real instruments will be recorded live or sampled from sample CDs, but in some instances they can also be 'borrowed' from other records. Consequently, chill-out can utilize almost any time signature, from the four to the floor to a more swing-orientated 3/4, and be produced adagio, andante or moderato. In terms of physical tempo, this can range from 80 BPM and move towards the upper limits of 120 BPM, but most tend to use a happy medium, staying around the 110–120 BPM mark. This, of course, isn't a demanding rule, but it is important not to go too fast as you need to relax to it!

To better explain how this theory is applied in practice we'll look at the creation behind a typical chill-out track. As always, any music is an entirely individual, artistic endeavour and the purpose of this analysis is not to lay down a series of 'ground rules' on how it should be created, rather its intention is to describe some of the general principles behind how to accomplish the characteristic arrangements, sounds and processing. Indeed, in the end it's up to the individual (i.e. you) to experiment and put a twist on the music that suits your own particular style.

The rhythms

There are no particular distinctions as to what makes a good chill-out drum loop; the only guide that can be offered is that the loop should remain relatively simple and exhibit a laid-back feel. The kick drum can lie on the beat, every beat, or it can be less common, appearing on the second and fourth, or first and third beats, or any division thereof. This is simply because the laid-back feel is often derived from the pattern and position of the snares in relation to the kick. Indeed, it's the rhythmic interplay between these two elements that creates the 'feel' of the drum rhythm. Additionally, closed and open hi-hat patterns are invariably employed in these rhythms to act as a form of rhythmic syncopation (i.e. not on the beat) to help the groove flow along.

Figure 12.1 Typical chill-out drum loop.

As Figure 12.1 shows, the closed hi-hats, unlike most other genres of dance, do not necessarily have to sit on every 16th division and, along with the open hats, they can play a pattern of their own. It is wise, however, to remove any closed hi-hats that occur on the same position as the open hi-hat, as this results in a frequency clash and can often result in a rhythm sounding too 'programmed'. Auxiliary instruments such as congas, bongos, toms, cabassas, triangles, shakers and tambourines also often make an appearance in chill-out and, similar to the hi-hat patterns, these act as a form of syncopation to help the rhythm flow.

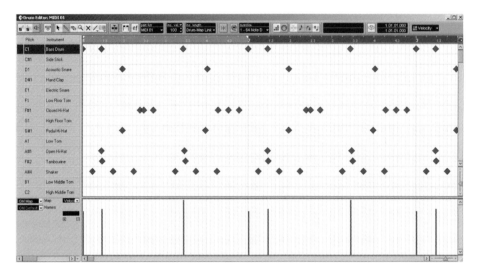

Figure 12.2 A more complex chill-out drum loop.

This, of course, is only a general guideline to the drum patterns used and it's fully open to artistic licence. The key is not to produce a loop that sounds rigid or programmed through experimenting by moving the kicks in relation to the snares to adjust the interplay between the two. If the tempo appears too fast, reducing the amount of snares or auxiliary instruments employed in the rhythm will often help slow it down and is preferable to physically slowing down the tempo, since this will affect the rest of the instrumentation laid on top. Additionally, as this genre occasionally follows a verse chorus structure similar to most 'pop' music to denote the change between the two, it's common to employ a couple of extra snare hits positioned together at the end of the bar to create a skip in the rhythm.

Of particular note, the velocities of each percussive instrument should be manipulated to help create a more flowing groove. Although this is entirely open to interpretation, the kick commonly follows the typical *strong–weak–medium–weak* syncopation. This is, of course, provided that four kicks are per bar. If there are three then it could follow a *strong–weak–medium* or perhaps a *strong–medium–strong* syncopation. What is common, however, is that the first kick of the bar remains the strongest so as to denote the beginning of a new bar. Even then, this is simply the more conventional and convention shouldn't always play a part in your music, so you should feel free to experiment by placing the accents at different divisions on the beat. By doing so, the rhythm of the piece can change quite severely, so it is worth experimenting.

With chill-out the sounds used for the rhythm will often play a vital role in obtaining the laid-back feel, and once the basic sounds have been laid down it's quite usual to apply effects to the individual elements. The hits can be sourced from anywhere, including other records, most drum machines (including the TR909 and TR808) or even the standard GM drum kits (particularly jazz kits). It is, of course, possible to synthesize a kit and this can often prove the best method, since the available parameters will allow you to modify each sound to suit any particular rhythm. As we've already discussed the basics behind synthesizing a drum kit, what follows is a short overview of the three most important elements (kicks, snares, hi-hats), with some tips on how the parameters can be used to create typical 'laid-back' timbres. Keep in mind that these are only tips and, above all, however you do it, if it sounds right then it most probably is.

Rhythm timbres

Generally, the kick is quite deep and 'boomy', as tight kicks tend to add some small urgency to the music. Thus, a good starting point is a 40 Hz sine wave with positive pitch modulation using a fast attack and a medium to longish decay period. There is usually little need to employ a 'click' at the beginning of the waveform, since this makes it appear sharper which isn't *generally* required but, as always, experimentation is essential. If possible, it's also worthwhile experimenting by setting the decay's transition to exponential, so that the pitch curves outwards as it falls, creating a 'boomier' sound. Additionally, it's often worth sending the kick to a reverb unit set to a medium pre-delay and short tail to help lengthen the timbre and prevent it from dropping off too sharply.

If you decide to use a kick sample or a GM module and there is too much of an attack on the sound, a compressor set so that it clamps down on the transient may produce the sound you require. That said, in the long run a preferred option would be to use a transient designer such as featured in the Waves *Transform Bundle* plug-in or SPL's hardware *Transient Designer*. Using these, the front of the kick can be removed to produce a more relaxed sounding timbre.

In direct contrast to the kicks, the snares often use a sharp transient stage, as these dictate the rhythmic flow of the loop and need to be apparent in the loop, more so if it's a particularly busy loop packed with auxiliary instruments. The typical snare can be created by using a triangle oscillator mixed with pink noise. A high-pass filter across these generally produces the most common chill-out snares, but it's also worth trying a notch or band-pass filter to see which you prefer. The amp's attack should be set fast, while the decay can be set anywhere from medium to long depending on the sounds that are already in the mix and how slurry you want the snare to appear.

Alternatively, if it's possible with the chosen synthesizer it's prudent to employ a different amp EG for both the noise and triangle wave. The triangle wave can be kept quite short and swift by using a fast decay, while the noise can be made to ring a little further by increasing its decay parameter. The further this rings out, the more slurred and 'ambient' it will become. If the snares are too bright, it can be removed with some EQ or, preferably, try replacing the triangle wave with a pulse and experiment with the pulse width.

The snares also often benefit from small amounts of reverb by sending, not inserting, them to a reverb unit. A typical snare room preset that's available on nearly every reverb unit around will usually suffice, but depending on the loop it may be worth reducing the decay slightly. Again, if samples are used in place of synthesis, the jazz snares on most GM modules produce a fairly good rendition, which may be more suited towards your track. Using a transient designer along with sending the snare to a reverb unit will often help you achieve the typical 'chilled' feel.

Some chill-out will benefit from a snare with a less prominent attack phase, which can be accomplished by using a transient designer to remove the attack phase. On top of this, it may also be worthwhile introducing a small amount of pink noise over the top of the snare to give the impression of a brush stick used to play. Alternatively, you could just use a brush kit from the GM standard or sample one from a jazz record. For obvious legal reasons I can't condone this latter approach, but a proportionate amount of dance artists do it.

The hi-hats can be created in two ways – with noise or ring modulation – but typically for chill-out, ring modulation tends to produce better results. This is accomplished by modulating a high-pitched triangle wave with a lower-pitched triangle to produce a high-frequency noise. As a starting point, set the amp envelope to a zero attack, sustain and release, with a short to medium decay. If there isn't enough of a noise present, it can be augmented with a white noise waveform using the same envelope. Once this basic timbre is constructed, shortening the decay creates a closed hi-hat, while lengthening it will produce an open hat.

As touched upon, not all chill-out will use MIDI programmed loops and some artists will borrow loops or at least elements of loops from previous records. Again for legal reasons I can't con-done this behaviour, but nevertheless it is a popular technique. These sampled loops are often imported into a sample slicing program and messed around until it sounds nothing like the original. Of course, you shouldn't just settle with rearranging parts and, after they're remodelled, it's worthwhile importing them into a wave editor and using a transient designer to model the transients and releases of the drum hits. Alternatively, another approach is to use two sampled loops together and mix them together in a sequencer. By time-stretching the loops so they share the same tempo, the loops are played simultaneously and any conflicting elements can be cut out from one of the loops.

Figure 12.3 Lining up a series of loops and removing elements to create a basic rhythm.

This technique is useful particularly if you wish to create polyrhythms that are difficult to accomplish any other way. On top of this, it's worth experimenting by moving one of the loops back in time to make some of the elements of the loop occur in different places.

With the loop laid down you can then look towards effecting and processing the rhythm, although roughly speaking, chill-out loops are not compressed for numerous reasons, including:

- They've already been sampled from a record and are already compressed.
- They are from a synth and are already compressed.
- They don't need to 'pump' musically.
- Second-order harmonic distortion from valve compression isn't usually required.

However, if the loop has been composed of numerous elements from different sources, compression may help to bring some of the levels under control. Typically, as a starting point set a relatively slow attack with a medium release and a ratio of about 4:1, and adjust the threshold so that the loop registers approximately 4 or 5 dB on the gain reduction meter. It's usually wise to ensure that the loop doesn't pump, so always experiment with the compressor's release to make its action as transparent as possible.

Effects-wise, although in many genres it's usually worth avoiding applying any effects to drum timbres, chill-out is often an exception to the rule. While it's unwise to wash the entire loop in cavernous reverb, if it's lightly applied to the snares and kicks, it can often present a more laid-back feel to the timbres. That said, any reverb or any effect for that matter must be applied cautiously. The drums form the backbone of the record and if an effect is applied too heavily it will often destroy the transients, reducing the bottom-end groove to mush.

> Note: On the CD you can hear the 'original' loop and the subsequent results from chopping up rhythms.

Melodic leads

Although it isn't unusual for a chill-out track to be based entirely around synthetic instruments, most do tend to contain real instruments. These can range from acoustic guitars playing Spanish rhythms, classical strings, pianos, wind or a combination thereof. Roughly speaking, these are best laid down before the bass, as much of the attention of the track is commonly directed towards the lead and the bass simply acts as the underpinning. Since most real instrument leads are commonly sampled from sample CDs or other records, it's much easier to form the bass around their progression rather than try to reprogram a lead to sit around a bass. Although it is possible to cut up and rearrange a lead in a sample splicing program to suit a pre-programmed bass, this can often result in the lead behaving unnaturally, so it's prudent to rearrange the bass instead.

Samples of real instruments do not necessarily have to be used (although in the majority they do sound better when used for leads) and, provided that you have a good tone module, it's possible to program realistic instruments through MIDI. If you take this approach, however, it's important to take note of the key and range you write the song in, because every acoustic instrument has a limit to how high and low it can play. For instance, with an acoustic guitar, the (standard) tuning is E, A, D, G, B, E, with this latter E a major third above middle C. This is incredibly important to keep in mind while programming, since if you exceed the limitations of an acoustic instrument the human ear will instinctively know that it's been programmed rather than played.

Acoustic guitars

The key to programming acoustic guitars is to take note of how they're physically played and then (provided the tone module produces an authentic sound) emulate this with MIDI. Firstly, you need to take into account the way that a guitarist's hand moves while playing. If the style is Hispanic then the strings are commonly plucked rather than strummed, so the velocity of each note will be relatively high throughout, since this will be mapped to control the cut-off (i.e. the harder a string is plucked, the brighter the timbre becomes).

Not all of these velocity plucks will be the same, however, since the performer will want to play in time with the track and if a note occurs directly after the first, it will take a finite amount of time to move the fingers and pluck the next string. This often results in the string not being accentuated as much as the preceding one due to 'time restraints'. Conversely, if there is a larger distance between the notes then there is a higher likelihood that the string will be plucked at a higher velocity.

In many instances, particularly if the string has been plucked hard, the resonance may still be dying away as the next note starts, so this needs to be taken into account when programming the MIDI. Similarly, it's also worth bearing in mind that a typical acoustic guitar will not bend more than an octave, so it's prudent to set the receiving MIDI device to an octave so you can use the pitch bend wheel to create slides. In between notes that are very close together, it may also be worth adding the occasional fret squeal for added realism.

If the guitar is strummed, then you need to take the action of the hand into account. Commonly, a guitarist will begin by strumming downwards rather than upwards and if the rhythm is quite fast on an upwards stroke it's rare that all the strings will be hit. Indeed, it's quite unusual for the bottom string to be struck, as this tends to make the guitar sound too 'thick', so this will need to be emulated while programming.

Additionally, all the strings will not be struck at exactly the same time due to the strumming action. Obviously, this means that each note on message will occur a few ticks later than the previous, which will depend on whether it's strummed upwards or downwards and the speed of the rhythm. To keep time, guitarists also tend to continue moving their hands upwards and downwards when not playing, so this 'rhythm' will need to be employed if there is a break in the strumming to allow it to return at the 'right' position in the arrangement. Finally, and some-what strangely, if a guitarist starts on a downward stroke they tend to come in a little earlier than the beat, while if they start on an upward stroke they tend to start *on* the beat. The reason behind this is too complex to explain, plus it would also involve me knowing why.

If you want to go even further and replicate trills or tremolo, it's worthwhile recording pitch bend movements onto a separate MIDI track and imposing them onto the longer notes of the original guitar track by setting them to the same MIDI channel. These pitch bend movements are recorded onto a separate channel since, once it's imposed onto the notes, it's likely that the pitch bend information will need further editing.

Note: On the CD you can hear a programmed acoustic guitar and view the associated MIDI file.

Wind instruments

Although fundamentally there are two types of wind instruments, brass and reed, in the concept of MIDI programming they're both programmed following similar principles, since they both rely on variations in air pressure to produce the timbre. As a result, if you're planning on making music that contains wind instruments it's prudent to invest in a breath controller. These are small devices that connect directly to the MIDI IN port and act in a similar to any wind instrument, although they do not produce any sound of their own. Rather, they measure changes in air pressure as you blow into them and convert this into CC2 (breath controller) messages that can then be used to control the connected MIDI synthesizer or sampler. These can be expensive, though, so if the use of wind instruments is only occasional, it's worth programming the typical nuances by hand into the sequencer.

Firstly, the volume and brightness of the notes are proportional to the amount of air pressure in the instrument. All good MIDI instruments will set the velocity to control the filter cut-off of the instrument, so the brightness can be controlled with judicious use of MIDI velocity. The volume of the note, however, is a little more complicated to emulate, since many reed instrumentalists will deliberately begin most notes by blowing softly before increasing the pressure to force the instrument to become louder. Essentially, this means that the notes begin softly before quickly rising in volume, and occasionally pitch. The best way to emulate this feel is to carefully program a series of breath controller (CC2) messages while using a mix of expression controllers (CC11) to control the volume adjustments. Alternatively, brass instruments will often start below the required pitch and slide up to it, so this is best emulated by recording live pitch bend movements into a sequencer and editing them to suit later.

On top of this, many wind instruments also introduce vibrato if the note is held for a prolonged period of time due to the variations in air pressure through the instrument. While it is possible to emulate this response with expression (CC11) controllers, generally introducing small pitch spikes at the later stages of the sustain can produce better results. It should be noted that these pitch 'spikes' only appear in the later stages of the note's sustain, though, and should not be introduced at the beginning of a note. Possibly the best way to determine where these spikes should appear is to physically emulate playing a wind instrument by beginning to blow at the start of the MIDI note on and when you begin to draw short of breath, insert some pitch bend messages. Alternatively, if the synth allows you to fade in the LFO, you can use this to modulate the volume by setting it to a sine wave on a slow rate, modulating the volume lightly. As long as the LFO fade-in time is set quite long, it will only begin to appear towards the end of a note.

On the subject of breathing, bear in mind that all musicians are human and as such need to breathe occasionally. In the context of MIDI, this means that you should avoid playing a single note over a large number of bars. Similarly, if a series of notes are played consecutively, remember that the musician needs enough time to take a deep breath for the next note (wind instruments are not polyphonic!). If there isn't enough time, the next note will generally be played softer due to less air velocity from a short breath, but if there is too short a space the instrument will sound emulated rather than real.

Finally, you need to consider how the notes will end. Neither reed nor brass instruments will simply stop at the end of the note but instead they will fade down in volume while also lowering in pitch as the air velocity reduces (an effect known musically as diminuendo). This can be emulated with a series of expression (CC11) messages and some pitch bend.

Note: On the CD you can hear a programmed brass and view the associated MIDI file.

As touched upon, some tracks do not employ real instruments and rely solely on synthetic, but if this is the case it's often worth avoiding any sharp aggressive synthesizer patches such as distorted saw-based leads, since these give the impression of a 'cutting' track, whereas softer sounds will tend to sound more laid back. This is an important aspect to bear in mind, especially when mastering the track after it's complete. Slow relaxed songs will invariably have the mid range cut to emphasize the low and high end, while more aggressive songs with have the bass and hi-hats cut to produce more mid range and make it appear insistent.

MIDI programming-wise, if synthetic timbres are used then many tracks tend to refrain from using notes less than a quarter in length, since shorter notes will make the track appear faster. Additionally, by using longer notes many of the tones can utilize a long attack, which will invariably create a more relaxed perception of the music. In fact, long attack times can form an important aspect of the music, and it's good to experiment with all the timbres used in the creation of this genre by lengthening the amp's/filter's attack and release times and taking note of the effect it has on the track.

Vocal chants are sometimes used in the place of leads, as these can be used as instruments in themselves. Typically, in many popular tracks these are sampled from other records or sample CDs, but it's sometimes worthwhile recording your own. Characteristically, condenser microphones are the preferred choice over dynamic as these produce more accurate results, but the diaphragm size will depend on the effect you wish to achieve. As touched upon in Chapter 8, many producers will use a large diaphragm for vocals, but if you're after 'ghostly' chants a small diaphragm mic such as the Rode NT1 will often produce better results due to the more precise frequency response. Of course, this is simply the conventional approach and, if at all possible, it's worthwhile experimenting with different microphones to see which produces the best results.

Vocals will also benefit from compression during the recording stage to prevent any clipping in the recording, but this must be applied lightly. Unlike most other genres, where the vocals are often squashed to death to suit the compressed nature of the music, chill-out relies on a live feel with the high frequencies intact. A good starting point is to set a threshold of −9 dB with a 2:1 ratio, a fast attack and moderately fast release. Then, once the vocalist has started to practise, reduce the threshold so that the reduction meters are only lit on the strongest part of the performance.

Similar to the drum rhythms, effects can also play a large part in attaining the right lead sound and you should feel free to experiment. Reverb is the main suspect here, but phasers and flangers can also work especially well if lightly applied. A good (if often overused) technique to produce ghostly vocal chants is to reverse the vocal sample, apply a large cavernous reverb and then reverse them back the right way around again. This way the reverb tail-off will act as a build-up to the vocal chant. The key is, as always, to experiment.

Bass

With the lead laid down, the bass is often programmed next to form the basic groove of the song. Generally, this tends to be quite minimal, consisting mostly of notes over an eighth in length to prevent the groove from appearing too fast. These are often derived from the principles encapsulated in the dub scene, staying quite minimal to allow space for the lead to breathe without having to compete with the bass for prominence. Indeed, it's incredibly important that the bass is not too busy either rhythmically or melodically and many chill-out tracks borrow heavily from the pentatonic scale used in dub and R'n'B. That is, they use no more than a five-note scale playing a single harmony, restricting the movements to a maximum five-semitone shift.

Naturally the bass should interact with the lead and, as touched upon in Chapter 2 on music theory, this can be accomplished through using parallel, oblique or contrary motions. In many cases, an oblique motion is the preferred option, since a majority of this music is driven by the melody and you don't want the bass to detract from it. Having said that, you should feel free to try out both contrary and parallel motions to see which produces the best results. What is important, however, is that the bass coincides with the drum rhythm. Bear in mind that if the drums have been effected with reverb, the subsequent tail-off between each hit will leave less room for the bass, so it will need to be less melodic or rhythmic. Conversely, if the drums are left dry, it's possible to employ a more rhythmic or melodic bass without creating any conflicts.

Figure 12.4 A typical chill-out bass line.

Typically, most chill-out basses are synthesized rather than real, since the timbre is often quite deep and doesn't want to attract as much attention as the lead. Analogue synthesizers usually provide the best results due to the uncontrollable yet pleasing phasing of the oscillators. Nonetheless, the beginnings of a typical timbre can be programmed on most synthesizers. A good starting point is to use a sine wave with a sawtooth transposed up from it by 5 or 7 cents. Unlike most of the timbres used throughout the production of chill-out, the attack should be well defined to prevent the bottom-end groove turning to mush, but there is usually very little decay, if any, to prevent it from becoming too plucky. Indeed, in much of the music of this genre the bass tends to hum rather than pluck to prevent it from drawing attention away from the chilled feeling. The best way to accomplish this type of sound is to use a simple on/off envelope for the amp envelope. This is basically an envelope with no attack, decay or release, but a quite high sustain. This forces the sound to jump directly into the sustain stage, which produces a constant bass tone for as long as the key is depressed. If this results in a bass with little or no sonic definition, a small pluck can be added by using a low-pass filter with a low cut-off and high resonance that's controlled with an envelope using a zero attack, sustain and release, but a medium decay. By adjusting the depth of the envelope modulating the filters or increasing/decreasing the decay stage, more or less of a pluck can be applied. If this process is used, however, it's prudent to employ filter key tracking so that the filter's action follows the pitch.

At this stage it's also prudent to experiment with the release parameters on the filter and amp envelope to help the bass sit comfortably in with the drum loop. In fact, it's essential to accomplish this rhythmic and tonal interaction between the drum loop and the bass by playing around with both the amp and filter's A/R EG before moving on. This is the underpinning of the entire track and it needs to be right for the style of music.

Since these waveforms are produced from synthesizers, there is little need to compress the results since they're already compressed at the source. Also, it's unwise to compress the drums against the bass to produce the archetypal 'dance music' pumping effect, as this will only accentuate the drum rhythm and can make the music appear too 'tight' rather than relaxed. Similarly, effects are usually avoided on the bass timbre because they can not only spread the timbre across the stereo spectrum, but they can also bring too much attention to the bass. Of course, that said, the entire track may be based around the rhythmic movements of the bass and if this is the case you should feel free to experiment with every effect you can lay your hands on!

> Note: On the CD you can hear the programmed bass and view the associated MIDI file.

Chords/pads

Slow evolving pads can play a vital role in a proportionate amount of chill-out. Since many of the tracks are thematically simple, a pad can be used to fill the gap between the groove of the record and the lead and/or vocals. What's more, slow evolving strings often add to the overall atmosphere of the music and can often be used to dictate the drive behind the track.

We've already covered the principles behind creating chord structures in Chapter 2, but as a quick refresher, the chords here will act as a harmony to the bass and lead. This means that they should fit in between the rhythmic interplay of the instruments so far without actually drawing too much attention (rather like the genre as a whole). For this, they need to be very closely related to the key of the track and not use a progression that is particularly dissonant. Generally, this can be accomplished by forming a chord progression from any notes used in the bass and experimenting with a progression that works. For instance, if the bass is in E, then C major would produce a good harmony because this contains an E (C–E–G). The real solution is to experiment with different chords and progressions until you come up with something that works.

Once the progression is down, it can then be used to add some drive to the track. Although many chill-out tracks will meander along like a Sunday stroll, if you place a chord that plays consistently over a large number of bars all of the feel can quickly vanish. This can sometimes be avoided by moving the chords back in time by a small amount so that they occur a little later than the bar. Alternatively, if they follow a faster progression, they can be moved forward in time to add some push to the music. It should be noted here, however, that if the pads employ a long attack stage, they may not actually become evident until much later in the bar, which can destroy the feel of the music. In this instance, you will need to counter this effect by moving the chords so that they occur much earlier.

Figure 12.5 A typical chord progression.

The instruments used to create these are more often than not analogue in nature due to the constant phase discrepancies of the oscillators, which creates additional movement. Thus, analogue or analogue emulations will invariably produce much better results than an FM synth or one based around S&S. What's more, it's of vital importance that you take into account the current frequencies used in the mix so far.

With the drums, bass, lead and possibly guide vocals playing along, there will be a limited frequency range where you can fit the chords in and if the timbre used is too harmonically rich it will be incredibly difficult to fit them into the mix without having to resort to aggressive EQ cuts. This will change the entire character of the chords and in some instances may make it inappropriate for the music, so it's much better to program the chords timbre to fit in now, rather than when approaching the mix. Ideally, during this construction stage you should look towards producing a 'mix' that sounds right at this stage and not rely on the mixing desk's EQ later to cure a host of problems. Keep in mind that the EQ on a mixing desk is very subtle and usually kept aside for subtle tonal changes. This obviously means that there are two programming possibilities for the chords. They either need to be quite lush to fill out any noticeable gaps in the mix or they need to be quite thin so they do not collide with the rest of the instrumentation.

If the mix is particularly busy in terms of frequencies, then it's prudent to build a relatively thin pad that can then be 'thickened' out later, if required, with phasers, flangers and reverb or chorus effects. This is accomplished by using pulse waves, as they have less harmonic content than saws and triangles and do not have the 'weight' of a sine wave (plus, of course, they sound more interesting!). Generally, only one pulse wave is required with the amp envelope set to a medium attack, sustain and release, but a fast decay. If this timbre is to sit in the upper mid range of the music then it's best to use a 12 dB high-pass filter to remove the bottom end of the pulse, otherwise use a low-pass to remove the top end and then experiment with the resonance until it produces a general static tone that suits the track. Following this, set an LFO to positive modulation with a slow rate on the pulse width of the oscillator and the filter to add some movement. Which waveform to use for the LFO will depend on the track itself, but by first sculpting the static tone to fit into the track you'll have a much better idea of which wave to use and how much it should modulate the parameters. If the timbre appears too 'statically modulated' in that it still seems uninteresting, use a different rate and waveform for the oscillator and filter so that the two beat against each other. Alternatively, if the timbre still appears too thin even after applying effects, add a second pulse detuned from the first by 3 cents with the same amp envelope but use a different LFO waveform to modulate the pulse width.

If the track has a 'hole' in the mix then you'll need to construct a wider, thicker pad to fill this out. As a starting point for these types of pads, square waves mixed with triangles or saws often produce the best results. Detune the saw (or triangle) from the square wave by 5 or 7 cents depending on how thick you need the pad to be and set the amp attack to a medium sustain and release, but no attack and a short decay. Using a low-pass filter set the cut-off to medium with a high resonance and set the filters' EG to a short decay and a medium attack, sustain and release. This should modulate the filters positively so that the filters sweep through the attack and decay of the amp but meet at the sustain portion, although it is worth experimenting with negative modulation to see if this produces better results for the music. This will produce a basic pad timbre, but if it seems a little too static for the music, use an LFO set to a sine or triangle wave using a slow rate and medium depth to positively modulate the pitch of the second oscillator. For even more interest, you could also use an LFO to modulate the filters' cut-off.

Effects can also play an important role in creating interesting and evolving pads. Wide chorus effects, rotary speaker simulations, flangers, phasers and reverb can all help to add a sense of movement to help fill out any holes in the mix. Again, as with most instruments in this genre, compression is generally not required on pads since they'll already be compressed at the source, but a noise gate can be used creatively to model the sound. For instance, by setting a low threshold and using an immediate attack and release, the pad will start and stop abruptly, producing a 'sampled' effect that cannot be replicated through synthesis parameters. What's more, if you set the hold time to zero and adjust the threshold so that it lies just above the volume of the timbre, the gate will 'chatter', which can be used to interesting effect.

Note: On the CD you can hear a programmed pad and view the associated MIDI file.

Arrangement

As touched upon in the musical analysis, chill-out is incredibly diverse, and pretty much anything with a slow tempo and a laid-back feel could be classified as part of the genre. Consequently, the arrangement of the music can follow any structure you feel suits. In fact, as this genre is so varied, it's difficult to point out globally accepted arrangements and the best way to obtain a good representation of how to arrange the track is to listen to what are considered the classics and then copy them. This isn't stealing, it's research and nearly every musician on the planet follows this same route. As a helping hand, however, we'll look at one of the popular structures used by this genre, which is based around the same principle as most popular music.

Most popular music is constructed of around five parts, consisting of the verse, chorus, bridge, middle eight and occasionally a key change near the end. The verse is the part of the song where the story is told and a good song will feature around three or four verses. The chorus is the more exciting part of the music which follows the verse and is where all the instruments move up a key and give the listener something to sing along with. The bridge is the break between the verse and chorus, and usually consists of a drum fill which leads onto the middle eight of the track. This, as its name would suggest, is usually eight bars long and is the break of the track that is often the most exposed segment to the dance musician's sampler. Finally, there's the key change and, although not all records will use this, it consists of shifting the entire song up a key to give the impression that it's reached its apex. If we break this down in a number of bars we're left with:

*   Verse 1 – *commonly* 16 bars.
*   Chorus – *commonly* eight bars.
*   Verse 2 – *commonly* 16 bars.
*   Chorus – *commonly* eight bars.
*   Verse 3 – *commonly* 16 bars.
*   Chorus – *commonly* eight bars.
*   Bridge – *commonly* one bar.
*   Middle eight – *commonly* eight bars.
*   Double chorus – *commonly* 16 bars.

This is, of course, open to artistic licence and some artists will play the first two verses before hitting the chorus. This approach can often help to add excitement to a track, as the listener is expecting to hear the chorus after the first verse. If this route is taken, even though there may be different vocals employed in the verses, if they are played one after the other it can become tiresome, so a popular trick is to add a motif into the second verse to differentiate it from the first.

Recommended listening

Ultimately, the purpose of this chapter is to give some insight into the production of chill-out and there is no one definitive way to produce the genre. Indeed, the best way to learn new techniques and production ethics is to actively listen to the current market leaders and be creative with processing, effects and synthesis. Chorus, delay, reverb, distortion, compression and noise gates are the most common processor and effects used within chill-out, so experiment by placing these in different orders to create the 'sound' you want.

While the arrangement and general premise of each track is generally similar, it's this production that differentiates it between all the others. With this in mind, what follows is a short list of some of the artists that, at the time of writing, are considered the most influential in this area:

*   Aphex Twin (*Selected Ambient Works*).
*   William Orbit (*Excursions in Ambience*).
*   Future Sound of London.
*   Ministry of Sound (*Chill-Out Sessions*).
*   Groove Armada.

Note: On the CD you can hear an excerpt from the chill-out track constructed through this chapter, along with the MIDI file to accompany it.

13 House

By 1981 they declared that disco was dead and there were no more up-tempo dance records. That's when I realized I had to start changing things to keep feeding my dance floor.

Frankie Knuckles

Although some house aficionados will refuse to admit it, the development of house music has much of its success accredited to the rise and fall of disco. As a result, to appreciate the history of house music we need to look further back than the 1980s and the development of the TR909 and TR808 drum machines; we also need to examine the growth of disco during the 1970s. This is because disco still forms a fundamental part of some of today's house music and in many instances older disco records have been scrupulously sampled to produce the latest house tracks.

Pinning down an exact point in time where disco first appeared is difficult, since a majority of the elements that make disco appeared in earlier records. Nonetheless, arguably it is said to have first originated in the early 1970s and was derived from the funk music that was popular with black audiences at that time. Some big name producers such as Nile Rodgers, Quincy Jones, Tom Moulton, Giorgio Moroder and Vincent Montana began to move away from recording the self-composed music, and started to hire session musicians and produce hits for artists whose only purpose was to supply vocals and become a marketable commodity.

Diana Ross became one of the first *disco* manufactured success stories with the release of *Love to Love You Baby* in 1975, believed by many to be the first disco record to hit and be accepted by the mainstream public. This 'new' form of music was still in its infancy, however, and it took the release of the motion picture *Saturday Night Fever* in 1977 before it eventually became a widespread phenomenon. Indeed, by the late 1970s over 200 000 people were attending discotheques in the UK alone and disco records contributed to over 60% of the UK charts.

As with most genres of music that become popular, many artists and record labels jumped on the wave of this new happening vibe and it was soon deluged with countless disco versions of original songs and other pointless, and poorly produced, disco records as the genre became commercially bastardized. As a result, disco fell victim to its own success in the late 1970s and early 1980s, with the campaign of '*disco sucks*' growing ever more popular. In fact, in one extreme incident Steve Vahl, a rock DJ who had been against disco from the start, encouraged people to bring their disco collections to a baseball game on 12 July 1979 for a ritual burning. After the game, a huge bonfire was lit and the fans were asked to throw all their disco vinyl onto the fire.

By 1981, disco was dead, but not without first changing the entire face of club culture, changing the balance of power between smaller and major labels, and preparing the way for a new wave of music. Out of these ashes rose the phoenix that is house, but it had been a large underground movement before this and, contrary to the misconceptions that are spread around, it had actually been in very early stages of evolution before disco hit the mainstream.

Although to many Frankie Knuckles is seen as the 'godfather' of house, it's true foundations lie well before and can be traced back to as early as 1970. At this time, Francis Grosso, a resident DJ at a converted church known as the Sanctuary, was the first ever DJ to mix two early disco records together to produce a continual groove to keep the party-goers on the dance floor. What's more, he is also believed to be the first DJ to mix one record over the top of another, a technique that was to form the very basis of dance music culture.

Drawing inspiration from this new form of mixing, DJ Nicky Siano set up a New York club known as The Gallery, and hired Frankie Knuckles and Larry Levan to prepare the club for the night by spiking the drinks with lysergic acid diethylamide (LSD/Acid/Trips). In return he taught both all about the basics of this new form of mixing records, and soon after they moved on to become resident DJs in other clubs. Levan began residency at The Continental Baths, while Knuckles began at Better Days, to soon rejoin Levan at The Continental Baths 6 months down the line. The two worked together until 1977, when Levan left the club to start his own and was asked to DJ at a new club named The Warehouse in Chicago. Since Levan was now running his own club he refused, but recommended Knuckles, who accepted the offer and promptly moved to Chicago.

Since this new club had no music policy, Knuckles was free to experiment and show off the techniques he'd been taught by Nicky Siano. Word quickly spread about this new form of disco and The Warehouse quickly became the place to be for the predominantly gay crowd. Since no 'house' records actually existed at this time, the term house did not refer to any particular music, but simply referred to The Warehouse and the style of continual mixing it had adopted. In fact, at this time the word 'house' was used to speak about music, attitudes and clothing. If a track was house it was from a cool club and something that you would never hear on a commercial radio station, whereas if you *were* house it meant you attended all the cool clubs, wore the 'right' clothing and listened to 'cool' music.

By late 1982/early 1983, the popularity of The Warehouse began to fall rapidly as the owners began to double the admission price as it became more commercial, so Knuckles decided to leave and start his own club known as The Powerhouse. His devoted followers went with him, but in retaliation The Warehouse was renamed The Music Box and the owners hired a new DJ named Ron Hardy. Although Hardy wasn't a doctor he dabbled in numerous pharmaceuticals and in turn was addicted to most of them, but was nonetheless a very talented DJ. While Knuckles kept a fairly clean sound, Hardy pounded out an eclectic mix of beats and grooves, mixing euro disco, funk and soul to produce an endless onslaught to keep the crowd up on the floor. Even to this day, Ron Hardy is viewed by many as the greatest ever DJ.

Simultaneously, WBMX, a local radio station, also broadcast late night mixes made by the Hot Mix Five. The team consisted of Ralphi Rossario, Kenny 'Jammin'' Jason, Steve 'Silk' Hurley, Mickey 'Mixin'' Oliver and Farley 'Jackmaster' Funk. These DJs played a non-stop mixture of British New Romantic music, ranging from Depeche Mode to Yazoo and Gary Numan, along with the latest music from Kraftwerk, Yello and George Clinton. In fact, so popular was the UK New Romantic scene that a third of the American charts consisted of UK music.

However, it wasn't just the music that the people tuned in for, it was the mixing styles of the five DJs. Using techniques that had never been heard of before, they would play two of the same records simultaneously to produce phasing effects, perform scratches and back spins, and generally produce a perfect mix from a number of different records. Due to the show's popularity, it was soon moved to a daytime slot and kids would skip school just to listen to the latest mixes. In fact, it was so popular that Chicago's only dance music store, 'Imports Etc', began to put a notice-board up in the window documenting all the records that had been played the previous day to prevent them from being overwhelmed with enquiries.

Meanwhile, Frankie Knuckles was suffering from a lack of new material. The 'disco sucks' campaign had destroyed the industry and all the labels were no longer producing disco. As a result, he had to turn to playing imports from Italy (the only country left that was still producing disco) alongside more dub influenced music. More importantly for the history of house, though, he also turned to long-time friend Erasmo Rivieria, who was currently studying sound engineering, to help him create reworks of the earlier disco records in an attempt to keep his set alive. Using reel-to-reel tape recorders, the duo would record and cut up records, extending the intros and breakbeats and layering new sounds on top of them to create more complex mixes. This was soon pushed further as he began to experiment by placing entirely new rhythms and bass lines underneath familiar tracks. While this undeniably began to form the basis of house music, no one had yet released a true house record, and in the end it was Jesse Saunders' release of *On And On* in 1984 that landmarked the first true house music record.

Although some aficionados may argue that artist Byron Walton (a.k.a. Jamie Principle) produced the first house record with just a portastudio and a keyboard, the track entitled *Your Love* was only handed to Knuckles for him to play as part of his set. Jesse Saunders, however, released the track commercially under his self-financed label 'Jes Say' and distributed the track through Chicago's 'Imports Etc'.

The records were pressed courtesy of Musical Products, Chicago's only pressing plant, owned and run by Larry Sherman. Taking an interest in this scene, he investigated its influence over the crowds and soon decided to start the first ever house record label 'Trax'. Simultaneously, however, another label 'DJ International' was started by Rocky Jones, and the following years involved a battle between the two to release the best house music. Many of these consisted of what are regarded as the most influential house records of all time, including *Music is the Key, Move Your Body, Time to Jack, Get Funky, Jack Your Body, Runaway Girl, Promised Land, Washing Machine, House Nation* and *Acid Trax*.

By 1987 house was in full swing; while still borrowing heavily from 1970s disco, the introduction of the Roland TB303 bass synthesizer, along with the TR909, TR808 and the Juno 106, had given house a harder edge as it became disco made by 'amateur' producers. The basses and rhythms were no longer live but recreated and sequenced on machines, resulting in a host of 303-driven tracks starting to appear.

One of these budding early producers was Larry Heard, who after producing a track entitled *Washing Machine* released what was to become one of the most poignant records in the history of house. Under the moniker of Mr Fingers, he released *Can U Feel It*, the first ever house record that didn't borrow its style from earlier disco. Instead, it was influenced by soul, jazz and the techno that was simultaneously evolving from Chicago. This introduced a whole idea to the house music scene as artists began to look elsewhere for influences.

One of these was Todd Terry, a New Yorker and hip-hop DJ. He began to apply the sampling principles of rap into house music. By sampling drum loops from old records and layering them together, he introduced harsher and heavier rhythms into the scene and released. *Three Massive Dance Floor House Anthems*, which pushed house music in a whole new direction. His subsequent house releases brought him insurmountable respect from the UK underground scene and he has duly been given the title of Todd 'The God' Terry.

Over the following years, house music mutated, multiplied and diversified into a whole number of different subgenres, each with their own names and production ethics. In fact, to date there are over 14 different subgenres of house, consisting of progressive house, hard house, deep house, dark house, acid house, Chicago house, UK house, US house, Euro house, French house, tech house, vocal house, micro house and disco House … and I've probably missed some too.

Musical analysis

The divergence of house music over the subsequent years has resulted in a genre that has become hopelessly fragmented and as such cannot be easily identified as featuring any one particular attribute. Indeed, it can be funky, drawing its inspiration from disco of the 1970s; it can be relatively slow and deep, drawing inspiration from techno; it can be vocal, it can be party-like or it can simply be pumping. In fact, today the word house has become somewhat of a catch-all name for music that is dance (not pop!) yet doesn't fit into any other dance category. The good news with this is that you can pretty much write what you want and as long as it has a dance vibe it could appear somewhere under the house label. The bad news, however, is that it makes it near impossible to analyse the genre in any exact musical sense and it is only possible to make some very rough generalizations.

Firstly, we can safely say that house music invariably uses a 4/4 time signature and is produced either allegretto or allegro. In terms of physical tempo, this can range from a somewhat slow (by today's standards) 110 BPM to a more substantial 140 BPM, but many of the latest tracks seem to stay around the 127 or more recently 137 'disco heaven' BPM. This latter value is referred to as such since this is equal to the average clubber's heart rate while dancing, but whether this actually makes the music more 'exciting' in a club has yet to be proven.

Fundamentally, house is produced in one of three ways: everything is sampled and rearranged; only some elements are sampled and the rest is programmed; or the entire track is programmed in MIDI. The approach taken depends entirely on what style of house is being written. For example, the disco house produced by the likes of Modjo, Room 5 and Daft Punk relies heavily on sampling significant parts from previous disco hits and dropping their own vocals over the top (Daft Punk's *Digital Love* and Modjo's *Lady* being prime examples). If you write this style of music then this is much easier to analyse, since it's based around the disco vibe. Generally, this means that it consists of a four-to-the-floor rhythm with a heavily syncopated bass line and the characteristic electric wah guitar. On the other hand, deep house uses much darker timbres (deep bass lines mixed with atmospheric jazzy chords) that don't particularly exhibit a happy vibe but are still danceable, while acid house relies heavily on the squawking TB303. In fact, while all house music will employ a four-to-the-floor pattern, the instrumentation used will often determine the style it comes under. Consequently, what follows is simply a guide to producing the basic elements of all types of house and since you know how your choice of genre already sounds, it can be adapted to suit what you hear.

House rhythms

When it comes to producing house rhythms, nearly all house producers will not use software sequencers, preferring the AKAI MPC or E-Mu SP1200 sampling drum machines. This is simply due to the solid timing that is requisite in house and can only be attained through using the internal hardware sequencers. Notably, of these the SP1200 has a maximum sampling rate of 12-bit, but this isn't of any concern since it often imparts a harsher sound to the rhythms, which is sometimes required. In fact, after programming a house rhythm and the subsequent timbres, it's often worth reducing the bit rate to see if it hardens up the beat for the subgenre of house you want to produce.

Generally, house relies heavily on the strict four-to-the-floor rhythm with a kick drum laid on every beat of the bar. Typically, this is augmented with a 16th closed hi-hat pattern and an open hi-hat positioned on every eighth (the off beat) for syncopation. Snares (or claps) are also often employed on the second and fourth beats underneath the kick.

Figure 13.1 The cornerstone of house loops.

This, of course, is only the start to a house loop and is based around the disco patterns from where it originated; congas, toms, bongos, tambourines and shakers are often added to create more of a feel to the rhythm, but this does depend on the subgenre. Where these are placed is entirely up to your own discretion, but generally they are best played live from the sampler's pads (or an attached keyboard) and left unquantized to keep the feel of the loop live. This helps in acquiring the sampled feel that's often exhibited, since more often than not the loop will have been sampled from previous records.

More importantly, though, house more than any other genre relies on plenty of drive. This can be applied in a variety of ways, but the most commonly used technique is to employ a slight timing difference in the rhythm to create a push in the beat. Typically, this involves keeping the kick sat firmly on all the beats but moving the *snares* forward in time by tick or two. Then, once the bass line is laid down, its timing is matched to these snares using a sequencer's match quantize function. Alternatively, to provide a heavier sound, a clap is laid on beats two and four (along with the snares), but these claps are then moved forward in time by a couple of ticks. This results in the claps' transient starting *just* before the snares, which not only helps to thicken out the snares but also adds drive to the rhythm.

Figure 13.2 Thickening out the snares (note how the claps are slightly ahead of time).

On this same theme, dynamics also play a vital role in acquiring the rhythmic push required in the genre. Usually, the kick drum will follow the *strong–weak–medium–weak* syncopation, while the closed hi-hats follow a less defined velocity pattern. The velocities of these will depend entirely on the feel you want to acquire, but as a general starting point the main emphasis often lies on the second and fourth 16th pulses of the bar, as illustrated below:

1	2	3	4	5	6	7	8	9	10	11	12	13	14	15	16
M	S	W	S	M	S	W	S	M	S	W	S	M	S	W	S
Kick				Kick				Kick				Kick			

Of course, this is simply convention, and convention shouldn't always play a part in music, so you should feel free to experiment by placing the accents at different divisions on the beat. By doing so, the rhythm of the piece can change quite severely. If you take this approach, however, keep in mind that the open hat sits on the third (offbeat) of the bar, so generally the closed hi-hat should remain weak here to prevent frequency clashes between the two.

For house kicks, the Roland TR909 is the most frequently used drum machine, but the Simmons SDS-5, Roland CR and E-Mu Drumulator are also used to produce the timbres. To my knowledge, however, only the TR909 has ever been emulated in software form, so unless you're willing to synthesize your own kick, sample another house record, use a sample CD or source and pay for one of these archaic machines, then you'll have little option but to use a 909.

The house kick is commonly kept quite tight by employing a short decay combined with an immediate attack stage. This is to help keep the kick dynamic, but also prevents it from blurring and destroying the all-important interplay between kick and bass. Typically, a 90 Hz sine wave produces the best starting point for a house kick, with a positive pitch EG set to a fast attack and a quick decay to modulate the oscillator. Although using a pitch envelope produces the best results, if the synth doesn't offer one then a self-oscillating filter will produce the requisite sound and the decay can be controlled with the filter's envelope.

In many cases, this kick will also be augmented with a square waveform to produce a heavier transient that can then be later crushed with a compressor. For this, the square wave requires a fast attack and decay stage so that it simply produces a 'click'; once this is layered over the original sine you can experiment by pitching it up or down to produce the required timbre. With this basic timbre laid down, it's then worth experimenting with the decay parameters of the square and sine. Generally, a concave decay will produce the typical 'housey' character, but for a deeper or darker house vibe a convex decay may be more appropriate.

Production-wise, it's worthwhile experimenting with small *controlled* amounts of distortion to give the kick a little more character in the loop, but this must be applied cautiously. If it's applied too heavily much of the bottom-end weight of the kick can be lost, resulting in a loop with little or no energy. Also, house kicks rely very heavily on compression to produce the archetypal hardened sound. This should be applied now rather than when the rest of the loop is in place, since in the context of a typical house loop, this would mean that the hi-hats, congas and toms are pumped with the compressor. This is generally avoided at this stage as the loop will be compressed along with the bass at a later stage. Instead, it's better to run just the 4/4 kick through the compressor and experiment with the settings to produce a 'hard' sounding timbre.

Although the settings to use will depend on the sound you wish to achieve, a good starting point is to use a low ratio combined with a high threshold and a fast attack with a medium release. The attack will crush the initial transient of the square, producing a heavier 'thump', while experimenting with the release can further change the character of the kick. If this is set quite long, the compressor will not have recovered by the time the next kick begins and this can often add a pleasing distortion to the sound. Once the kick has been squashed, as a final stage it's often worthwhile using the PSP VintageWarmer[1] to add more warmth to the overall timbre.

Note: On the CD you can hear the kick through GM, then programmed and then compressed.

Unlike the kicks, in the majority of cases the house snare is not sourced from the TR909 but the E-Mu Drumulator due to its warmer, rounded sound. Similar to the kick, however, the decay on the snare is kept particularly short to keep with the short, sharp dynamics of the loop. This characteristic sound can be produced through using either a triangle or square wave mixed with some pink noise or saturation (distortion of the oscillator), both modulated with an amp envelope using a fast attack and decay. Typically, pink noise is preferred over white as it produces a heavier snare, but if this proves to be too heavy for the loop then white noise may be a better option. Much of this noise will need removing with a high-pass filter to produce the archetypal house snare, but in some cases a band-pass filter may be more preferable, since this allows you to modify the low and high content of the snare more accurately.

If at all possible, it's worthwhile using a different envelope on the main oscillator and the noise, since this will allow you to keep the body of the snare (i.e. the main oscillator) quite short and sharp, but allow you to lengthen the decay of the noise waveform. By doing so, the snare can be modified so that it rings out after each hit, producing a timbre that can be modified to suit all genres of house. Additionally (and provided that the synth allows it), it also permits you to individually adjust the decay's shape of the noise waveform. Generally, the best results are from employing a concave decay, but as always it's best experimenting. Some house snares will also benefit from some judicious amounts of pitch bend applied to the timbre, but rather than apply this positively, negative will invariably produce results that are more typical to house music, as it often produces a sucking/smacking sound.

If you use this latter technique, it's sometimes prudent to remove the transient of the snare in a wave editor and replace it with one that uses positive pitch modulation. When the two are recombined, it results in a timbre that has a good solid strike but decays upwards in pitch at the end of the hit. If this isn't possible then a viable alternative is to sweep the pitch from low to high with a sawtooth or sine LFO (provided that the saw starts low and moves high) set to a fast rate. After this, small amounts of compression so that only the decay is squashed (i.e. slow attack) will help to bring it up in volume so that it doesn't disappear into the rest of the mix.

Note: On the CD you can hear the snare through GM, programmed and then compressed.

Notably, it's only the timbre for the kick and snare that truly define the house sound and, while in the majority of cases the remaining percussive and auxiliary instruments such as the open and closed hi-hats, congas, cowbells, claps, tambourines, shakers and toms are sourced from the TR808 or TR909, almost any MIDI module will carry sounds that can be used. Plus, many modules today carry the ubiquitous TR808 and TR909 percussion sound-sets as part of the collection anyway. Normally, none of these should be compressed as it can reduce the high-frequency content, but having said that this is simply convention and you should always be willing to 'break the rules', so to speak.

With the drum loop complete, it's worthwhile cycling it over a number of bars and experimenting with the synthesis parameters available to each instrument to create a rhythm that gels together. This includes lengthening or shortening the decay parameters, pitching the instruments up or down and applying *subtle* effects such as reverb. The basic principle here is to make the loop sound as good as possible before moving onto the subsequent processing. If it doesn't seem to gel properly together or seems wrong, then a common problem is that too many elements are playing at once. Keep in mind that, although house prides itself on funky rhythms, it doesn't necessarily have to be a complex loop to achieve this and sometimes simply removing one element can suddenly make the loop work.

Once the loop is 'working' together, a common practice is to sample the loop into audio and then run it through a compressor and noise gate. Whether the gate or compressor comes first is a matter of individual taste, but generally the rhythm is compressed and then gated. As always, there are no definitive settings to use when compressing, as it depends entirely upon the sound your after. Having said that, a good starting point is to set the ratio quite high with a low threshold and use a fast attack with a medium release. Finally, set the make-up gain so that the loop is at the same volume when the compressor is bypassed and begin experimenting with the release settings. Shorter settings will force the compressor to pump and as a result make the kicks more prominent in the loop, while longer settings will be less evident. There is no universal answer for the 'correct' settings, so you'll have to try out numerous configurations until you reach the sound you need. Ideally, the best compressors to use for this should be either valve or opto, as the second-order harmonic distortion they introduce plays a large part in capturing the sound, but if this is not possible, following the compressor with the PSP VintageWarmer should help things along nicely.

Following compression, it quite usual to feed the results into a noise gate and use this to modify the loop further. One of the key elements of house loops is that the sounds remain short and sharp, as this makes the entire loop feel more dynamic, since long sounds tend to mask the gaps between each hit, which lessens the impact as a whole. This also helps it to accentuate the often funky bass rhythms that lie underneath, as these tend to use quite long release times to help 'funk' them up. As with compression, the settings to use on a noise gate are up to your artistic interpretation, but as a start, set the threshold so that it's just lower than the kicks and snare, use a fast attack with a medium hold time and then experiment with the release setting. The faster this is set, the more the sounds will be cut, so you'll need to find a happy medium where they do not cut off too soon but do not decay for too long either. If vocals are to be employed in the mix, it's prudent to set the threshold lower so that the open hats are also gated slightly. In house these are often at a similar frequency to the vocals and if they're left too long they can mix with the vocals, which results in both drums and vocals losing their dynamic edge. Finally, remember that house music is a popular genre to remix, so it's sensible to keep a note of all compression or effects settings along with the original MIDI files. Most studios will create these 'track' sheets and it's wise to do so too, since many remixers will expect a track sheet if they're to remix the music.

> Note: On the CD you can hear the house loop played through MIDI and then sampled and processed, along with associated MIDI files.

Of course, it would be naive to say that all house beats are created through MIDI and a proportionate amount of rhythms are acquired from sampling previous house records. Indeed, these will have probably been sampled from other records, which will have been sampled from previous house records, which will have been sampled from other records, which will probably have been … well, you get the picture. For obvious legal reasons I can't condone this type of approach – it's illegal – but that's not to say that dance artists don't do it and so in the interests of theory it would only be fair to cover some of the techniques they use.

Almost every house record every pressed to vinyl starts with just a drum loop, so sourcing one isn't particularly difficult, but the real skill comes from what you do with it after it's in the sampler. Fundamentally, it's unwise to just sample a loop and use it as is. Even though there's a very high chance that what you're sampling is already a sample of a sample of a sample of a sample (which means that the original artists don't have a leg to stand on for copyright infringement), you might be unlucky and they may have actually written it in MIDI. What's more, it isn't exactly artistically challenging, so after sampling a loop it's always worthwhile placing it into a sample slicing program (such as Wavesurgeon) and moving the parts around a little to create your own variation.

> Note: A demo of Wavesurgeon is included on the CD.

Of course, this is the most simplistic approach and a number of artists will take this a step further by first employing a transient designer to alter the transients of the loop before placing it into a sample slicing program. Using this technique, the loop does not have to be sourced from a house track as any drum break can be manipulated to create the snappy sounds used in house.

Along with working with individual loops, many house producers also stack up a series of drum loops to create heavier rhythms. Todd Terry and Armand Van Helden are well known for dropping more powerful sounding kicks over pre-sampled loops to add more of a heavy influence. If you take this approach, however, you need to exercise care that the kick starts in just the right place, otherwise you'll experience a phasing effect. Possibly the best way to circumvent this is to cut the bar into four segments where the kick occurs and then trigger it alongside the new kick at the beginning of each segment to produce a continual loop that stays in time. This also permits you to swap the quarter bars around to create variations.

Note: On the CD you can hear the complete loop from the example track.

Bass

After the drums, the bass is commonly laid down, since many genres of house rely a great deal on the groove created from the interaction between the two. As this genre of music has its roots firmly embedded in disco, most house basses tend to follow the atypical disco funk pattern, which can range from the rather simplistic walking bass to more complex patterns produced by real bassists improvising with the drum rhythms.

Figure 13.3 Disco's infamous walking bass line.

The bass line in Figure 13.3 is only a very simple example, but house basses do not have to be particularly complex and in many instances you'll find that even an incredibly unsophisticated bass can have additional groove injected into it by simply changing the duration and timing of *some* notes, as in Figure 13.4.

Figure 13.4 By changing the duration and timing of some notes, new grooves can be created.

It's important to note that the complexity of the bass often depends on the lead instruments that are to lie over the top. Although some house music relies on a particularly funky bass groove, the overlying instruments are kept relatively simple by comparison, whereas if the overlying elements are more complex, then the bass line should remain relatively simple. This is incredibly important to comprehend, since if both the bass and overlying instruments are playing complex grooves it can easily undermine the record, resulting in a miscellany of melodies that are difficult to decipher. Thus, when producing house you need to decide whether the centrepiece of the music is created by the bass *or* the overlying elements. For instance, the piano in Laylo and Bushwacka's *Love Story* produces the centrepiece of the music, while the bass acts as an underpinning and so is kept relatively simple. Conversely, tracks such as Kid Crème's *Down and Under*, Stylophonic's *Soul Reply* and Supermen Lovers' *Starlight* use a more funky bass, which is complemented with simple melodies playing over the top.

Figure 13.5 The bass rhythm from *Starlight* (notice how the bass occurs before the beat for additional groove).

Note: Some house basses, particularly those used in acid or deep house, will follow a simpler pattern, similar to techno. Refer to Chapter 14 on techno for more information.

As with the rhythms, the timbre used for the basses in house can vary wildly from one track to the next. On the one hand, it can be from a real bass guitar, while on the other it could consist of nothing more than a low-frequency sine wave pulsing away in the background, to anything in between. Like much of house music, this makes it incredibly difficult to pin down any particular timbre, so what follows are a few synthesis tips to create the fundamental structure of most house basses. After you have this, experimentation is the real key and it's worthwhile trying out new waveforms, envelope settings and modulation options to create the bass you need.

The foundation of most synthetic house basses can be constructed with a sine wave and saw-tooth or pulse oscillator. The main waveform (the sine) provides a deep body to the sound, while the secondary oscillator can either provide raspy (saw) or woody (square) overtones. If the sound is to be quite cutting and evident in the mix then a saw is the best choice, while if you want to be more rounded and simply lay down a groove for a lead to sit over then a square is the better option. Listening carefully to the sound they produce together, begin to detune the secondary oscillator against the first until the timbre exhibits the 'fatness' and har-monic content you want. Typically, this can be attained by detuning +5 or +7 cents, but as always let your ears decide what's best.

Nearly all bass timbres will start immediately on key-press, so set the amp's attack to zero, and then follow this with a fast decay, no sustain and a short release. This produces a timbre that starts immediately and produces a small pluck before entering the release stage. If the sound is too short for the melody being played, increase the sustain and experiment with the release until it flows together to produce the rhythm you require. If the bass sounds like it's 'running away', try moving it a few ticks forward or back and play with the amp's attack stage. To add some movement and character to the sound, set the filter cut-off to low pass and set both the filter cut-off and resonance to midway, and then adjust the envelope to a fast attack and decay with a short release and no sustain. Normally, this envelope is applied positively, but experi-ment with negative settings as this may produce the character you need. If possible, it's also worthwhile trying out convex and concave settings on the decay slope of the filter envelope, as this will severely affect the character of the sound, changing it from one similar to a Moog through to a more digital nature and onto a timbre similar to the TB303. In most house tracks, the filter follows the pitch (opens as the pitch increases), so it's prudent to use filter positive key-follow too.

This creates the basic timbre, but it's practical to begin experimenting with LFOs, effects and lay-ering. Typical uses of an LFO in this instance would be to lightly modulate the pitch of the primary or secondary oscillator, the filter's cut-off or the pulse width if a pulse waveform is used. On the effects front, distortion is particularly effective and an often used effect in house is phasing/-flanging, but if you choose this latter option you will have to exercise care. In all dance mixes the bass should always sit in the centre of the stereo spectrum so that not only do both speakers share the energy, but also any panning applied during mixing is evident (this is covered in Chapter 17). Consequently, if heavy flanging or phasing is applied then the bass will be smeared across the stereo image and the mix may lack any bottom-end cohesion. Of course, if the bass is too thin or doesn't have enough body then effects will not rescue it, so it may be sensible to layer it with another timbre or in some cases five or six others. Above all, bear in mind that we have no expectations of how a synthesized bass should sound, so you shouldn't be afraid of stacking up as many sounds as you need to build the required sound. EQ can always be used to remove harmonics from a bass that is too heavy, but it cannot be used to introduce harmonics that are not already present.

Alongside the synthetic basses, many house tracks will employ a real bass. More often than not, these are sampled from other records or are occasionally taken from sample CDs rather than programmed in MIDI. However, provided that you have (a) plenty of patience, (b) a willingness to learn MIDI programming and (c) a good tone or bass module (such as Spectrasonics' Trilogy), it is possible to program a realistic bass that could fool most listeners. This not only prevents any problems with clearing copyright, but it also allows you to model the bass to your needs.

The key to programming any real instrument is to take note of how they're played and then emulate this action with MIDI and a series of CC commands. In this case, most bass guitars use the first four strings of a normal guitar, E–A–D–G, which are tuned an octave lower, resulting in the E being close to three octaves below middle C. Also, they are monophonic, not polyphonic, so the only time notes will actually overlap is when the resonance of the previous string is still dying away as the next note is plucked. This effect can be emulated by leaving the preceding note playing for a few ticks while the next note in the sequence has started. The strings can either be plucked or struck and the two techniques produce different results. If the string is plucked, the sound is much brighter and has a longer resonance than if it were simply struck. To copy this, the velocity will need to be mapped to the filter cut-off of the bass module, so that higher values open the filter more. Not all notes will be struck at the same velocity, though, and if the bassist is playing a fast rhythm the consecutive notes will commonly have less velocity, since he has to move his hand and pluck the next string quickly. Naturally, this is only a guideline and you should edit each velocity value until it produces a realistic feel.

Depending on the 'bassist' they may also use a technique known as 'hammer on', whereby they play a string and then hit a different pitch on the fret. This results in the pitch changing without actually being accompanied with another pluck of the string. To emulate this, you'll need to make use of pitch bend, so this will first need setting to a maximum bend limit of two semitones, since guitars don't 'bend' any further than this. Begin by programming two notes, for instance an E0 followed by an A0, and leave the E0 playing underneath the successive A0 for around 100 ticks. At the very beginning of the bass track, drop in a pitch bend message to ensure that it's set midway (i.e. no pitch bend) and just before where the second note occurs drop in another pitch bend message to bend the tone up to A0. If this is programmed correctly, on playback you'll notice that as the E0 ends the pitch will bend upwards to A0, simulating the effect. Although this could be left as is, it's sensible to drop in a CC11 message (expression) directly after the pitch bend, as this will reduce the overall volume of the second note so that it doesn't sound like it has been plucked. In addition to this, it's also worthwhile employing some fret noise and finger slides. Most good tone modules will include fret noise that can be dropped in between the notes to emulate the bassist's fingers sliding along the fret board.

Whether you decide to use a real bass or a synthesized one, after it's written it's an idea to compress the previous drum loop against the bass. The principle here is to pump the bass to produce the classic bottom-end groove of most house records – they tend to pump like crazy. This is accomplished by feeding both bass and drum loops (preferably with the hi-hats muted) into a compressor and then set the threshold so that each kick registers approximately −6 dB on the gain reduction meter. Use a ratio of around 9:1 with a fast attack and set the gain make-up so that it's at the same volume level when the compressor is bypassed. Finally, set the release parameter to 200 ms, and then experiment by increasing and reducing this latter parameter. The shorter the release becomes, the more the kick will begin to pump the bass, becoming progressively heavier the more that it's shortened. Unfortunately, the only guidelines for how short this should be set are to use your ears and judgement, but try not to get too excited. The design is to help the drums and bass gel together into a cohesive whole and produce a rhythm that punches along energetically. On the same note, it should not be compressed so heavily that you lose the excursion of the kick – you have to reach a happy medium!

Melodies and motifs

With the bottom-end groove working, the lead instruments can be dropped over the top. As touched upon when discussing basses, the lead instruments melody will depend entirely on the type of house being produced. Funky bass lines will require less active melodies to sit over the top, while less active basses will require more melodic elements. Additionally, since these leads are often sampled from other records, it's naive to suggest that they are programmed with MIDI and therefore it's impossible to offer any guidelines apart from to listen to the latest house records to see where the current trend is.

This also means that there are no definitive sounds that characterize the genre and as such practically everything should be seen as fair game. In fact, a proportionate amount of house producers feel the same, as many have sampled heavily from previous records (particularly disco) to produce the sound. With that said there are some timbres that have always been popular in house, including the hoover, plucked leads and pianos from the DX series of synthesizers. These have all already been discussed in detail in the chapter on sound design, so here we'll look at the general synthesis ideals.

Fundamentally, synthetic house leads will more often than not employ sawtooth, triangle and/or noise waveforms to produce a harmonically rich sound that will cut through the mix and can be filtered if required. Depending on how many are employed in the timbre, these can be detuned from one another to produce more complex, interesting sounds. If the timbre requires more of a body to the sound then adding a sine or pulse wave will help to widen the sound and give it more presence. To keep the dynamic edge of the music, the amplifier's attack is predominantly set to zero so that it starts upon key-press, but the decay, sustain and release settings will depend entirely on what type of sound you require. Generally, it's unwise to use a long release setting, since this may blur the lead notes altogether and the music could lose it's dynamic edge, but it's worth experimenting with the decay and sustain while the melody is playing to the synth to see the effect it has on the rhythm. As lead sounds need to remain interesting to the ear, it's prudent to employ LFOs or a filter envelope to augment the sound as it plays. A good starting point for the filter EG is to set the filter cut-off to low pass and set both the filter cut-off and resonance to midway, and then adjust the envelope to a fast attack and decay with a short release and no sustain. Once this is set, experiment by applying it to the filter by positive and negative amounts. On top of this, LFOs set to modulate the pitch, filter cut-off, resonance and/or pulse width can also be used to add interest. Once the basic timbre is down, the key is, as always, to experiment.

For chords, the most common instrument used in house is the Solina String Machine, which has made an appearance on hundreds of house records, including those by Daft Punk, Air, Joy Division, Josh Wink, STYX, New Order, Tangerine Dream, Roger Sanchez, Supermen Lovers and nearly every disco track ever released. Although this synth is now out of production, a similar timbre can be created in any analogue style synth. Begin by using three sawtooth waveforms and detune two from one of the oscillators by + and −3. Apply a small amount of vibrato using an LFO to these two detuned saws and then use an amplifier envelope with a medium attack and release and a full sustain (there is no decay, since sustain is set to full and it would have nothing to drop down to). Now, experiment with the sawtooth that wasn't detuned by pitching it up as far as possible without it becoming an individual sound (i.e. less than 20 Hz) and if possible use two filters – one set as a low pass to remove the low-end frequencies from the two detuned saws and a high pass to remove some of the high-frequency content from the recently pitched saw.

Once this basic sound is created, it may be prudent to make it appear as though it's been sampled and subsequently effected. To do this, its worthwhile copying the timbre to three sequencer tracks and removing all the bottom-end frequencies of the first track, all of the mid range from the second and everything but the high frequencies on the third with a good EQ unit. Following this, insert a flanger or phaser across the audio track that only has the high frequencies and then mix the three tracks together until it produces a timbre that sounds appropriate for the track. Depending on the music so far, this may also involve applying the flanger (or phaser) to the mid-range or the low-range track, or perhaps even all of them. In this case, the basic principle is to have the effects sweep the frequencies present in each track by differing amounts to construct a big textural timbre. Once complete, export these three files into a single audio file and then place it into a sampler and create a chord structure to suit the music.

Even if this sound is not used, effects can play an important role in creating interesting and evolving pads for the genre. Wide chorus effects, rotary speaker simulations, flangers, phasers and reverb can all help to add a sense of movement to help fill out any holes in the mix. A noise gate can also be used creatively to model the sound. For instance, by setting a low threshold and using an immediate attack and release, the pad will start and stop abruptly, producing a 'sampled' effect that cannot be replicated through synthesis parameters. What's more, if you set the hold time to zero and adjust the threshold so that it lies just above the volume of the timbre, the gate will 'chatter', which can be used to interesting effect. Overall, compression is generally not required on pads since they'll already be compressed at the source, but if distortion or heavy filtering or EQ is used then it may be prudent to follow them with a compressor to prevent them from clipping the desk.

With all the parts programmed we can begin to look at the arrangement. As touched upon in the musical analysis, house is an incredibly diverse genre, so there are no definite arrangement methods. Nevertheless, as much of house relies on the sampling ethics introduced by Todd Terry, a usual approach is to arrange the instruments in different configurations to produce a series of different loops. For instance, a loop could consist of just the drums, another of the drums and bass mixed together, another consisting of the drum, bass and leads, and another made from just the leads. Each of these are then sampled (or exported from an audio sequencer) and set to a specific key on a controller keyboard connected to the sampler. By then hitting these keys at random you can produce a basic arrangement, plus if one-shot mode is disabled on the sampler it's possible to create the stuttered mix effect used on some house tracks.

Note: On the CD you can hear the stuttered effect in use on the example track.

That said, as with all genres this is subject to change, so the best way to obtain a good representation of how to arrange the track is to listen to what are considered the classics and then copy them. This isn't stealing, it's research and nearly every musician on the planet follows this same route. This can be accomplished using a method known as 'chicken scratching'. Armed with a piece of paper and a pen, listen back to the track and on every bar place a single scratch mark on the paper. When a new instrument is introduced or there is a change in the rhythmic element, place a star below the scratch. Once you've finished listening back to the song, you can refer to the paper to determine how many bars are used in the track and where new instruments have been introduced. You can then follow this same arrangement and if required change it slightly so that it follows a progression to suit.

Recommended listening

Ultimately, the purpose of this chapter has been to give some insight into the production of house and, as ever, there is no one definitive way to produce the genre. Indeed, the best way to learn new techniques and production ethics is to actively listen to the current market leaders, and be creative with your processing, effects and synthesis. In fact, in many cases it's this production that makes one track better than another, and if you want to write house you need to be on the scene and listen to the most influential tracks. If you *listen* closely to these they will invariably reveal a significant amount about how they were made. Keep in mind that the timbres are unimportant since this genre does not necessarily rely on a particular sound; it's more to do with *feel* and *groove*. With this in mind, what follows is a short list of some of the artists that, at the time of writing, are considered the most influential in this area:

- Daft Punk.
- Room 5.
- The Supermen Lovers.
- Modjo.
- Stardust.
- Kid Crème.
- Stylophonic.
- Some of Moby's work.
- Laylo and Bushwacka.
- Roger Sanchez.

And of course, some disco artists who *could* be used as a source of inspiration:

- Alec Costandinos.
- Bohannon.
- Bootsy Collins.
- Cerrone.
- Chic.
- Deodato.
- Donna Summer.
- Eddie Kendricks.
- Evelyn King.
- France Joli.

- George McRae.
- Gino Soccio.
- Giorgio Moroder.
- Gregg Diamond.
- Gwen McRae.
- Harold Melvin and the Blue Notes.
- Heatwave.
- LaBelle.
- Larry Levan.
- Lime.
- Love Unlimited Orchestra.
- Meco Monardo.
- Patrick Adams.
- The Salsoul Orchestra.
- Shalamar.
- Sister Sledge.
- Slave.
- Sylvester.
- Sylvia.
- Tavares.
- Thelma Houston.
- Tom Moulton.
- Van McCoy.
- Vicki Sue Robinson.
- Vincent Montana Jr.

Note: On the CD you can hear an excerpt from the house track constructed through this chapter, along with the MIDI file to accompany it.

14 Techno

Techno is a complete mistake. It's like George Clinton and Kraftwerk stuck in an elevator with only a sequencer to keep them company.

Derrick May

To the uninitiated techno is used to describe any electronic dance music and, although this was initially true, over the years it has evolved to become a genre in its own right. Originally, the term techno was coined by Kraftwerk in an effort to describe how they mixed electronic instruments and technology together to produce 'pop' music. However, as the following years were riddled by numerous artists taking the idea of technology on board to write their own music, the true foundations of where techno as we know it today originated are difficult to pinpoint accurately.

To some, the roots of this genre can be traced back to as early as 1981 with the release of *Shari Vari* by A Number of Names, Donna Summer's (and Giorgio Moroder's) *I Feel Love* and Cybotron's *Techno City*. To others, it emerged in the mid 1980s when the 'Belleville Three' collaborated together in Detroit. These three high school friends – Juan Atkins, Kevin Saunderson and Derrick May – used to swap mix tapes with one another and religiously listen to the Midnight Funk Generation on WJLB-FM. The show was hosted by DJ Charles '*Electrifying Mojo*' Johnson and consisted of a 5-hour mix of electronic music from numerous artists, including Kraftwerk, Tangerine Dream and George Clinton.

Inspired by this eclectic mix they began to form their own music using cheap second-hand synthesizers such as the Roland TR909, TR808 and TB303. The music they produced was originally labelled as '*house*', and both May and Saunderson freely admit to gaining some of their inspiration from the Chicago clubs (particularly The Warehouse and Frankie Knuckles) and the house music they played. In fact, Derrick May's 1987 hit *Strings of Life* is still viewed by many as house music, although to Derrick himself and many other aficionados it was an early form of Detroit techno.

This form of music was characterized by its mix of dark pounding rhythms mixed with a soulful feel and a stripped-down vibe. This latter stripped-down feel was a direct result of the limited technology available at the time. Since the TB303, TR909 and TR808 were pretty much the only instruments obtainable to those without a huge budget, most tracks were written with these alone, which were then recorded direct to two-track tape cassettes.

It wasn't until late 1988, however, until techno became a genre in its own right, when Neil Rushton produced a compilation album labelled *Techno – The New Dance Sound of Detroit* for

Virgin Records. Following this release, techno no longer described any form of electronic music, but was used to describe minimalist, almost mechanical, house music. Similar to most genres of dance, this mutated further as more artists embraced the ideas and formed their music around it. By 1992 and the evolution of the new 'rave generation', techno bore almost no relationship to the funky beats and rhythms of house music as it took on more drug-influenced hypnotic tribal beats.

As technology evolved and MIDI instruments, samplers, sequencers and digital audio manipulation techniques became more accessible, techno began to grow increasingly complex. While it still bore a resemblance to the stripped-down feel of Detroit techno, consisting solely of rhythms and perhaps a bass, the rhythmic interplay became much more complex. More and more rhythms were laid atop one another and the entire studio became one instrument with which to experiment.

Of course, Detroit techno still exists today, but it has been vastly overshadowed by the tribal beats of 'pure' techno developed by numerous artists, including Thomas Krome, Redhead Zync, Henrik B, Tobias, Carl Craig, Kenny Larkin and Richie Hawtin. Each of these artists has injected their own style into the music while keeping with some of the original style set by their contemporaries.

Musical analysis

Techno can be viewed as dance music in its most primitive form, since it's chiefly formed around the cohesion and adaptation of numerous drum rhythms. Although synthetic sounds are also occasionally employed they will, more often than not, remain atonal, as it's the abundance of percussive elements that remain the most vital aspect of the music. In fact, in most techno tracks any additional synthetic instruments are not often used in the 'musical' form to create bass lines or melodies; instead, the genre defines itself on a collection of carefully programmed and manipulated textures rather than melodic elements.

Fundamentally, this means that it's produced with the DJ in mind and, in fact, most techno is renowned for being 'DJ friendly', being formed and written to allow him (or her) to seamlessly mix all the different compositions together to produce one whole continuous mix to last through the night. As such, techno will generally utilize a four-to-the-floor time signature, but it isn't unusual to employ numerous other drum rhythms written in different time signatures, which are then mixed, effected and edited to fit alongside the main 4/4 signature. Tempo-wise, it can range from 130 to 150 BPM and, although some techno has moved above this latter figure, it is in the minority rather than the majority.

Techno is also different from every other genre of music covered so far, since it does not rely on programming and mixing in the 'conventional' manner (if there is such a thing). Rather it's based around using the entire studio as one interconnected tool. While a sequencer (hardware or software) is still used as the centrepiece, it's commonly only used to trigger numerous drum rhythms contained in the connected samplers and drum machines. Each of these rhythms has been previously edited and manipulated with effects to produce new variations, which are then layered with others or dropped in and out of the mix to produce the final arrangement.

These rhythms are layered on top of one another so that they all interact harmonically to produce interesting variations of the original patterns. As more of these patterns are laid together, they create an often syncopated feel as the rhythmic harmony becomes progressively more complex. The mixing desk is then used not to mix the rhythms together in a conservative manner, but as a creative tool with EQ employed to enhance the interesting harmonic relationships created from this layering, or to prevent the cohesive whole from becoming too muddy or indistinct.

This method of working is analogous to subtractive synthesis, whereby you build a harmonically rich sound and then employ filters to shape the results. With techno, you construct a hectic, yet harmonically rich, rhythm by layering loops, and then proceed to deconstruct them with filters and EQ until you're left with some interesting harmonic interplay and rhythm. This produces incredibly complex beats that could never be 'programmed' through MIDI or played live and that also subtly move from one interesting rhythmic interaction to another with some live (or automated) tweaking of EQ or additional effects.

The obvious place to start with any techno is to begin by shaping the loops and occasionally they may be programmed in MIDI, but are more often sourced from sample CDs or other records. These are then subsequently diced, sliced and manipulated with programs such as Wavesurgeon to create rhythms that are very different from the original. However you decide to create the loops, the individual rhythms begin quite simply, and the beginnings consist of nothing more than a kick drum, snare, closed and open hi-hats, along with an occasional cymbal crash employed every four or so bars to mark the end of a musical segment.

Figure 14.1 The beginnings of a techno loop.

Although this generally provides a good starting pattern, any further loops should be programmed differently and/or use different timbres, so that when the rhythms are overlaid with one another a more complex sound and rhythm evolves. To accomplish this, it's often worth experimenting by programming tribal rhythms to mix with the previous loop. How these are programmed is entirely up to your own creativity, but a good starting point is to employ side sticks, splash cymbals, tom drums and more snares that can dance around the main starting loop. An example of a typical tribal loop is shown in Figure 14.2.

Figure 14.2 A tribal techno loop.

Note that in the example in the figure, very few of the instruments actually land on the beat, rather they are all offset slightly. This helps them to combine more freely with the initially creat-ed loop, creating thicker textures as the sounds play slightly earlier or later than the first loop, thus preventing the loops from becoming too metronomic. Also, note how this second loop does not contain any kicks. This is simply because if each consecutive loop contained a series of offbeat kicks, the rhythm would lose it's four-to-the-floor feel, while if the kicks were all laid on the beat of the bar they would all amalgamate, producing a mix with a very heavy bottom end. The principle is to create a series of loops, each of which focuses on a different timbre to carry the loop forward.

These two loops could be further augmented with another tribal loop, this time much more complex than the last, containing a number of closely paired instruments that all complement the previous two loops and create an even more complex drum arrangement.

In figure 14.3, note how few of the instruments land on the beat. This prevents the rhythm from becoming too metronomic and allows these new timbres to mix with the previous two loops to thicken out those timbres already present. Additionally, it is also worthwhile experimenting by mixing loops with different time signatures together. For instance, two 4/4 loops mixed with two 3/4 loops can produce results that cannot be easily acquired any other way.

Once a number of MIDI files have been created in this manner, they can be exported/imported as audio and sliced, diced and equalized to modify the sounds further to create abstract timbres suitable for techno. Although for this example we've only discussed three loops, you should feel free to program and layer as many loops together as you feel are necessary to accomplish the overall sound. This can range from using just three to layering over eight together and then progressively reducing the content of each until you are left with a harmonically rich and inter-esting sound. The layering of these should also not just be restricted to dropping them dead

Figure 14.3 Another tribal techno loop to mix with the previous two.

atop one another, and in many cases moving one of two of the loops forward or back in time from the rest can be used to good effect.

Once a few basic loops have been created, effects and aggressive EQ cuts or boosts are employed on each individual element of a sliced loop to create interesting timbres. Although the premise of techno is to keep a 4/4 kick pumping away in the mix, other percussive elements are commonly heavily affected, processed and equalized to produce timbres that add to its tribal nature. Although any effects can produce good results, of particular note the *Sherman Filterbank* is almost a requisite for creating strange evolving timbres. That said, the effects and processing applied are, of course, entirely open to artistic licence as the end result is to create anything that sounds good, but what follows is a list of possibilities to start with:

- Use a transient designer to remove the transients of a snare or hi-hat and then increase the sustain.
- Reverse a sample, apply reverb with a long tail and then reverse the sample so it plays the correct way again (albeit with a reversed reverb effect).
- Use a noise gate to shorten sounds in the loops.
- Apply heavy compression to squash the transients of some sounds.
- Apply heavy compression but only to the sustain (i.e. use a long attack).
- With the aid of a spectral analyser, identify the frequencies that contribute to the body of a sound and reduce their volume while increasing the volume of those surrounding them.
- Merge two timbres (such as a snare and hi-hat) together and use EQ to remodel the sound.
- Pitch-shift individual notes up and by extreme amounts.
- Apply heavy chorus or flangers/phasers to singular hi-hats or snares.
- Write a hi-hat pattern, export it as audio (if required), apply heavy delay and then cut the resulting file up to produce a new pattern.

- Apply heavy delay/chorus/flanging or phaser to an entire drum loop and cut segments out to mix with the rest of the loops.
- Constantly time-stretch and compress audio to add some digital clutter and then mix this with the other loops.

The main principle of applying all these processes is to generate timbres that are tonally different from every other loop, so that layered together they combine to create an interesting sound. Using a mixing desk's EQ and low-, high- or band-pass filters, you can then amplify or filter this area and slowly evolve its progression over the length of the mix with some automation. This live 'tweaking' forms a fundamental part of the music, so if the rhythms are contained in a software audio sequencer, it's of paramount importance to use an external controller so you can modulate the various parameters in real-time. Obviously, the more parameters this controller offers, the more creative you can become so, ideally, you should look towards a controller that has numerous options. Novation's *ReMote 25* can be particularly useful, since this offers a plethora of real-time controllers, consisting of knobs, faders, a touch pad and a yoke (a control stick that can be used to control both pitch and modulation simultaneously). All of these can then be recorded as CC data in real-time, permitting you to edit the automation further if required.

Naturally, techno relies on compression to produce the harsh, heavy beats as much as any other dance genre, but its production ethics are slightly different. Whereas in much of dance you usually wish to refrain from pumping the hi-hats or additional percussion by compressing the 4/4 kick singularly, it isn't uncommon in this genre to actually pump these percussive elements on some of the individual loops to create new sounds. Additionally, the compressor is often treated as an effect as well as a processor, and it isn't considered unusual to access a valve compressor as a send 'effect'. With this set at low threshold and high ratio, the returned signal can be added to the uncompressed signal, while experimenting with the attack and release will produce a series of different timbres. If you do not have access to a valve compressor, then sending the signal out to the *PSP VintageWarmer* can often add more harmonics to the signal similar to using a valve compressor.

Of course, once the beats are finally laid down, with all the frequency, EQ and pitch interaction, the loop will need compressing to prevent any potential clipping and in techno this is applied very heavily indeed. Generally, valve-based compressor are used, since these tend to pump musically. The most commonly used compressors for this are the *Joe Meek SC 2.2, UREI LA 3* or the *UREI 1176 LN* due to the amount of second-order harmonic distortion they introduce. If you don't have access to a good valve compressor, then after compressing it with a standard unit it's certainly worth throwing the loop through the *PSP VintageWarmer* to recreate the requisite sound. In fact, even if it is put through a valve compressor it's often worth placing it through the VintageWarmer anyway. The amount of compression to apply will, as always, depend heavily upon the timbres used, but as a general guideline start by setting the ratio to 12:1, along with an attack of 5 ms and a medium release of approximately 200 ms. Set the threshold control to 0 dB and then slowly decrease it until both the kick and second loudest elements of the loop (commonly the snare) register on the gain reduction meter by at least 5 dB. To avoid the volume anomaly (i.e. louder invariably sounds better!), set the make-up gain so that the loop is at the same volume as when the compressor is bypassed and then start experimenting with the release settings. By shortening the release the loop will become progressively heavier the more that this is shortened. Unfortunately, the only guidelines for how short this should be set are to use your ears and judgement, but try not to get too excited. Keep in mind that it should not be compressed so heavily that you lose the excursion of the kick, otherwise the loop will lack any real punch.

As previously mentioned, techno commonly consists of drums alone, but some may also include a bass rhythm to help the music groove along. In these instances, the bass is kept incredibly simple so as not to detract from the fundamental groove created by the drums. In other words, the bass commonly consists of nothing more than a series of 16th, eighth or quarter notes (sometimes consisting of a mix between them all), with no or very little movement in pitch.

Figure 14.4 A typical techno bass line.

In the example in Figure 14.4, the bass remains atonal but movement is provided by lengthening and shortening the bass notes, while the velocity controls the filter cut-off, allowing the groove to move in and out of the drum rhythms. Some tracks may also employ some pitch movement in the bass, but if this is the case then the timbre used is invariably mono and makes heavy use of portamento, so that that any overlapping notes slide into one another.

Figure 14.5 Using portamento on a techno bass.

Note how, in the example in Figure 14.5, a secondary note overlaps the first. This creates a timbre that rises and falls during its period, and it's this type of movement that is fundamental to a techno bass. Since the bass will generally only consist of one bar that is continually looped over and over, it's the harmonic and timbral movement that plays a primary role in attaining the groove. This is not only accomplished by using portamento, but also by adjusting various synthesis and effects parameters as the bass plays alongside the drum track. The basic belief here is to manipulate the frequencies contained in the bass so that they augment the frequencies in the drum track. That is, it adds to the harmonic relationship already created through manipulating the drums to create a cohesive whole that pulses along. The bass should still remain a separate element to the drum track but, nonetheless, any frequency-dependent movement should be to bring further interest to the harmonic interaction with the drums than to bring attention to itself.

For the bass timbre, the TB303 is the most widely used instrument since this is what was originally used by the techno originators; however, more recently, any analogue synthesizers (or DSP equivalents) are used so long as they contain enough frequencies to be swept with a filter so that they interact with the rhythms. If you want to stay with the roots of techno, though, the TB303 is the bass to use, but this timbre can be created in almost any subtractive synthesizer. A good starting point is to employ both a saw and square oscillator and set the amplifier's EG to zero attack and sustain with a medium release and a fast decay. Use a low-pass filter with both the resonance and cut-off set midway, and then adjust the filter's envelope to a short attack, decay and sustain with no release. Finally, as ever, experiment with all these settings until you obtain the sound you require.

For those that are a little more adventurous, try combining four sawtooth waves, or a couple of saws and a sine wave, to add some bottom end if required. If a sine wave is used, detune this by an octave below the other oscillators and then proceed to detune each saw from one another by +/−3, +/−5 and +/−7. The amp envelope for all the waveforms is commonly set to a fast attack with a medium to long decay and no sustain or release. A filter envelope is not commonly used, as this adds harmonic movement through the period of the sound that may conflict with the already programmed/manipulated rhythms and the oft preferred option is to keep it quite static and employ filter movement manually to suit the constantly changing frequencies in the rhythms. That said, if a pitch envelope is available in the synth, it may be prudent to positively or negatively modulate the pitch of the oscillators to add some small amounts of movement. Once this basic timbre is laid down, it's prudent to experiment with the attack, decay and release of the amp's EG to help the bass sit comfortably in with the kick drum loop. In fact, it's essential to accomplish this rhythmic and tonal interaction between the drum loop and the bass before moving on. Techno relies on a very sparse collection of 'instruments' and the interaction attained here will form a majority of the record.

As the harmonic movement and interaction with the bass and rhythms provide the basis for most techno, it's also prudent to experiment by applying effects to the bass timbre to make it richer sounding. While most effects should be avoided since they tend to spread the sound across the image (which destroys the stereo perspective), small amounts of controlled distortion can help to pull the bass out of the mix or give it a much stronger presence. Similarly, a conservatively applied delay effect can be used to create more complex sounding rhythms.

One final aspect of techno is the addition of sound effects and occasionally vocals. While the sound effects are generated by whatever means necessary, from sampling and contorting anything with effects and EQ, the vocals very rarely consist of anything more than a short sample. The verse and chorus structure is most definitely avoided and in many cases only very small phrases are used, which are often gleaned from the old 'speak and spell' machines of the early 1980s. This particular machine isn't a requirement (with its increased use in techno, the second-hand prices of these units have increased considerably) and the same effect can be obtained from most vocoders so long as the carrier consists of a saw wave and the modulator is robotically spoken.

When it comes to the arrangement of techno, it's important to understand that it does not follow the typical dance structure. Instead, it relies totally on the adaptation and interrelationship between all of the elements together. This consists of dropping beats in and out of the mix, along with the bass, vocals and sound effects (if used), but it mostly centres on the real-time application of filters, effects and EQ. The plan is not to create a track that builds to a crescendo or climax, but rather stays on one constant rhythmical level that warps from one rhythmically interesting collective to another. Indeed, it's careful use of filters and effects that creates the final arrangement, not by introducing new melodic elements. The overall goal is to create the impression that all the sounds are interweaving with one another at different stages in the music. This helps to prevent monotony, but also averts the building sensation often introduced by adding new melodic elements into the mix.

Recommended listening

Ultimately, the purpose of this chapter has been to give some insight into the production of techno and, as ever, there is no one definitive way to produce the genre. Indeed, the best way to learn new techniques and production ethics is to actively listen to the current market leaders and experiment wildly with the mixing desk, effects and processors. With this in mind, what follows is a short list of artists that, at the time of writing, are considered the most influential in this area:

- Derrick May.
- Juan Atkins.
- Kevin Saunderson.
- Eddie Fowlkes.
- Richie Hawtin (Plastikman).
- Carl Craig.
- Kenny Larkin.
- Stacey Pullen.
- Jeff Mills and Mike Banks.
- James Pennington.
- Robert Hood.
- Blake Baxter.
- Alan Oldham.

Note: On the CD you can hear an excerpt from the techno track constructed through this chapter.

15 Trip-hop

Trip-hop is British hip-hop that lacks the lyrical skills of the US counterparts, but British kids have got the musical side.

James Lavelle (Mo' Wax label)

Trip-hop is, in the context of dance music, a relatively new genre that evolved out of Bristol in the early 1990s. During this time, American rap was the predominant musical style that was taking Europe by storm. Since this genre of music requires a heavy involvement in the entire hip-hop scene (hip-hop is actually a culture of which rap is only a part), British DJs and musicians contorted it further. The principle elements behind the construction of American hip-hop were fully embraced with the emphasis remaining on slow, laid-back heavy beats mixed with the gratuitous sampling of old records, but the vocals were left out. This resulted in a genre that was termed British hip-hop by the media, but at this stage it lacked any real diversity, consisting mostly of slow 'tripped out' beats and bass lines mixed in with samples of old jazz records.

In 1991, the release of Bristol-based Massive Attack's *Blue Lines* album marked the first serious release of British hip-hop and also revealed its close connections with American hip-hop (the single *One Love* displaying a remarkably similar feel to its relative). It wasn't until 1994, however, when British hip-hop was coined as trip-hop by UK's *Mixmag* magazine with the release of Massive Attack's second album entitled *Protection*, alongside the appearance of Portishead and Tricky. Portishead defined a new style of trip-hop music labelled 'lo-fi' through a mixture of Beth Gibbons' brooding vocals mixed amongst samples of 1960s and 1970s jazz music, which were left with a predominantly 'raw' feel.

In fact, this raw approach to music was deliberately encompassed by Portishead as they made an effort to record all their instruments into old analogue tape recorders rather than straight to digital media. Tricky's style was somewhat different again, branded by low-pitched singing and an overall cleaner sound, but like Portishead and Massive Attack, the style often exhibited a slow, almost depressing feel. Although these three acts did not necessarily aim to create music that was particularly dark, it just so happened that the brooding attitude of the music often oozed a dismal feeling. Part of this may have been attributed to the fact that they had all worked in the same circle. Massive Attack and Tricky originally produced music together under the moniker of 'The Wild Bunch', and Portishead's founder Geoff Burrow aided Massive Attack in producing *Blue Lines*.

On the back of this relatively new genre, more trip-hop artists began to emerge, each putting their own distinctive twist on the music. Artists such as Red Snapper, Howie B, Baby Fox,

Lamb, Sneaker Pimps and the Brand New Heavies mutated the genre by mixing it with break beat, ambience, house and acid jazz. The vocals became more upbeat and lively to encapsulate a wider audience, resulting in trip-hop being associated with more energetic music rather than the dark and gloomy vibe. Indeed, because trip-hop is often associated with a dark brooding atmosphere, many artists do not appreciate being placed under the trip-hop tag and will describe their music as 'illbient', 'ambient hip-hop', 'British hip-hop' or 'jazz hop'.

Musical analysis

Actually defining trip-hop for musical analysis is difficult because, as mentioned, most artists will flatly deny that they produce this particular style of music. In fact, only Massive Attack, Portishead and Tricky don't seem to mind being labelled as producing the genre. Nevertheless, it can be roughly summarized that trip-hop is commonly produced using an eclectic mix of acoustic and electronic instruments combining ideas from R'n'B, hip-hop, dub, ambient, industrial and jazz. This means that it often features instruments such as acoustic pianos, Rhodes pianos, upright basses, clavinets, horns and flutes, along with electric and acoustic guitars. Principally, these are combined to produce an often nostalgic or dark ambience, which is helped along further with haunting vocals and samples taken from vintage radio and films.

On the subject of samples, in keeping with its original roots of hip-hop, many of the instrumental riffs, melodies and drums are commonly sampled for old records. It's this approach that is often accredited to the creation of 'lo-fi', since these samples are not respectively 'cleaned up' and are often left dirty and gritty, even to the point that the vinyl crackle is left evident in the background of the music. This approach has meant that even if the sounds are recorded from a live instrumentalist (which is progressively becoming more common), it's quite usual to dirty them up a little (as mentioned, Portishead in particular are renowned for recording all their instruments to old analogue tapes before submitting them to digital media for editing and mixing). This helps to retain the 'old' feel that is often crucial to the production of the genre.

When it comes to the equipment used by the artists, many of them are particularly nonchalant about what's used, but Portishead are notoriously cagey about both their production techniques and equipment. Nevertheless, producing this style of music does require you to use the 'right' type of instrumentation and effects to produce the atypical feel of the genre. The first of these is access to a large collection of old vinyl records, particularly jazz, a decent record player and a sampler. For legal reasons, it's immoral for me to suggest that you should rely on sampling from records to produce music, but it's important to note that much of this genre relies so much on sampling that many record companies now ask which records have been sampled when the music is submitted.

Of course, these samples are very rarely used 'as is' and it's common to manipulate them in wave editors, or more specifically, sample slicing programs such as Wavesurgeon. This is predominantly the case with the drum rhythms as these, more than any other aspect, are commonly sampled to help attain the feel of the genre. Generally, it's the break beat of the record that is sampled (i.e. the middle eight of the original record, where the drummer goes off on one for a couple of bars), which is then cut up, rearranged and, if required, the tempo is reduced or increased to between 100 and 120 BPM to form the basis of the requisite relaxed feel. That said, it should be noted that while sampling and rearranging breaks is the most common form of creating a loop, it's not the only method employed. The rhythms can also be programmed in drum machines or sequencers. It all depends on the artists working on the project and their methodology, so we'll examine the principles behind generating new grooves with both applications.

The most popular process, as touched upon, is to sample break beats from old jazz records which can then be sliced diced and rearranged in sample slicing programs such as Wavesurgeon, or they can be sliced and have each individual sample mapped to a key in the sampler to be played live from a keyboard. Although sampling from vinyl will often produce quite dirty timbres, it's often the case that they're not filthy enough, so when first sampling the break beats it's advisable to sample at a much lower rate than CD quality (most trip-hop musicians use the E-Mu SP1200 or the Casio FZ due to the poor sampling quality). The typical resolution is 12-bit, 22 kHz, but reducing the bit rate further to 8-bit may provide even better results depending on the source. Alternatively, for those using sample CDs in WAV or AIFF format you can often lower the rate in most wave editors, or Wavesurgeon offers a 'quantize' feature that adds sufficient grit to any samples imported into it.

Note: A demo of Wavesurgeon is included on the CD.[1]

While sampling at a lower quality or using Wavesurgeon's quantize feature will grit up the beats to a good degree, it's worth manipulating the sounds further using EQ, filters, transient designers and distortion to acquire the sounds you need. The use of these effects is purely down to experimentation, but it's often worth inserting a compressor directly after them so you can go as mad as you like without the fear of overloading the mixer's inputs/outputs.

Of course, sampling beats from old records brings up numerous copyright issues, so it's sometimes prudent to program your own drum timbres and contort them with effects into sounds that are more suited to the genre. These timbres can be sourced from almost anywhere, including the jazz kit from GM modules or more commonly a mix of the Roland TR808, TR909 and the E-Mu Drumulator. As always, however, you can program all these timbres in any capable synthesizer and then experiment with the various parameters on offer before applying effects to construct the atypical 'trip-hop' kit.

The trip-hop kick drum can range from being quite low and boomy to quite bright and snappy; it depends on how you want the loop to appear. As a good starting point, the kick can be produced with a 100 Hz sine wave that's modulated with a positive pitch envelope set to an immediate attack and a medium decay. If you need it to be a little tighter, then reduce the decay and place a small transient click at the beginning of the waveform. A good way to produce this click is to sample any of the GM drum kicks and, using a wave editor, remove the tail end of the sample before dropping it over the top of the programmed kick. Alternatively, it can be produced through using a square wave with an immediate attack and very short release for the amp envelope. Once programmed, this can be laid over the top of the sine wave and can be tuned up or down to produce the basic timbre. It's important to note here that you're only after a fairly basic kick drum timbre at this point, since after the rhythm has been programmed it will be affected further to produce the finished article.

---

[1] Wavesurgeon. © 1997–2003 Square Circle Software. All rights reserved.

Moving onto the snares, these tend to be quite bright and snappy, which is often accomplished by filtering and augmenting the timbre with small amounts of controlled distortion, but the basic sound can be produced with nothing more than a triangle wave mixed with a white noise waveform. An amp envelope set to an immediate attack with no sustain or release and a very short decay is used to shape the triangle, while a second amp envelope is used to shape the noise waveform. This uses the same attack, sustain and release parameters, but by setting the decay a little longer than the triangle, the noise will ring out beyond the triangle and gives you independent control over the triangle and noise to create the timbre you're after. As the final icing on the cake, a band-pass filter can be used to remove the low-end frequencies along with some of the high-end noise to produce a typical trip-hop snare. On the other hand, on occasion, once the snares have been created, the tail is removed so that only the initial transient is left, producing a click rather than a thwack. Massive Attack and Portishead have both used this technique.

This same sample cutting technique is also commonly employed on the hi-hats to produce a timbre that ticks rather than hisses. In fact, some trip-hop tracks have actually used samples of the ticking of a clock to produce the timbre. These can of course be synthesized quite easily using any synth and, although ring modulation does produce better hi-hat timbres, for trip-hop white noise is perfectly sufficient. To create this, simply select white noise as an oscillator and set the filter envelope to a fast attack, sustain and release with a medium to short decay. Once this is done, set the filter to a high pass and use it to roll off any frequencies that are too low to create a hi-hat. Once this basic timbre is down, lengthening the decay parameter can create open hi-hats and shortening it will produce closed hats. By setting this very short you can also produce the typical ticking timbre. Alternatively, if samples are used it's prudent to either use the amplifier's decay envelope in the sampler or use a wave editor to remove the subsequent tail of the hats.

Whether the grooves are sampled or programmed, the most fundamental aspect of creating the typical groove lies with the pattern. All trip-hop derives its feel from hip-hop and jazz, and so relies on a relatively sparse drum arrangement with the laid-back feel produced by the interplay between the kick and the snare. As a consequence, the often used kick/snare/kick/snare pattern is usually (but not always) avoided and generally more rhythmic kick patterns are utilized, augmented with the occasional snare. Of course, closed hi-hats, ride cymbals and pedal hi-hats are also employed, along with the occasional open hat or crash cymbal, all of which add some syncopation and steadiness to the rhythm.

As mentioned, snares are not always used in the production of trip-hop and any 'clicky' percussive timbre can be used in its place. What is important, however, is that the entire loop is quite reserved in terms of instruments and patterns, and in many cases these are kept incredibly simple to allow room for the delicate vocals, chords and lead instruments to play over. Also, it's important to bear in mind that many of these elements will sit just off the beat rather than strictly quantized on it, so that the rhythm appears to be played live. In fact, the real key to producing these rhythms is to keep the drums as simple as possible and play them live from a workstation or keyboard and then use functions such as iterative quantize to prevent the pattern from becoming too wayward.

Velocity also plays a large role in the creation of the rhythm to keep the realism and the most usual method of producing a live beat is to implement the *strong–weak–medium–weak* syncopation on the kick drums, although the *actual* velocities may stray wildly from this guideline depending on the sound *you* wish to achieve. The snares too are often subjected to different velocities, with the first and final snares using the highest values, and those in between using a mixture of different values.

Figure 15.1 A trip-hop rhythm.

Figure 15.2 A trip-hop rhythm using snares.

On top of this, many trip-hop tracks also often employ small snare rolls in the rhythms to add to the flow of the groove. These are often played monophonically, however, and very close together so that the consecutive snares remove the tails of those that precede them. This can be accomplished by setting the synth or sampler to monophonic operation, or alternatively they can be programmed in audio by physically cutting off the tails of the snares and positioning them one after the other. For added interest these can also be subjected to pitch bend so that the snares roll downwards in pitch. This downwards roll is often preferred to pitching upwards, since a movement down the scale tends to sound less uplifting.

Figure 15.3 A typical trip-hop snare roll augmented with pitch bend.

As always, these are only generalizations and are fully open to further experimentation. Indeed, after programming a typical loop it's quite usual to use a series of effects and processors to seriously grunge up the audio. The first of these techniques involves speeding up the tempo of the rhythm from the usual 100–120 BPM to 150–160 BPM. This loop is then sampled at a low bit rate, typically 12- or 8-bit, set across the key range of the sampler and then played in the lower registers so that not only does the tempo slow down, but the pitch also becomes significantly lower. Alternatively, the loop could be sampled at the original tempo and then time-stretched numerous times by extreme amounts each time. The more that this is applied, the more the loop will begin to degrade, as each subsequent time-stretching algorithm will impart some degradation into the audio.

Another technique involves playing the loop through a speaker system and miking up the speakers using a poor quality microphone and pre-amp. This can often add sufficient dirt to anything that is recorded this way and you can experiment by moving the placement of the microphone. You may also capture small amounts of incidental sounds such as traffic outside using this method, which may actually enhance the overall lo-fi effect. Similarly, recording the loop down to an old (and hopefully knackered) analogue tape machine will introduce adequate hiss and sound degradation that can then be re-recorded into the audio editor. So long as any of these techniques are followed by a compressor, you should feel free to experiment wildly.

On the subject of compression, trip-hop drum loops benefit hugely from incredibly heavy compression and in many instances this contributes a great deal to the overall sound. Although there are no definitive settings to use, as it depends on the loop and the timbres being used, a good starting point is to set a very low threshold with a ratio of approximately 10:1. Once these are set, experiment with the attack and release settings and, while generally the idea is not to create a loop that pumps, you need to compress the loop as hard as possible by trying out various settings so that the loop becomes as 'heavy' as possible.

Note: On the CD you can hear the trip-hop rhythm used on the example track.

With the basis for the drums laid down, it's quite usual to follow these with the chords or leads rather than the bass. Since trip-hop draws it's inspiration for bass lines from dub (and in some cases hip-hop), they remain relatively simple and only act as a basic underpinning for the chords and lead to interweave with. As a consequence, it's usually prudent to lay down the chord structure first, followed by the leads and vocals, before finally developing the bass to sit around them.

Generally, the chords and most of the overlying instruments are usually written in a minor rather than major key since, as we've already touched upon in music theory, these provide a more serious and sometimes dark feel to the music. Naturally, coming up with the actual chord structure is up to your own artistic licence, but generally they do not involve large amounts of movement through the pitch range and mostly tend to stay rather static using simple inversions rather than jumping manically from one key to another. A good starting point for these is to write a chord in A minor and then create some inversions based around this. That said, it is often worth moving between consonant and dissonant chords to produce a cacophonous feel to the music. Portamento is also useful here, so experiment by overlapping notes slightly and using it so that the pitch sweeps up and down between the notes of the consecutive chords. This will invariably produce more laid-back results than a chord sequence that triggers suddenly on each new note.

The timbres used for the chords can range from samples of a real string section (ensure that you write these in the right key – violins, for example, cannot play any lower than the G below middle C!) to synthesized pads that are swept with a low- or high-pass filter. If you decide to take the latter approach, the timbre to use is down to your discretion, but we can nonetheless look at the construction of a basic pad, which can then be contorted further to suit your music.

Typically, the pad in trip-hop is fairly rich in harmonics, allowing it to be swept with a low-pass filter if it's necessary to add some weight, or alternatively swept with a high-pass or band-pass filter if the idea is to produce a ghostly backing. A good starting point for this is to use either two sawtooth waveforms or a sawtooth and a pulse wave. Detune one oscillator from the other by 5 or 7 cents and then set the amp envelope to use a fast attack with a short decay, medium sustain and a fast release. If you want to sweep the filter (keep in mind that a swept low-pass filter will use a proportionate amount of frequencies in the mix), then set the filter's envelope to a slow attack and decay with no sustain or release, and set this to positively modulate the filter's cut-off. Otherwise, use a high-pass or band-pass filter or, to keep the timbre static (at this point), use the same settings as the amp envelope with a slightly longer decay on the filter and set the filter key-follow to positively track the pitch. If the pad is not being swept and consists of long sustained notes in the chord structure, then it's prudent to add some movement to the timbre by using an LFO to augment the pulse width or the resonance of the filter. The waveform, rate and depth to use on the LFO will obviously depend on what you wish to achieve, but generally a square, triangle or sine wave will produce the most 'natural' results.

Although any effects should generally be avoided this early, since they tend to occupy a significant amount of space within a mix, if the pad is to play a large role in the music then it may be worth applying them now to produce the finished timbre. Typical effects for this include heavy chorus or light reverb, but it's worth experimenting with other effects to produce the results you want. One of the classic effects to use on a pad in trip-hop is a vocoder. By employing the track's drum loop as a modulator and using the pad as a carrier wave, the drums will impose a new character onto the pad, forcing it to flow in and out of the mix with the drums. Alternatively, you can make a copy of the drum track, move it forward in time and then use this to modulate the pad through the vocoder. This results in the pad 'pulsing' between the kicks of the original loop, creating a forced syncopation. And of course, as a final stage, the pad will need to be sampled at 8- or 12-bit to produce the archetypal sound of the genre.

With the chord structures down, the vocals and lead melodies can be laid over the top. Roughly speaking, the vocals are usually recorded before any lead instrumentation, since these instruments are often kept very simple to prevent detracting the attention from the vocals. In fact, the vocals play such a fundamental role in trip-hop that it's rare to hear of a track that doesn't feature any. Most usually, mellow, laid-back female vocals are used, but occasionally they are performed by male singers, although if this is the case they tend to be more rap (i.e. poetic) based than actually sung.

The typical microphone for recording trip-hop is a large diaphragm model such as the AKG C414, which is amplified with a valve pre-amp to add the requisite warmth, but it should be noted that these are not necessarily a requirement. Indeed, since this genre relies on a gritty, dirty feel, the vocals can be recorded with pretty much anything so long as they're decipherable. Even a cheap dynamic microphone from the local electronics store recorded direct into a portable cassette recorder such as a walkman can produce great results.

If this latter approach is used, then generally you will not need to use a compressor while recording, as the cassette tapes will reach magnetic saturation before clipping is introduced, but if you're recording into a digital system then compression is vital to prevent any clipping. At this stage it should be applied lightly, since it cannot be undone later, and a good starting point is to set a threshold of −8 dB with a 3:1 ratio, a fast attack and moderately fast release. Then, once the vocalist has started to practise, reduce the threshold so that the reduction meters are only lit on the strongest part of the performance.

Once the vocals have been recorded (whether to analogue tape or hard disk), they will invariably require more compression to even out the levels so that they don't disappear behind other instruments in the mix. As always, the settings to use will depend on the initial performance, but to begin with use a threshold setting so that you have −7 dB on the reduction meter, along with a ratio of 3:1, a fast attack and moderately fast release (approximately 200 ms). If you're after a more commercial type of sound, then try increasing the ratio to 5:1, but exercise caution and use your ears to decide whether it's right or not. Keep in mind that trip-hop is all about generating atmosphere and much of this is derived from the vocals. If these are too heavily compressed, the ratio between the peak and average signal level will be seriously compromised, which can remove all emotion from a recording. If they sound a little thin in the context of the mix, then small amounts of reverb may help, but similar to compression this should be kept to a minimum to retain the feel of the music. In fact, reverb should be applied so lightly that its application is only noticeable when it's removed. Alternatively, making a second copy of the vocal track and pitching this up or down by five semitones can be used to produce harmonies which can help fill out the vocals.

With the vocals sat in the track, it's much easier to hear where there are gaps in both frequency spectrum and arrangement, and these can be filled out with the lead instrumentation. As already mentioned, these very rarely play complicated riffs so as not to draw attention to themselves, but simply add to the atmosphere of the music. As a result, in the majority of tracks they consist of single hits at the beginning of the bar or extremely simple melodies which appear as though they've been played ad lib over the chords and vocals. The old adage of 'less is more' is certainly the case with most trip-hop. Fundamentally, these tend to be played with real instruments rather than synthetic, and a proportionate amount of tracks will use clavinets, organs, horns, theremins or flutes, but more commonly Rhodes pianos and electric guitars that are treated heavily to tremolo effects (i.e. pitch modulation), phasing, flanging, distortion and a whole host of effects. In actual fact, as these leads play such simple riffs, it's the effects applied that add to the atmosphere of the music and so experimentation is the key.

Figure 15.4 A typical Rhodes piano lick.

Most artists rely on guitar pedals to produce most of the effects and, ideally, it's wise to follow suit. Since these are essentially produced to be used live on stage, they don't have a particularly low noise floor, so not only are they powerful effects, they're also pretty noisy. You can pick up guitar pedals new for around £50 and they can change hands for as little as £20 on the used market. Generally, you'll need tremolo, flanger, reverb and delay pedals to begin with, as these are all suited and used in the production of the genre. You can, of course, use digital or plug-in effects units in their place, but these tend to be too clean so you'll need to grunge them up a little.

Due to the sparseness of trip-hop, delay is the most commonly used of these effects, but this can be subsequently soiled by sending the audio out to a delay unit set to a high feedback and sampling the subsequent delays. Once you have these, import them into a separate track in an audio sequencer and, using a mix of EQ and distortion, automate the EQ to gradually remove both the high and low frequencies on each delay, while also increasing the distortion applied to each. Export this as another audio file and then use the scissors tool (or similar) to cut up the delays and place them where required in the arrangement. If the sequencer does not allow automation (or you're using hardware), then you can simply record the EQ and distortion tweaks to audio cassette (or DAT) and then edit the results in the sequencer.

The final element of trip-hop is the bass, but before proceeding any further it should be noted that not all tracks will feature a bass line, and many that do will use an upright bass that can also act as a lead instrument. Trip-hop is a descendant of hip-hop mixed with elements of jazz and ambient, and as such it is totally unnecessary to fill up the entire frequency spectrum and arrangement with sounds. In fact, the relatively large spaces that are left between instruments help in creating the all-important atmosphere of the music. This is important to keep in mind when writing the bass, since it can be quite easy to get carried away and produce a bass with far too much movement.

Fundamentally, trip-hop draws the basses from dub and so on the whole they are kept relatively simple, as they merely act as the underpinning of the track and should not form a major melodic element in themselves. This is mainly to prevent it from drawing attention away from the vocals. Consequently, a bass line will often consist of nothing more than a few eighth or perhaps quarter notes per bar that will tend to stay at one pitch or move by no more than three, five or seven semitones. Simplicity is the solution here. More importantly, since the bass is often played live, when programmed in MIDI the notes should very rarely occur at the beginning of each bar and it's much more natural for it to begin just a few ticks *before* or *after* to offset it from the rest of the record. This not only emulates the nuances of a real bassist, but the drums will appear to pull or push the record forward, helping to create a more flowing groove. Creating this relationship can form an essential part of the music, but whether to sit it before or after the beat is entirely open to artistic licence and after programming it's worth experimenting by moving it before or after the beat to see which produces the best results.

Figure 15.5 A typical trip-hop bass rhythm.

For the timbre, real bass guitars, particularly upright and acoustic, are usually chosen due to the resonant sounding strings, which help it cut through the mix. These sounds are best sourced from a sample CD or bass-specific module, such as *Spectrasonics' Trilogy*, since recording live guitars is difficult without good equipment and this subject is beyond the scope of this book, as it requires a whole new set of production ethics. Nevertheless, provided that you have a good bass module it is perfectly possible to emulate a real guitar with MIDI, and while it does involve a lot of work, it is only as much as is involved with recording a real bass guitar – it just doesn't take as much equipment. The details of emulating a bass with MIDI have already been discussed in Chapters 10 and 11, but for the benefit of those who've jumped straight to this chapter you can either play the book style version of Dungeons and Dragons by jumping back and forth depending on what information you want, or just read on.

The key to programming any real instrument is to take note of how they're played and then emulate this action with MIDI and a series of CC commands. In this case, most bass guitars use the first four strings of a normal guitar, E–A–D–G, but these are tuned an octave lower, resulting in the E being close to three octaves below middle C. Also they are monophonic, not polyphonic, so the only time notes will actually overlap is when the resonance of the previous string is still dying away as the next note is plucked. This effect can be emulated by leaving the preceding note playing for a few ticks while the next note in the sequence has started. The strings can either be plucked or struck, and the two techniques produce different results. If the string is plucked, the sound is much brighter and has a longer resonance than if it were simply struck. To copy this, the velocity will need to be mapped to the filter cut-off of the bass module so that higher values open the filter more. Not all notes will be struck at the same velocity, though, and if the bassist is playing a fast rhythm the consecutive notes will commonly have less velocity, since the player has to move their hand and pluck the next string quickly. Naturally, this is only a guideline and you should edit each velocity value until it produces a realistic feel.

Depending on the 'bassist' they may also use a technique known as 'hammer on', whereby they play a string and then hit a different pitch on the fret. This results in the pitch changing without actually being accompanied with another pluck of the string. To emulate this, you'll need to make use of pitch bend, so this will first need setting to a maximum bend limit of two semitones, since guitars don't 'bend' any further than this. Begin by programming two notes, for instance an E0 followed by an A0, and leave the E0 playing underneath the successive A0 for around 100 ticks. At the very beginning of the bass track, drop in a pitch bend message to ensure that it's set midway (i.e. no pitch bend) and, just before where the second note occurs, drop in another pitch bend message to bend the tone up to A0. If this is programmed correctly, on playback you'll notice that as the E0 ends the pitch will bend upwards to A0, simulating the effect. Although this could be left as is, it's sensible to drop in a CC11 message (expression) directly after the pitch bend, as this will reduce the overall volume of the second note so that it doesn't sound like it has been plucked.

In addition to this, it's also worthwhile employing some fret noise and finger slides. Most good tone modules will include fret noise that can be dropped in between the notes to emulate the bassist's fingers sliding along the fret board. The pitch bending is best emulated by first programming the notes to overlap slightly and then recording movements of the pitch bend wheel live and editing them in the sequencer. Again, since the pitch bend range is set to two semitones, only the MSB value will need editing if the live recording didn't go according to plan.

> Note: On the CD you can hear a programmed 'realistic' bass and view the MIDI file.

Of course, you don't particularly have to use a real bass guitar and, as ever, it's open to artistic licence; pretty much any low-frequency sound can be used provided that it sounds dark, moody and atmospheric. As we all have different ideas of what constitutes a moody, atmospheric bass, there are no real generic timbres and practically any synthesizer can be used to program the basic sound. As a recommendation, the *Access Virus, Novation Supernova* or the *Novation Bass Station* (VSTi) are particularly suited towards creating deep atmospheric bass sounds.

Fundamentally, a synthesized bass for trip-hop should exhibit plenty of low-end presence (dub influence), yet at the same time not be so powerful that it takes up too much of the mix (hip-hop influence). To accomplish this, a good starting point is to use three oscillators: one set to a sawtooth waveform, the second set to a sine wave and the third using a triangle wave. Transpose the triangle wave up an octave and then adjust the amp's EG to a fast attack, decay and release but a high sustain. This will allow the bass note to sustain while the key is depressed (many trip-hop rhythms depend on long notes rather than short stabby ones). If a resonant pluck is needed for the beginning of the note, using a low-pass filter set the cut-off quite low with a high resonance and use a filter EG with a zero attack, sustain and release with a medium decay. Finally, set the filter envelope to full positive modulation and then experiment by detuning the sine wave downwards from the saw. The further away this is detuned, the deeper and more 'full' the bass will become. If the notes are sustained for a lengthy period of time, then it may also be worth applying very small amounts of LFO modulation to the filter or pitch of one or all of the oscillators depending on the type of sound you want to achieve.

As with most other instruments in this genre, the bass can also benefit from effects, but these must be chosen and applied carefully. Any stereo effects such as flangers and phasers will often spread the sound across the image (which will, in turn, move the bass from centre of the mix and consequently destroy the stereo perspective), so generally only mono effects should be applied. This can include distortion, EQ and filters, but if using any of these it's useful to place a compressor directly after so that you can experiment by overdriving the signal without fear of actually overdriving the desk (although this can be put to creative uses if it's an analogue desk).

Note: On the CD you can hear the bass through GM and programmed synthesis.

As touched upon throughout, the most fundamental aspect of creating trip-hop is the dirty/gritty character that the whole mix exhibits. While this is certainly not an excuse for poor mixing, it is a good reason to be experimental and push things further to attain the character of the sounds. For instance, record everything to analogue tape rather than direct to hard disk or create wild feedbacks by sending an audio signal to one channel of a stereo effect, then return the signal back into two inputs of the desk and feed one of these back out to the other channel of the effect. Alternatively, feed an effect as usual but return the outputs into a normal mixing channel and then feed these down another aux send back into the effect. Also, feel free to experiment with very heavy EQ to thin sounds right down or use filters such as the Sherman Filterbank to warp sounds beyond comprehension. So long as each of these processes is followed by a compressor to keep the levels under some control there should be no restrictions. What's more, even if the elements of the record are *not* sampled, they should nevertheless sound as though they are. This means gratuitous use of bit reduction and sample reduction, along with sampling the vinyl crackle from a record and applying this over the top of the mix. For added hiss, a popular method is to record a source at a relatively low level and then increase the gain so it comes up to nominal level, thus also increasing the noise floor. Furthermore, you may wish to play your final track through a pair of poor quality speakers and record the sound output with a poor quality microphone (as suggested for individual sounds); by importing this sound back into your sequencer you will have further material to edit and add back to your original track, creating areas of lo-fi double-tracked sound.

Once the basic elements are programmed, you can begin to lay the arrangement down. Since trip-hop relies heavily on vocals the music tends to be structured in a similar manner to most popular music songs, consisting of a verse/chorus structure. As such, it can be roughly broken down into four distinct parts, consisting of the verse, chorus, bridge and middle eight. If we break this down in a number of bars we can derive that the order of a song and subsequent bars are as follows:

- Verse 1 – *commonly* 16 bars.
- Chorus – *commonly* eight bars.

306   The Dance Music Manual

- Verse 2 – *commonly* 16 bars.
- Chorus – *commonly* eight bars.
- Verse 3 – *commonly* 16 bars.
- Chorus – *commonly* eight bars.
- Bridge – *commonly* one bar.
- Middle eight – *commonly* eight bars.
- Double chorus – *commonly* 16 bars.

This is, of course, open to artistic licence and some artists will play the first two verses before hitting the chorus. This approach can often help to add excitement to a track, as the listener is expecting to hear the chorus after the first verse. If this route is taken, even though there may be different vocals employed in the verses, if they are played one after the other it can become tiresome, so a popular trick is to add a motif into the second verse to differentiate it from the first. Also, keep in mind that this is only a general guideline and as always the best way to gain a good representation of how to arrange a trip-hop track is to actively listen to what are considered the classics and then copy them. This isn't stealing, it's research and nearly every musician on the planet follows this same route.

Recommended listening

Ultimately, the purpose of this chapter has been to give some insight into the production of trip-hop and there is, of course, no one definitive way to produce the genre. Indeed, the best way to learn new techniques and production ethics is to actively listen to the current market leaders and be creative with processing, effects and synthesis. With this in mind, what follows is a short list of some of the artists that, at the time of writing, are considered the most influential in this area:

- Tricky.
- Massive Attack.
- DJ Shadow.
- Portishead.
- Red Snapper.
- DJ Food.
- DJ Cam.
- DJ Kicks.
- Thievery Corporation.
- DJ Krush.
- Coldcut.
- Moloko.
- DJ Vadim.
- Herbalizer.
- 9 Lazy 9.

Note: On the CD you can hear an excerpt from the lo-fi track constructed through this chapter, along with the MIDI file to accompany it.

Part 3
Mixing and Promotion

16 Mixing

The ideas for *A Huge Ever Growing Pulsating Brain That Rules From the Centre of the Ultra World* were there weeks before, but the mix itself took 20 minutes.

Alex Patterson

Mixing is often seen by many as the most complex procedure involved in producing a dance record; however, the truth of the matter is 99.9% of the time a poor mix isn't a result of the mixing process, but a culmination of factors before anything even reaches the mixing desk. If you've taken care producing the track throughout, then when it eventually comes to mixing you'll find that the music has more or less mixed itself and all that is required is gentle use of EQ, volume and effects to polish the results.

The mixing desk is simply a tool used to position instruments in the mix and allow you to add a *very* small final sparkle. So, before you even touch the desk the only problem should be that you can't hear some of the instruments too clearly. Apart from that, the mix should otherwise sound just as you imagined it would. Thus, before we even touch upon the use of faders, EQ and effects in mixing, if you want to produce a great mix you need to begin by going back and examining what you have so far.

Generally, a poor 'pre-mix' is created by one or all of the following:

- Poor recording, programming or choice of timbre/sample.
- Poor quality effects, or use of effects when programming.
- Poor arrangement or MIDI programming.

We've touched upon the importance of all of these in earlier chapters, but at this stage you should go back and check everything one final time. It's prudent to solo each instrument and ensure that the timbre really is suitable for the track and that it sounds the best it possibly can. If you're unhappy with any of the timbres, reprogram or replace them. It is vital at this stage that you don't settle for anything less than the very best – it may be an overused analogy but you wouldn't expect to make a great tasting cake if some of the ingredients were out of date!

Perhaps most important of all, though, before you approach the mix you need to ask yourself one final yet vitally significant question about your track. Close your eyes, listen back to the music and ask yourself: 'Can you feel it?'

Above the arrangements, sounds, processing and effects, this is the ultimate question and the answer should be a resounding (and honest!) yes. Dance music is ultimately about 'groove', 'vibe' and 'feel', and without these fundamental elements it isn't dance music. A track may be beautifully programmed and arranged, but if it has no 'feel' then the whole endeavour is pointless.

Mixing theory

At the outset, it's important to understand that mixing is a creative art and as such there are no right or wrong ways to go about it. Every engineer will approach a mix in their own distinctive way – it's their music and they know how they want it to sound. A unique style of mixing, provided that it sounds transparent, will define your creative style as much as the sounds and arrangement do. Despite this, there are some practices that you can employ that will put you on the right track to producing transparent and detailed mixes.

The first step to creating any great dance mix is to understand the theory behind mixing and how this can be adapted in a more practical sense to suit your own particular style. This means that you need to comprehend how the soundstage of a mix can be created and adapted, how frequency ranges are grouped together and how to come to terms with our natural hearing limitations. We'll look at each of these in detail.

Hearing limitations

Our hearing is far from perfect because not only do we perceive different frequencies to be at different volumes, but the overall volume at which we listen to a mix will also determine the dominant frequencies. This inaccurate response can be traced back to when we used to brandish clubs and live in caves. Since talking (or grunting) to each other in a cave resulted in vast amounts of reverberation, our hearing adapted itself to concentrate on the frequencies where speech is most evident and decipherable in this situation – approximately 3–4 kHz. This means that, at conversation level, our ears are most sensitive to sounds occupying the mid range and any frequencies higher or lower than this must be physically louder in order for us to perceive them to be at the same volume.

> Note: At normal conversation levels it's 64 times more difficult to hear bass frequencies than it is the mid range and 18 times more difficult to perceive the high range.

If, however, the volume is increased beyond normal conversation level, the lower and higher frequencies gradually become (albeit perceivably) louder than the mid range. This again is related to when we lived in caves and played a part in our survival mechanism. It allowed us to understand when something was shouted rather than spoken, as not many people are in the habit of casually telling you there's a hungry looking bear right behind you!

In the 1930s, two researchers – Fletcher and Munson from Bell Laboratories – were the first to accurately measure the uneven response of the ear and so it is referred to as the 'Fletcher–Munson contour control curve'. It's absolutely vital that you keep this contour control in mind while mixing, since it means that the bass energy produced will depend entirely on the mix volume. For example, if you balance the bass elements at a low monitoring level then there will be a huge bass increase at higher volumes. Conversely, if you mix at high levels there will be too little bass at lower volumes.

Mixing appropriately for all three volumes is something of a trade-off, as you'll never find the perfect balance for all listening levels, but generally whenever mixing dance music you should always check that it sounds okay at low and medium volumes and great at high volume. That is, of course, assuming that the clubs are your market – they don't play music at low or medium levels.

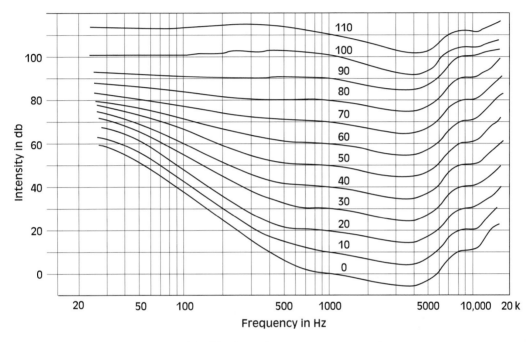

Figure 16.1 The Fletcher–Munson contour control curve.

Frequency bands

The second step is to become familiar with how frequencies are generally banded together, as these bands play a large role whenever we speak about sub bass, bass and mid range, etc. This isn't as difficult as it may initially sound; as we've been subjected to these frequencies since we were born, you just need to understand how they are grouped together. To help you decipher this, what follows is a list of how frequencies are generally banded and described.

- **Sub bass (under 50 Hz)**. At frequencies this low it's impossible to determine pitch, but nonetheless is commonly occupied by the very lowest area of a kick drum and bass instruments. Notably, most loudspeaker and nearfield loudspeaker monitors cannot reproduce frequencies this low reliably, and in most genres of dance music all signals this low will be rolled off to prevent loss of volume in hi-fi systems or damaging the bass response in club PA systems.
- **Bass (50–250 Hz)**. This range is typically adjusted when applying the bass boost on most home stereos, and where most of the bass is contained in all dance music mixes. EQ cuts (or boosts) around this area can add definition and presence to the bass and kick drum.

- **Mid-range muddiness area (200–800 Hz)**. This frequency range is the main culprit for mixes that are described as sounding muddy or ill-defined and is often the cause of a mix that is fatiguing to the ear. If too many sounds are dominating in this area, a track can quickly become tiring and irritating.
- **True mid range (800–5000 Hz)**. As previously touched upon, the true mid range is where loudspeakers produce most of their energy and human hearing is particularly sensitive to these frequencies. This is because the human voice is centred here, as are TVs and radios; thus, even small boosts of a decibel will be perceived to be the same as boosting 10 dB at any other frequency. Subsequently, you should exercise extreme caution when adjusting frequencies in this area.
- **High range (5000–8000 Hz)**. This range is typically adjusted when applying the treble boost on most home stereos and is where the 'body' of hi-hats and cymbals often reside. This area is sometimes boosted by a couple of decibels to make sounds artificially brighter.
- **Hi-high range (8000–20 000 Hz)**. This final frequency area often contains the higher frequency elements of cymbals and hi-hats. Some engineers also apply a small shelving boost at 12 000 Hz to make the music appear more hi-fidelity. This often adds extra detail and sheen without introducing any aural fatigue. It does, however, require plenty of care, as boosting anywhere in this region can pronounce any background hiss or high-frequency noise.

There's much more to applying EQ to a mix than understanding the separate frequency bands, and to become truly competent at engineering it's also vital to train your hearing so that you can identify timbres and the corresponding frequency in hertz. Unfortunately, this isn't something that you can pick up from just reading about it and will only come with practical experience and careful listening.

Creating a soundstage

Finally, you need to be able to envisage the 'stage' on which the instruments will be placed. When approaching any mix it's usually best to imagine a three-dimensional room – the soundstage – on which you can place the various instruments.

Sounds placed on this stage can be positioned anywhere between the left or right 'walls' using a pan pot; they can be positioned at the front or back, or anywhere in between, using volume and the frequency content of the sound will determine whether it sits at the top of the stage (high frequencies), the middle (mid-range frequencies) or the bottom (low frequencies).

Using this as a template, the concept behind mixing is to ensure that each sound occupies its own unique space within this room so that it can not only be heard, but so that it also fits in well with everything else. To do this we can break the soundstage into three distinct areas: the front to back, the horizontal and the vertical. We'll begin by examining the front to back.

Front-to-back perspective

One of the primary auditory clues we receive about how far away we are from a sound source is through the intensity of air pressure that reaches our eardrums. As we've touched upon in Chapter 3, sound waves spread spherically outwards in all directions from any sound source and the further these have to travel the less intense the sound becomes. In other words, the further we are from a source of sound, the more the sound waves will have dissipated, resulting in a drop of volume.

Figure 16.2 A chart defining the general ranges taken up by instruments in dance music.

Note: The intensity of sound is based around the inverse law which states that:

Sound pressure decreases proportionally to the square of the distance from the source.

Roughly translated this means that each time the distance from the sound doubles, it'll become roughly 6 dB quieter.

If we interpret this into the context of a mix we can say that the louder an instrument is, the more 'in your face' it will appear to be. However, although many dance mixes appear to have everything right in your face and at the front of the soundstage, this is far from the case.

Indeed, this depth perception is the first aspect to take into consideration when producing a good mix.

If every instrument were placed at the forefront, all the volumes would be at equal gain and this would produce a cluttered image as every instrument fights to be at the front. What's more, the mix would appear just two-dimensional because, when listening to music, our minds always work on the basis of comparison. That is, for us to gain some perception of depth there must be some sounds in the background so we can determine that some sounds are at the front of the mix and vice versa.

This means that you need to determine which sounds should be upfront and which should be progressively further back in the mix, but this isn't as difficult as it may sound. Dance music, by its very nature, is based on rhythm and groove because that's what we dance to. Thus, in all genres of dance it makes sense that both the drums and bass should take soundstage priority over every other instrument.

The most obvious way to provide this depth perspective is with volume adjustments, as it's natural for us to believe that the louder a signal is, the closer it must be. This only applies to sounds that we are familiar with, though, and for dance music it's quite usual to use unnatural or unrealistic timbres. We can overcome this problem by taking advantage of how our ears and mind perceive a sound depending on the acoustic properties it projects.

As touched upon in Chapter 3, the higher the frequency of a sound, the shorter the wavelength becomes; therefore, we can assume that if a high-frequency sound has to travel over a long distance, many of these frequencies will dissipate along the way. This means that if a high-frequency sound is playing a good distance away from a listener, some of the high-frequency content will be reduced.

This effect can be particularly evident with passing cars whose stereo systems take up more space than the engine and shift enough air to blow out a candle at 50 paces. You hear the low frequencies while the car is at a distance and as it approaches the higher frequencies become more pronounced until it passes by, whereby the higher frequencies begin to decay as it moves off into the distance. We can therefore emulate this effect with some creative EQ. By applying cuts of just a few decibels at the higher frequency ranges of a sound, it will be perceived to be more distant when listened in association with other sounds with low-frequency content. Alternatively, by increasing the higher frequencies using enhancers or exciters, a timbre can appear much more upfront.

Note: This frequency-dependent effect is an important aspect to consider when working with a compressor, since if a fast attack is used the transient will be clamped down, thus reducing some of the high-frequency content and pushing it to the rear of the mix! This is often the cause of numerous problems, as most usually you'll want to compress to prevent clipping or introduce 'punch' into a full mix. While you could always increase the gain to bring the timbre to the front of the mix, it would not work as well as if they had all the high-frequency content intact. The only suitable way to avoid this is to use a long attack on the compressor or employ a multi-band compressor and set it to compress the lower frequency content only, leaving the higher frequencies unaffected.

Another characteristic of the front-to-back perspective is derived from the amount of reverberation the signal has associated with it. All sound will have some natural reverberation as the reflections emanate from any surrounding surfaces, but the amount and stereo width of these reflections will depend on how far away the source of the sound is. If a sound is far away, then the stereo width of the reverberations will dissipate as they travel through the air, but they will also be subjected to more reverberation. This is important to bear in mind, since many artists wash a sound in stereo reverb to push it into the background and then wonder why it doesn't sound quite 'right' in context with the rest of the mix.

From this we can also determine that if you need to place an instrument at the rear of a mix, it's wise to use a mono reverb signal with a long tail and remove some of the higher frequencies. This will emulate the natural response we've come to expect from the real world, even if that particular timbre doesn't occur in the real world. Most good reverb units will allow you to remove higher frequencies with an on-board filter but if not, by returning the reverb effect into a normal mixing channel you can use the channel's EQ to remove the high frequencies this way.

Of course, sounds that are up close or at the front of a mix will have little or no reverb associated with them, but in some instances you may wish to apply reverb to thicken out the timbre. On these occasions it should be applied in stereo, but you should also aim to keep the stereo width of the reverb contained to prevent it from occupying too much of the left and right perspective. Also, it should use a short tail and a pre-delay of approximately 50–90 milliseconds to separate the timbre from the effect to prevent it from washing over the attack stage, otherwise it may force the instrument to the back of the mix.

Note: Always keep in mind that applying effects too heavily can make sounds very difficult to localize and it's absolutely vital that each instrument can be pinpointed to a specific area within a mix, otherwise it will appear indistinct and muddy.

Horizontal perspective

The next consideration in a mix is the horizontal plane – that is, the distance and positioning of sounds between the left and right walls of the virtual room. The major audio clue that helps us derive the impression of panning and stereo can be determined by:

- the volume intensity between sounds; or
- the timing between sounds.

The principle behind altering the volume of sounds to produce a stereo image was first realized by Alan Blumlein in the early 1930s. An inventor at EMI's Central Research Laboratories, he researched the various ways in which the ears detect the direction from which a sound comes. Along with deriving the technique for creating a stereo image in gramophone records, he also figured that, to maintain realism in a film, the sound should follow the moving image.

This technique was first employed in Walt Disney's *Fantasia*, when sound engineers asked Harvey Fletcher (of the same Fletcher–Munson curve) if he could create the impression of sound moving from left to right for the movie. Drawing on Alan Blumlein's previous work, he came to the conclusion that if a sound source is gradually faded in volume in one speaker and increased in the other, it will gradually move from one to the other. The engineers at Disney adapted this idea into practice by using a potentiometer to vary the volume between two speakers and labelled the process 'Panoramic Potentiometer', hence the term pan pot.

Although this volume intensity difference between two speakers is still the most commonly used method for panning a sound around the image, we can also receive directional clues from the timing between sounds. Otherwise known as the Directional Cues, Precedence or Haas effect, this process takes advantage of the Law of the First Wavefront, which states that: '*If two coherent sound waves are separated in time by intervals of less than 30 milliseconds, the first signal to reach our ears will provide the directional information.*'

In layman's terms, if a direct sound reaches our ears anywhere up to 30 milliseconds before the subsequent reflections, we can determine the position of a sound. For example, if you were facing the central position of the mix, any sound leaving the left speaker would be delayed in reaching the right ear and vice versa for the left ear, an effect known as interaural time delay (ITD). Considering that sound travels at approximately 340 m/s, this effect can be emulated by setting up a delay unit on a mono signal and delaying it by a couple of milliseconds, producing the impression that the sound has been panned.

For this effect to work accurately, we also need to consider that our ears are on the side of our heads and therefore it gets in the way of the frequencies from opposite speakers (an effect known as head-related transfer function). This means that, provided we are facing the centre of the stereo image, some of the higher frequencies emanating from the left speaker will be reduced before they reach the right ear, simply because sound doesn't travel through our heads, it has to travel around it. This effect can be simulated by cutting a decibel at around 8 kHz from the delayed signal.

Naturally, to accomplish these types of effects means that you need to use mono sounds within the mix and this subject relates to one of the biggest mistakes made by practising musicians. With all of today's tone modules, keyboards and samplers featuring stereo outputs, it's quite easy to fall into the trap of using stereo sounds throughout an arrangement, but this only leads to a mix that lacks any real definition.

A stereo image is formed by spreading any sound to both the left and right speakers using one of two methods. It uses either effects or the process of layering two different sounds together, the results of which are then panned left and right so that they play across both speakers. While this always makes them sound much more interesting in isolation, it's only to persuade you to part with your money for the synthesizer.

If you made a mix completely from stereo files they will all collate together in the mix and occupy the same area of the soundstage. Of course, you could narrow the stereo width of each file with the pan pots so that they don't all occupy the same position, but this approach is not particularly suitable for making a transparent mix.

The soundstage for any mix should be transparent enough so that you can picture the mix in three dimensions with your mind's eye and pinpoint the exact position of each instrument. Since many dance mixes are quite busy, encompassing anything from eight to 12 different elements playing at once, if these were all stereo it would be difficult, if not impossible, to find a pan placement for each. This often leads to an inexperienced engineer resorting to unnecessary EQ in a futile attempt to carve out some space for the instruments or start to 'creep the faders'.

Note: Creeping faders is typical of most inexperienced engineers and is the result of gradually increasing the volume of each track so that it can be heard above other instruments. For instance, they may increase the gain of vocals so that they can be heard above the bass, and then increase the drums so they can be heard above the vocals, and then increase the bass so it can be heard with the drums and then move back to the vocals and, well, you get the picture. Eventually, the mixer is pushed to its maximum headroom and the mix turns into loud, incomprehensible rubbish.

Unnecessary EQ and volume creep can be avoided by utilizing mono sound sources, since if two instruments share the same frequency range they can be separated by panning one sound to the left and the other to the right. There's little need to spatially separate them by panning them as far as possible in each direction and in many cases a panning space of approximately '2 hours' will allow both instruments to be heard clearly. By 'hours' we're referring to the physical position of a pan pot on a mixing desk, i.e. 1 o'clock, 2 o'clock, etc. Thus, for a 2-hour panning the right pan would read 1 o'clock and the left would read 11 o'clock.

Note: This physical pan pot measurement is only mentioned in a theoretical sense and you should always pan with your ears and not your eyes. All mixing desks are not equal and where one may place the sound at a certain position at a certain setting, the same setting on a different mixer may place the sound in a different position.

There is yet another reason why you should avoid using stereo files throughout and this is based upon our perception of stereo. Similar to how different volumes aid us in gaining some perception of depth within a mix, the same is true of stereo.

It's in our nature to turn and face the loudest, most energetic part of any mix, which in many mixes is the kick drum. If this happened to be a stereo file, it would be spread between the left and right speakers, and the main energy would be spread across the sound stage. If, however, mono samples were used and the kick were placed centrally, we can perceive the position of other sounds much more clearly, which increases the overall sense of space within the mix.

On top of this, when working with mono files on a stereo system we also perceive the volume of a sound by its position in the soundstage. This means that if a signal is placed dead centre the energy is shared by both speakers and it will appear louder than if the same signal, at the same gain, were placed in the left or right speaker alone.

Typically, this perceived volume difference can be as much as 3–6 dB, so many professional and semi-professional mixing desks (and some audio sequencers) will implement the panning law. Using this, any sounds that are placed centrally are subjected to either 3 or 6 dB of attenuation, which can be set by the user.

Note: Not all mixing desks will implement this panning law, so after panning the instruments to their respective positions, it is not unusual to readjust the respective volumes again.

Perhaps most important of all, if you wish to pan instruments faithfully, your speaker system must be configured correctly in relation to your monitoring position. Ideally, you should be positioned at one point of an equilateral triangle with the loudspeakers positioned at the other two points. This means that speakers should be positioned an equal distance apart from one another and your listening position to ensure that the signal from each reaches your ears at the same time.

If there is just a few inches difference between these three points, the sound from one speaker could be delayed in reaching your ears by a couple of milliseconds, which results in the stereo image moving to the left or right. What's more, you also need to ensure that the volume from each speaker is the same when it reaches your current listening position – even differences as small as 1 or 2 dB can shift the image considerably to one side.

Note: A possible solution to producing an accurate soundstage would be to mix through headphones, but this should be avoided at all costs. While they are suitable for listening to a mix without disturbing the neighbours, they overstate the left and right perspective because they're positioned either side of your head. Keep in mind that when listening to a mix from loudspeakers, sound from the left speaker reaches your right ear and vice versa for the left speaker. If you mix using headphones alone, you can easily over- or understate the stereo field.

Finally, any stereo mix must also work well in mono. Most mixing desks will offer a mono button that, when used, sums the channels together into a mono signal and this is important in preventing any phasing that, while not immediately evident in stereo playback, can result in a comb filtering effect. While this may seem worthless when all of today's hi-fi systems are stereo, most TV and radio stations still broadcast in mono, most clock/transistor radios are also mono and many club PA systems are mono.

This means that if you've relied entirely on stereo while mixing, then parts of the mix can disappear altogether. This doesn't mean you have to destroy the stereo mix, because by switching the mix to mono the image will simply be collated into the centre. When in mono, the general tones of the instruments and their volumes should remain constant but, if not, you will need to re-examine the mix and check for what is causing mono compatibility problems. In the majority of cases this is caused by using stereo files or the application of too many stereo effects.

Vertical perspective

The final perspective of a mix is the top to bottom of the soundstage, with higher frequency elements sitting towards the top of the stage and lower frequencies sat towards the bottom. Much of this positioning will already be dictated by the timbres used in the mix. For example, basses will tend to sit towards the bottom of the mix and hi-hats will naturally sit towards the top. However, it should be noted that these, and any other timbres, will contain frequencies that do not necessarily contribute to the sound when placed into a mix.

A typical case in point is with the basses used in dance music, as they do not consist entirely of low-frequency elements, but also some mid-range elements and possibly some high frequencies too. If you were to tonally adjust this type of sound by removing all of the mid-range frequencies, the bass would move more towards the bottom of the soundstage, making some space for instruments to sit in the mid range. This 'corrective' adjustment is one of the most fundamental aspects of mixing and is accomplished with the EQ on a mixing desk's channel.

EQ is a frequency volume control that allows you to reduce or increase the gain of a band of frequencies contained within a sound, but it can be used for much more than simply making space in a mix. Applied cautiously, it can be used to make insipid timbres much more interesting, add more definition to the attack portion of a sound to pull it to the front of a mix, or prevent the bass from overpowering and 'muddying' the entire lower frequency range of the mix. However, if you apply it without knowing exactly what effect it will have on a sound and those surrounding it, a mix can fall apart within minutes and you end up with nothing more than a muddy, ill-defined mess that doesn't translate at all well on a typical club PA system.

To understand the concept behind using EQ, we need to revisit Chapter 3, which explained that all sounds are made up of a number of frequencies that are a direct result of the fundamental and its associated harmonics at differing amplitudes. Also, that it's this predetermined mix of harmonics that helps us establish the type of sound we perceive, whether it's an earth shaking bass, a screaming lead or a drum kick.

We only require all these frequencies present if the sound is played in isolation, though, and if it's mixed in with a number of other instruments we only need to hear a few basic frequencies because our hearing/perception will persuade itself that the others must be present – they're just masked behind the other instruments.

This natural occurrence is one of the main keys when mixing, since if frequencies that we don't necessarily need are removed, it'll make more room for the important frequencies of other instruments that we do need to hear. This means that we need a way of pinpointing a specific group of frequencies within a sound that we can remove or enhance, and in 'professional' desks this is accomplished with a parametric EQ.

Fundamentally, a parametric EQ consists of three controls: a frequency pot, a bandwidth pot and a gain pot. The frequency pot is used to select the centre frequency that you want to enhance or remove, while the bandwidth pot determines how much of the frequencies either side of the centre frequency should be affected. Finally, the gain pot allows you to increase or reduce the volume of the chosen frequencies.

To explain these uses, suppose you had a lead sound that contained frequencies ranging from 600 Hz to 9 kHz, but the frequencies between 1 and 7 kHz were not required. Using the frequency pot you could home in on the centre of these (4 kHz) and then set the bandwidth wide enough to affect 3 kHz either side of the centre frequency. By then reducing the gain of this centre frequency, the frequencies 3 kHz either side would be attenuated too.

The size of this bandwidth is often referred to as the Q (Quality) and the smaller the Q number, the larger the 'width' of the bandwidth.

Such a precise bandwidth as shown in Figure 16.3 is entirely theoretical and in the real world the two extremes of the bandwidth are not attenuated at right angles. This is because, similar to a synthesizer's filters, EQ has a transition period or 'response curve' for it to be more natural. Typically, this is around 3 dB per octave, but can sometimes be 6 dB per octave.

The problem with this approach is that if you apply a particularly heavy cut or boost and the 'Q' does not remain constant, the bandwidth will increase in size as you boost or cut further.

Figure 16.3 Theoretical EQ.   **Figure 16.4** Practical EQ.

The action of a constant Q is best explained with an example. Suppose that the bandwidth were set to remove 2 kHz either side of the centre frequency at an EQ boost of 3 dB, if you were to then boost by another 3 dB, the bandwidth would affect more of the frequencies either side of the centre frequency.

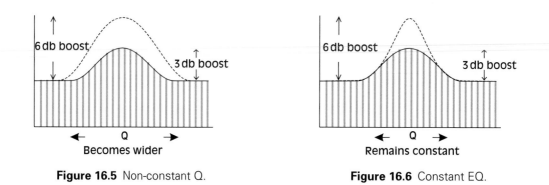

Figure 16.5 Non-constant Q.   **Figure 16.6** Constant EQ.

This action is known as a 'non-constant Q' because the Q changes as you increase or decrease the gain at the centre frequency. This creates an unnatural sound, so many professional desks and external EQ units will vary the response curve as the gain is increased or decreased to maintain a constant bandwidth. This produces a sound that appears much more natural to our ears and is the sign of a well-designed EQ unit.

'Q' should only be used when discussing the theory behind EQ and I wouldn't advise taking it any further than this book. In a studio situation you're communicating with musicians not techno freaks, so it's usual to talk about the Q in octaves, otherwise you could find yourself laughed out of the studio. For those with an unhealthy academic interest in how the Q value can be transformed into octaves, you can use the following mathematical theorem:

$Y = (4 \times Q^2 + 1)^{0.5}$
Octave $=$ Log base 2 of $((Y + 1)/(Y - 1))$

But for those who simply can't be bothered, it's worth memorizing the values in Table 16.1, as they cover the most commonly used Q settings while mixing.

Table 16.1 'Q' settings and equivalent octave ranges

'Q' setting	Octave range
0.7	2
1.0	$1\frac{1}{3}$
1.4	1
2.9	$\frac{1}{2}$
5.6	$\frac{1}{4}$

Note: Generally, a bandwidth of a quarter of an octave or less is used for removing problematic frequencies, while a width of a couple of octaves is suitable for shaping a large area of frequencies. For most work, though, and especially if you're new to mixing, you should look at always maintaining a non-constant Q of 1.0. This is sometimes referred to by many engineers as the 'magic Q', since it commonly avoids any masking problems (we'll look at this in a moment) and sounds much more natural to the ears.

As of yet, we've only concerned ourselves with one form of EQ – parametric – but in addition to this there are also shelving EQs. These will, or at least should, be available on most semi-professional and all professional desks, and consist of low-pass and high-pass filters along with low-shelf and high-shelf. Because these all share similar names, they're often confused with one another, but they perform different functions.

Both low-pass and high-pass EQs are based on exactly the same principle as the low- and high-pass filters used in synthesizers. The low-pass filter can be used to attenuate any higher-frequencies that lie above the cut-off point while a high-pass filter is used to attenuate any lower frequencies that lie below the cut-off point (see Chapter 3). With a shelving EQ, however, you can home in on a specific frequency and cut or boost all the frequencies that lie beyond it. For instance, by using a low-shelf EQ set at 40 Hz you could cut or boost all the frequencies that lie below 40 Hz.

In some respects, this action isn't too different from low- and high-pass filters and is the reason why the two are often confused with one another. As an example, a high-pass filter could be used to cut all the frequencies below 40 Hz but you would obviously have no control over how much gain reduction (or amplification) is applied to the affected frequencies.

The specifications of the mixer will depend on how much control you have over the filters and shelves but, generally, a shelving filter will have two controls: one to select the frequency (often called the knee frequency) and one to adjust the gain of the frequencies above (or below) the shelf. Some of the more expensive desks may also offer the opportunity to adjust the transition slope of all these filters and this can sometimes be important when we need to 'roll off' frequencies.

Since many speaker systems are incapable of producing frequencies as low as 40 Hz, they should be removed altogether, and the attenuation of frequencies above or below the cut-off point is often termed as 'rolling off'. Thus, if you were asked to roll off all frequencies below 40 Hz, you would employ a low-shelf EQ and position the cut-off point at 40 Hz. Or at least you would if the transition slope of the EQ were more or less vertical. As we've seen, EQ units will use a transition slope to make the effect appear more natural to our ears, but we don't necessarily need this when rolling off frequencies that are not required, so it is useful to reduce this if it happens to be particularly long (6 dB per octave and over).

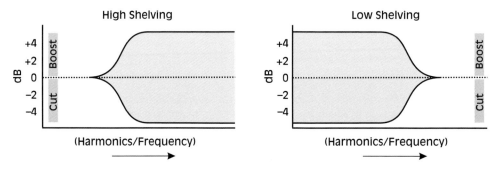

Figure 16.7 The action of shelving filters.

Naturally, all this control requires dedicated pots on a mixer and additional circuitry to implement it, so many of the cheaper mixing desks will not employ shelving or high-/low-pass filters and in some instances may not even offer any parametric EQ. Indeed, it's fair to say that the cheaper the desk, the less EQ control you'll be offered and this can be broadly categorized into three areas: fixed frequency, semi-parametric and graphic.

- *Fixed frequency*. Fixed frequency EQ is possibly the worst type to have on any mixing desk as it's the least flexible, but it is also the cheapest to implement so many cheap consumer desks will employ it. Fixed frequency EQ consists of two, sometimes three, rotary pots that control the low, mid and high frequencies. To accomplish this they commonly use low-/high-pass filters with fixed cut-off frequencies at 12 kHz and 75 Hz for the high and low bands, respectively. If there is a third pot then this controls the mid-band and is generally a notch filter with a centre frequency that's fixed at around 2 kHz and has a fixed bandwidth of approximately two octaves (see Table 16.1 for the respective Q). This type of EQ makes precise, frequency-specific shaping impossible and unless you have an external parametric EQ unit it is far from suitable for creating a good mix.
- *Semi-parametric*. Semi-parametric EQs (sometimes known as sweep EQs) utilize a similar design as the fixed frequency EQs with fixed low/high filters at 75 Hz and 12 kHz respectively, but rather than have a fixed mid-band, only the bandwidth is fixed and you are free to sweep the centre frequency. This type of control is typical of mid-priced consumer mixing desks, but while offering more freedom to select frequencies the fixed bandwidth means that it isn't versatile enough for creative EQ and it can be particularly difficult to produce a good mix with this limitation.

- *Graphic.* Graphic EQs do not appear on mixing desks but are sometimes used by mastering engineers to set the overall tonal colour of a full mix. They're also possibly the most instantly recognizable of all EQ units as most home stereos now have them fitted as standard, appearing as a number of faders, each of which offers control over a specific frequency band. The size of the bandwidth covered by each fader obviously depends on how many faders it has, so the more faders that are available the better the EQ will be, since each bandwidth will be smaller. The reason these are referred to as graphic EQs is because you can often tell what type of music people listen to by the way the equalizer is configured, hence it's rather graphic in nature.

With the introduction of computers and audio sequencers, this graphic nature is no longer limited to graphic EQs, as many software mixing desks and some of the latest digital desks will mix a parametric EQ with the visuals of a graphic EQ to produce a paragraphic EQ. This gives all the freedom of a parametric EQ with the visual ease of using a graphic EQ. The immediate benefit of this is obvious: rather than having to determine the current EQ settings through a series of often cryptic pot positions, you can actually see the frequency range being affected.

That said, keep in mind that your potential audience do not see EQ settings, they only hear them, so you should do the same. One of the biggest problems with mixing on any 'visually enhanced' software desk is that you have the tendency to use your eyes more than your ears and this can be disastrous. If it looks right, it may not sound right, yet if it sounds right, it most probably is.

Practical mixing

It should go without saying that the first step to creating a good mix is to ensure that you can actually hear what you're doing reliably. If you don't hear the frequencies within a mix accurately then you certainly can't EQ or effect them accurately. This means that you *must* use loudspeaker studio monitors to scrutinize a mix rather than placing your trust in everyday hi-fi speakers.

No matter how 'hyped' they are by manufacturers for having an excellent frequency response, all hi-fi speakers deliberately employ frequency-specific boosts and cuts throughout the sonic range to make the sound produced by them appear full and rounded. As a result, if you rely entirely on these while mixing you'll produce mixes that will not translate well on any other hi-fi system. For instance, if your particular choice of hi-fi speakers feature a bass boost at 70 Hz and a cut in the mid range, you're obviously going to produce a mix based on this and if it's then played on a system that has a flatter response there will be less bass and an increased mid range.

It's impossible to recommend any particular make or model of studio monitor because we all interpret sound differently, but generally you should look for a model that makes any professionally produced music you like sound moderately lifeless. If you can make your mixes sound good on these, then it's likely to sound great on any other system. Then again, you shouldn't aim for a totally flat response and it's usual to look for monitors that are a happy medium between coloured and flat. Indeed, most dance musicians will deliberately use monitors with a vaguely coloured response, since 99.9% of people who are going to listen to it will do so in a club, car or on a typical home stereo system.

Even with a good monitoring system it's important to also note that their actual positioning will have an effect on the frequencies they produce. Placing any monitors close to a corner or wall will increase the perceived bass by 3, 6 or 9 dB due to the boundary effect. This is similar in some respects to the boundary effect that's exhibited when singing too close to a microphone – there's an artificial rise in the bass frequencies.

With monitors this can be attributed to the low-frequency waveforms emanating from the rear of the monitor onto the wall or corner directly behind it. This low-frequency energy culminates with the energy produced by the monitor itself and doubles the perceived bass. As a result, a general rule of thumb is that unless they're specifically manufactured to be used close to a wall (i.e. they have a limited bass response and use the wall to produce the low-frequency energy), they should be placed at least 3 feet away from any walls. Of course, this may not always be possible and if it isn't then you'll have to learn to live with the increased response by listening to as many commercial mixes as you can on them and training your ears to accept the response they produce.

On that note, before you even approach any mix you should take time out to listen to some of your favourite commercial mixes. We hear music every day but we very rarely actually *listen* to it, and if you want to become competent at mixing you need to differentiate between hearing and listening. This means that you need to sit down and play back your favourite commercial mixes – at a moderate volume (just above conversation level) – and try to pick out the action of the bass, the drums, vocals, leads and any sound effects. Listen for how the melodies have been programmed and ask questions such as:

- Is there any pitch bend in the notes?
- Does the drum loop stay the same throughout the track or does it change in any way?
- What pattern are the hi-hats playing?
- Do they remain constant throughout?
- What effects have been used?
- Where is each instrument placed in the mix?

Although many of these may not seem related to mixing, listening this closely for small nuances in drum loops, hi-hat patterns and basses, etc. will train your ears into listening closely to any music and will help you to not only begin identifying frequencies, but also increase your awareness of arrangement and programming techniques.

Most important of all, though, despite the huge array of audio tools available today, the most significant ones are those that are stuck to the side of your head. Continually monitoring your mixes at excessive volume or listening to any music loud will damage the ear's sensitivity to certain frequencies. A recent survey conducted showed that those in their late teens and early twenties who constantly listened to loud music had more hearing irregularities than people twice their age. Also, keep in mind that dabbling in pharmaceuticals for recreational use leads to a heightened sensory perception. Although they may have played a large role in the dance scene they don't help in the studios. Apart from losing hours from fits of laughter about terms such as ring modulation, the heightened sensory perception makes (a) everything sound great or (b) everything sound wrong, so you spend the next 10 hours twiddling EQ on something that doesn't need it.

Volume adjustments

The first step in creating a mix is to begin by setting the relative volume levels of each track, and this means taking the musical style into account. You need to identify the defining elements of the genre and make sure that these are placed at the very front of the mix. Typically, for all dance music this means that the drums and bass should sit at the front of the mix with every-thing else that must be located centrally sat behind them. This includes the vocals, as more often than not they will not be the truly defining part of the music.

It's generally best to begin mixing by starting with these defining points, so it's prudent to mute every channel bar the drum tracks and commence mixing these. As the kick drum is the most prominent part of a drum loop, set this so it's at unity gain, and then introduce the snare, claps, hi-hats, cymbals and other percussion instruments at their relative levels to the kick.

At this point don't concern yourself with panning, EQ or effects, but try to find a happy medium so that the drum levels are close to how you imagine the finished product to sound. Follow this by adding the bass, then the vocals (or the next important part of the mix) and gradually add each audio track in order of priority, adjusting the volume levels until all the instruments are playing together in the mix. It's vital that you're not concerned with masking, EQ or effects at this stage; you need to make the mix work quickly before you start jumping in with com-pressors, EQ or effects, as these can quickly send you off on a tangent and you can lose the direction of the mix. If you're experiencing troubles setting the volume levels, then a quick trick is to switch the mix to mono, as this allows you to hear the relative volume levels more accurately.

Note: This form of 'additive' mixing is only one way to approach a mix and some users feel more at ease using the 'subtractive' method. This means that all the faders are set at unity gain and each is reduced to obtain the appropriate levels required. Which to use is entirely up to your own digression and depends on which you feel most comfortable with.

Panning

With all the relative volume levels set, you can then begin to pan the instruments to their appropriate positions. This is where your own creativity and scrutinizing of previous mixes of the same genre will come into play, as there are no 'strict' rules for creating a good sound-stage. Of course, this isn't going to be much help to those new to mixing, so what follows is an outline guide to where instruments are usually positioned in most dance genres (see Table 16.2).

Apart from placing the kick, bass and vocals in the centre of the soundstage, note that the posi-tionings suggested are only a *very* general guide and the panning of instruments will define your own particular style of mixing. Indeed, the best solution to panning instruments is to have a damn good reason as to why you're placing a timbre there in the first place. Always have a plan and don't just throw sounds around the image regardless.

Table 16.2 General guide to positioning of instruments

Instrument	Pan position/description
Kick	Positioned centrally so that both speakers share the energy
Snare	Positioned from 1 o'clock to 2 o'clock or centrally
Hi Hats	Positioned at the far left of the soundstage with a delayed version in the right
Cymbals	Positioned centrally or from 1 o'clock to 2 o'clock
Percussion	Positioned so that the different timbres are in the left and/or right of the soundstage
Bass	Positioned centrally so that both speakers share the energy
Vocals	Positioned centrally since you always expect the vocalist to be centre stage
Backing vocals	Occasionally stereo, so they're spread from 2 o'clock to 4 o'clock
Synth leads (trance)	A stereo file positioned fully left and right with perhaps a mono version sat centrally
Synthesized strings	Positioned at 4 o'clock or stereo spread at 9 o'clock and 3 o'clock
Guitars	Positioned at 3 o'clock
Pianos	Commonly stereo with the high notes in the left speaker and low notes in the right
Wind instruments	Positioned in the left or right speaker (or both) at 3 o'clock and 9 o'clock
Sound effects	Wherever there is space left in the mix!

Note: In direct contradiction, in many dance mixes the positioning of instruments is rarely natural, so you should feel free to experiment too. In fact, in many cases it may be important to exaggerate the positions of each instrument to make a mix appear clearer and more defined. Nevertheless, you shouldn't position a sound in a different area of the soundstage just to avoid any small frequency clashes with other instruments. In these circumstances you should try to EQ the sound beforehand, then if it still doesn't fit consider panning it out of the way.

After the instruments have been panned it's highly likely that the relative volumes will have to be adjusted, so you'll need to go back and readjust these so that the instruments are all at their appropriate volumes before you move onto EQ.

EQ

Although there are no rules on how to use EQ, when it comes to mixing there are some generalizations that can be made. Firstly, any EQ is essentially acting like a filter and this means that, like any other type of filter, it introduces resonance at the cut-off point and phase shifting. In other words, it applies small amounts of comb filtering into the audio, which sounds like small amounts of distortion. This is where the quality of the EQ unit being used plays a major role, as less competent/cheaper EQ units will introduce more comb filtering than the more expensive units.

As a result you need to exercise care while equalizing sounds so that it doesn't become too apparent; this is why many engineers will always advise that you should look to cutting frequencies rather than boosting them. This is because boosting any frequency will make the phase distortion much more noticeable. To support this theory further, in the 'real' world our ears are used to hearing reductions in frequencies rather than boosts, since any frequencies are generally reduced by walls, objects and materials.

Note: Yet another reason to cut rather than boost lies with the recording of the audio itself. To increase the signal-to-noise ratio, any audio should be captured as 'hot' as possible and if an audio file is already close to clipping, EQ boosts may push the audio to distortion.

As mentioned, these are only generalizations and so, in direct contradiction, on occasion boosting the frequencies that are lower in volume than the fundamental can be used as a creative sound design tool. Indeed, one of the most common misconceptions about EQ is that you should never really have to use it if the sound is recorded correctly in the first place.

This viewpoint stems from when EQ was first used as a means of correcting the frequency response of a recording to prevent it from compromising the recorded sound. This was because recording in the early days was carried out with poor quality gear when compared with today's standards and therefore frequencies had to be adjusted to reproduce the original recording.

While many older engineers still stand by this view today, it means absolutely nothing to the dance musician, since a proportionate amount of the timbres used will be synthetic anyway. Thus, apart from using EQ to help instruments sit together and create space within a mix, it should also be seen as a way of creatively distorting timbres.

Since EQ is essentially a filter it can be used to reduce certain frequencies of a sound, which will have the immediate effect of increasing the frequencies that have not been touched. As an example of this, we'll use a spectral analyser to examine the frequency content of the bass sound that was created for the house track (Figure 16.8).

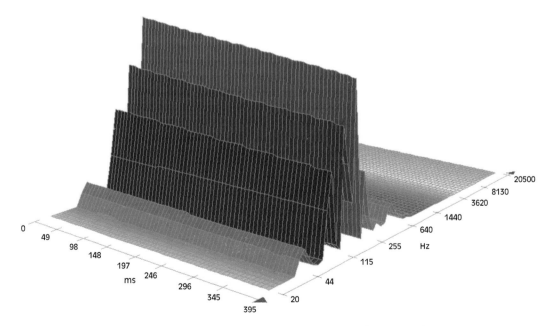

Figure 16.8 A spectral analysis of the house bass sound.

Notice how there are a number of peaks in the timbre. If you were to listen to this sound you would primarily hear these four loudest frequencies. However, using EQ creatively you can turn down these peaks to hear more of the entire frequency spectrum of the bass. It then becomes much more harmonically complex and can sustain repeated listening, because the closer you listen to it the more you can hear. Making a sound appear as complex as possible by reducing the peaks is a common practice for many big name producers. This could be further accentuated by applying small EQ boosting to the lower peaks of a sound.

On the other hand, instead of complexity, simple sounds can also work equally well, accomplished by cutting all but these four peaks, and this cutting can also offer another advantage. As it is unlikely that you would be able to perceive the lowest frequencies (harmonics) contained in the sound, there is little need to leave them in and by removing them you would make more room for instruments in the mix.

Note: You shouldn't simply just grab a pot and start creatively boosting aimlessly, since you need to consider the loudspeaker system that the mix is to be played back on. All loudspeaker systems have a limited frequency response and produce more energy in the mid range than anywhere else. This means that if you decide to become 'creative' on the bass or mid range, you could end up increasing energy that the speaker cannot reproduce faithfully, resulting in an overzealous or muddy mid range or, worse still, a quieter mix overall!

The main use for EQ within mixing, however, is to prevent any frequency masking between instruments. As touched upon in Chapter 3, the fundamental and subsequent harmonics contribute to making any sound what it is, but if two timbres of similar frequencies are mixed together some of the harmonics are 'masked', resulting in the instruments sounding different in the mix than in isolation.

While this can have its uses in sound design (hocketing, layering, etc.), during mixing it can cause serious problems, since not only is it impossible to mix sounds that you can't hear, but the frequencies that mix together can make both sounds lose their overall structure, become indistinct or, worse still, the conflict will result in unwanted gain increases.

Naturally, knowing if you have a problem with frequency masking is the first step towards curing it, so if you're unsure whether some instruments are being masked, you can check by raising the volume of every track to unity gain and then panning each track in turn to the left of the stereo field and then to the right. If you hear the sound moving from the left speaker through the centre of the mix and then off to the right, it's unlikely that there are any conflicts. This, however, is about the only generalization you can make for mixing and the rest is entirely up to your own artistic preferences. Because of this, what follows can only ever be a very rough guide to EQ and to the frequencies that you may need to adjust to avoid any masking problems. As such, they are open to your own interpretations because, after all, it's your mix and only you know how you want it to sound.

Drums

Drum kick

A typical dance drum kick consists of two major components: the attack and the low-frequency impact. The attack usually resides around 3–6 kHz and the low-end impact resides between 40 and 120 Hz. If the kick seems very low without a prominent attack stage, then it's worthwhile setting a very high Q and a large gain reduction to create a notch filter. Once created, use the frequency control to sweep around 3–6 kHz and place a cut just below the attack, as this has the effect of increasing the frequencies located directly above.

If this approach doesn't produce the results, set a very thin Q as before, but this time apply 5 dB of gain and sweep the frequencies again to see if this helps it pull out of the mix. Taking this latter approach can push the track into distortion, so remember to reduce the gain of the channel if necessary.

If the kick's attack is prominent but it doesn't seem to have any 'punch', the low-frequency energy may be missing. You can try small gain boosts around 40–120 Hz, but this will rarely produce the timbre and the problem more likely resides with your choice of drum timbre. A fast attack on a compressor may help to introduce more punch as it'll clamp down on the transient, reducing the high-frequency content, but this could change the perception of the mix so it may be more prudent to replace the kick with a more significant timbre.

Snare drum

Generally, snare drums contain plenty of energy at low frequencies, which can often cloud a mix and is not necessary, so the first step should be to employ a shelving (or high-pass) filter to remove all the frequency content below 150 Hz.

The 'snap' of most snares usually resides around 2–10 kHz, while the main body can reside anywhere between 400 Hz and 1 kHz. Applying cuts or boosts and sweeping between these ranges should help you find the elements that you need to bring out or remove, but roughly speaking cuts at 400 and 800 Hz will help it sit better, while a small decibel boost (or a notch cut before) at 8 or 10 kHz will help to brighten its 'snap'.

Hi-hats and cymbals

Obviously, these instruments contain very little low-end information that's of any use and if left in can cloud some of the mid range. Consequently, they benefit from a high-pass filter to remove all the frequencies below 300 Hz.

Typically, the presence of these instruments lies between 1 and 6 kHz, while the brightness can reside as high 8–12 kHz. A shelving filter set to boost all frequencies above 8 kHz can bring out the brightness, but it's advisable to roll off all frequencies above 15 kHz at the same time to prevent any hiss from breaking through into the track. If there is a lack of presence, then small decibel boosts with a Q of about an octave at 600 Hz should add some presence.

Toms and congas

Both these instruments have frequencies as low as 100 Hz, but are not required for us to recognize the sound and if left can cloud the low and low mids; thus, it's advisable to shelve off all frequencies below 200 Hz. They should not require any boosts in the mix as they rarely play such a large part in a loop, but a Q of approximately half an octave applied between 300 and 800 Hz can often increase the higher end, making them appear more significant.

Bass

Bass is the most difficult instrument to fit into any dance mix, since it's interaction with the kick drum produces much of the essential groove, but it can be fraught with problems.

The main problems with mixing bass can derive from the choice of timbres and the arrangement of the mix. While dance music is, by its nature, loud and 'in your face', this is not attributed to using big, exuberant, harmonically rich sounds throughout. As we've touched upon, our minds can only work in contrast, so for one sound to appear big, the rest should be smaller. Of course, this presents a problem if you're working with a large kick and large bass as the two occupy similar frequencies, which can result in a muddied bottom end.

This can be particularly evident if the bass notes are quite long, as there will be little or no low-frequency 'silence' between the bass and kicks, making it difficult for the listener to perceive a difference between the two sounds. Consequently, if the genre requires a huge, deep bass timbre, the kick should be made tighter by rolling off some of the conflicting lower frequencies and the higher frequency elements should be boosted with EQ to make it appear more 'snappy'. Alternatively, if the kick should be felt in the chest, the bass can be made lighter by rolling off the conflicting lower frequencies and boosting the higher elements.

Naturally, there will be occasions whereby you need both heavy kick and bass elements in the mix, and in this instance the arrangement should be configured so that the bass and kick do not occur at the same point in time. In fact, most genres of dance will employ this technique by offsetting the bass so that it occurs on the offbeat. For instance, trance music almost always uses a 4/4 kick pattern with the bass sat in between each kick on the eighth of the bar.

If this isn't a feasible solution and both bass and kick must sit on the same beat, then you will have to resort to aggressive EQ adjustments on the bass. Similar to most instruments in dance music we have no expectations of how a bass should actually sound, so if it's overlapping with the kick, making for a muddy bottom end, you shouldn't be afraid to make some forceful tonal adjustments.

Typically for synthetic instruments, small decibel boosts with a thin Q at 60–80 Hz will often fatten up a wimpy bass that's hiding behind the kick. If the bass still appears weak after these boosts you should look towards replacing the timbre, since it's dangerous practice to boost frequencies below these, as it's impossible to accurately judge frequencies any lower on near fields. In fact, for accurate playback on most hi-fi systems it's prudent to use a shelving filter to roll off all frequencies below 60 Hz.

Of course, this isn't much help if you're planning on releasing a mix on vinyl for club play, as many PA systems will produce energy as low as 30 Hz. If this is the case, you should continue to mix the bass, but avoid boosting or cutting anything below 40 Hz. This should be left to the mastering engineer, who will be able to accurately judge just how much low-end presence is required. As a very rough guide for *theoretical* purposes alone, a graphic equalizer set to −6 dB at around 20 Hz gently sloping upwards to 0 dB at 90 Hz can sometimes prove sufficient enough for club play.

If the problem is that the bass has no punch, then a Q of approximately half an octave with a small cut or boost and sweeping the frequency range between 120 and 180 Hz may increase the punch to help it to pull through the kick. Alternatively, small boosts of half an octave at 200–300 Hz may pronounce the rasp, helping it to become more definable in the mix. Notably, in some mixes the highest frequencies of the rasp may begin to conflict with the mid-range instruments and, if this is the case, then it's prudent to employ a shelving filter to remove the conflicting higher frequencies.

Provided that the bass frequencies are not creating a conflict with the kick, another common problem is the volume in the mix. While the bass timbre may sound fine, there may not be enough volume to allow it to pull to the front of the mix. The best way to overcome this is to introduce small amounts of controlled distortion, but rather than reach for the atypical distortion unit it's much better to use an amp or speaker simulator.

Amp simulators are designed to emulate the response of a typical cabinet, so they roll off the higher frequency elements that are otherwise introduced through distortion units. As a result, not only are more harmonics introduced into the bass timbre without it sounding particularly distorted, but you can use small EQ cuts to mould the sound into the mix without having to worry about higher frequency elements creating conflicts with instruments sat in the mid range.

If the bass is still being sequenced from a tone module or sampler, then before applying any effects you should attempt to correct the sound in the module itself. As we've already covered in Chapter 7, we can perceive the loudness of a sound from the shape of its amplitude envelope and its harmonic content, so simple actions such as opening the filter cut-off or reducing the attack and release stage on both the amplitude and filter envelopes can make it appear more prominent. If both these envelopes are already set at a fast attack and release, then layering the kick from a typical rock kit over the initial transient of the bass followed by some EQ sculpting can help to increase the attack stage, but at the same time be cautious not to overpower the track's original kick.

Although most genres of dance will employ synthetic timbres, on occasion they do utilize a real bass guitar and, if so, a slightly different approach is required when mixing. As previously mentioned, bass cabs will roll off most high frequencies, reducing most high-range conflicts, but they can also lack any real bottom-end presence. As a result, for dance music it's quite usual to layer a synthesizer's sine wave underneath the guitar to add some more bottom-end weight.

Another potential problem with real bass guitars are the finger or string noises. These can result in conflicts in the midrange so it's prudent to remove any frequencies above 300 Mz. More importantly, though, unlike tone modules, where the output is compressed to even level, bass guitars will fluctuate wildly in dynamics and these must be brought under control with compression. Keep in mind that, no matter what the genre of dance, the bass should remain even throughout the mix. If it fluctuates in level, the whole groove of the record can be undermined.

If, after trying all these techniques, there is still a 'marriage' problem between kick and bass, then it's worth compressing the kick drum with a short attack stage so that it's transient is captured by the compressor. This will not only make the kick appear more 'punchy', but it will allow the initial pluck of the bass to pull through the mix. That said, you should avoid compressing the kick so heavy that it begins to ring, as this will only muddy up the bottom end of the mix.

As ever, the compression settings to use are entirely dependent on the kick in question, but generally a good starting point is a ratio of 8:1 with a fast attack and release, and the threshold set so that every kick activates the compressor. Ultimately, though, as all bass sounds are different, the only real solution is to experiment by EQ boosting or cutting around 120–350 Hz, or by applying heavy compression to any conflicting instruments.

Vocals

Although vocals take priority over every other instrument in a pop mix, in most genres of dance they will take a back seat to the rhythmic elements of a dance mix. Having said that, they must be mixed coherently, since while they may sit behind the beat, the groove relationship and syncopation between the vocals and the rhythm is what makes us want to dance. Subsequently, you should exercise great care in getting the vocals to sit properly in the mix.

Firstly, it should go without saying that the vocals should be compressed so that they maintain a constant level throughout the mix without disappearing behind instruments. Generally, a good starting point is to set the threshold so that most of the vocal range is compressed with a ratio of 9:1 and an attack to allow the initial transient to pull through unmolested.

The choice of compressor used for this is absolutely vital and you must choose one that adds some 'character' to the vocals. Most of the compressors that are preferred for vocal mix compression are hardware units such as the LA-2A and the UREI 1176, but the Waves RComp plug-in followed by the PSP VintageWarmer can produce good results. This is all down to personal choice, though, and it's prudent to try a series of compressors to see which produces the best results for your vocals.

> Note: It's worth noting that, even when compressed, it isn't unusual to automate the volume faders to help the vocals stay prominent throughout. It's quite common to increase the volume on the tail end of words to prevent them from disappearing into the background mix. If you take this approach, you may need to duck any heavy breath noises, but you should avoid removing them between phrases, otherwise they could sound false.

Once compressed, vocals will, or should, rarely require any EQ as they should have been captured correctly at the recording stage and any boosts now can make them appear unnatural. That said, small decibel boosts with a half-octave Q at 10 kHz can help to make the vocals appear much more defined, as the consonants will become more comprehensible. If you take this 'clarity' approach, however, it's prudent to use it in conjunction with a good de-esser. This will remove the sibilance boosting that 10 kHz produces, but will leave the body of the vocals untouched. This must be applied cautiously, otherwise it could introduce lisps.

Alternatively, if the vocals appear particularly muddy in the mix, an octave Q placing a 2 dB cut at a centre frequency of approximately 400 Hz should remove any problems. If, however, they seem to lack any real energy while sat in the mix, a popular technique used by many dance artists is to speed (and pitch) the vocals up by a couple of cents. While the resulting sound may appear 'off the mark' to pitch-perfect musicians, it produces higher energy levels that's perfectly suited towards dance vocals and, besides, not many clubbers are pitch perfect.

It's also sensible to add any of the effects you have planned for the vocals. As these play an important part in the music they must fit now rather than later, as any instruments that are introduced after them should have their frequencies reduced if they conflict with the vocals or effects. For instance, it's quite typical to apply a light smear of reverb to the vocals (apply it cautiously, though!) to help them produce a more natural tone, but if this were applied later in the mix you may find yourself equalizing the effect to make room for other instruments, which can make for unnatural results.

Synthesizers/piano/guitars

The rest of the instruments in a mix will (or should) all have fundamentals in the mid range. Generally, you should mix and EQ the next most important aspect of the mix and follow this with progressively less important sounds. If vocals have been employed there will undoubtedly be some frequency masking where the vocals and mid-range instruments meet, so you should look towards leaving the vocals alone and apply EQ cuts to just the instrument. Alternatively, the mid range can benefit from being inserted into a compressor or noise gate, with the vocals entering the side-chain so that the mid range dips whenever the vocals are present. This must be applied cautiously and a 'duck' of 1 dB is usually sufficient; any more and the vocals may become detached from the music.

Most mid-range instruments will contain frequencies lower than necessary, and while you may not actually be able to physically hear them in the mix they will still have an effect of the lower mid-range and bass frequencies. Thus, it's prudent to employ a shelving filter to remove any frequencies that are not contributing to the sound within the mix. The best way to accomplish this is to set up a high-shelf filter with maximum cut and, starting from the lower frequencies, sweep up the range until the effect is noticeable on the instrument. From this point, sweep back down the range until the 'missing' frequencies return and stop. This same process can also be applied to the higher frequencies if some are present and do not contribute to the sound when it's sat in the mix.

Generally, keyboard leads and guitars will need to be towards the front of the mix, but the exact frequencies to adjust will be entirely dependent on the instrument and mix in question. Nevertheless, for most mid-range instruments, it's worth setting the Q at an octave and applying a cut of 2–3 dB while sweeping across 400–800 Hz and 1–5 kHz. This often removes the muddy frequencies and can increase the presence of most mid-range instruments.

Above all, try to keep the instruments in perspective by asking yourself questions such as:

- Is the instrument brighter than hi-hats?
- Is the instrument brighter than a vocal?
- Is the instrument brighter than a piano?
- Is the instrument brighter than a guitar?
- Is the instrument brighter than a bass?

What follows is a general guide to the frequencies of most sounds that sit in the mid range along with the frequencies that contribute to the sound. Of course, whether to boost or cut will depend entirely on the effect you wish to achieve.

Piano

- 50–100 Hz: add weight to the sound.
- 100–250 Hz: add roundness.
- 250–1000 Hz: muddy frequencies.
- 1000–6000 Hz: add presence.
- 6000–8000 Hz: add clarity.
- 8000–12 000 Hz: reduce hiss.

Electric guitars

- 100–250 Hz: add body.
- 250–800 Hz: muddy frequencies but may add roundness.
- 1000–6000 Hz: allows it to cut through the mix.
- 6000–8000 Hz: add clarity.
- 8000–12 000 Hz: reduce hiss.

Acoustic guitar

- 100–250 Hz: add body.
- 6000–8000 Hz: add clarity.
- 8000–12 000 Hz: add brightness.

Synth leads/strings/pads

- 50–100 Hz: add bottom-end weight.
- 100–250 Hz: add body.
- 250–800 Hz: muddy frequencies.
- 1000–6000 Hz: enhance digital crunch.
- 6000–8000 Hz: add clarity.
- 8000–12 000 Hz: add brightness.

Wind instruments

- 100–250 Hz: add body.
- 250–800 Hz: muddy frequencies.
- 800–1000 Hz: add roundness.
- 6000–8000 Hz: add clarity.
- 8000–12 000 Hz: add brightness.

Processing and effects

We've already looked at the various effects in previous chapters, so rather than reiterate it all again here it should suffice to say that you should refrain from using any effects during the mixing process (bar the vocals) and only once the mix is together should you consider adding any effects. Even then, you should only apply them if they are truly necessary.

Always keep in mind that empty spaces in a mix do not have to be filled with reverb or echo decays and a good mix works on contrast. When it comes to effects and mixing, less is invariably always more. Any effects, but particularly reverb, can quickly clutter up a mix, resulting in a loss of clarity. Since there will probably already be plenty going on, adding effects will only make the mix busier than it already is, and it's important to keep some space in between the individual instruments. Indeed, one of the biggest mistakes made is to employ effects on every instrument when in reality only one or two may be needed throughout. Therefore, before applying any effects it's prudent to ask yourself why you're applying them – is it to enhance the mix or make a poorly programmed timbre sound better? If it's the latter, then you should look towards using an alternative timbre rather than try to disguise it with effects.

Above all, remember the golden rules when using any effects:

1   Most effects should not be audible within a mix. Only when they're removed should you notice the difference.
2   If an effect is used delicately it can be employed throughout the track, but if it's extreme it will have a bigger impact if it's only used in short bursts. You can have too much of a good thing and it's better to leave the audience gasping for more than gasping for a break.

Common mixing problems

Frequency masking

As touched upon earlier, frequency masking is one of the most common problems experienced when mixing, whereby the frequencies of two instruments are matched and compete for space in the soundstage. In this instance, you need to identify the most important instrument of the two and give this the priority while panning or aggressively equalizing the secondary sound to make it fit into the mix.

If this doesn't produce the results, then you should ask yourself if the conflicting instrument contributes enough to remain in the mix. Simply reducing the gain of the offending channel will not necessarily bring the problem under control, as it will still contribute frequencies which can still muddy the mix. Rather, it's much more prudent to simply mute the channel altogether and listen to the difference it makes. Alternatively, if you wish to keep the two instruments in the mix, consider leaving one of them out and bringing it in later during another 'verse', 'chorus' or after the reprise.

Clarity and energy

All good mixes work on the principle of contrast – that is, the ability to hear each instrument clearly. It's all to easy to get carried away by employing too many instruments at once in an effort to disguise weak timbres, but this will very rarely produce a great mix. A cluttered, dense mix lacks energy and cohesion, so you should aim to mix so that you can hear some silence behind the notes of each instrument; if you can't, start to remove the non-essential instruments until some of the energy returns.

If no instruments can be removed, then aim to remove the unneeded frequencies rather than the objectionable ones by notching out frequencies of the offending tracks either above or below where the instruments contribute most of their body. This may result in the instrument sounding 'odd' in solo, but if the instrument must play in an exposed part, then the EQ can be automated or two different versions could be used.

Additionally, when working with the groove of the record, remember that the silence between the groove elements produces an effect that makes it appear not only louder (silence to full volume) but also more energetic, so in many instances it's worthwhile refraining from adding too many percussive elements. More importantly, though, good dance mixes do not bring attention to every part of the mix. Keep a good contrast by only making the important sounds big and upfront, and leave the rest in the background.

Prioritize the mix

During the mixing stage always prioritize the main elements of the mix and approach these first. It's all too easy to spend a full day 'twitching' the EQ on a hi-hat or cymbal without getting a good balance on the most important elements first. Always mix the most important elements first and you'll find that the 'secondary' timbres tend to look after themselves.

Mixing for vinyl and clubs

When mixing a record down that will be pressed onto vinyl for club play, you'll need to exercise more care in the frequencies you adjust. To begin with, any frequencies above 6 kHz should not be boosted and all frequencies above 15 kHz should be shelved off. On top of this, it's also prudent to run a de-esser across each individual track.

These techniques will help to keep the high-frequency content under some control, since if the turntable's cartridges are old (as they usually are in many clubs) it will introduce sibilance into the higher frequencies. Additionally, the frequencies on the kick and bass should be rolled off at 40 Hz, while every other instrument in the mix should have all frequencies below 150 Hz rolled off. While this may sound unnatural for CD, it makes for a much clearer mix when pressed to vinyl and played over a club PA system.

Relative volume

Analytical listening can tire your ears quickly, so it's always advisable to avoid monitoring at loud volumes, as this will only quicken the process. The ideal standard monitoring volume is around conversation level (85 dB), but you need to keep the Fletcher–Munson contour control in mind during mixing and, after every volume or EQ adjustment, reference the mix again at various gain levels. It can also be beneficial to monitor the mix in mono when setting and adjusting the volume levels, as this will reveal the overall balance of the instruments more clearly.

EQ

EQ can be used to shape all instruments, but you can only apply so much before the instrument loses its characteristics, so be cautious with any EQ. Always bypass it every few minutes to make a note of the tonal adjustments you're making, but remember that, while an equalized instrument may not sound correct in isolation, what really matters is that it sounds right when run with the rest of the mix.

Cut EQ rather than boost

Our ears are used to hearing a reduction in frequencies rather than boosts, since frequencies are always reduced in the real world by walls, objects and materials. Consequently, while some boosts may be required for creative reasons, you should look towards mostly cutting to prevent the mix from sounding too artificial. Keep in mind that you can effectively boost some frequencies of a sound by cutting others, as the volume relationship between them will change. This will produce a mix that has clarity and detail.

Don't use EQ as a volume control

If you find yourself having to boost frequencies for volume, you should not have to boost by more than 5 dB. If you have to go higher than this then the chances are that the sound itself was poorly recorded or the wrong choice for the mix.

Remember the magic Q

A 'Q' setting of $1\frac{1}{3}$ octaves has a bandwidth that's generally suitable for equalizing most instruments and often produces the best results. That said, if the instrument is heavily melodic or you're working with vocals, wider Q settings are preferred and a typical starting point is about two octaves. Finally, drums and most percussion instruments will benefit from a Q of half an octave.

Shelf EQ

Shelf equalizers are generally used to cut rather than boost because they work at the extremes of the audio range. For instance, using a shelving filter to boost the low frequencies will only accentuate low-end rumble, since there's very little sound this low. Similarly, using a shelf to boost the high range will increase all the frequencies above the cut-off point and there's very little high-frequency energy above 16 kHz.

Fix it in the mix

'Fix it in the mix' is the opinion of a poor engineer and is something that should never cross your mind. If a timbre is wrong, no matter how long it took to program, admit that it's wrong and program/sample one that is more suitable.

Table 16.3 Frequencies for mixing

Frequencies	Musical effect	General uses
30–60 Hz	These frequencies produce some of the bottom-end power, but if boosted too heavily can cloud the harmonic content, introduce noise and make the mix appear muddy.	• Boosts of a decibel or so may increase the weight of bass instruments for drum 'n' bass. • Cuts of a few decibels may reduce any booming and will increase the perception of harmonic overtones, helping the bass become more defined.
60–125 Hz	These frequencies also contribute to the bottom end of the track, but if boosted too heavily can result in the mix losing its bottom-end cohesion, resulting in a mushy, 'boomy' sound.	• Boosts of a decibel or so may increase the weight of kick drums and bass instruments, and add weight to some snares, guitars, horns and pianos. • Cuts of a few decibels may reduce the boom of bass instruments and guitars.
125–250 Hz	The fundamental of bass usually resides here. These frequencies contribute to the body of the mix, but boosting too heavily will remove energy from the mix.	• Small boosts here may produce tighter bass timbres and kicks, and add weight to snares and some vocals. • Cuts of a few decibels can often tighten up the bottom-end weight and produce clarity.
250–450 Hz	The fundamentals of most string and percussion instruments reside here, along with the lower end of some male vocalists.	• Small boosts may add body to vocals, kicks and produce snappier snare timbres. It may also tighten up guitars and pianos. • Cuts of a few decibels may decrease any muddiness from mid-range instruments and vocals.
450–800 Hz	The fundamentals and harmonics of most string and keyboard instruments reside here, along with some frequencies of the human voice. Cuts are generally preferred here, as boosting can introduce fatigue.	• Small boosts may add some weight to the bass elements of instruments at low volumes. • Cuts of a few decibels will reduce the body sound, and may help to add clarity to the mix.
800 Hz–1.5 kHz	This area commonly consists of the harmonic content of most instruments, so small boosts can often add extra warmth. The 'pluck' of most bass instruments and click of the drum kick's attack also reside here.	• Boosts of a decibel or so can add warmth to instruments, increase the clarity of bass, kick drums and some vocals, and help instruments pull out of the mix. • Small decibel cuts can help electric and acoustic guitars sit better in a mix by reducing the dull tones.

Table 16.3 (*Continued*)

Frequencies	Musical effect	General uses
1.5–4 kHz	This area also contains the harmonic structure of most instruments, so small boosts here may also add warmth. The body of most hi-hats and cymbals also reside here, along with the vocals, BVs and pianos.	• Boosts of a decibel or so can add warmth to instruments and increase the attack of pianos and electric/acoustic guitars. • Small decibel cuts can hide any out-of-tune vocals (although they should be in tune!) and increase the breath aspects of most vocals.
4–10 kHz	Finger plucks/attacks from guitars, the attack of pianos and some kick drums and snares, along with the harmonics and fundamentals of synthesizers and vocals, reside here.	• Boosts of a few decibels can increase the attack on kick drums, hi-hats, cymbals, finger plucks, synthesizer timbres and pianos. It can also make a snare appear more 'snappy' and increase vocal presence. • Small decibel cuts may reduce sibilance on vocals, thin out guitars, synthesizers, cymbals and hi-hats, and make some sounds appear more transparent or distant.
10–15 kHz	This area consists of the higher range of vocals, acoustic guitars, hi-hats and cymbals, and can also contribute to the depth and air in a mix.	• Boosts of a few decibels may increase the brightness of acoustic guitars, pianos, synthesizers, hi-hats, cymbals, string instruments and vocals.
15–20 kHz	These frequencies often define the overall 'air' of the mix but may also contain the highest elements of some synthesizers, hi-hats and cymbals.	• Boosts here will generally only increase background noise such as hiss or make a mix appear harsh and penetrating. Nonetheless, some engineers may apply a shelving boost around this area (or at 10 kHz) to produce the Baxandall curve.

17 Mastering[1]

A failure will not appear until it has passed final inspection.

Albert Einstein

Mastering can be viewed as the final link in the music-making chain. After any track has been mixed and processed, it will need to be mastered. This is not something that you should attempt yourself and it's difficult to stress enough the importance of employing a professional mastering engineer.

All record companies will have their music mastered professionally regardless of the proficiency of the original artist and it is fully recommended that if you value your music, you should do the same. Simply because you produced and mixed the music does not mean you are the perfect choice to master it, and it is essential to appreciate that there is a world of difference between mixing and mastering.

Whereas recording and mixing music is a definite skill that requires complete attention to every individual element of the track, mastering involves looking at the bigger picture subjectively. This subjective viewpoint is something that is practically impossible for the producer (i.e. you) to accomplish due to the involvement throughout the entire project.

The process mastering involves depends on the quality of the mix in question, but if it has been mixed competently then it will usually involve equalization to set the overall colour of the mix, compression to add more punch and presence, and loudness maximization to increase its overall volume. The result is that the music will sound much more professional and will compare well with every other record played on the radio, CD or on vinyl.

This overall balance is vital if you plan to release music for commercial sale, as the public will expect the record to have a professional polish similar to other commercial releases – if the recording sounds professional, you'll sell more records. This is more complicated than it first appears, though, and to accomplish it proficiently requires experienced ears, technical accuracy, knowledgeable creativity and a very accurate monitoring environment.

Of course, some may suggest that you could master it yourself and then test the music on a number of systems to check that it is correct, but this takes time and no matter how experienced you

---

[1] This chapter was written with the help of Steve Marcus, a mastering engineer for EMI.

are, it is doubtful you would achieve the same result as that of a professional mastering engineer. That said, many artists will master their own music if it's only to be submitted to a web site in MP3 format, to a magazine for a review (such as Future Music[2]) or to a professional Internet record label such as Peoplesound or MP3.com. Consequently, some knowledge of what is involved in the mastering process and how it is employed is useful for these circumstances, so, for this chapter, we'll look at the basic *theory* behind it all.

To master any record you'll need the appropriate tools for the job, and these are available as all-in-one units in both hardware and software form. The most celebrated hardware mastering device is the TC Finalizer, which has all the mastering tools you could require in one compact single rack unit. Alternatively, digital audio workstations can be used to master music, and this is usually accomplished with an established wave editor and a number of individual mastering plug-ins such as Steinberg's mastering tools or all the mastering effects required in one unit such as Izotope's excellent Ozone.

Note: A demo version of Ozone V2.0 is available on the CD.

More importantly, though, no matter how good the equipment used, it is extremely easy to mis-use it and, if used incorrectly, it will deteriorate the sound of music and distort the audio. This distortion is not to the extent that you can hear it, rather it makes the mix sound more muddled or it reduces its stereo perspective.

Mastering theory

As we've previously discussed in earlier chapters, the order of processing can dramatically affect the overall sound; thus, in mastering each processor should be connected in a sensible order. Although this can depend on the music being mastered and the mastering engineer, the most usual order consists of:

- Noise reduction.
- Equalization.
- Mastering reverb (although this is uncommon).
- Dynamics.
- Harmonic exciters.
- Stereo widening.
- Loudness maximizers.

We'll look at each of these in turn.

Noise reduction

Although any extraneous noise should have been removed during the recording stages, occasional clicks, pops, hiss or hum can slip through, so it is important to listen to the mix thoroughly

---
[2] Future Music is part of Future Publishing.

and carefully on both monitors and headphones to see if there are any present. This is where a mastering engineer will have an immediate advantage, as they will use subwoofers and loud-speakers with an infrasonic response and, without these, microphone rumble, deep sub basses or any jitter noise introduced with poor quality plug-ins can easily be missed.

If there is any audible noise present it will need removing before the track is mastered. Stray transients such as clicks or pops will prevent you from increasing the overall volume, while if there is any hum, hiss or noise present, if you increase the gain it will also increase the volume of the noise.

Ideally, the best way to remove any noise problems in a recording is to re-record the offending part, as this will inevitably produce much better results than attempting to reduce them at the master-ing stage. If this is not possible, then you will have no choice but to use noise reduction algorithms or close waveform editing. This latter method is especially suited for removing pops or clicks that are only a couple of samples long and involves using a wave editor to zoom right into the individual samples and then, using a pencil tool or similar, reduce the amplitude of the offending click. This approach is only suitable for extraneous noises that occur over a very short period of a few sam-ples, however, and if the noise is longer than this it is advisable to use dedicated reduction algo-rithms. These have been designed to locate and reduce the amplitude of short instantaneous transients typical of clicks or pops and are available in most wave editors or as a third-party plug-in. Depending on the amount of pops and the mix in which they're contained, *Waves Click Remover* or *Waves Crackle Reduction* are both suitable for reducing or eliminating the problem.

If hum and/or hiss are present, then these are much more difficult to deal with and often require specialist plug-ins to remove. As the frequency of tape hiss (also the result of a poor soundcard A/D or low recording levels) consists of frequencies ranging from 8 to 14 kHz, removing these using a parametric EQ can result in the higher frequencies of the mix also being removed. Thus, it is important to use plug-ins specifically designed to remove it, such as *Steinberg's De-Noiser*. Again, the reliability of using these types of plug-ins will depend on the source and if not used with some degree of caution they can result in flattening the overall sound of the mix.

AC hum is a little easier to deal with than hiss, as much of the 'noise' is centred around it's fundamental frequency, which lies around 60–80 Hz. Using a parametric EQ set to a thin band-width with a maximum cut can reduce most hum, but there may also be additional harmonics at 120, 180, 200, 300, 350 and 420 Hz, and removing any of these can compromise the overall tonal balance of the music, muddying the lower and mid-range frequencies. Thus, if a parametric EQ does not remove the problem, it's worth using a dedicated plug-in such as *Waves X-Hum*.

Clicks, pops, hiss and hum can also be removed, sometimes much more reliably, by using plug-ins or wave editing features that profile the noise. Often referred to as FFT filtering, if there is a section of the track that consists of just the noise it can be sampled by the FFT filter to create a noise profile that can then be applied over the entire mix to remove the extraneous noise.

Waves X-Noise and *Sonic Foundry's Noise Reduction* plug-ins are typical examples of this, but some wave editors may also offer this function. Notably, these do not always produce the desired results and, in some instances, they can remove frequencies that are essential to the music, resulting in a mix that has no bottom-end weight or a flat top end.

Above all, any form of noise reduction is not something that should be approached lightly. Reducing any noise within a final stereo mix will often affect frequencies that you may wish to keep and this is especially the case with removing hiss or hum. In fact, if there is an obvious amount of noise that cannot be removed reliably using FFT noise profiling or plug-ins, then you should seriously consider re-recording the performance, see if a professional mastering engineer can resolve the problem or drop the project and start a new one. It is much better to write a new record rather than attempt to release one that's below your best capabilities.

Equalization

Assuming that there is no noise present, or that it has been removed proficiently, the next step in the mastering process is equalization (EQ). This should not be confused with the use of EQ during the mixing process, as at this stage the track(s) should already sound pretty much correct. Instead, the general principle behind applying mastering EQ is to achieve an overall tonal balance that is sympathetic to our ears.

The quality of the EQ being used obviously plays a large part in attaining this balance and many mastering engineers will choose EQ units that have a pleasing resonant quality, helping to give the mix a warm tone. These are not necessarily complex parametric or multi-band graphic EQ units, as any precise EQ adjustments should have been made during the mixing process. Rather, during the mastering process wide Q settings with gentle transition slopes (sometimes as low as 3 dB per octave) are used as these are more pleasing to our ears.

Note: When mastering, Q settings of 0.4–0.9 are the most popular.

Generally, mastering EQ can be broadly split into three different sections: the bass, the mid range and the highs.

Bass

The only 'real' EQ that's required on the bass end of a mix is a low-shelf filter to remove any frequencies below 40 Hz. Frequencies lower than this cannot be reproduced by many loudspeakers, but by sending a signal this low to them they will still attempt to replicate them. This inevitably results in the bandwidth of the speaker system being restricted, resulting in the overall volume of the mix dropping. By setting a high-pass shelving filter to remove everything below 40 Hz, the sub-harmonic content of the sound will be removed, not only opening up the bandwidth of the speakers for a higher volume, but also resulting in an improvement in the definition of the bass frequencies.

The bass frequencies may also appear quieter than commercial records, but it is important not to use EQ boosts in an attempt to add anything that isn't already present and the same goes for any maximizers. Although bass-specific maximizers such as Waves MaxBass can be used to introduce additional harmonics based on those already present, these are best avoided at this stage as compression can be used to introduce additional low-end punch to the music. If, however, there is a distinct lack of definition in the low-end frequencies, they will need to be pulled out using EQ.

In most dance mixes the kick's main energy will lie around 100–200 Hz, but perceivably boosting this will only result in muddying the bottom end; rather it's the transient that needs to be more defined and this commonly resides around 1000–2500 Hz. Thus, you will need to notch out the frequency that lies just below the transient so that it brings it out of the mix slightly. Possibly the best way to find these transient frequencies is to set a very narrow Q with a gain boost of around 6 dB. By then sweeping through the frequency range you should be able to locate the required transient and, once located, you can reduce the gain to around −4 dB and sweep to the frequencies just below this.

Mid range

The mid range is the most common source of problems when mastering, so you need to listen carefully to these frequencies and then identify any problems they may have. The best way to accomplish this is to compare your current mix with a professional CD of the same genre using both your ears and a good spectral analyser. Take note of the frequency dips, if there are any, and then try to employ these to your mix to see if it makes a beneficial difference.

Similar to mixing, the mid-range problems can be divided into three categories: it's too muddy, too harsh or too nasal.

If the sound is too muddy, then using a fairly wide Q place a cut around 180–225 Hz of a few decibels so that the frequency range between 100 and 300 Hz is dipped with the largest cut centred around the current frequency setting. Then, listening carefully, reduce the width of the Q until the muddiness dissipates.

If the overall sound of the mix appears too harsh, using the previously described principle place a cut at around 1.8 kHz so that there is a transition slope down from 1 kHz to the centred frequency and then a transition slope up to 3 kHz. Then, once again start reducing the width of the Q setting until the harshness is removed.

This same process applies for a mix that sounds too 'nasal', but place a cut at 500 Hz with a Q that will create a slope from 250 Hz back up to 1000 Hz and slowly reduce the width of the Q.

Above all, bear in mind that wide bands will sound more natural to our ears, so try to keep the Q quite wide unless you're making surgical corrections to a specific frequency. If you find that you have to use a very narrow Q or too much of a cut to remove the problem, then the trouble is from the mixing itself and you should return to the mixing stage to repair it.

High range

Finally, carefully listen to the high-end frequencies in the mix. In the improbable event that these are too harsh or bright, then wide cuts placed at a centre frequency of 14 kHz can reduce the problem, but it is more likely that they will not be bright enough when compared to commercial CDs.

Although brightness could be added to the signal by placing a wide Q with a boost of a few decibels at 12–16 kHz, as touched upon in the previous chapter, any gain boosts are usually avoided, with many engineers preferring to use harmonic excitement to brighten them. Not all engineers will use harmonic excitement, preferring to leave the signal unaffected, and in this instance they'll employ the Baxandall curve.

This is created by using a wide Q setting on a parametric EQ and boosting by 6 or 8 dB at 20 kHz. Due to the width of the Q, it introduces a gradual transition curve from 10 kHz upwards, which is not only similar to the ear's natural response, but also introduces small amounts of harmonic distortion similar to valves. Both these factors contribute to a pleasing sound that is often preferred to using harmonic enhancers.

While the majority of these adjustments should be made in stereo to maintain the stereo balance between the two channels, occasionally it may be more beneficial to make adjustments on one channel only. This is useful if the two channels have different frequency content from one another or if there is a poorly recorded/mixed instrument in just one of the channels. In these instances, subtlety is the real key and the best approach is to start with extremely gentle settings, slowly progressing to more aggressive corrections until the problem is repaired.

Also, when it comes to this form of surgical EQ, the most important rule of thumb is to avoid 'itchy fingers' and listen to the material carefully before making any adjustments. You need to identify exactly what the problem is before applying any adjustments and even then only to the sections that require it. If the 'problem' frequency sounds fine while it plays along with the rest of the track but appears wrong in the breakdown, then it's sensible to just correct it during the breakdown rather than attempt to change the frequencies throughout the mix. Ideally, there should be more things right with the mix than there is wrong, and if you concentrate too heavily on a fault you could end up messing up the rest of the frequencies while trying to repair just one small problem. Like mixing, this type of EQ should also be applied decisively and rapidly. Although it is difficult to perceive on a home monitoring system, changes of less than 0.5 dB are audible and our ears can quickly become accustomed to the changes, so it's vital that you compare the recent EQ movements with the original mix.

Note: As a general rule of thumb you should play the same piece of the track being mastered over and over for up to 10 minutes before switching to the unequalized comparison and listening to that for another 10 minutes to hear the changes that are being made. On this same note, this should also be referenced against no expense spared professionally mastered music of the same genre, both aurally and visually, using a spectral analyser.

Boosts and cuts

Although frequency boosts are generally best avoided altogether to retain a natural sound, any EQ adjustments will affect the perception of the overall tonal balance of the whole mix and so must be applied carefully. For example, even applying a 0.5 dB boost at, say, 7 kHz could have the same result as dipping 300–400 Hz by 2 dB. Always bear in mind that all the frequency ranges on a full mix will interact with each other and even small changes can have an adverse affect on every other frequency, so whenever you make a change you will have to reconsider all the others too.

More importantly, the combination of all these EQ settings will determine the overall dynamics of the song and this is one of the most critical aspects of any mix – it must sound correct for the style of music. When anyone listens to a mix, the first thing they'll hear is the overall EQ and if this is wrong for the genre of music then the track will fall flat. This is where a mastering engineer's expertise and subjective viewpoint plays yet another role. Every genre of music is approached and balanced differently, and there is no definitive guide to how any one genre should be balanced.

As an example, consider two entirely different genres of music; techno and chill-out. A techno record will generally need to sound sharp and aggressive to suit the style. This can be accomplished by cutting the lower and high-range frequencies, resulting in the mid range becoming more apparent. Conversely, chill-out will be quite relaxed and nowhere near as aggressive, so a cut in the mid range will accentuate the low and high frequencies, making it appear gentler to the ear. Similarly, euphoric trance is encapsulated by the often filter-swept lead motif, so this would need to be pulled to the forefront of the mix by dipping the low and high frequencies, whereas in hip-hop the bottom end is bigger and the top end is more aggressive, thus the mid range is usually attenuated.

As with mixing, the key behind setting a good tonal balance with EQ is developing an ear for the different frequencies and the effects they have, but it's worth employing a spectral analyser so you can also see how the movements are affecting the overall balance of the music. Additionally, using an analyser it's possible to measure the frequencies of a professionally mastered track and then compare these with the results you have on your own mix.

A general guide to mastering EQ frequencies

- **0–40 Hz**. All these frequencies should be removed from the mix using a high shelving filter.
- **40–200 Hz**. These frequencies contain the bottom end of the mix and are not usually attenuated or boosted. If this range has no punch or depth, or appears 'boomy', then it is common practice to use compression to cure the problem.
- **100–400 Hz**. These frequencies are often responsible for a 'muddy' sounding mix and the effect can be reduced by cutting the gain.
- **400 Hz–1.5 kHz**. These frequencies can contribute to the 'body' of the sound, and both compression and EQ gain can be used to introduce punch and impact to the mix.
- **800 Hz–5 kHz**. These frequencies often define the clarity of the instruments. Perceived increases applied around here will occasionally pull out the fundamentals of many instruments, making them appear more defined.
- **5–7 kHz**. These frequencies are usually responsible for sibilance in a recording and can be removed with a parametric EQ set to a narrow bandwidth to cut the problem, or alternatively compression can be used.
- **3–10 kHz**. These frequencies are accountable for the presence and definition of instruments. Introducing perceived 1 or 2 dB increases here can augment the presence of some instruments, while cutting can reduce the effects of sibilance or high-frequency distortion.
- **6–15 kHz**. These can contribute to the overall presence of the instruments and the 'air' surrounding the track. Introducing small gain boosts here will increase the brightness but may also boost noise levels.
- **8–15 kHz**. Creating a boost at these frequencies will add sparkle and presence to a mix (although it will also boost hiss if there is any), while cutting them will help to smooth out any harsh frequencies.
- **20 kHz and above**. This is beyond the limit of human hearing, but setting a wide Q of around 1 and boosting (not perceived boosting!) by a few decibels at this frequency will create the Baxandall curve, creating a more ear-pleasing effect on the mix.

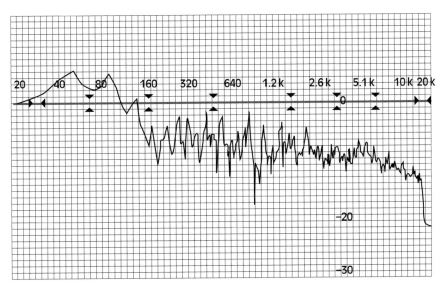

Figure 17.1 Spectral analysis of the results of mastering EQ on dance music.

Note: Figure 17.1 is only an example of typical settings and should not be regarded as the definitive graph to use. Rather they are just guidelines and in the end you should rely on your ears and judgement as to what sounds right for your mix.

Mastering reverb

Mastering reverb (applying reverb across an entire mix) is rarely required in most mastering situations as it should have been applied properly during the mixing stages, but in certain cases it may still be required. Typically, this is when instruments don't seem to gel with one another, creating a series of holes in the mix, or the mix seems to be lacking the full body of other similar mixes.

Although from an ideal viewpoint you should return to the mixing stage and repair these problems, it may not always be possible, so lightly (and carefully!) applying reverb over the entire mix can help to blend the music together to produce a better result.

Of course, the amount of mastering reverb that should be applied is totally dependent on the mix, the spacing between the instruments and the amount of reverb that has been applied to each during the mixing stage, but there are some general guidelines that can be followed.

First and foremost, the reverb must be of a very high quality as any tail noise created by poor algorithms will only serve to destroy the mix further. Indeed, if you don't have access to a good reverb unit or plug-in, then it isn't not worth applying, as it isn't worth turning an average mix into a poor one just for the sake of trying to introduce some cohesion between the instruments.

We've already looked at the uses and parameters of reverb in Chapter 5, and essentially the same methods and ideologies also apply during the mastering process, although a little more prudence should be applied. Also, unlike all other effects in the mastering chain, reverb is applied as a send effect rather than an insert as you don't want to wash over the mix completely, but only add small amounts to the original signal. A typical return to start with is around 15% wet signal against 85% dry, but as always this depends entirely on the mix in question. Nevertheless, using this setting you can set the other reverb parameters and then back off the wet signal until it suits the mix.

Following the mix fader, it's worthwhile adjusting the decay time. Broadly speaking, you should aim to set this so that the tail is a little longer than the reverb that's already been applied during the mixing stage, so that the instruments mix together more coherently. Typically, for most styles of dance music this would equate to around 2–8 milliseconds; any longer and the mix could appear too 'washy' and undefined. This, however, is related to the room size and as a general starting point a square room (if the option is available) with a relatively small size to begin with will inevitably produce the best results. It's important to note that the size of the room will also determine the diffuse settings you should use, so for a small room this should be set quite low so as not to create an inaccurate impression.

As you're working with a stereo reverb unit, you need to emulate the diffuse effect that results from the left and right reverberations reaching the ears at different times. If you set this high but use a small room, you would essentially be creating the effect of a small confined room with walls that are far apart! Although on some mixes this inaccurate impression has been known to work quite well, ideally it is best to keep small diffusion settings with small rooms and vice versa. More importantly, any adjustments with the diffusion can have a significant effect on the stereo aspect of the mix, so always check the mix in mono too.

Applying reverb to a signal may also introduce or enhance any sibilance or low-end artefacts, so to remove these it's preferable to use the reverb's low-pass and high-pass filters. On average, for most dance mixes a high cut-off applied at 2–4 kHz should prevent the introduction of sibilance, while a low cut-off set at 80–120 Hz will prevent the low body of the mix from becoming too muddy.

Possibly the most important aspect of applying reverb, though, is to continually bypass the effect to check that you're not getting too carried away with its application. Ideally, as with mixing, reverb should be applied so that when it is present it isn't particularly noticeable, but when removed it becomes immediately perceptible.

Dynamics

Following EQ (and perhaps mastering reverb), many engineers will compress the entire mix in an effort to add volume, punch, clarity and possibly emotion. This cannot be accomplished by simply strapping a standard compressor across the entire mix, however, since the loudest parts of the track will drive the compressor, forcing the entire mix to pump with each kick.

While all good dance mixes should pump with some energy, much better results can be gleaned from using multi-band compression. Using these, it's possible to alter the dynamics at different frequency bands, which can result in a clearer sounding mix. For example, compressing the high end of a mix heavily can flatten the sound of a mix, so it's best applied lightly while, on the other hand, the lower ranges of a mix are usually compressed more heavily, not only to even out the sound but also to help the mix pump along and add extra volume to some frequency bands of the mix.

This frequency-specific volume is an important aspect of most dance music, especially on the lower frequencies, and should not be underestimated. Almost every record company will want their music to measure up to the competition in terms of volume, as it's a long proven fact that if you play the same musical passage to an audience at different volume levels, most will prefer the louder playback. However, while it would seem sensible to reduce the dynamics of a mix as much as possible to increase the volume, there are limits to how far you should push it.

A typical home hi-fi system will reproduce about 80 or 90 dB of the dynamic range, but this is going to be pointless if the music is squeezed into just a few decibels. In fact, we naturally expect to hear some dynamics when listening to music and if this is eliminated it isn't going to sound real, no matter how good the hi-fi system reproducing the music is.

Furthermore, while louder music will inevitably capture the attention of the listener, it will only be for the first minute or two. While the track may appear hard-hitting at first, if it's too 'hot' then it can quickly become fatiguing to listen to and most listeners will turn off even if they like the track. In fact, any ruthless dynamic restriction is only really considered useful for playback on radio stations.

Note: All radio stations will limit the signal for two reasons:

1   They need to keep the music's level at a specific maximum so as not to over-modulate the transmission.
2   They want to be as loud as possible to hold an audience's attention.

This means that no matter what the input level is, the output level will always be constant. The inevitable result is that, even if your mix happens to be a couple of decibels louder than every other mix, when it's played on the radio it would be at the same volume as the one played before or after.

Incidentally, some artists will argue that by compressing a mix heavily at the mixdown/mastering stage the radio station cannot squash the signal any further, so there is no possibility of any broadcasters destroying the signal. While there is some logic behind this, setting the average mix level too high through heavy compression can actually create more problems because the station's processors might confuse the energy of the music with the signal peaks and attempt to squash it even further, totally destroying what's left of the mix. This will seriously reduce the overall punch of the music, making it appear lacklustre when compared to other dance records.

Consequently, you will often have to settle for a compromise between the output volume and leaving some dynamic range in the mix. As a very general rule of thumb, the difference between the peak of the kick and the average mix level (known as the *peak to average ratio*) should never be less than 4 or 5 dB. This will enable you to increase the overall gain while also retaining enough dynamic variation to prevent the mix from sounding 'flat' on a home hi-fi system. Alternatively, if you really want a loud mix with less punch for radio airplay, then it may be worthwhile releasing two different versions of the record. The promotional records can be compressed heavily for radio airplay, while the record store version can be mastered with more punch for clubs or home equipment.

There's more to multi-band compression than simply using it to add some punch to a dance track, and a manipulation of dynamics around the vocals or lead melody can introduce additional impact. This is often what gives music that extra listenable edge, although acquiring this dynamic feel often requires a fresh pair of experienced ears too, as it involves listening out for the mood, flow and emotion it conveys.

As an example, we'll examine an imaginary house track:

	......I love to party......			to dance the				night away...........				I just love to party			
Beats	**1**	2	3	4	**1**	2	3	4	**1**	2	3	4	**1**	2	3	4

Of course, this isn't a particularly complex vocal line, but notice that the accents occur on the downbeat of each bar. If you were to apply compression heavily throughout the mid range (where the vocals reside), then this vocal line could easily become:

	......I love to party...			**to dance the**				**night away...........**				**I just love to party**			
Beats	**1**	**2**	**3**	**4**	**1**	**2**	**3**	**4**	**1**	**2**	**3**	**4**	**1**	**2**	**3**	**4**

By compressing the vocal area hard throughout, the dynamics of the track have been lost and the resulting feel has been totally removed. However, by carefully compressing and only applying light amounts to specific parts of the vocals, it could result in:

	......I love to			**party**......		**to** dance the night				**away**..............				I just love **to party**			
Beats	**1**	2	**3**	4	**1**	2	3	4	**1**	**2**	3	4	**1**	2	**3**	**4**

As the accents show, you're not just working on the vocal but the mix that's also sat behind it, so you're also affecting the music. This can help to add a dynamic edge as the music will move dynamically with the vocals, enhancing the general feel and emotion of the song.

Of course, this compression has to be precise if it's applied, so the compressor's parameters must be set carefully. If the attack parameter is set too fast then the transients of the words and music will be compressed meaning that both will be softened, which you need to avoid. Similarly, if the release is set too long the compressor probably may not recover fast enough and distort the next accent but if it's set too short the sound could also distort. It's a very careful balance.

More importantly, any compression will result in you having to increase the make-up gain on the passage being compressed, and as we've discussed, loudness can have an adverse affect on our judgement. Consequently, even applying accent compression in the wrong place will still lead you to think it sounds better, so it's important to keep the gain the same as the uncompressed passages to ensure that it isn't just pure volume that's making the passage sound better.

When it comes to setting up a multi-band compressor for a mix there is no set approach and there are certainly no generic settings to use, as it depends entirely on the mix in question. All the same, there are some very general guidelines you can follow to initially set up a compressor across a mix.

- On a three-band compressor, start by setting each band of the compressor to 2:1 and then adjust the frequency of each band so that you're compressing the low range, mid range and high range.
- On a four-band compressor, adjust the crossover frequencies so that one band is compressing the bass frequencies, two are compressing the mid range and one is compressing the high range.
- On a five-band compressor, adjust the crossover so that two bands are concentrated on the bass, two are compressing the mid range and one is compressing the high range.

The crossover frequencies will depend on the mix being squeezed, but try to ensure that the vocals are not compressed with either the low range or high-frequency ranges of the compressor. By doing so, a single (or double on a five-band) compressor concentrated on the vocals alone can be adjusted to add the aforementioned dynamic movement to a mix.

Applying this multi-band compression to parts of the mix will upset the overall balance of the mix, but rather than start equalizing again, adjust the output gain of each compressor to try to gain a more level mix. It's important at this stage to realize that overall track volume shouldn't be an immediate concern. Ideally, you should be aiming to produce a mix that has plenty of bottom-end energy with a well-balanced mid and high range.

If the bass end of the mix is short of any bottom-end weight, then it is worthwhile lowering the threshold further and raising the ratio until the bass end comes to the forefront. It should be noted, though, that the mid range may simply need less compression to lift up the bass and it's a careful mix of all the bands that creates a fuller mix.

For instance, if the mid range is compressed heavily it will be pushed to the front of the mix, overtaking both the high and bass ranges, so rather than compress the bass even more it may be more prudent to lower the compression on the mid range. As a general guideline, for most dance mixes, it's usual to compress the bass heavily and use lighter settings on the mid range, as this can help the mix appear less muddied.

Both the attack and release parameters can also affect the overall tonal balance of a mix, so these must be adjusted with caution. Usually, on the bass frequencies you can get away with using a fast attack time, as it consists purely of low-frequency waveforms with few or no transients. Also, the release setting should be set as fast as possible so the compressor can recover quickly, but remember that a compressor can track the different states of a low frequency waveform so be cautious not to set it too fast.

Generally, the faster the release is set, the more the bass end will pump, so it should be set so that it pumps slightly but not too much as to destroy the feel of the mix. For the mid range, it is sensible to use longer attack times than the bass to help improve the transients that are contained here, while the release setting should be set as short as possible without introducing any gain pumping. Determining pumping at mid range frequencies can be difficult, so it is preferable to use a compressor that allows you to solo the current band to hear the effect you're imparting onto the mix.

Finally, the high range should use a relatively fast attack to prevent the transients becoming too apparent, and as the waveform of these is often considerably shorter than lower frequencies you can also get away with using short release times. As with the mid range, however, do not set this release too short, otherwise you may introduce pumping, which will ruin the overall dynamics contained here.

As a very general starting point, what follows is a list of typical settings for three-, four- and five-band compressors:

Three-band compression

Band 1: to tighten up the bottom end of a mix

- Frequency – 0–400 Hz.
- Ratio – 4:1.
- Threshold – 3 dB below the quietest notes so that the compressor is permanently triggered.
- Attack – 10–20 ms.
- Release – 140–180 ms.
- Gain make-up – increase the gain so that it is 3–5 dB above pre-compression level.

Band 2: to tighten up and add some 'weight' to the mix

- Frequency – 400 Hz–1.5 kHz.
- Ratio – 2:1.
- Threshold – just above the quietest notes so that it is triggered often but not continually.
- Attack – 10–20 ms.
- Release – 100–160 ms.
- Gain make-up – increase the gain so that it is 1 dB above pre-compression level.

Band 3: to increase clarity of instruments and reduce 'harsh' or 'rough' artefacts

- Frequency – 1.5–15 kHz.
- Ratio – 2:1.
- Threshold – just above the quietest notes so that it is triggered often but not continually.
- Attack – 5–20 ms.
- Release – 100–130 ms.
- Gain make-up – increase the gain so that it is 1 dB above pre-compression level.

Four-band compression

Band 1: to tighten up the bottom end of a mix

- Frequency – 0–120 Hz.
- Ratio – 4:1.
- Threshold – 3 dB below the quietest notes so that the compressor is permanently triggered.
- Attack – 10–20 ms.
- Release – 150–180 ms.
- Gain make-up – increase the gain so that it is 3–5 dB above pre-compression level.

Band 2: to tighten up the mix in general and add warmth to instruments and vocals

- Frequency – 120 Hz–2 kHz.
- Ratio – 2.5:1 or 3:1.
- Threshold – just above the quietest notes so that it is triggered often but not continually.
- Attack – 20–30 ms.
- Release – 100–160 ms.
- Gain make-up – increase the gain so that it is 1–3 dB above pre-compression level.

Band 3: to increase the clarity of instruments

- Frequency – 2–10 kHz.
- Ratio – 3:1.
- Threshold – 2 dB below the quietest notes so that the compressor is permanently triggered.
- Attack – 10–20 ms.
- Release – 100–180 ms.
- Gain make-up – increase the gain so that it is 2 dB above pre-compression level.

Band 4: to increase the clarity of hi-mid to high-frequency instruments

- Frequency – 10–16 kHz.
- Ratio – 2:1 or 3:1.
- Threshold – just above the quietest notes so that it is triggered often but not continually.
- Attack – 5–25 ms.
- Release – 100–140 ms.
- Gain make-up – increase the gain so that it is 1 dB above pre-compression level.

Five-band compression

Band 1: to tighten up the bottom end of a mix

- Frequency – 0–180 Hz.
- Ratio – 5:1.
- Threshold – 3 dB below the quietest notes so that the compressor is permanently triggered.
- Attack – 10–20 ms.
- Release – 130–190 ms.
- Gain make-up – increase the gain so that it is 3–5 dB above pre-compression level.

Band 2: to tighten up the rhythm section and the mix in general

- Frequency – 180–650 Hz.
- Ratio – 2.5:1 or 3:1.
- Threshold – just above the quietest notes so that it is triggered often but not continually.
- Attack – 10–20 ms.

- Release – 130–160 ms.
- Gain make-up – increase the gain so that it is 1–3 dB above pre-compression level.

Band 3: to add some weight to the mix and increase its energy

- Frequency – 650 Hz–1.5 kHz.
- Ratio – 3:1.
- Threshold – 1 dB below the quietest notes so that the compressor is permanently triggered.
- Attack – 15–25 ms.
- Release – 100–130 ms.
- Gain make-up – increase the gain so that it is 1–3 dB above pre-compression level.

Band 4: to increase the clarity of mid- to high-frequency instruments

- Frequency – 1.5–8 kHz.
- Ratio – 2:1 or 3:1.
- Threshold – just above the quietest notes so that it is triggered often but not continually.
- Attack – 5–10 ms.
- Release – 100–140 ms.
- Gain make-up – increase the gain so that it is 1–4 dB above pre-compression level.

Band 5: to reduce unwanted artefacts such as a 'rough' or 'harsh' top end

- Frequency – 8–16 kHz.
- Ratio – 2:1 or 3:1.
- Threshold – set so that it is triggered occasionally on the highest peaks.
- Attack – 5 ms.
- Release – 100–140 ms.
- Gain make-up – increase the gain so that it is 1 dB above pre-compression level.

As always, these are not precise settings and are only listed as a general guide to begin with. With these settings dialled into the compressor, you will need to experiment by increasing and lowering the make-up gain, the threshold, attack, release and the ratio while listening out for the difference each has on the mix. Also, remember to bypass it often and compare it with the uncompressed signal, ensuring that you're not destroying the dynamics of the mix.

Harmonic exciters

Following compression, some engineers may send the mix out to a harmonic exciter to add some extra sparkle to the mix. This is a moot point amongst many engineers, as some regard any harmonic excitement as sounding too artificial, but nonetheless a proportionate amount of music released today will have been 'cooked' with some form of enhancement to increase the overall resolution of the music.

The main reason behind applying any form of psychoacoustic enhancement to a mix stems from the introduction of modern recording techniques. As we now have the ability to record and re-record music, the clarity and detail of music is usually compromised, particularly in the higher frequencies.

This results in the music's frequency content seeming dreary, dull and lifeless. By using an enhancer, these dulled or missing frequencies can be restored or replaced, in effect making the music appear clearer, brighter and crisper. While this seems relatively simple in principle, in application it is much more complicated, as there are various enhancers available and each use different methods to brighten the mix.

The first exciters were developed by the American company Aphex in 1975 and were, according to the story, discovered by accident. A stereo amplifier kit was put together incorrectly, ending up with only one channel working, the other giving a thin distorted signal. When both channels were mixed together, the resulting signal came out sounding cleaner, brighter and generally more enhanced.

This principle was researched and developed before eventually coming to the conclusion that adding controlled amounts of specific harmonic distortion to an audio signal can actually enhance it. Originally, the first Aphex exciters were only available for rent to studio at around £20 per minute of music, but due to the overwhelming demand they soon became available for purchase. These commercial units work by allowing you to select the low-, mid- or high-range frequencies, which are then subsequently treated to second- and third-order harmonic distortion. This distortion is then added back to the original audio before the whole signal is treated to small amounts of phase shifting.

This form of synthesizing additional harmonics from an original signal is only used by the Aphex exciters and to many engineers they are the preferred choice, but there are some who don't like the idea of applying any distortion to a mix and so use enhancers instead. As Aphex registered the name Aural Exciter to its family of exciters, all other manufacturers refer to their exciters as '*psychoacoustic enhancers*', and these are available from Behringer, SPL and BBE to name a few.

Each of these utilize slightly different forms of enhancement, many of which are closely guarded secrets by the manufacturers. This could be because nobody can quite explain why enhancers manage to fool our senses into believing something special is happening. And although I would love to explain the details of why psychoacoustic enhancements sound so special, (a) it would take up most of this book to explain and (b) it would also involve me knowing why. Nonetheless, I can say that they are all based around some form of dynamic equalization and phase alignment.

Much of what we determine from a sound is derived from the initial transient, but during the recording process and subsequent processing the phase can be shifted, which results in the transient becoming less noticeable. By realigning the phase of these transients using an enhancer they become more defined, which results in extra clarity and definition.

Most usually enhancers and exciters are single-band, either allowing you to specify the frequencies you wish to enhance or simply enhancing all of the frequencies at once. However, some enhancers, such as Izotope's Ozone, also offer individual control over numerous bands of 'enhancement', allowing you to process different frequency bands separately. Generally, these types of enhancers introduce second-order harmonic distortion into the original signal, which makes them suitable for adding the typical tube emulation throughout the frequency range.

Ultimately, while applying an extra sonic sheen to music will make pretty much anything sound better, they should be used conservatively. The results they produce can be extremely addictive and our ears can grow accustomed to the processing quickly. This can result in over-processing (often referred to as 'overcooking' or 'frying'), which results in a sound that can be very fatiguing to the ear. Thus, like all other effects, they should be used sparingly.

Stereo widening

Following the excitement effect, a proportionate amount of mixes will also be sent through a stereo widening effect to increase the overall image of the mix. This is a very simple effect and requires little explanation, as usually they will only offer a percentage parameter. By increasing this, the width of the stereo image will be perceived to be wider than it actually is.

This works on the principle that sounds that are shared in both left and right channels can appear to be in the middle of the mix rather than panned to either side, even if they are. However, if you were to subtract one of these channels from the other then the phase would be adjusted, resulting in a wider stereo effect.

Like any other mastering process, overuse of stereo widening effects can have a detrimental effect on the music, so it must be applied carefully. This is because as you widen the spectrum further left and right, the phase adjustments can create a 'hole' in the middle of the mix, and as the bass, kick drum and vocals are usually located here, it can result in a distinct lack of body, or in extreme cases their total removal. To avoid this, some mastering modules will allow you to adjust frequency bands individually, and if this is the case, it's worthwhile widening the high frequencies but applying less to the mid range and the bass consecutively.

Loudness maximizers

The final link in the chain is the loudness maximizer. The basic premise of this is to reduce the volume level of any peaks so you can increase the relative volume of the rest of the mix without fear of overloading the signal. This is similar to the operation of a limiter in most respects, in that they establish a strict dynamic maximum, but, unlike a typical limiter, loudness maximizers are designed to sound more natural by 'rounding' off any peak signals rather than cutting them dead. This is the preferred approach to increasing the relative volume of a mix, as normalizing on an audio editor can introduce unwanted artefacts.

As previously discussed, when audio is normalized, the file is first scrutinized to locate the loudest waveform peaks and then the volume of the entire section of audio is increased until these peaks reach the digital limit. Theoretically, this should raise the volume without introducing any unwanted artefacts, but any digital recording will sound much better if the waveform was captured as a loud signal in the first place. If you artificially increase the gain using normalization it will often introduce quantization noise. What's more, as it raises the volume in direct relation to the peaks, while it may look louder, remember that we do not perceive volume by the peaks in a signal, but through the RMS.

Using loudness maximizers, the distance between the peak and main body of the music is limited, which increases the volume more substantially. This does, however, mean that the more maximization that is applied to a mix, the more the dynamics will be reduced, so like any other processor they must be used cautiously. Ideally, you want to keep a peak to average ratio of at least 4 or 5 dB, or perhaps more if you want the kick drum to really stand out (keep a large excursion).

Due to their simple nature, most maximizers will consist of only a few controls: a threshold, a make-up gain control (often known as a margin), a release and the option to use either brick-wall or soft limiting. These functions are described below.

Threshold

Similar to a limiter and compressor, the threshold on a loudness maximizer is used to set the level where the limiting will begin. By reducing this, more of the signal will be limited, whereas increasing it will reduce the amount of the signal being limited. The amount that this should be set is determined by the mix in question, but as a general rule of thumb it should be adjusted so that only the peaks of the waveform (usually the kick drum) are limited by a couple of decibels, while ensuring that the excursion between the kick and the main mix is not compromised to less than 4 dB. For most competent mixes (i.e. those that have a good signal level) the threshold will usually be set around −1 to −4 dB.

Make-up gain (a.k.a. margin)

This is similar to the make-up gain on a typical compressor, and is used to set the overall output level of the mix after 'limiting' has taken place. Broadly speaking, it isn't advisable to increase this so that the output level is at unity gain, rather it is more prudent to set this to 1 dB lower just in case any further processing has to take place.

Release

Again, the release parameter is similar to that on a compressor and determines how quickly the limiter will recover after processing the signal. Ideally, this should be set as short as possible to increase the volume of the overall signal, but if it is set too short the signal will distort. The best way to set the release parameter is to adjust it to its longest setting, and gradually reduce it while listening to the mix. As soon as the mix begins to distort or slice, increase the time so that it's just set above these blemishes. You'll usually find that the more limiting that is applied to an audio file, the longer this release setting will need to be.

Brickwall/soft limiting

Most mastering limiters or loudness maximizers are often described as being brickwall or soft, and some will offer the choice to switch between the two modes. Fundamentally, if the maxi-mizer is particularly good then these will both produce similar natural results, but soft limiting may appear more transparent than brickwall depending on how hard the limiter is being pushed.

When using a soft limiter, the overall level can exceed unity gain if it is pushed too hard, but if not it will produce results that are generally seen as more natural and transparent than brick-wall. Alternatively, using a brickwall setting, no matter how hard the limiter is pushed it will not exceed unity gain, but if it is pushed too hard the limiter may have a difficult time rounding off the peaks of the signal, resulting in a sound that's best described as 'crunchy' or 'digital'. The choice on which to use is dependent on the mix, but a soft setting will often produce better results on a mix.

General mastering tips

- *Avoid the temptation to master your own music.* All mastering engineers are impartial to the music and they will provide a subjective viewpoint on the music that you will not be aware of due to your involvement in the project. Even if you are not willing to employ a professional mastering engineer, ask a knowledgeable friend or fellow musician to master it for you. Another person's interpretation can produce results that would never have crossed your mind.
- *If at all possible use full-range, flat monitors.* Both vinyl and CD are capable of producing a frequency range that is beyond the capabilities of all near-field monitors, and if you can't hear the signal you can't judge it accurately. Mastering engineers will use monitors that are incredibly flat which, although they sound a little bland, guarantee that the music will sound appropriate on all sound systems.
- *Monitor at a fixed volume.* Adjust the monitoring volume so it is at a fixed, comfortable volume. Loud monitoring volumes will tire your ears more quickly and affect the frequencies of the mix due to the Fletcher–Munson contour control. Once the volume is set, listen to some professionally mastered CDs and compare them to your own mix.
- *Always reference your track with a professionally mastered track.* This is possibly the best way to train your ears. By continually switching between your own music and a professionally mastered track you will be able to keep some perspective on your current work.
- *Always perform A/B comparisons.* Our ears become accustomed to tonal changes relatively quickly, so after applying any processing, perform an A/B comparison with the unaffected version.
- *If it isn't broke, don't try to fix it.* One of the biggest mistakes made by many musicians is to immediately jump in and start adjusting the EQ and/or multi-band compression without actually listening to the mix first. Itchy fingers will inevitably produce a poor master, so listen to the mix thoroughly before touching anything. The less processing that is involved during the mastering stage, the better the overall results will be.
- *Avoid using presets or wizards.* All music is different, so always use your ears rather than someone else's idea of what the EQ or compression should be. Also, bear in mind that the approach is different from one track to the next, so don't think that mastering settings from a previous track will work on the next.
- *Do not finalize anything as you're recording to external hardware such as DAT.* Unlike computers, DAT machines do not have an undo function and if you apply the wrong settings at this point there is no going back.
- *Do not normalize a mix.* All normalizing algorithms look for the highest peaks and then increase the volume of the surrounding audio, but the ear doesn't respond in the same way. Rather, it judges the loudness of music by the average level.
- *Do not apply any processing that is not vital to improving the sound.* Simply because some may mention that a certain process is used in mastering, do not take it for granted on your own mix. Every DSP process comes at the cost of sound quality. The less processing that is applied to a mix during the mastering stage, the better it will inevitably sound.
- *If possible, use a VU meter to measure the levels of the mix.* Although VU meters are considered old hat compared to the latest digital peak meters, VU meters are much more suitable for

mastering. A digital peak meter only measures the loudest peak of a signal, thus if you have a drum loop measured by one it will seem as though the level is particularly high, even though it may be quiet. Conversely, a VU meter generally measures the overall loudness (RMS).

- *Keep the bit rate as high as possible and only dither (if required) before it is recorded.* Use as many bits as possible in the mix until it is ready to be copied to the final media. If you continually work with 16-bit throughout the recording and mixing, further processing such as raising the gain will truncate the overall sound. Always bear in mind that the more bits that are contained in a signal, the better the signal-to-noise ratio, with less chance of distortion being introduced.

- *Use a top quality soundcard with the best software you can afford.* Poor quality soundcards will reduce the overall quality of the end results, and poor dithering algorithms will reduce the quality further.

- *Make sure that the headroom is optimal.* Ensure that the faders on the mixing desk are not too far below their optimum 0 dB and that the main mix output signal is around 0 dB (or −15 dBFS for digital signals). Keeping them too far below this can often result in compensating for it by adding more volume at the amplifier, which can result in a loss of transients at the recording stage.

- *Learn your mastering processors.* Learn the effect that each processor has on a mix, and unless you're experienced create a number of takes using a mix of gentle and aggressive settings. Write these settings down, copy the audio to CD and then play it on as many systems as possible. All engineers have learnt the most from previous mistakes, so don't be afraid of making them. It's much better to make a mistake in private than publicly.

- *Avoid using a noise gate.* Although noise gates are useful for removing any potential noise during silent passages, they can also result in clipping as the sound stops or starts. It is more prudent to leave any potential noise in, as this can be removed by applying the noise reduction software over just the noisy part and can also be used to create a noise profile if it is present throughout the track.

- *Avoid using rapid fades.* If there is noise present at the end of a recording, don't try to avoid it by rapidly fading the music down to nothing, as it may result in cutting off the end of an instrument. A better approach is to leave the recording running for a few seconds longer and then use noise reduction software to remove the problem. As with the above tip, any noise present here can also be used as a noise profile.

- *Always check your mastered mix in mono.* If the track is destined for radio play, it is imperative that it sounds well in mono as well as stereo. Although a majority of radio stations broadcast in stereo, slight signal discrepancies can result in the mix being received in mono. If the mix has not been checked for mono compatibility it could sound drastically different.

- *Consider decreasing the tempo of the master.* It isn't unusual for some radio stations to increase the tempo of a track so they can cram in more commercials throughout the day. This is only by a few beats per minute, but nevertheless if a song is destined for radio airplay it is sometimes worth time-stretching the track to a slower tempo. By doing so, when the radio station speeds the track up it will be back at its original tempo.

- *Use the best CD burner that you can afford.* If the final media for a mastered mix is CD, then it is vital to use a high-quality CD burner. These are not only more reliable but will reduce the number of errors that can be introduced onto the disc. Additionally, although many CD burners can record at speeds of 40× and above it is advisable not to record at speeds greater than 4×. Apart from preventing the chance of buffer underrun (CDs are recorded in one continuous pass and if the computer cannot deliver data fast enough, an interruption will occur and destroy the recording), if the burn rate is too fast, crackling and distortion can appear on quieter passages of music.

- *Use quality blank CDs.* This may seem like common sense, but many musicians use cheap quality CDs in an effort to save some money. Silver and gold dye on CDs will invariable perform better than the dark green, and if you value your music it is worthwhile spending a little extra for a better recording quality.
- *Do not use CD labels.* The typical sticky paper labels should be avoided because they can often increase the CD's error rate. This is not usually perceptible on many CD players, but it will degrade with age and low error rates are required by mastering plants. Ideally, you should use a specific CDR pen and try to write outside the area where the data is recorded.
- *Avoid using 80-minute CDs.* Despite the manufacturer's claim that 80-minute discs are as reliable as 74-minute CDs, they have the tendency to introduce distortion, so stay with 74-minute CDs.
- *Use a professional mastering plant.* If the record is for the public market (hi-fi players, vinyl, etc.) you should very seriously consider taking the recording to a professional – the difference can be remarkable and worth the additional outlay. Mastering is a precise art and you really do need golden ears to produce an accurate master copy.

18 Publishing and promotion

I signed away Voodoo Ray for a hundred pounds so that I could buy a new drum machine. I had no idea that the record would be so big.

A Guy Called Gerald

On completion of a dance track (or album), you're obviously going to want to release it to the general public. The first step in this is to copyright the music recording and performance. This is to ensure that, wherever the music is played, you'll be paid or credited for it, and is defined by UK law as *'the right granted to the creators of original literary, dramatic, artistic and musical works to ensure that copyright owners are rewarded for the exploitation of their works'*. Basically, this means that as long as you taken the appropriate steps to copyright your creation, you have the right to determine what people can do with what is essentially 'your property' and allows you to prevent anyone from:

- Broadcasting the work through radio or television.
- Copying the work onto any medium.
- Distributing the work to the public.
- Renting the work out to anyone.
- Adapting any part of it – for instance, remixing it.
- Performing in a public place.
- Using it as 'background' music on films or corporate videos.

Although the current UK law states that, as soon as you produce an original piece, it is automatically copyrighted to you, it's vital that you take steps so that you can offer physical proof that it *is* your work before you sign it over to a record company or release it commercially. The most common perception of how you should keep this evidence is to write a copyright statement on the media, seal this into an envelope and post it to yourself. As long as you don't open it, you have proof from the date stamped on the postage. However, while this approach is still commonplace amongst many musicians, it is sometimes not considered suitable evidence of copyright ownership in a UK court of law. Therefore, a much more sensible approach is to write the copyright statement on the media and mail it to a solicitor or bank and ask them to keep it (unopened) for you, as this proves to be much more reliable in court (should the copyright ownership be questioned).

Note: A copyright statement simply consists of:
© Artists names, Year, All rights reserved

Also, you can only copyright the recording of the music (which lasts for 50 years after first publication), and the melodies and words used in the music (which lasts for 70 years after the death of the last surviving composer). This means that you cannot claim copyright to the musical arrangement or the song title!

Once your material is copyrighted, you can then look to releasing it onto the general market. Generally, this is possible in one of three ways: you can sign the record over to a record label, press and release it yourself as a white label, or promote and sell it yourself over the Internet. Each of these requires a different approach, so we'll look at each of these options in detail, starting with the most obvious route, signing the recording over to a record label.

Record companies and contracts

Sending a 'demo tape' to a record company has to be the most infamous way of acquiring a record deal. Even though this approach is looked upon with a certain amount of scepticism today, it cannot be denied that many artists still use this route and, as a result, some of them have been signed on the strength of the demo. Before we go any further into this, though, there are some home truths that you should be aware of if you have big plans to sign to a large label and become an overnight 'megastar':

1   Signing to a large label is *not* a guarantee that you will ever reach superstar status.
2   The record company will *expect* the rights to everything you ever plan to do.
3   The record company will *expect* you to become who they want.
4   The record company will expect you not to change your musical style midway through the contract, but reserve the right to turn away any music you produce and *tell* you to make it more 'commercially viable'.
5   The record company will *expect* you to do exactly what they say, when they say.
6   The record company will *expect* you to appear on *Top of The Pops* and other humiliating music programmes.
7   They will invest money in what *they* want, not what you want, and will *expect* you to pay for these expenditures out of your royalties.
8   You will be expected not to make any demands or requests to the record company whatsoever.
9   They reserve the right to drop you if they see fit, but you do not have the option to drop them if their performance is terrible.
10 It costs between £1.5 and 2 million to launch an artist into the 'public' eye and the record company will *expect* you to do *everything* to earn this money back, and then some, before you receive any royalties.

If you're still interested in being signed to a large label for superstar status, my advice would be to seek urgent medical attention or psychiatric help. Instead, if you want to take the record label route, it's much more advisable to forget stardom and aim for the smaller labels that specialize in the genre of music you write. These are much more reliable, make fewer demands and will appreciate you for who you are.

Even taking this route, however, the days of sending 'demo-quality' CDs have long since died and they will expect the music to be professionally mixed or, at the very least, have some good production values behind it. Imagine if you owned a record label and you received an overtly loud, muffled, poorly defined mix on a CD. It would more likely end up as the office Frisbee or drink mat than a record deal, no matter how good the ideas on it may be. Therefore, if you're serious about sending a demo to a label and are not fully confident with mixing, then it should be put in the hands of a qualified engineer.

Typically a studio will charge around £50–90 per hour, and provided that the track only requires mixing and perhaps vocals, it should take around 12 hours. However, before visiting any studio you should ensure that the arrangement is complete; attempting to make it up on the spot while the studios clock is ticking can be incredibly stressful and the attitude that a further instrument could be added during studio time can prove to be an expensive endeavour. Bear in mind that a typical session musician will charge anything from £50 to £100 an hour!

Note: Note that even employing a session musician to play a tambourine can cost in excess of £50 per hour. Don't think that you can play one yourself either. As simple as it looks, it is difficult.

Most studios will record the mix direct to Digital Audio Tape (DAT), but also ask for them to put the mix onto CD. Many engineers will do this free, but if not they will only charge a couple of pounds and it is worth having more than one copy of a mix in case of accidents. Plus, demos that are delivered to a label on CD are more likely to be heard than one that arrives on DAT, as not many offices or cars have access to a DAT machine.

When it comes to posting the mix, ensure that you put your name, address and phone number on absolutely everything you send, along with a publicity shot of yourself which is in keeping with your musical direction. This way, the label can immediately tell what music you write and where you get your influences. What's more, it is also worth including a single sheet of A4 paper containing everything that you want to say about yourself and your music, but this should be kept short and to the point.

Note: The addresses of most record companies can be found printed somewhere on their latest release, or alternatively in a publication known as the White Book; this contains the names and addresses of most reputable labels.

The basic premise behind sending in a demo is to give the company an idea of your talent, not your entire songwriting history. Also, you should start by sending only two or three songs, ensuring that your best music is first. This way, and assuming that they do actually listen to your demo, the first 20 or 30 seconds are the ones which matter the most; if that doesn't grab them they may listen to 20 seconds of the second, but if that doesn't work then it's game over.

This means that the musical arrangement will also need to be closely analysed, especially for club-based dance music. As it isn't unusual for club music to consist of the drum track alone for the first 16 or 32 bars, if you're submitting to an A&R department, this should be shortened so that the main elements of the music, that would commonly start around 2–3 minutes into the track, start within the first 20 or 30 seconds. Ideally, the intro should be kept particularly short and new melodies should be introduced as soon as possible to maintain attention. That said, care does need to be taken that you don't introduce too much at once, it's a fine line that has to be trodden carefully.

You should also prepare yourself for a negative response, if you receive one that is. Larger labels very rarely respond unless they're interested in signing you, but smaller companies are more likely to either phone or write to you informing you of why they've rejected your music. Indeed, rejection is something that you will become accustomed to, unless you happen to be a musical genius, but it's important to bear in mind that it's nothing personal, they're simply business judgements that are made according to the current market status.

If you are accepted, then you may be notified by post, but more usually by phone. Naturally, you will be tempted to say yes to anything they have to offer, but remember that fools rush in. With smaller labels, this call will probably be from the director/owner of the label, but with larger companies it will almost certainly be the head of the A&R department. At this stage, do not agree to anything contractual over the phone and insist that you meet them before you agree to anything. This is important as some companies will send the contract by post and expect you to sign and return it, but without meeting the people in charge you don't have the opportunity to negotiate a better deal.

When the time comes to meet the director or head of A&R, make an effort to look presentable, reputable and trustworthy. Though you may have been offered a contract on the strength of the music alone, you haven't signed at this point and first impressions count. It's vital that you adopt a professional and businesslike attitude. These companies exist to make money and you need to prove that you're dedicated enough to be both reliable and sensible, and this also means turning up on time. The excuse of 'I missed the bus' isn't particularly concerting. They will have literally hundreds of artists lining up waiting to get on their books and if you present a poor or unreliable image they'll simply go for someone else.

While it's doubtful that you'll be asked to sign any contracts at the first meeting, it is sensible to take a solicitor who has dealt with the music industry before to negotiate the terms of a future contract. This not only helps to ensure that you turn up on time, but if the label does present a contract, the solicitor can look through it for you. As tempting as it may be to save money at this point by reading up on the law yourself, I *strongly* recommended that you employ one to check that the terms are acceptable. A typical record contract is crammed full of numerous clauses and if any escape your attention you could end up losing not only money but the rights to your current music and any future releases.

Royalties

When presented with a contract, the part you're most likely to be interested in are the royalty rates, as these dictate how much money you will earn from each sale. Most labels will offer either a publishing deal or simply choose to license the music, and the royalties you receive will depend on whether you're being offered a licensing deal or a publishing deal.

Publishing deals generally pay a much higher percentage for royalties, but in return you're giving them the rights to release any future material you produce over a specific number of years that is stipulated in the contract. This is commonly set over 5 years, but some labels are known to sign for a period of 10 depending on the past sales performance of the artist. As these types of contract are long term they should be scrutinized very carefully by a qualified music-savvy solicitor, as a missed clause on your behalf can result in a long, drawn-out legal battle against the label if they're not performing up to your expectations.

Despite the temptation of being on a label's books for a number of years, from personal experience it is not worth signing these types of deals because you will be tied to the label. Instead, if they are offered the label involved are clearly confident that your music will sell, they will be open to negotiation and they could be persuaded to sign a licensing deal instead. This is a more secure route to take, not only allowing you to assess the label's competence, but also giving you the freedom to walk away and find another label for your next release if it all goes horribly wrong.

Licensing deals are less complicated and are more common on the dance music scene. Essentially, by signing such a contract your music is licensed to the record company for a specific period of time, in which they can do with it whatever they see fit. These are most usually signed for a period of 1 year, although it can sometimes stretch to 2 or 3 depending on how popular the label thinks the track may become. It can therefore be generally accepted that the longer the time period you're expected to sign, the more potential the record company believes the track will have and this can open up some bargaining power. In fact, it is often worthwhile asking for a few changes in the contract if there is something that you're not happy with.

Note: Many labels don't fully understand their own contracts and they are occasionally only too happy to change some aspects as long as it isn't changing the percentage of the royalties you or they receive. An important aspect to look out for here is a clause stipulating that they have the right of first refusal on any other tracks you produce. This means that you could be forced by a legally binding contract to hand over any subsequent releases to the same label regardless of their previous performance.

Alongside the royalties received from a record company, you may also receive payments from the Music Publisher, PRS, MCPS, AURA, PAMRA and PPL. Some of these payments will be made directly to the record company who will pass these, along with your royalty payments, onto you or your manager, while others will pay them directly to you, depending on the circumstances.

To better understand the roles that each of these companies play and how it all ties together in the real world, we'll use an example by envisaging two artists, James and Jon. They have just finished engineering a track and featured an artist (Emma) to perform the vocals on the record. So the artists can continue to make music rather than deal with the recording industry, they've also employed a manager, who we'll call Steve.

Unless there is a specific written contract between the band members, the manager or the record company stipulating that they are all entitled to an equal share of *all* proceeds, the members may be able to claim different sums of money from different organizations, as well as the record company, depending on the circumstances. Figure 18.1 outlines how this works.

- *Record company.* The record company will pay any royalties due to you from record sales or licensing deals. If you have a manager, he/she will receive the payments and then distribute

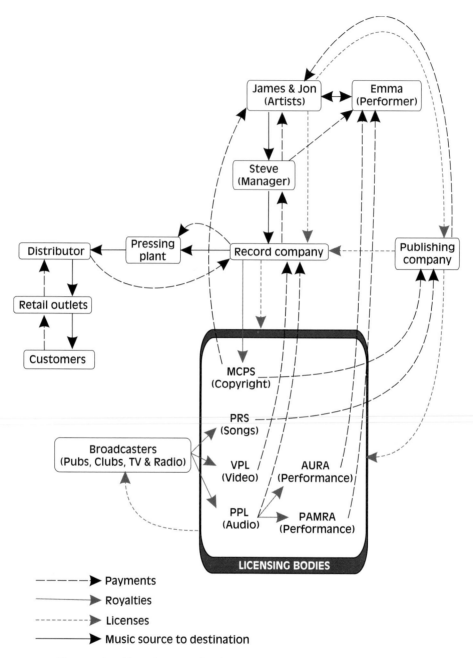

Figure 18.1 How the recording industry licenses and collects monies.

them to you and any other group members while keeping an agreed percentage for his/her managing duties.

• *Publishing company.* Publishing companies work very closely with record companies and deal with releasing the recording and tracking the national royalty systems around the world on behalf of the original author of the work. For this they take a percentage of any of the royalties, paying the rest to the original author.

- *Phonographic Performance Limited.* The PPL license records to radio stations, clubs and pubs. This is because every time the record is played to a collective public audience, whoever is playing the music must pay for the privilege of playing your music. The PPL then collects these royalties and pays it direct to the record company involved and to any UK featured artists. For this service, the PPL keeps a percentage of the royalties collected.
- *Video Performance Limited.* Similar to the aforementioned PPL, the VPL licenses the music video to be played by broadcasters (such as MTV), pubs and clubs. Any royalties collected from this are paid direct to the record company involved, and for supplying this service the VPL keeps a percentage of the royalties collected.
- *Association of United Recording Artists.* If a piece of music has been engineered by a professional producer or it features an artist (session musicians, etc.) who is not a member of your group, then AURA may become involved if the featured artist is not from the UK. These collect royalties from the PPL and pay them direct to the producer or featured artist. It is up to the producer or featured artist to join this association and, although membership is free, AURA keeps a percentage of the royalties that are collected.
- *Performing Artists' Media Rights Association.* PAMRA is similar to AURA and collects any royalties that are due to the performers. Like AURA, it is up the performing artist to join, but while membership is free, they keep a percentage of royalties for offering this service.
- *Performing Rights Society.* If the song is going to be performed in public by a third party, then the original writer of the song is due royalties every time it is performed. These royalties are collected by the PRS and paid directly to the author of the song. Like AURA and PAMRA, it is up to you to join if you wish to collect. For the rights to 15 songs, membership costs £400, with the PRS also taking a small percentage of the royalties. Joining the PRS is rare for dance musicians, since it is rarely performed by another party.
- *Mechanical Copyright Protection Society.* Once signed to a label, the record company will make a number of copies of the record for distribution and sale. These are referred to as 'mechanical copies' and all the artists involved are due 'recording royalties' based on the value of the number of copies made. The MCPS collect these royalties, which is currently 8.5% of the wholesale value, and pay them direct to you. Again, it's up to you to join this organization and costs £50 for membership, along with a percentage of the royalties collected.

Naturally, there is no universally accepted amount that you should receive from any of these companies, but especially from the record company, as it depends entirely on the size of the label, your past performance and how the label works their accounts. Most labels will operate on an 'all-in' basis, which is derived from the suggested retail or the wholesale price. An all-in royalty rate means that if a third-party producer, musician, instrumentalist or songwriter was employed during the recording process, their payment is derived from your own royalties rather than the record company's.

Whether the royalties are based around the suggested or wholesale price is also an important consideration. The suggested retail price (SRP) is usually double the wholesale price, so if you have a 40% wholesale-based royalty rate then you'll get paid the same as a 20% retail-based royalty. This percentage is quite normal for a royalty rate and if it's any less you should try to negotiate a better deal.

Packaging

One of the biggest mistakes any artist can make is asking that their single or album is released in commemorative or novelty packaging. While it is true that leather cases or steel boxes sell more copies due to their novelty value, they are expensive to manufacture and this additional cost will come out of your royalties.

In fact, though jewel cases and booklets aren't actually part of your music, they're essential for delivering it to the audience and you pay for their manufacture, not the record label. This is usually described by many labels as a 'sales allowance' and, depending on the label involved, the charge can be anywhere from 8% to 20% of your net royalties, although this can quickly rise to above 50% if you're adamant that your recording is released in novelty packaging. This allowance is also used to cover any damage that the packaging might suffer from on its way to the shops. Along with this sales allowance, most labels will also take a percentage of royalties for 'recoupables'. These are one-off expenses that are incurred by the label for promotional purposes. This includes tour support, personal advances, photo sessions, artwork, publicists, copies given away for promotion and music videos.

Today, more than ever, videos provide a huge promotional tool and the current MTV generation need videos to associate with the music. The price of making a music video varies tremendously from as little as £5000 to over £2 million and you will be expected to pay a percentage towards making it, which will be subtracted from any royalties. That said, it is very doubtful that any record label would allow you to make a video for more than £10 000 if you're a relatively new artist and, in some cases, they may insist on a video budget of less than £5000. Again, the percentage you pay towards a video will vary from company to company, but you should insist/negotiate on not paying more than 40% towards it. What's more, you should insist/negotiate that no more than 10% of your recordings are given away as promotional items. By providing promotional copies to record stores, a sales team can purchase both shelf and window space, but if you let them they'll give away a huge amount of records to secure a good space in the window and/or shelves, which ultimately results in loss of income for you.

Note: As a word of warning, all major record labels will not allow you to find your own supplier of CD cases, artwork or pressing plants, even if they are cheaper than the amount being paid by the record company (and, of course, ultimately you). Whatever the label decides to do, you do not have the opportunity to audit and, however much the label decides to spend, you'll have to pay for it from your own royalties!

More importantly, ensure and then double (and triple) check that some of the promotional copies are sent to any related magazines *before* they arrive in the shops rather than after. Most magazine reviewers will not review a record that is freely available in the shops, so if they receive them after they are available they will usually not be reviewed, reducing your potential sales. This is a surprisingly common occurrence and an apology from the label for overlooking this 'small problem' will not help flagging sales.

To prevent this from happening, many artists take this into their own hands by requesting 20 or 30 copies of the record before they go into the final manufacturing stage. Often referred to as test pressings, you can use the excuse that it allows you to listen to what the mastering engineer has done to your record and so you can hand out copies to friends and relatives. In truth, you're sending them out to magazine reviewers yourself to ensure that they receive copies to review before they're released onto the general market.

One final issue that is vital to fully comprehend is the advance payment. All record companies, no matter how big or small, should offer an advance on signing of the contract. This is not only the label's proof that they are committed to promoting your record, but also that you're paid for the work. Naturally, you shouldn't expect a sum amounting to millions like those often reported by the press; for this you need to have a long history of best sellers and a good reputation with the public, and besides these huge amounts are before tax and are always blown out of all proportion by the media.

For a first release, a label may realistically offer between £200 and £5000 depending on the size of the label and how popular they believe your music is going to be. Advances are refundable from the profits, though, which means that you will not receive any royalties until the sum total of your royalties from sales have exceeded the advance. It is worth ensuring that there are no 'hidden' clauses in respect of this advance too, by making sure that it is only recoupable from the profits. If they are not, the label may have the right to recoup the money if they don't release the track or if the record does not sell as well as they initially expected it to.

Promotion

Once the final details have been accepted by both company and artist, the promotional campaign will start and the recording will hit the shops. The first 3 months are the most crucial stage, and many record companies and shops will judge the selling potential of the record within this time period. If only a few copies are sold, it's very likely that the shop will return them to the record company and ask for a percentage of what they paid for the record back.

Most of the high street stores settle for a 25% or 50% return rate, but many of the smaller independent shops will demand a 100% return if they can't sell it. Consequently, whenever a record is released all record companies set up what's known as a reserve royalty account. This allows the label to withhold anywhere from 40–80% of your net earnings and hand them out over a 2-year period, preventing them from making a loss by paying you too much.

This can cause some substantial problems if you've signed more than one record or a publishing deal, since they will view it as one financial deal rather than a number of individual ones. Thus, if your first release is a failure but the second or third is a hit, they are within their rights to recoup any losses from the failures by taking a larger percentage from the hit. In a worst case scenario, this could result in you earning nothing whatsoever.

Ultimately, here we've only looked at the most common pitfalls and recording contracts can contain many more clauses. They are a tedious read but it is vital that you go over every aspect with a good solicitor, because if you make a mistake now it will stay with you for the rest of your recording career. Also, as you become more successful it is advisable to hire a good manager to take care of everything for you, so you can continue making music rather than dealing with all the paperwork. Until you have a manager, it's almost certainly worth joining the Musicians' Union, as they can offer solid advice relating to both contracts and publishing deals. They can be contacted at www.musiciansunion.org.uk.

Note: The Mechanical Copyright Protection Service can be contacted at www.mcps.co.uk
The Performing Rights Society can be contacted at www.prs.co.uk
The Phonographic Performance Limited can be contacted at www.ppluk.com
The Performing Artists Media Rights Association can be contacted at www.pamra.org.uk
The Association of United Recording Artists can be contacted at www.aurauk.com

Independent release

While sending off a demo to a record label is still viewed by many as the best way to get your recording released onto the general market, actually being accepted by a label is becoming increasingly difficult. Indeed, many labels will not entertain the idea of signing an artist if they have no previous history of releases, and even if they are willing to sign you, they pretty much have all the bargaining power, giving them the freedom to dictate much of the contract you'll be expected to sign. Consequently, an increasing number of musicians are releasing their singles independently. This does mean that it's you who has to do all the legwork involved, from sorting out the artwork and distribution along with incurring all the costs, but on the plus side you aren't tied down to a contract and there are no record company executives to answer to.

Releasing your own music isn't as complicated or difficult as many are led to believe, and although it does take some financial capital to get it off the ground, any earnings are yours to keep. Additionally, if it does well, it also means that you have a history of a release, giving you some bargaining power if you ever decide to sign to a record label.

Over recent years, CDs have invariably become the ultimate format on which to record and distribute music. They're cheap and easily available, but this also happens to be the problem. With the explosion of the DJ culture, vinyl turntables are the definitive system to play dance music and very few clubs or DJs will take a dance mix that arrives on a CD seriously. Apart from the fact that it doesn't particularly show any real commitment to the 'scene', vinyl reproduces the lower frequencies much better than a CD. This obviously means that if the music is bass driven, as dance music is, vinyl is always going to be the number one choice for dance.

This does, however, pose some problems if you've mixed down with CD in mind, because there's a tendency to overcompensate to create a deep bass. If this is then pressed to vinyl, the bass can be so loud that the stylus can skip along the record. What's more, the louder the bass the wider the groove needs to be on the record, and this can force you to reduce the overall length of the track.

> Note: Broadly speaking, a typical dance record with a heavy bass should be no longer than 10 minutes at 33 RPM or 8 minutes at 45 RPM. It is possible to go a minute or so longer than this at both speeds, but it may result in a drop of volume or, in some cases, distortion in the high-frequency elements.

Also, consider that at the beginning of a record the outer circumference of the groove is around 36 inches, but as it continues towards the centre it continually reduces in size until it reaches around 13 inches. This means that as the record progresses there is less and less space to hold all the frequencies, so the audio quality will reduce, commonly introducing distortion in the higher frequencies. Consequently, it's sensible to avoid boosting any frequencies above 8 kHz while also rolling off everything above 16 kHz. It's also worth running any high-frequency sounds through a good de-esser while mixing for extra security. Both these techniques have to be applied cautiously, though, as once the mix has been committed to vinyl there's no going back if it's wrong. Therefore, I strongly advise that if you want to submit a mix to vinyl, you send it to a professional mastering house.

Although we've already covered the general principles behind mastering in the previous chapter, without a doubt there is absolutely no substitute for employing a professional mastering engineer. Even though there are numerous programs available that promise to produce 'professional' results, which some can, none of them replace a mastering engineer's experience.

These guys have years of experience behind them and will not only give a different perspective on your music which you may not have considered, but also make sure it's mixed correctly for vinyl. On top of this, there may be no need to adjust a mix that's been prepared for a CD, it can just be sent to the mastering house as is. That said, it is worth checking with the mastering house before you send any material, as they may need the CD preparing in a particular way. Also, you should include any important details such as track numbers, the length of each song in minutes and the total length of a side including the pauses between the tracks, provided there is more than one.

Most mastering houses today will accept CDs in data form, often called PMCD (Pre-Master Compact Disc), but they will also accept Exabyte, U-matic 1630 and perhaps Digital Audio Tape. This latter medium is becoming less popular as a recording standard, as many reputable mastering engineers do not consider this to be of a high enough standard. This is because DAT can suffer from drop-outs and other tape-related problems that can make the mastering engineer's job even more difficult, which can result in one of two things:

1 They'll charge you a hefty sum of money to 'repair' the problem on-site.
2 They'll charge you for the time they've spent listening to the DAT before deciding that they have to return it to you.

The prices mastering houses charge vary from plant to plant, but at the time of writing a typical price for mastering and preparing two tracks (remember there's usually a B-side!) is around £250–400 depending on the reputation of the mastering engineer and house involved. On the subject of a B-side to the record, it's a wise idea to use this for the instrumental version and include plenty of exposed segments consisting of just the drum loops, along with instrumental riffs and vocals. This way you open up more options for the record:

- If the A-side does incredibly well it will not only increase the demand for it, but it also allows any producers to remix the record to promote it further.
- If the A-side does poorly, a producer or well-known DJ may decide to remix it using the elements included on the B-side, increasing its chances of becoming a hit record. If it does well, a label will look to sign it and you'll receive royalties for every sale – a very profitable venture.
- A producer may drop different vocals over the top of the B-side which could promote the record further.

Once the track has been mastered, the plant will often return a copy of the master back to you in a format that you request, allowing you to verify the quality, but this copy should not be sent to the pressing plant. The mastering engineer will, or at least should, have recorded the mastered version onto Exabyte or U-matic 1630, which is then sent directly from the mastering house to the pressing plant to produce the record. This is because the quality required to produce the 'glass master' must be of an excellent quality and CD or DAT are not of a high enough standard. This does, of course, depend on the pressing plant involved and some will accept these lesser quality formats simply to get their hands on your money.

Note: Glass masters are so called because they are destroyed during the production process. After the recording has been copied onto the glass master it can be transferred onto a stamper, a round metal form that is then used to process the lacquer to produce the final record. This glass/stamper/lacquer process generally costs around £65 per side, but as always the prices will vary from company to company.

Once this is completed the plant will press a small amount of records known as test pressings or TPS to give you the idea of how the finished product will sound. Again, these are not free and you'll be expected to pay anything from £2.00 to £4.00 per 'test pressing'. Many plants will refuse to make just one TPS, insisting that you have a minimum of 10. Nevertheless, with these test pressings, you can then begin to look towards distribution deals, publishing deals and, more importantly, registering yourself with the Mechanical Copyright Protection Society.

Since you're looking to press and release a record, you're becoming a small record label, so you should register with the MCPS for a number of reasons. Firstly, you'll need to check that no one else has used the name that you want to use. This mistake has been made many times, with the most notable example being made by the Chemical Brothers when they decided to change their name to the Dust Brothers. Rather than check with the MCPS first they simply used the name, which understandably resulted in the 'original' Dust Brothers being a little more than upset.

Secondly, the MCPS will ensure that your record is licensed and protected so that if it becomes a hit, you'll receive mechanical royalties. For providing this service, they are usually paid 8.5% of the total income from every unit sold. You do not have to pay this if you press and release less than 500 records (provided that you reach an agreement with them first) or for records that are used for promotion. These promotional copies must, however, be prominently marked with a non-removable, non-erasable notice carrying the words '**PROMOTIONAL COPY – NOT FOR SALE**' on the sleeve *and* packaging *and* the record itself – you cannot simply just attach this label to the dust jacket. Plus, of course, they must be supplied free to the DJ, reviewer or club!

After registering, you can begin to look towards getting a distribution deal using the 10 or so test pressings you've received. The reason behind looking for a distributor now rather than when all the records have been pressed is that you may be able to secure a pressing and distribution deal. This will depend on the quality of the record; if your chosen distributor believes that it has a good sales potential they may offer to pay 50% towards the price of pressing. This investment will obviously be recouped by taking a percentage of profits from every sale, along with their distribution percentage and further percentage for organizing the manufacturing. Though this probably sounds like a poor deal for you it is the best deal to make, as the distribution company will be heavily motivated to sell your records so that they can recoup their expenditure. Additionally, they will also be obliged to sort out the bar-codes that help record shops detail their sales, something that is very difficult to organize yourself.

Many distributors may also want you to sign a contract giving them exclusive rights to distribute your record for a minimum of 1 year. Although a relatively small press of 2000 records may not sell over this time period, if they do, you will be tied to the same distributor until the end of the contract. This means that you should take your time to find a good one.

Finding a good distributor can make the difference between your record reaching the best independent DJ and high street shops or ending up with a backstreet dealer frequented by 'wedding DJs'. Consequently, it is worth shopping around, phoning the companies and asking who exactly they deal with, and if they have any previous history of distributing 'hit' dance records.

Typical questions to ask are:

- Who do they distribute to?
- Have they ever distributed a hit record?
- Do they have plenty of contacts?
- Are they in contact with any radio stations, producers or DJs?
- Are they contactable 24 hours a day?
- What publicity plans, if any, do they have?
- If you let them distribute the record, when would they plan to release it?

This latter question is important in establishing whether they have their finger on the pulse of the music industry. Obviously, you'll want to release your record as soon as is possible, but if they plan to release it during Christmas or at the same time as one of their other, more popular, artists, then it will be lost in the flood of releases or not receive their utmost attention.

Enthusiasm, faith and an understanding of the music you produce are the most important qualities of any good distributor, and this is something you should be able to ascertain after speaking with them for 10 or so minutes. Remember that it's your future being placed in their hands, so ask them to inform you of their release schedules and, assuming that they have heard your recording, ask what they think of it. A good distributor will not be obsequious but offer solid advice on the selling points of your music and let you know of any ideas they have that may increase its sales potential. The price for all this commitment is usually around 20–30% of your earnings, but don't underestimate the power of negotiation. If you adopt a diplomatic attitude you may be able to reduce this to 15% or 25% respectively. Whatever happens, though, it's unwise to sign any contracts allowing them to take over 30% of the income.

> Note: If you do not want to become involved with a distributor, preferring to handle it all yourself, then you can use a vinyl broker instead. These companies are solely concerned with just pressing the vinyl and configuring the bar-codes and, as they have a good relationship with pressing plants, they can usually get records pressed cheaper than if you approached a plant yourself.

Assuming that you have a distribution deal or have decided to approach it all yourself and have approved the quality of the test pressings, the plant will continue to process the amount of records that have been ordered. Generally, 130-gram lacquer is the most commonly used for 12-inch records, but some plants may offer 180-gram. This grade produces a heavier bass, a sharper high end, a clearer stereo image and suffers from less distortion, but it costs approximately twice as much. It is possible to get very good quality from 130-gram, but you stand a better chance of quality if the vinyl is virgin and not recycled. A proportionate amount of pressing plants will use recycled lacquer unless you state otherwise, so it's always advisable to ask before you commit yourself.

> Note: Virgin lacquer may be a couple of pence more expensive per unit than using recycled, but the difference in audio quality is worth the extra expenditure – with pressing you get what you pay for.

A typical price for pressing a single 12-inch, 130-gram record is around 60 pence, but very few plants will press less than 500 and on average many will not press less than 1000 copies at a time. Although many unestablished artists would prefer to press an initial run of 500 records, the less that are pressed, the more expensive it is and, ideally, you should look to earn enough from pressing records to finance pressing your next release. This means that you're generally better off making a minimum press run of 1000 or, if you can afford it, a run of 2000.

Another option that may be open to you is the choice of coloured vinyl rather than black. It's a fact that coloured vinyl sells more copies, but it is considerably more expensive. Blue, red and yellow vinyl can cost as much as £1.50 per unit, while white vinyl or a customized colour can cost in excess of £3.00 per unit. Bear in mind, though, that the choice of colour can govern how many records you should press. Most pressing plants will reduce the price of each unit with the more records you order, so while it is recommended that you press over 1000 black records so you can recoup the money you're investing, if a different colour is used you should look at pressing more than 2000 to make them cheaper.

Artwork

The very name white label should give some clue as to how much artwork is involved on the record's sleeve. A majority of white labels have little more than the name of the artist and a contact number or, more commonly, for a record that contains an illegal sample, an email address. While I cannot condone the use of illegal samples in any record, a proportionate amount of white labels that have made their way into clubs and onto the charts have contained illegal samples, and by only including an email address at least you have the option of not replying if a record company gets in touch to press charges against you. That said, in a majority of cases if the recording is doing exceptionally well and the label who own the original sample believes that it has a big sales potential, they may wish to release it commercially, but be warned, it gives them the upper hand when it comes to negotiating a deal.

Many pressing plants will offer to design and print the artwork on your behalf, but this obviously costs extra and it's usually a better idea to actually supply the plant with your own artwork. As long as it's quite simple, the plant will only charge you a small 'cover preparation' or 'film charge', which is used to prepare the printers to create the labels and then charge 3 or 4 pence per label, provided there are no colours involved. Alternatively, if money is tight, a cheaper option is to leave the labels from the plant blank, design your own, print them out and then stick them on the sleeve yourself – if you have the patience to stick labels on over 1000 records. Remember that time is money to everyone, including you, so calculate the time it will take to prepare the labels and balance it against what the pressing plant may charge – it may actually work out cheaper to use the pressing plant!

Note: While a majority of white labels contain little more than the name of the track and some contact details, a trick employed by some musicians is to design a label that not only contains these details, but also gives the impression that it's an American import. The egos of many DJs should not be underestimated and the elitism of owning an 'import' can immediately raise the hype, so that DJs are more interested in playing it on the dance floor.

Most plants will also give you the option between an inner or outer sleeve. An inner sleeve is the thin 'tracing paper' type cover you often find inside normal record sleeves to protect the vinyl from being scratched by the outer sleeve, and these cost around 6 pence each. The outer sleeve, often referred to as a 'dust jacket' or 'disco bag', is constructed from cardboard; thus, it is more substantial and normally comes in black only. Obviously, this means that they are more expensive, costing between 15 and 20 pence each, but they do offer more significant protection for the vinyl and are preferred by many DJs for transporting their records.

Thus, assuming that you decide to press 2000 copies of your recording and you have no manufacturing deals, the price so far would be:

- Mastering – £200 × 2 = £400.00.
- Lacquer – £65 × 2 = £130.00.
- TPS – £4.00 × 10 = £40.00.
- Film charge – £8.00.
- Labels – 4 pence × 2000 = £80.00.
- Dust jacket – 20 pence × 2000 = £400.00.
- Pressing – 60 pence × 2000 = £1200.00.
- **Total: £2258.00**.

Once you've received the completed order of vinyl, you'll need to look into promoting the single before it hits the record shops. Simply releasing it without any publicity to back it up isn't going to help it sell, as few people will walk into a record shop and buy a track without knowing anything about it. There are two options available for this: you can either employ a press company to deal with it all or you can attempt to create some press yourself.

In an ideal situation you would employ a press company. These people, who often work in league with distributors, have all the contacts needed to publicize your record and will get your name in every magazine or paper that really matters. To accomplish this, they will ask for a minimum of 100 free copies of the record so that they can send them out to the people that are important – editors and journalists – who will, hopefully, give you rave reviews, increasing your sales potential when the record finally hits the shops. If you do decide to take this route, then finding a good publisher is as vital as finding a good distributor, and it is worth shopping around to find a company that deals in the genre of music you create.

In finding a good publishing company, typical questions to ask are:

- Who have they dealt with in the past?
- Which magazines or papers are they in contact with?
- What are they prepared to do to publicize your record?
- Will they take out adverts in music papers?

The latter question should result in a resounding no if the press agency is reputable, as unless you're an established artist adverts have been proven to be a waste of time and money. Employing a publishing company obviously isn't free and, generally, it can cost between £500 and £1000.

Whether you decide to use a press company or not, you should *seriously* consider employing a DJ promotion agency. These form an essential aspect of dance music promotion and without them it is doubtful that any record will ever get of the ground. Essentially, they have various lists of clubs and well-known DJs that are suited towards the genre of your record and will mail it directly to the ones that matter. Both clubs and DJs get the records for nothing, which means that you have to supply them for nothing too, but in return they fill out reaction sheets that report how each record is going down on the dance floor. This gives you a good idea of how well the track is doing on the circuit, allowing you to estimate its popularity.

Obviously, this means that you should go with established agencies that have the addresses of the best clubs and DJs for the type of music you produce; you don't want your records ending up in the hands of wedding and party DJs. Agencies such as *Waxworks* or *Rush* are both well respected (and whom I fully recommend), as both these have large lists and will advise you on how many records you should supply them. Obviously, this service is not free and they charge £2–3 per record and have a minimum mail out of 100. They will also expect payment upfront rather than after they've been mailed – just in case you go 'out of business'.

> Note: It certainly isn't advisable to miss this stage out in the hope of saving a little money. Hanging around in clubs in the hope of catching DJs and giving them a copy is a time-consuming affair, time that could be put to better use writing another record. Let the professionals handle the promotional aspects.

This brings the price so far to:

- 2000 130-gram black vinyl records, with sleeve and labels – £2258.00.
- DJ promo agency – £300 and 100 records.
- **Total: £2558.00 and 1900 records**.

On average, most record shops will pay approximately £2.00 for a 12-inch single, so provided that you could sell them all, it would equal a profit of £1442.00. Alternatively, if you only pressed 1000 records, the profit would be considerably lower, equalling around £382.00. However, while the latter still provides a profit, it isn't much to put towards a second release and also remember that many record shops operate on a no-sale return policy. This means that if they can't sell your record, they'll want their money back!.

> Note: As an extra precaution against losing money, a number of artists also send their labels to overseas distributors, effectively licensing the record to them. The benefit of this approach is twofold: it provides a larger potential audience and they'll press any additional recordings themselves. On some occasions, they may also be willing to pay you a licence fee as an advance of the records, but if the record doesn't sell many may well expect this back.

In an ideal world the record would fly out of the shops and the money you make can go towards releasing the next single until ultimately you have a number one hit, you start your own record company and start signing other artists or another label offers you a sub-licensing label deal. This has happened with many small independent labels, where a larger record company such as Sony or Virgin help not only to finance your work, but also use their marketing influence to reach a much wider audience, in return for a slice of your profits, of course. This kind of affiliation is more common than many realize and a proportionate amount of 'independent' labels depend heavily on the backing of large companies. So, if the premise that the end of large labels is on the horizon from the evolution of self-promotion on the Internet and illegal MP3 downloads, the smaller labels will fold first!

Some of the main distributors in the UK for dance are:

- Intergroove – 020 8838 2000.
- 3MV – 020 7378 8866.
- Kudos – 020 7482 4555.
- Pinnacle – 01689 870622.

Distributing via the Internet

While white labels invariably provide the most promising way of getting your track noticed, especially if you produce dance, if money is tight then Internet promotion may prove a more viable option. Indeed, the lines between web designer and musician are drawing closer, as more and more musicians are developing their own web sites and nearly all have a web site, even if its only purpose is to keep fans informed of when and where the next gigs are.

Designing a web site is a professional business and as it's essentially acting as the face of your music it must be well developed, easy to navigate and exude a professional feel. This latter aspect is especially vital if you plan to charge users to download the music – if the site looks cheap and tacky, it isn't going to instil a lot of confidence in the buyer.

Note: Developing a web site is beyond my area of expertise, but James Froggatt, a professional web designer and musician, has kindly donated a chapter on how to create a professional site, which can be located at the end of the book (Chapter 20).

Once a web site has been created, it has to be promoted. The first step towards this is to register a domain name, and this should be kept short, sweet and memorable. An IP address that uses long, drawn-out page names or squiggles is not only difficult to remember but a pain to type in and is considered tawdry by many web users. More importantly, do not use banners or pop-ups on your own site. These are often voted as the most annoying part of the web and many users avoid any sites that feature them like the plague. Similarly, it is not worthwhile paying for a web site to advertise yours with a pop-up window, as a proportionate amount of knowledgeable Internet users will employ software that prevents them from appearing.

Once you have a domain name, it can be submitted to a search engine. This is where meta-tags (text that is contained in the header of a web site) can help, as some search engines look for these while they're searching. Additionally, a number of search engines will allow you to add your URL to their engine by clicking on the appropriate link, but it is also advisable to purchase software that is specifically designed to submit URLs to most search engines. Wolf Submit Pro is considered by many as the best software for this, but there are various other recommended software packages such as Addme.

Of course, there's much more to registering a domain name and simply placing it in a search engine, because unless you already have a fan base no one is going to be able to locate the site. Simply typing in MP3, independent bands or dance into a search engine isn't going to result in just your site being displayed, there will be thousands to choose from. To prevent this it's a much better idea to create a site dedicated to a subject that your potential fans might be interested in, or something a little more fun, such as Flash script-based games with your music playing in the background. Alongside this, you could place the band history and full MP3 downloads on a separate page, with a link to them from the main web site. This way, if anyone finds the music interesting they can jump to the 'real' page.

Note: If you take this approach, remember that not everyone has access to a broadband connection, so don't attempt to place a full track in the background, just choose the best snippets and loop them. As a general rule of thumb, it's advisable that any web page should load faster than you can hold your breath, using a typical V90 56 K modem as a reference speed.

Furthermore, try typing in some keywords that relate to your genre of music and find out who the top bands are. Once you have their address, contact them and ask if they would be willing to exchange links with you. Similarly, join or start your own webring. This is a group of sites all dedicated to the same subject that are interconnected together with links. Essentially, this forms a virtual chain, allowing a visitor to go forwards or backwards through it, visiting each site in turn. You can find out which webrings deal with your genre of music, or start your own by visiting www.webring.org.

Notably, it isn't advisable to set up a new webring if there is one already covering your interests, and most webrings will insist that a number of factors are met before you can join. Consider translating the web site and the vocals in the music, if there are any, into different languages such as French and German too. This opens up the site to more visitors, which will hopefully lead to more clicks on your site, and by translating the songs to different languages, there is a better chance of them being played on a foreign radio station. For instance, many stations in France and Germany insist that a certain percentage of the music they play is in the native tongue.

Note: An interesting new development that may also prove useful is to check out recommend-it.com. By registering you can include a button on your site that allows any visitors to automatically send an email to their friends, letting them know how good the site is. To top it off, by doing so they're automatically entered into a competition to win prizes.

Although all these methods will go towards promoting your site and music, it isn't a good idea to charge visitors to download any music unless you are well established. Though offering small samples of the music before charging to download the full track may seem sensible, it will easily deter potential customers. A much better approach is to allow users to download all of your music for free, but ask them for contributions if they do. Most people will take anything if it's offered free and not only does it allow them to evaluate all of your music, if they like it they'll tell friends about it, which helps spread the music. As an additional incentive, you could create a member's only area for those that make contributions, giving them inside info on the band, access to 'limited edition' tracks or the chance to remix one of your tracks. This latter approach can obviously be beneficial to both parties. Above all, remember that when starting out, promotion is substantially more important than royalties from sales, and the more promotion you receive the bigger the chance that a label may snap you up.

To further promote your music it is also worth seriously considering signing to an Internet label (iLabel). These not only have much more financial clout and promotional influence, but some also pay royalties based around how many times your music is downloaded or listened to. This, however, can be fraught with complications similar to when signing with a normal record label, so it is sensible to shop around. What to look for depends on what you want to gain from the signing, but there are some things you should look out for.

First and foremost, find out how long the iLabel has been online and how popular it is. Sites such as Peoplesound.com and MP3.com have been around for years and are well known throughout the music population through their advertising in magazines. The label needs to be both big and professional enough to attract visitors for you to sell your music. Naturally, registering with any iLabel is useless if no one can find your music, so it's an idea to look for one that features plenty of genre categories. A web site that encapsulates every genre of dance music under one name in the search isn't going to help those looking for a specific style of music. Using narrower categories will help to funnel potential listeners towards your song. Also, ensure that the site categorizes these genres as accurately as possible. After sitting through a download, you expect it to at least be the right genre.

Another important consideration is what, if any, software do you need to access and download from the site. A few iLabels use their own proprietary download system, resulting in a visitor having to download and install software before they can even access the music. Others will not let you download a small snippet of music without requiring you to fill in various forms first, while some put stringent requirements on the version of browser that's needed to access the site and others are so laden with graphics that those with a slow Internet connection have to wait 5 minutes before they can even begin to browse the site. Any of these can deter potential customers. Many visitors will simply be browsing and only those who are incredibly patient or enthusiastic will put up with having to mess around to access the music, and even their patience can wear thin if it's all too convoluted.

On this same subject, the overall design of the site is important. It's a safe bet that they have all been programmed by professional designers, but some are complex to navigate and finding a specific genre of music can be a drawn-out experience, almost bordering on torturous. Ideally, they should be quick to navigate and you should be able to find the genre within a minute of entering the site. More crucially, the site should also reflect the style of your music. If you're producing music that's considered 'cool', you need to register with a site that projects a cool image. Moreover, ensure that the web site vets all music before it can be posted to the site. All successful record companies only sign and release music that they think will sell and make an effort to ensure that they are professionally recorded and mixed. The Internet doesn't have any of these filters and there can be nothing more annoying than spending time downloading an

MP3 that promised to be good but turned out to be self-indulgent rubbish. To avoid this and provide some security of quality, iLabels, such as Peoplesound.com, vet the recordings first and so stand a larger likelihood of being visited.

If you already own a web site then its worth checking that the iLabel will allow you to place a link to your own web site. Some iLabel sites only allocate a single page to each artist and it is difficult, if not impossible, to fit all the information about yourself on one page. By including a link to your own page, any potential customers can jump to it to find out more about you. If you don't already own a site and have no plans to start one, then look for an iLabel that offers promotional pages that permit you to advertise upcoming gigs, new album releases or merchandise. What's more, some iLabels will even offer web space with an address that you can use and these may be the best option if you have no home page.

While in a few cases this is all supplied free, many iLabels will charge you for the services they provide by taking a percentage of any royalties you are due. The amount varies wildly from site to site, with large sites that have hundreds of visitors every day taking a higher percentage than one that only attracts a couple of hundred a week. This begs the question as to whether you will make more money signing to a large site that pays fewer royalties that is frequented by lots of visitors against a site that is not as popular but pays a higher royalty rate. Obviously this is something you'll have to consider carefully, but a few factors that could help you decide are the terms of the iLabel's contract.

Most reputable sites will not ask for exclusive rights, but signing to those that do means that you can't post the music to any other sites, thus limiting the potential market. On top of this, if a major record label discovers you, the iLabel may not let you terminate the contract with them until it has run its course. Few sites impose these kinds of restrictions, but it is worth examining all the contractual clauses before you agree. Finally, consider that some sites will not send you a cheque for royalties until they have grossed above a certain amount, while others will send them religiously every month. Both these methods are a potential double-edged sword, as if you're paid only when the royalties have grossed over a specific total it could be a while before you see any return, but if they're sent religiously every month you may be cashing cheques for only a few pounds.

Above all, although promotion forms a vital part in getting your music heard, a much more important factor is the quality of the music you produce. If it's good enough, you can guarantee that people will hear about it and a number of artists have been signed to labels through an MP3 demo alone.

Music charts and play lists (UK only)

The ultimate promotional tool for any artist is being placed on BBC Radio One's play list and this is the holy grail of most record labels. In fact, labels want their artists on this list so much that the producers of Radio One are wined and dined continually.

These producers meet at the beginning of every week and listen to all the latest releases to decide which make the A, B and C play lists for that week. If a single makes it onto the A list then it's played three or four times a day, every day. If it makes the B list then they recommend that it should be played once a day and if it's on the C list it's entirely up to the DJ hosting the show. Once you're on any of these lists you stand a much better chance of making it to the shops that go towards making up the official UK music charts.

The official music chart is compiled by a market research group known as Gallup. The figures they use to compile these charts are obtained from the sales records from just under 1000 shops spread across the country. When anyone buys a record from one of these shops the sale is registered and stored on a computer. Every Saturday at the close of business, Gallup chooses a random 300 of these chart return shops and from them compiles the charts for radio play on a Sunday evening.

19 Remixing and sample clearance

Everybody is influenced by someone else. Thus, while complete originality is impossible individual style is not.

Leonard Bernstein

There can be little doubt that remixing plays a major role in much of today's music industry. It is now regularly used as a viable form of promotion even if the original record isn't dance, yet while many artists still scorn it as 'untalented theft' or the 'plague of today's music scene', remixing has played a vital role throughout music history, since we've all been inspired by previous releases. Even classical musicians such as Mozart, Bach and Beethoven could be seen as remixers, since it wasn't considered unusual for classical composers to borrow entire sections of another's composition and rework it into their own style.

Although many may view this as stealing, it's a natural progression that has formed the basis of all today's music. Artists have always derived the foundation of their own music from what has gone before: disco formed the foundations of house music, rap formed the basis of lo-fi, and trance borrowed heavily from techno and house. Even pop music has encapsulated the most fashionable pieces of every genre, creating a hybrid of all previous musical styles, resulting in what it is today.

The roots of remixing in the context of dance music, however, can be traced back to 1976, when a New York DJ, Walter Gibbons, altered Double Exposure's single *Ten Percent*. He blended the track into a disco fusion, which went on to become the first ever commercial 12-inch vinyl single, something that to many was seen as the first ever modern remix. Building on this foundation, when the first electronic drum machines and samplers appeared on the scene, DJs such as Kevin Saunderson and Ron Hardy began to chop up vocal and instrumental passages, mixing them with four-to-the-floor loops, manipulating them into something more danceable.

From these humble beginnings remixing has since become an industry all unto itself, with DJs such as Armand Van Helden, Joey Negro, David Morales, Timo Maas and Grammy winners (who established a remixer category in 1997) Deep Dish now worshipped as club legends. Remixing is also beneficial to everyone involved as the author, record company and remixer will profit from it. The author and their record company are entitled to some of the proceeds while the remixer has the opportunity to produce a potential club anthem from another artist's palette.

Having someone else ideas as a starting point may leave the impression that remixing is easy but it's far from straightforward. There's much more to it than simply rearranging the supplied parts in a sequencer or programming a new rhythm section. In fact, transforming any song into the next club anthem is a definite skill that requires years of practice. Since it commonly involves making the music suitable for the dance floor, it requires plenty of familiarity with the dance music movement. Plus, since you're essentially turning a 4-minute pop mix into a massive 8-minute club hit, it also requires its own style of production and arrangement.

Session tapes

The beginnings of any remix start with the source material which, in the majority of cases, is resourced from copies of original session tapes recorded by the studio involved in producing the original track. This is obviously the best approach since having the individual tracks allows contortion of the parts without having to worry about finding exposed parts of the original that you can 'borrow' and utilize in your own mix.

Today, nearly all producers keep the original samples from a recording and, provided you have some charm and a little luck, you may be able to get your hands on the samples from the studios, along with written permission to remix the material from the record company involved. Even with this written permission, however, the studios may charge you to rent the session tapes from them, or simply refuse to let you have them at all.

This isn't as unlikely as it probably sounds, since they may not trust you with the original tapes, the original artist may implicitly refuse to release them, the producer may not want to let them go, if it's an old recording they may have been lost, or they could have been destroyed in a studio fire. Although this latter occurrence is rare, with all the electrical equipment contained within a professional studio it's certainly not exceptional. Deep Purple's *Smoke on Water* was conceived and written while they watched the smoke from their burning studio travel across the lake next to it. If, however, the session tapes are available then you can expect to receive the original recording on one of four formats, outlined below.

DASH

An acronym for Digital Audio Stationary Head, these use two formats, either quarter-inch tape that uses eight digital tracks to record a stereo signal or half-inch tape capable of recording up to 48 digital tracks. Price-wise they can cost as much as a three-bedroom house in the most expensive area of your town and therefore it's unlikely that you would receive a recorded session on this format. That's not to say it doesn't happen, though, and some remixers (including myself) have received this format and usually end up taking them to a professional studio to have them copied onto a more acceptable medium.

PCM

PCM derives its name from the method it uses to record, Pulse Code Modulation. This form of recording is also employed in the A/D converters in most of today's computer soundcards. PCM machines use three-quarter-inch video cassettes to record, but as the machines themselves are so exceptionally expensive to purchase, few studios use them for recording and so it is unlikely (although not improbable) that you'll receive it on this format.

DAT

DAT recorders are perhaps the most common recording method employed today, and every professional studio will have access to a number of them. They record on a very narrow, slow-moving tape and achieve a 48 kHz bandwidth by using two heads on a rotating cylinder. The tape, when 'loaded', is always contacting the head even while using fast forward or rewind, and so allows you to 'scrub' (forward and rewind the audio while hearing it) to locate parts that you want to use quickly. This is almost certainly the format you'll receive the remix parts on.

ADAT

Alesis Digital Audio Tape works on a similar principle to DAT machines, although they can record eight multi-track channels using SVHS video cassettes. This makes it the most popular format for studio recording, as it not only offers digital quality but its costs significantly less than an equivalent analogue machine and allows individual instruments to be stored in separate tracks onto a single tape for mixdown. Moreover, ADAT can also be connected digitally to a computer for editing using the ADAT optical interface, occasionally referred to as Light-Pipe. This is another format you'll likely receive as a remixer. Some labels will transfer from ADAT onto CD for those who do not own an ADAT machine, but they may charge for this service.

Bootlegs and Internet remixing

If you're just starting out then permission is going to be incredibly difficult to obtain, as most large labels are not willing to let little-known artists remix their artists; for that, they want big name remixers. Consequently, many aspiring remixers choose to illegally remix a track in the hope that, upon a small unofficial vinyl press and release to club DJs, it will become a massive club hit and a record company will search them out. These illegal remixes, known by the industry as 'dubs' or 'bootlegs', are often made by creatively sampling the original stereo recording.

Note: A proportionate amount of remixes that have done particularly well during the past few years have been dubs, since the general agreement is that it's better to 'ask for forgiveness than beg for consent'. If permission is refused before the track is even constructed, it could result in a potential floor-filler disappearing forever.

For example, Richard 'X' created an instant hit when he mixed a certain old soul tune with Gary Numan's *Are Friends Electric?* to create *Freak Like Me* for the lovely Sugarbabes. This type of remixing is becoming increasingly popular, since it doesn't require you to be a musician, you just need a keen ear and a little luck. In fact, many top DJs are renowned for spending weeks slicing and dicing yesteryear's popular tracks, removing the drums and other elements and replacing them with more 'club' style timbres to create special versions of tracks for their sets. These are often created with just the original track, a laptop running a sequencer and/or wave editor, and a few VSTi synthesizers and samplers.

While it would, of course, be irresponsible to encourage these illegal activities, some larger labels will request a dub remix to gain an idea of what you intend to do before they will release the session tapes, so it's only practical to discuss the techniques used.

At the time of writing there are no processors, hardware or software that can reliably remove a single instrument or vocal from a complete mix, so many dub remixes are constructed from a mixture of exposed parts (commonly at the intro and middle eight) and hard-line equalizing to home in and, hopefully, expose specific instruments.

The familiar parametric EQ found on most mixing desks isn't suitable for this form of aggressive tonal cutting, though, requiring the use of specialist processors such as the *Electrix EQ Killer*. This is an extremely aggressive three-band EQ unit, allowing you to pinpoint and extract specific instruments. It doesn't provide the perfect solution, however, since its performance at stripping frequencies from a mix depends on the type of music you're working with and isn't suitable for extracting vocals. Rather, for these, it's more prudent to search record stores or the Internet and file-sharing programs such as Kazaa for 'a cappellas'. These sometimes appear on the B-side of a record and are a vocal-only mix which can be gratuitously sampled, contorted and rearranged to sit on a new arrangement alongside the exposed instruments from the original mix.

> Note: Please note that downloading or copying a cappellas is illegal, and something I cannot condone – you should purchase the original record!

If this sounds too much like hard work, or you prefer to stay on the right side of the law, the Internet provides plenty of downloadable session tapes from smaller labels and lesser known artists simply for the sake of promotion. Often referred to as 'sample packs' and usually in MP3 format, they allow you to download and remix their work, which they can later judge and often release if it's first-rate.

However you choose to acquire the original parts, the most important aspect of starting any remix project is deciding which parts should remain and which should be discarded. Naturally, this judgement is most likely already made for you if you're creating a dub remix, as there will only be certain elements exposed in the mix that you can use. Alternatively, if you're approaching a remix using the original multi-track recordings you have to make a conscious decision. Nonetheless, both will require you to leave just enough of the original to deem it a remix, while at the same time not including too much of the original so that it simply becomes a trivial variation.

Fundamentally, a great remix retains the key elements of the original that are mixed with your own original twist to suit the dance floor. Consequently, the first step is to understand what the key elements of the original are and this means you should be prepared to listen to the original persistently until you can pinpoint the focal points of the music.

Usually, this is down to the main hook of the record, the instrumental melody or vocal lines that you just can't get out of your head. However, in many instances there is more than just one hook line in any popular record; in fact, there are usually two, three or sometimes four, all of which manage to indent their way into our subconscious minds, but in such a way that we only perceive that there is one. As a result, while you may feel that the main instrumental hook or vocal is the one of importance to keep while remixing, our minds work on a much higher level and simply hearing one of the smaller, less frequent hooks can often remind us of the original record too. To gain a better understanding of this we need to look at the common tricks employed by producers and songwriters when producing a 'pop' record.

Although dance music is often scorned as being far too repetitive, all pop records use the same amount of repetition to drive the message of the track home, only in a more devious way. One of the most classic techniques employed is to subliminally reinforce the message of the music by replacing the vocals with instrumental copies. For instance, many tracks will open with the vocal melody being played on an instrument; this will then be replaced with the vocals during the verse and 'hook' chorus, before being replaced again with an instrumental version of the 'hook' vocals during the breakdown. In some cases this instrumental hook then appears again at the end of the track to finally drive the message home.

These instrumental 'clones' are not always played on the same instruments, though, so as not too appear too repetitive; rather they may be played on piano, followed by a wind section, then a string section to finish. This is where we, the listeners, derive the main melody or hook from a record, but there are also other hooks that, although not driven home as much, can still form a proportional part of the music we hear.

Alongside the main hook of the record there will be any number of secondary hooks that are used to maintain listener interest. These are commonly short riffs, motifs or small vocal ad libs that appear between the lines of a song and are created to fill in the gaps. These ad-lib hooks and vocals are often ignored because we're concentrating on the main hook, but they are still subconsciously imprinted into our minds as part of the record. Just using these small ad libs and synthesizing the remainder of a track can produce a remix that still remains immediately recognizable without having to worry whether the main hook will suit the genre you have in mind.

At this stage, its also worthwhile envisaging the final arrangement, as this will often determine which parts of the original are to be kept aside. There's usually little need to worry too much about how the finished product will sound, but rather have a good idea of how the arrangement will fit together using some of the parts from the original recording. It's also at this song mapping stage that you can determine what genre of music the remix is aimed at: is it going to be trance, house, garage, etc.?

As with all forms of music, this isn't a fixed method and every artist will use their own methods and ideals; however, it's only on very rare occurrences, or bootlegs, that you will be given total freedom from musical style. If a record company requests a remix from an artist, they have been carefully chosen by the A&R department because of the style of music they have produced in the past.

It's not contemptuous to believe that some big name remixers are approached simply because of the name that would be attached to a recording for the sole purpose of profit margins but, optimistically, they are solicited simply through their previous stylistic endeavours. By requesting Oakenfold to produce a remix, they wouldn't expect a hip-hop recital; instead, they would expect a club-style hit typical of his previous releases. After all, there is little point in taking a record that is already a hit in R'n'B circles and reworking it in the same style – the public market is the same and in many instances only the die-hard fans of the original will want to purchase the remix. The result is that it hasn't been promoted, but is merely an over- (or under-) produced original aimed at the same listeners.

Another benefit from using this form of 'song mapping' is that it can often help to envisage the remix in more ways than one. Along with helping with figuring out which of the original samples are going to be used, it can also offer an idea for the general tempo of the piece. As already discussed in earlier chapters, the tempo can divulge a great deal about any piece of music, especially in dance music circles. As the original will most probably require some time-stretching to suit the new tempo, it's an idea to stick with your chosen tempo to avoid introducing unwanted digital distortion from continually stretching a piece of audio each time you change your mind.

These days it can be difficult to encapsulate a tempo for any remix as they're constantly being re-invented, but roughly speaking most tend to fall between 120 and 145 BPM. Very few remixes with a tempo over this have ever reached mainstream clubs and even fewer will be accepted by a record company. This is because the vast majority of regular club-goers are not able to dance to a rhythm at this speed while still managing to look cool, and if nobody can dance to it then the DJ isn't going to play it. An 180 BPM drum 'n' bass remix of Madonna's latest hit may capture the imagination of some, but this is only a small percentage when compared to the millions of clubbers who go dancing every weekend. On this same note, creating a remix that constantly switches between, say, 4/4 and 3/4 signatures or has a constantly fluctuating tempo will only serve to confuse and alienate listeners further. While it may seem inequitable to suggest that the public are fickle, unimaginative and short of attention span, there have been plenty of popular artists who have released 'experimental' music only to have it fail miserably because it didn't have the familiar elements the fans want.

Note: Most unsuccessful musicians and remixers will undoubtedly argue this point, insisting that they create mixes to please only themselves, yet the most successful are the ones who know what the clubbing generation want and how to deliver it. Morales, Oakenfold, Deep Dish, BT, Negro, Tiesto and Maas have continually proven these points, and as a result are currently considered some of the best in club land.

Of course, there is certainly nothing wrong with innovation, but unless you're already a well-known remixer with a fan base it certainly doesn't have the same promotional power as the more conventional four-to-the-floor club mixes. It is important to have a distinctive style but it's equally as important not to get too carried away. Dance music is an exact science and close scrutiny of the current market is crucial. This is one of the key reasons why DJs usually produce the best club remixes.

Another important factor as far as the tempo is concerned lies with that of the original recording. If you're remixing on behalf of a record company, then the chances are that you'll be given the original working tempo of the track, or occasionally you may even receive the original track sheets. These are written records usually taken by the tape ops at the time of the original recording, and note the start times, EQ settings, channel assignments and general working tempo. Having these to hand obviously makes remixing much easier and if you're not automatically supplied with one, it is worth asking if there is one available. There are no guarantees that there will be, though, as the filing system in most studios usually consists of throwing them into the nearest cardboard box and some don't even bother compiling track sheets at all.

If you are provided with a tempo of the original it still shouldn't be taken for granted that it will be this quoted tempo when you begin working with it. A mistake often made is not taking into account the clock in the sampler or computer being used for the remix, and this can often lead to small discrepancies in tempo as the track progresses. While it would seem natural that something recorded at, say, 120 BPM would remain the same on everything, all computers and samplers utilize a crystal clock that controls the rate they execute their instructions and not all crystals operate at exactly the same speed. Additionally, the operating system being used can also affect this clock, resulting in a stated tempo of 120 BPM being anything between 1 and 2 BPM faster or slower. This is significant if you decide to approach a remix using the cut, drag and drop method – a system mainly used as an ideas generator, consisting of importing the original track into a sequencer, cutting it up into bars and then moving the segments around.

Consequently, it's worth calculating the tempo yourself, even if you've been supplied with the original tempo. Most of today's sequencers offer various methods for calculating the tempo of any piece of music. Generally, you can import a segment of the original track, select four or eight bars of the sound file and then tell the sequencer how many bars are present. The sequencer will then determine the BPM and adjust the tempo to suit. This method isn't applicable to all audio sequencers and some, such as Steinberg's Cubase SX, require you to tap in the tempo of a piece of music using the keyboard before it calculates it. Using this method can be a little hit and miss, so a preferred method is to calculate the tempo manually.

To begin with you need to place the original track into a wave editor and cut out a four-bar loop for importing into the sequencer. It's usually best if you choose a segment where there is a drum solo featuring just the kick, but any part where the kick is present will be suitable, as this lets you see exactly where the beats occur, allowing you to determine a four-bar loop. It's vital that the loop starts with a kick and ends just before the final kick so that it can loop over and over with no glitches. Once achieved, this loop can then be imported into a sequencer set to constantly cycle over these four bars, whereby you can manually adjust the sequencer's tempo until the audio file loops clearly. Alternatively, if you know the length of the loop in seconds and the number of beats in a measure, then it's possible to use mathematics to calculate the tempo:

60 divided by Loop Length in seconds = N
N × 8 (assuming you have 2 bars) = BPM
(the 8 is for two four-beat measures)

Note: Some sequencers offer functions which can perform this calculation for you!

While this is the widely accepted ideal for determining the tempo of any piece of music, if the original was recorded before the 1980s the process can be a little more difficult. Before this and the intervention of step-time drum machines, tracks often speeded up naturally as they progressed and frequently the producer would also slow down the tape machine by 8.5% on the final mix.

The reason behind this is that when the track is played back at normal speed the mix becomes a semitone higher in pitch, in effect tightening up the playing and adding an additional polish. This clearly makes calculating the tempo throughout the entire song a complex procedure, as it can fluctuate by as much as 10–15 BPM throughout the track. If this is the case, then it's advisable to import each loop from the original that you plan to use in the remix and time-stretch each constituent loop to the tempo that the remix will be.

Time-stretching is available on most audio sequencers, audio editors and samplers, but the resulting quality relies entirely on the algorithms that are used and the material being stretched.

Time-stretching adjusts the length of a sample (effectively changing its BPM) while also leaving the pitch unchanged. It does this by cutting or adding samples at various intervals during the course of the sample so that it reaches the desired length, while, to a certain extent, smoothing out the side-effects of this process on the quality and timbre of the sound. This is quite obviously a complex, processor-intensive process and is not suitable for extreme stretching. For instance, stretching an 87 BPM loop into 137 will undoubtedly introduce unpleasantness to the rhythmic flow of a loop, producing a side-effect often referred to as digital clutter.

Occasionally, this digital clutter can be used to good effect, but on riffs and more especially vocals it's unadvisable to stretch them by any more than 25 BPM. In fact, vocals, more than any other instrument, can be adversely affected by the process, even if it's only stretched by a small amount. The problem with stretching vocals derives from when vibrato is used while singing. We're become accustomed to the human voice, so even small time-stretching adjustments can become immediately obvious, more so if there is an element of vibrato involved. With the advancements in technology, there are various methods to avoid vocals sounding deliberately affected, often involving spending days cutting vocals into individual syllables to make them fit a new groove.

We've already seen the uses of sample splicing software, such as Wavesurgeon, in earlier chapters as a form of reconfiguring and rewriting sampled loops, but it can be equally useful as a form of stretching audio. By placing cuts manually at specific points throughout a riff or vocal line it's possible to change the tempo without introducing any unwanted digital artefacts. The basic principle is that if you need to stretch a vocal from 80 to 140 BPM, you can thinly slice the audio and, after importing them into a sampler or audio sequencer at this new tempo, the decay of each sliced sample will be stretched or reduced to prevent gaps or any overlapping.

Yet another advantage of taking this approach is that, when a loop is sliced, each slice is allocated a different MIDI note gradually ascending in semitone steps. These MIDI notes, or the audio slices if you prefer, can then be rearranged to produce alternative vocal lines or riffs from the original.

Clearly, this type of editing can be an incredibly time-consuming affair and in the majority of cases doesn't work too well if there are vibrato or legato phrases in a vocal performance. To get around this, a common practice is to 'comp' the stretched vocals with the original vibrato or legato parts. Comping plays an important part of any production process and consists of mixing numerous vocal takes together to produce the finished article. In the context of remixing, this often involves cutting each word of a vocal phrase into separate segments and stretching each to suit the mix. Using this method, if there are any vibrato, legato or held notes they can be paired up with the stretched parts and cross-faded together to produce a more natural sound.

Failing this, another popular method involves pitch-shifting, rather than time-stretching, the audio to get it in time. By importing the vocals into a sampler, you can use the pitch bend wheel of an attached controller keyboard to find the best speed. Clearly, this affects not only the speed but also the pitch; however, by using a sequencer's controller editor you can try various pitch bend values until you find the right speed, and then use a good quality pitch-shifter to move the vocals back to the correct pitch. This often removes the additional digital 'clutter' introduced by more conventional time-stretching.

On some occasions, while most popular remixes will retain the vocals from the original, there isn't the need to use all of them. In most forms of music they commonly form the most influential and

Figure 19.1 Slicing vocals into smaller segments for stretching.

dynamic part, and many producers will treat them with the utmost care, but in dance music and remixing the part they play is negligible. Pure dance music commonly has very little lyrical content, dealing in just one thing – feel. Vocals in dance are simply a form of distraction from the main event – the underpinning groove, the rhythm that keeps everyone up on the floor, the DJ in work and the record companies in profits. Subsequently, when remixing popular music that consists of the typical vocal verse/chorus for the dance floor, there is little need to keep all the vocals intact and running throughout the remix. In reality, it's common practice to lift just the most prominent parts of the vocal performance, i.e. those that make the track instantly identifiable, and disregard the rest.

Note: Armand Van Helden's number one remix of Tori Amos' *Professional Widow* is an excellent example of this technique. He ignored a large proportion of the vocals, choosing to keep just a few small, unaltered vocal snippets, practising the 'minimal vocals for minimal attention spans' technique.

Having said that, even if you cut up vocals or use the various methods to transform and stretch them, there is a limit as to how much of a tempo adjustment you can make from the original. This is not solely down to the vocals, as stretching riffs can also introduce unwanted artefacts, so after auditioning the original parts the song's tempo must be determined in order to have a starting point from which to derive the remix's tempo. If a song is already 120 BPM a house remix will be a pretty easy stretch, assuming, that is, you actually want to make the record faster, but if the song is 90 BPM, reaching this tempo is going to be much more difficult. It's significant, therefore, to consider the remix's tempo carefully, because the less you time-stretch (or time-compress) a track, the better it's going to sound.

As an example, suppose that you've been given an original vocal recording with a tempo of 80 BPM and that you're working at 140 BPM. Rather than attempt to time-stretch the vocals by 60 BPM to suit the working tempo, it would be more sensible to time compress them to 70 BPM. This is half your woking tempo and is only 10 BPM from the original vocals. One bar of the vocals would then run over two bars of the music, which can often help contribute to the length of the remix.

That said, keep in mind that not all songs have been written close to a dance tempo and some hooks that make the original what it is will not suit any other tempo. As we've seen, using the aforementioned techniques it may be possible to make anything fit any tempo if you're willing to put the additional time into it, but stretching some hooks can ruin the integrity of the original song. To become a good remixer you have to respect the track you're remixing and it isn't worth destroying it just because you want it faster.

Of course, not all remixes use unaltered vocals and there are many popular techniques for twisting vocals, and riffs, in remixes, although in many cases the techniques employed will depend entirely on the artist that's being remixed. What follows is by no means an exhaustive list of possibilities as the market is constantly changing, but currently these techniques are proving to be the most popular:

- Import large sections of vocals into sample splicing software such as Wavesurgeon and use this to slice the vocals/melodies into individual phrases. Once completed, use either MIDI note commands to a sampler or the actual audio and rephrase the line. This technique is more popular with melodies as the timbre remains the same as the original, yet the melody is different.
- On this same note, assign each vocal/melodic phrase to a specific key in a sampler and then use a keyboard to play different pitches, layering them over one another to create chords.
- Vocals and melodies can also benefit from low-pass, high-pass and band-pass filters. By assigning a controller keyboard's mod wheel to control the filter it allows you to filter specific phrases live. A variation on this technique that's also popular is to use different filters every few phrases.
- Although the stutter effect is becoming rather clichéd, it is helpful in bringing attention to the vocals or riff. These can be created by setting a sampler to play a sample only while the key is held and then sending 32nd or 16th MIDI note on-commands to the sampler.
- An alternative to the previous method is to import a riff or vocal into an audio sequencer and then place cuts every 16th before deleting every other 16th. This recreates the typical gating effect.
- To bring more attention to the vocals, an often practised method is to create a vocal phrase that is physically impossible to sing. The methods used to create this are varied, but often include the use of auto-tuners set to adjust the key of some phrases by a large amount so that the side-effects are noticeable.
- Time-stretching can also help towards creating phrases that in the real world would be impossible. By stretching some vocal phrases by a large amount and leaving others unaffected the vocal line can take on a whole new meaning. Similarly, using cross-fading techniques, try to loop part of a vocal phrase for longer than would be physically possible but so that it still sounds natural.
- Reverse parts of a vocal phrase while leaving others playing normally.

- Effects play a large role in creating interesting melodies and vocals. Simply double-tracking the vocals and offsetting one from the other by just a few clicks can create an interesting effect.
- Reverse reverb can also work particularly well, especially if you do not want to affect the vocals too much. This can be accomplished by reversing the vocal phrase then applying reverb with a large hall setting, or similar, applying it to the waveform and then re-reversing the vocals so that they're the correct way again. Using this method the reverb will build up slowly to the vocal line, creating a sweeping effect.
- Specialist effects, such as Steinberg's GRM Tools versions 1 and 2, Voice Machine or Spektral Designs Ultra Voice, can be used to create strange results. The Pitch Accumulation from GRM Tools v1 is a popular effect in many remixes, creating a culmination of pitching and filtering.
- EQ should also not be underestimated; rather than save it solely for mixing, use it to remove the bottom end from vocals to create a 'telephone' or 'megaphone' vocal by rolling off all frequencies above 3 kHz and below 1 kHz.
- If you have access to a hardware reverb unit, use a long decay and apply it to the vocal line. While playing back the vocal, record only the reverb's return signal and then offset this against the vocal phrase.

These are only suggestions and, as always, experimentation and close scrutiny of the current market are essential for creating great effects. A visit to a local record store frequented by club DJs with £30 in your pocket will help you follow the latest manipulation trends with remixing.

Before you do begin to splice, dice or mangle vocals in any way, it's worth checking with the record company first, or if it's an illegal dub remix, avoiding over-processing altogether. Whereas some styles of music such as happy hardcore almost invariably use 'chipmunk' style vocals and the current trend is leaning towards additional processing with effects such as Antares Auto Tune, a proportionate amount of labels do not approve of their 'selling-point vocals' being mutated into something totally incomprehensible. Getting too carried away with vocal manipulation may not only annoy the label, who coincidentally will probably ask that they remain unaltered anyway, but it can also upset the original artist and alienate the audience.

Arranging

The method for arranging a remix varies from producer to producer, but while it's quite probable that the familiar verse/chorus will be used in the original mix, this does not mean to say that the remix should follow suit. Indeed, for clubs it's quite rare for any track to follow this structure and most will follow the atypical 'club' structure that, incidentally, is very similar to classical music. In other words, you begin with a theme followed by a variation, before eventually returning back to the original theme. Although how this style of arrangement is approached varies wildly, a typical example is detailed in Table 19.1.

This type of structure often forms a vital part of any remix, since most need to be at least 8 minutes long and the original track will more than likely be half of this. Additionally, even though Table 19.1 is an example and isn't a strict rule, it is considered a rule to keep both the intro and outro as drums only and leave them 32 bars in length. Most DJs prefer to mix beats than mix tuned instruments from one track to another (known as harmonic mixing), since they have to adjust the pitch accurately on the fly, plus it can also make some tuned instruments sound

Table 19.1 Example of a club arrangement

Part of arrangement	Length in bars	Analysis
Intro	32	Commonly consists of nothing more than a building of the drum track to allow the DJ to mix the track in with another record.
Body 1	32	A melodic instrument is introduced, most usually the bass to establish the groove of the record, which may be followed 16 bars later with another melodic instrument. This latter instrument is rarely part of the original track, but a motif programmed to sit around the original parts
Drop 1	16 or 32	Most percussive elements bar the kick are commonly dropped from the mix, followed by the introduction of a cut-up main hook line or vocal to establish (or give some clue) that it *is* a remix to the clubbers.
Body 2	32 or 64	All the previous elements may return, playing alongside the cut-up hook. Small ad libs either from the original or programmed are sometimes introduced after 16 bars, mixed amongst snare rolls, cymbal crashes and sound effects. Vocals from the original are sometimes also introduced here, but more often than not they have been cut up so the entire vocals are not revealed, adding to the tension of the music.
Drop 2	16 or 32	The main drop of the track is often similar to the first, but occasionally the kick is removed and the main hook may be removed or filtered down. This could be followed with a 16-bar snare roll with the filter cut-off on the main hook opening, creating a building sensation.
Body 3	32 to 64	All instruments return to the mix, alongside the full main hook or vocal line. Further ad libs may be added after 16 bars, alongside skips in the rhythm, small breaks, rolls and cymbal crashes.
Breakdown	32	Typically, all the instruments are dropped gradually from the mix, signified by drum rolls or cymbal crashes, so that only the drum track remains.
Outro	32	Similar to the intro, this commonly consists of nothing more than the drum track to allow the DJ to mix the track in with another record.

utterly terrible. By keeping with a percussive intro and outro, they only have to mix the beats so therefore are more likely to play the mix.

More importantly, to create an 8-minute mix from only 4 minutes of music, many of the bars are continually repeated, and so to maintain interest and keep the mix exciting, production techniques play a vital role. This includes using effects to create more interesting timbres, cross-fading between samples, creating stutter effects, altering transients (with a transient designer) as the track plays and using filter sweeps to add movement to static areas of the arrangement. Sudden breaks in the rhythm, snare rolls, small breaks, skips, scratches, sound effects and slight changes in the drum pattern, melodies and motifs are all used to add tension and a dynamic edge to the music. As ever, experimentation often produces the best results.

Contracts

If you've been commissioned to remix a tune, you're obviously going to need some kind of agreement with the label involved to ensure that you both fully agree what you're being paid for. Although remix contracts differ from label to label, they will all consist of a number of clauses that must be fulfilled if your remix is to be commercially released.

Firstly, as a remixer you will need to agree on the type of remix that the label requires; this means that you both need to concur on the genre of music and a suitable length for the track. Hopefully, the label will have asked you to perform the remix based on the same genre of music as your previous releases, if you've had any, but a suitable length of track can vary immensely.

The widely accepted length for a typical club remix is around 8 minutes, but it is important that you know exactly how long the label want it to be, because if it's over or under the specified length they may refuse to pay you. Additionally, you will need to confirm with the label how many versions of the remix they want from you, since it's not uncommon to be asked to supply not only the club mix with vocals, but also an instrumental version.

You need to agree on the terms of delivery to the label too. Most will only accept your completed mix on DAT, but CD is becoming a more widely accepted format. Also, there will be a rather annoying clause that states that the mix must be of a 'commercially acceptable standard'. All record companies adopt this and it's nigh on impossible to persuade them to remove it from the contract. It is particularly vague but essentially it gives them the right to refuse your remix if they don't believe there is an acceptable market for it, or its been poorly mixed, produced or mastered.

As previously touched upon, it's also essential to check what media the label will use to deliver the original track to you. Although larger labels will only supply a remix in ADAT format, smaller labels may put it to CD, but more often these commonly supply it on DAT with the master tape recorded track by track with the time code. The benefit of this is that it's then possible to accurately recreate the track in a hard disk editor or software sequencer, but of very little use if you're working with just a sampler.

Finally, you will also need to be in agreement on when the master tapes will be delivered to you and, of course, the deadline that they want your completed remix back. Deadlines can vary enormously, with some labels expecting you to turn around a professional remix in a matter of weeks, while others may offer a few months. It's rare that any remixer is given longer than 4 months to complete a mix, so before accepting a project you should ensure that you can perform it fast enough so as not to annoy the label and ruin your chances of being offered any further projects.

Assuming that you have come to a mutual accord on the terms and conditions of completing and delivering a remix, you need to reach an agreement on how much you should be paid. It's rare for any label to offer royalty-based payments and more often than not a label will pay you a one-off fee – usually half upfront and then the other half if the remix is accepted.

If it isn't accepted then you could be expected to pay back the previous payment, so this is something you should check before taking on the venture. The amount you can expect to be paid is dependent on the label, the artist you're remixing and your past performance. Little-known remixers can be paid as little as £500, and in some cases a label may ask you to remix for free, only agreeing to pay you when the remix is returned and if they want to release it. This may not seem so great, but as you produce more and more remixes, your name gets around and you can begin to charge thousands. It's not uncommon for a well-known remixer to charge in excess of £15 000 to remix a successful artist who's signed to a big label.

No matter how large, or small, the payment it isn't advisable to try and barter for more money, unless your remix happens to be selling incredibly well. If it is, try contacting the label asking for royalties, but depending on the remix this can be a bit of a grey area. Many remixes usually find their way onto the B-side of a record, and you could have a hard time trying to ascertain that it's your remix that's selling the record and not the artist who's featured on the A-side.

Clearing samples

If the remix has been created without the authorization from a record company and it is doing particularly well in a club, then you may need to look at releasing it commercially. Ideally, the record company that released the original track should be your first port of call, because these will often charge less for the use of their samples than if you signed to a rival label.

The subject of clearing samples has often been overly exaggerated as being a long-winded, drawn-out, expensive process, but this isn't necessarily the case. Since the first case of illegal sampling in 1991, when Gilbert O'Sullivan successfully sued Biz Markie, sampling other artists' records has become an art form in itself and numerous companies now exist to clear samples on your behalf.

Generally, from experience, most record companies and artists are only too happy to work out a deal and allow you to use their samples, simply because they will receive an income for doing absolutely nothing. This is especially the case if you can make an old out-of-date record hip again. However, be warned that if the record that has been sampled was originally a hit, then the record company and artist will take a bigger slice of any royalties. Similarly, the record company holds all the cards when it comes to negotiations, because you broke the law in the first place by illegally sampling and releasing a track featuring their artist or music.

This same principle applies even if you haven't created a remix but have sampled an artist to create your track. In these latter cases you have to be diplomatic, as the record company will quite obviously try to take all the proceeds from the record by arguing that the samples form a major part of the record. For instance, say that you've constructed a track but only relied on a few small guitar hits in the record. The record company will begin by arguing that they contribute, say, 90% towards the track, whereby you can argue that they only contribute, say, 40% to the music. They will then return by saying that it contributes 70% to the music, where you can reply that they only contribute 50% and so forth. Notably, you will have to be amenable, since if the record company are not happy with your dealings they can simply refuse to release the track while you watch an instant club hit disappear into obscurity.

A much better way is to avoid this type of negotiation by writing the track but to clear the samples before pressing to vinyl and releasing it. This has the immediate benefit that the record company doesn't know if it will be a hit or not, so you stand a better chance of getting a good deal. If the record is already released and doing well in the clubs, then the sample's owners know it's a hit and will try and squeeze you for as much money as possible.

If you go for sample clearance before release, 80% of the time the record company will offer clearance so long as their terms are accepted, but it's unwise to attempt to clear this yourself and you should look towards using a sample clearance company. Typically, these firms charge around £300–400 for all the legwork involved, but they stand a much better chance of getting you a good deal when it comes to paying for the sample's clearance. This is especially the case if you've sampled from a record that's originally signed to a label from France, Germany or America. From some very personal experience, these guys are notoriously difficult to deal with, and the cigar-chewing fat cats are equally adept at chewing you up and spitting you out again unless you know exactly what you're doing.

Of course, it is perfectly possible to clear samples on your own so long as you adopt a professional nature, but you will have to clear two copyrights: the mechanical rights and the publishing rights. The mechanical rights, sometimes referred to as the 'recording rights' or 'master rights', concern the physical recording of the music, which in many cases belongs to the record company, since they'll have put up the money for the studio time.

The performance rights belong to the composer of the music, but this is usually handled by a specific music publisher, who deals with the performance of the recording on behalf of the original artist. The first step to finding out exactly whom these are is to speak to the Mechanical Copyright Protection Society (www.mcps.co.uk). This organization has a list of every UK record label and music publisher, and will have all the information on who owns the mechanical and performance copyrights. Alongside this, they will give you the contact details and offer advice on the best ways to deal with the situation.

If the record has been released through a large label then clearing a sample is quite clear-cut, as most companies have adopted EMI's online sample clearance contracts. Using these, you simply fill in an online form and wait for a response from the clearance team. On the other hand, if the record is particularly old (but still in copyright) or was released by a smaller label, then you may find yourself having to go to the middle of nowhere to ask for permission.

This can be particularly complicated if more than one artist was involved in the original project, as you'll have to obtain permission from each one of them, or their closet relatives if they've passed away. Typically, they will give permission since it will earn them money, but it's important to note that they do not have to give approval or offer any reasons why they won't. If they do consent, then you'll have to come to a written agreement on how much they receive in royalties and, in many cases, how much they want as an upfront payment.

The amount of money asked for varies and depends mostly upon the popularity of the artist. Famous artists such as The Beach Boys, The Beatles and Michael Jackson expect at least 80% of royalties and demand an incredibly substantial cash sum upfront. As a guideline, when I went for sample clearance on the mechanical rights to use a lick of Sting's guitar, it would cost an upfront advance of £10 000 along with 70% of any royalties, while Barry White asked me for an almost unbelievable £60 000 per *word* and half of any and all royalties. Needless to say, his vocals never touched my sequencer.

On the other side of the coin, little-known disco records from yesteryear usually demand £300–800 along with 20–30% of all royalties, depending on how much of the original is used in your music. On top of this, you also have to pay for the publishing rights of the artist and again this varies depending on the artist. Commonly, they will only ask for 20–30% of all publishing royalties.

Note: In some cases you'll be confronted with the dilemma of the record company saying yes and the artist saying no, and if this is the case, you're pretty much stuck. By law, you cannot include a sample in your music if the original artist has refused permission, but if the artist has given consent and the record company has refused then there is a work-around. Since the record company only own the mechanical rights, you're free to recreate the sample yourself or alternatively employ a company or studio to recreate the sample for you.

Above all, if you do plan to sample another artist's record for use in a commercial track, then it's prudent to join the Musicians' Union. This costs around £80 a year, but they will offer to look over any contracts or agreements that you have and ensure that legally you are protected just in case the record company or original sampled artist change their minds. This has been known to happen and it's always better to be safe than sorry.

20 Designing an audio web site

With the Internet increasingly becoming a viable marketing tool for dance musicians to promote their music, many musicians are building their own web sites. This, however, can be a difficult endeavour to undertake unless you have a good basic understanding of how web sites are constructed and designed. In light of this, James Froggatt, a professional musician and web designer, has contributed this chapter, which explains how to design and construct a web site to promote music.

How the web works – the basics

This chapter goes through much of the theory behind making a great web site and covers many of the key considerations. For the purpose of this book, there is little room to go through the entire use of HTML (HyperText Markup Language – the language used to create web pages), as this would require an entire book alone. Instead, this chapter provides the basic theory and knowledge required to create a fully working, audio-enabled (downloadable audio) web site in a step-by-step fashion.

The code used in the example web site will be de-mystified as we go with descriptions explaining what the code used actually does. The code provided may be used and altered as you wish to create your own personalized web sites.

To understand how a good web site is designed and developed, some knowledge of the technology behind getting the web page from the server onto the user's screen is essential.

HTML and web browsers

As mentioned, web pages are written in a language called HTML. This is a series of instructions that tells the browser how to display the web page. The browser is an application, such as Internet Explorer, Safari (Mac OS), Netscape or Opera, that sits on a web-enabled computer and interprets these HTML instructions in order to display the results on the screen of the browser.

When surfing the web, by typing a URL (Uniform Resource Locator), such as http://www.google.com in the browser window, the browser is instructed to send a request to a server. In its own way it says: 'Server, wherever you are in the world, I'd like you to send me this page so that I can view it.' Using this example, the URL sends a request to the Google server, which then sends the HTML and associated graphics back to the browser that made the request through the Internet. Your browser then interprets the HTML and pieces the page together ready for viewing. It sounds fairly simple and it is. The difficulties for a web designer, however, lie in the fact that many people have different Internet connection speeds and there are many web browsers that may be being used to view the page.

Because HTML is an interpreted language, different browsers interpret the same HTML differently. This means it's important to take care with your web page design at the outset. If not carefully designed, a web page may be interpreted so badly by some browsers that it becomes non-functional for the end-user. Employing masses of web technologies and enhancing your web site with heavy use of additional languages such as JavaScript or overdoing the Macromedia Flash is likely to increase the probability that the site will not capture the largest audience by slowing it down or making it unusable on some viewing interfaces, particularly for people using old browsers and slow Internet connections to view the pages. Many problems arise due to the fact that some web technologies are now exceeding the capacity of the web to deliver them efficiently, simply due to the lack of global roll-out of broadband or faster connections (particularly in the UK), and many people are still using out-of-date web browsing software.

The most common web browsing software applications are Internet Explorer and Netscape, but there are many others, such as Opera, and specialized browsers, such as Lynx, that cater for special needs users. At the time of writing the latest versions of these packages are Netscape 7.1, Internet Explorer 6, Opera 7.11 and Lynx 2.8.3. With the latest version installed, how often do you upgrade your browser? Fortunately, people into dance music production will generally appreciate the benefits of updating software, because the last thing you need when you're under the dim blue light of your studio at 3 a.m. and 'in one' with a groove, having not eaten for 10 hours, is the computer suddenly stopping working. While dance music producers know that updating software with the latest version will help keep the computer stable, potential viewers of your web site (the public) may not, and as a consequence many web users will not be using up-to-date browsers. What we do know is the most common browser is Internet Explorer, with Netscape following behind, with a whole spectrum of versions of these browsers being used. From personal experience, a proportion of people are still using a version of Netscape 4.7x, which when developing a relatively complex web site that is fully functional on the latest browsers is an example of an archaic browser that can reduce a web site developer to tears!

Because of the various browser differences and the likelihood that most users will not be using up-to-date browser software, it's necessary to decide whether to design a site that caters for the 'elite broadband latest browser software' crew or the '56K dial-up modem Netscape 4.7 crew'. Whilst the difference between the two crews is a technological abyss, deciding whether to cater for the latter crew as well as the elite is an important starting point for the site's design.

Attempting to design a web site that works comfortably on Netscape 4.7, for example, will restrict how web pages function and look, simply because this version of the Netscape software series was written years ago and the web has moved on since.

To cater for most users, ensure your page works perfectly on Internet Explorer 5 and Netscape 6 versions. Preferably, a site should also work, at least so that the content can be read clearly, on Netscape 4.76. Alternatively, you can just design for the latest versions of various browsers and put along with your design a note instructing users to update their browser to the latest version.

Web design software packages

There are many good software packages available that will speed up and facilitate the creation of web sites. Probably the most commonly used of these is Dreamweaver created by Macromedia (www.macromedia.com) and Version 3 will suit a person wishing to create a basic web site. Macromedia Dreamweaver 4 Ultradev is also an excellent package that has many more facilities than previous versions, including the ability to create database-driven web sites with relative ease. The real beauty of this software package is the ability to create a web site using drag and drop facilities and see the HTML code being written automatically. This, however, does not negate the need to learn HTML, as quite often the automatic code the software creates will not be 100% perfect. A site created this way may appear to work on the latest browsers, but may look terrible on an older browser.

If you intend to create quite a few web sites, there is no substitute for purchasing this software and learning how to use it. It will save weeks if not months of what would otherwise be the lengthy procedure of hand-coding HTML. Dreamweaver is a great web site creation and learning tool.

Design considerations and techniques

The following is a series of considerations and techniques that are useful to be aware of when creating a web site. They are just outlines designed for general awareness and do not need to be understood fully here, as they are dealt with in actual demonstrations that follow.

Keeping it fast

> My web site will be the flashiest, most graphically beautiful on the web; it's bound to be a success …

Modern web technologies allow anybody to create great looking web sites, but it is all too commonplace for such web sites to actually serve pretty much every purpose other than delivering the content that they should have been originally designed for. These technologies are advertised to speed up productivity and most do, but what they do not necessarily tell you is the truth about the web experience the person viewing your pages will actually receive from your hard web site designing work. The reason for this is that it is not actually possible to give this information accurately, because many web users will in fact experience something different. There are many factors to consider during the process of producing a beautiful looking web site that actually fulfils its purpose. For example, some users may be fortunate enough to have a broadband (high-speed) Internet connection and enjoy viewing the page and other media, such as streaming audio, video or Macromedia Flash animations, at maximum speed. Other users may be connecting to the Internet via a dial-up connection that uses a 56K modem, while other, very unlucky users may even have a slower connection than that. All these people will experience different download times, which may mean that if a site is graphically rich and therefore slow, they may even give up trying to view the site before it has fully downloaded, so it's best not to assume that most people have broadband or other faster connections: while you may use that type of connection, there are many people that can't get it (most of the British countryside and city outskirts in the UK) who surely also deserve to see your site. More to the point, they must be able to see your site. Who knows, a big record company manager may live in a village some-where who, if only they hadn't given up on your site being so slow, may have heard your music and signed you up.

If you are to get ahead of the competition and get your page viewed, you must create web pages that cater for, at least to a reasonable degree, the lowest common denominator of access speeds. This is good news for non-technical people, because many of the web technologies pitched today are all bells and whistles and are not particularly useful for making successful web sites. More often than not, too much technology on a web site serves only to clutter the content, confuse the viewer and slow down the web site loading speed, if not render it totally non-functional on some browsers.

At the end of the day, whether you make your site predominantly based on HTML or create a really sexy Macromedia Flash style site is up to you, and this will depend on whether you wish to hit a specific target audience or make your web site available to the largest audience possible.

Here are a few guidelines for speeding up your site:

- Ensure all graphics are web optimized. This means that they are the smallest possible file size and will still look good on the web. Graphic design packages such as Adobe Photoshop have functions specifically for this purpose, so check your graphics software manual for information on sizing graphics files for the web.
- Use thumbnail images to display several graphics on a single page. These are dimensionally small images that are optimized to be as smaller file size as possible without them losing the desired clarity. When the user clicks on them, they link to a page displaying the full-sized, full-resolution image.
- Remove all the spaces between the tags from the final HTML code so that it reads as a con-tinuous page, but remember to keep a copy with the spaces intact as this file will be easier to refer to when editing. To demonstrate this, below is an example piece of HTML code.

```
<html>
 <head>
 <title>Dance Music Production</title>
 </head>

 <body bgcolor="#FFFFFF" text="#000000">
 </body>
</html>
```

The above code will be interpreted more slowly than the following, although the code is exactly the same.

```
<html><head><title>Dance Music Production</title></head><body
bgcolor="#FFFFFF" text="#000000"></body></html>
```

- Be very careful with the use of Macromedia Flash. How many web sites have you come across with a naff Flash animation on the front page with the text 'Skip Intro' underneath it? Did you wait to see the animation before pressing 'Skip Intro'?

Being consistent

When a user hits your page, they should feel at home, this is your 'home page'. Have a short introduction about what the site is about and any important information that the user needs to know whilst visiting. Do not give a lengthy life description. Think about the information the user wants to retrieve from your site. The first page must capture their attention, so it should give an indication of all the information that will be found in your site.

Once the first page has been created the entire site is practically finished, because the first page will serve as a template for all the other pages. The only differences should be the content and graphics. Common elements like the page header, menu and footers are most likely to remain the same throughout all pages. This gives a consistent look and feel. This, along with the applied CSS (Cascading Style Sheets – more about this later) to format the text within the site, will make a user feel that they are still within the same site, even though they will be viewing different pages.

Learn from other web sites you like

When setting out to design a web site, the common approach is to look at the competition or for inspirational pages on the web. In the case of dance music, you would be looking to produce a web site that has details about you, your co-workers, possibly your studio, and quite possibly downloadable demos of your work. Go and find web sites that do a similar thing and see how they function. What do you like? What don't you like? At this stage it's useful to view such pages via a 56K modem connection to give a true perspective of the lowest common denominator user.

When you find a few sites you like, a great way to learn how web sites are put together is to view other people's HTML code and see how they did it. To do this, go to a web page that you're interested in and, if you are using Internet Explorer 5, you can go to the **View** menu at the top of the browser and select **Source**. This will then show you the source code for that site. If you use other browser software, you will find out how to do this by consulting the help file within your specific browser. Another great method for learning how to build web sites is to go to **File** and select **Save As** (in Internet Explorer) or equivalent. This allows you to save the HTML and all associated graphics for offline browsing. Using this method you can then view examples of working web pages in Dreamweaver and see from the code how and why the page works.

Laying out your web site – fixed or relative width tables

Creating a web site is rather like putting a tabloid newspaper together. A web site is made up of a number of tables and rows. Within the rows are table data cells that contain the actual content of your site.

There are two ways that HTML can be used to tell the browser how to create a page layout: tables or framesets, or a combination of the two. The most common is to create the web page layout from tables (as framesets can be problematic on some browsers). Tables contain rows that contain table data cells. These three things work together to create a layout. Tables can function as fixed or relative table types.

With fixed tables, the horizontal width of the web page (that is, the table) is constant and is set as a pixel amount. This means that if the page is viewed on a high-resolution monitor, there may be space around the sides that is unused. This may be considered good or bad depending on the overall design. An example of a fixed width web site is http://www.bt.com/ (correct at the time of writing). You will notice that as you resize your browser window, the page does not change its dimensions, it remains at a fixed width. Further examples (correct at the time of writing) are http://www.tiscali.co.uk/, http://www.mp3.com/, http://www.winamp.com/, http://www.real.com/ and http://www.bbc.co.uk.

Relative width tables fill up a percentage of the browser window's width even if the browser window is resized and are generally more useful. This width will commonly be set to 100%, meaning the web page will fill 100% of the browser window, although this can be any value down to zero. Examples of web sites using relative width tables (correct at the time of writing) are http://www.btinternet.com, http://www.thereikischool.co.uk/, http://www.google.com and http://www.alexa.com/.

Generally, people will have their screen resolution set at anything from 800×600 pixels in dimension to higher resolutions of 1280×1024 and beyond. To cater for all users, it's recommended that you design your pages to fit an 800×600 pixel page and ensure that the user does not have to scroll horizontally at this resolution when viewing the page through a fully opened browser window.

Use of frames

Frames are a way to display multiple HTML documents in the same window. For example, frames may be set up to split the browser window into four sections, each displaying a different web page. Although at face value frames seem like quite a good idea, generally they are not and it is recommended they are not used, as some browsers do not interpret them properly.

Image types – GIF or JPG?

Several graphical file formats are suitable for display on the web, but for the purpose of this book only two are worth mentioning. These are GIF and JPG. Other graphics formats will rarely be used on the web.

The GIF image format uses a form of compression known as LZW (Lempel Zev Welch), and is ideally suited for images that are not complex and use few colours. An example of such an image is a graphical menu button, as this is likely to contain only a few colours and some text. GIF images can be anything from two (black and white) to 256 colours. These images can be optimized (reduced in size) for the web using most graphics packages, such as Adobe Photoshop.

A feature of GIF images is interlacing. If a GIF image is interlaced, during the downloading process to a browser window, the whole image will appear in low resolution whilst the complete picture is downloading and is displayed at its full resolution when the download has finished. This benefits the viewer of the web site if a large GIF is used, as the viewer will have an immediate impression of what the image will look like whilst it's downloading. On the other hand, if a non-interlaced GIF is used, it will not appear in full until fully downloaded.

An important aspect of GIF images is the ability to create transparent GIFs. With a transparent GIF any colour within the graphical composition can be rendered transparent, so that any images or text in the web page can be seen through the image. The transparent GIF is a most useful ally and the humble transparent 1 × 1 pixel GIF will prove to be one of the most useful tools in creating web sites that work, as explained later.

Finally, GIF images can be animated. This can be a very healthy alternative to Macromedia Flash, as they do not require specific plug-ins to work. The disadvantage is that an animated GIF can be quite large; it is a sequence of multiple GIFs wrapped into one file. They can be set up to loop continuously or just play once. Packages for creating animated GIFs (some of which are free) can be downloaded from the Internet.

The Joint Photographic Experts Group (JPG or JPEG) format, unlike GIF compression, is good for full-colour photographic images (24-bit). When creating JPG images, rather like GIF images, the size of the final file can be adjusted. With GIFs this is achieved by lowering or increasing the number of colours anywhere in the range 2–256 (ideally the lowest possible number of colours is chosen). With JPG, selecting the degree of compression changes the file's size. The more highly compressed the image, the lower the file size, but subsequently the image's quality degrades. A tip with JPG compression is to add a small amount of blur effect to the original image before you convert it into a JPG, as this can reduce the file size dramatically.

When working to optimize images for the web it's important to keep backups of the original files, because once a file has been optimized the process cannot be reversed. This is very similar to bit-crushing audio without saving the original.

Using thumbnails

Whichever graphical formats you decide upon, the file size of the image is the most critical. As already mentioned, it is important to use thumbnails (that have a smaller file size) whenever possible, as these improve the speed of the web pages' download times. Thumbnails are dimensionally small duplicates of the original image to display on the web. Linking from the thumbnail to the bigger version allows the web site viewer to receive a web page that is downloaded very quickly, whilst having the option to view the image in full when clicking on it.

The 1×1 transparent GIF

One of the most useful pieces of knowledge in web site development is knowing how to use a 1 pixel by 1 pixel transparent GIF to control the layout of your page. Knowing how to use this correctly within a design will save months of head scratching, as it will trick browsers into displaying web pages correctly. As shown in the demo web pages that follow, the transparent GIF is inserted into all table data cells that have no content, i.e. spacer cells within the layout. This simple spacer image resolves many of the bugs and errors in specific browsers. Just trust me: to explain how and why would require an essay. This is probably the most valuable information of all in producing a fully functioning page layout, so ignore this advice at your peril! You will see the 1×1 transparent GIF used extensively in Tutorial 2.

Controlling text using Cascading Style Sheets (CSS)

Cascading Style Sheets (CSS) is a formatting language that is used to control the style of a web page. CSS can be used to control, amongst other things, the fonts and margins of a web site.

A CSS file is a small file that sits on the server along with the web pages it relates to and contains information about the style of the pages' text. In its most basic form, a CSS file defines five features of a web site's formatting:

- What font styles are used.
- What colours are used.
- Whether text is bold, italic, underlined.
- What colours are used for links.
- How the links behave when the mouse is held over them.

CSS is excellent because it standardizes the 'look and feel' of text across the entire web site. Whenever text is added, a small line of code in the HTML file tells it to look at the CSS file for details of what to do with the new text. Because this ensures the consistency of the text formatting, the page will instantly look more professional. A word of warning though: CSS in its most basic of uses is excellent and is commonly used in many web sites, but complex CSS can cause problems, as it is not completely compatible on all browsers. Although Internet Explorer 3.0 and Netscape Navigator 4.0 versions apparently support CSS, they do not implement it in the same way. On a positive note, Internet Explorer 5 and Netscape Navigator 6 implement most aspects of CSS. If using CSS (which is highly recommended), restrict it to the basics: these aspects of CSS are the most used and useful anyway. The examples later in this chapter use CSS in its basic form to demonstrate its power.

One of the major advantages of CSS is that it enables the characteristics of the text to be changed across an entire site by changing the content of only one file, removing the need to go through the entire HTML code and change the style of text line by line. Furthermore, CSS has great advantages for people with disabilities that use your site, as they will be able to turn off the CSS setting in their browser and can apply their own style sheet as an alternative. In this way the user has control over your site's display, allowing practically anybody to use your site in a way that suits them personally.

Using ALT tags

ALT tags are crucially important if you are to allow people with disabilities access to your web site. Importantly, if you use ALT tags correctly, visually impaired people should be able to use your site effectively. ALT tags are an 'alternative' to graphics and some browsers such as Lynx will allow the user to utilize your site without even seeing the screen, as text-to-speech algorithms may be used to read the content of the site. ALT tags should give a description of what the image is conveying. If the image is a spacer image, such as the 1×1 transparent GIF, then the ALT statement within your HTML code should read ALT="". This will then be read as an image that is irrelevant to the actual displayed content and ignored.

ALT tags also add more 'keywords' to your web site, which may help search engines to index it correctly.

Using comments in HTML code

Comments are pieces of text within the HTML code that are ignored by the browser. They serve no purpose but to help you navigate through your own HTML code.

An example of a comment is:

```
<!-- Beginning of 2nd nested table - this will hold the main contents
of the Web Page -->.
```

This entire line will be ignored if inserted into HTML and will not be interpreted.

Using the tutorial files

The web site building demonstrations that follow are in two parts, Tutorial 1 and Tutorial 2. The first runs through key principles and techniques used in building a web site, and the second part builds a complete web site step by step, utilizing and reiterating knowledge from the first set of demonstrations to add clarity.

For the purpose of this demonstration, we will create a web site called *Dance Music Production* in Tutorial 2. Before beginning it's necessary to plan what content will be included and, because this is a demonstration, we will keep it simple. We will go through all the practical aspects of building a web site, from planning the content, to the design and layout of the site, adding audio files, buying a domain name and server space, and finally uploading the site to the web so that the world can view it.

The site's design in Tutorial 2 uses one of the most common and user-friendly layouts for a web site, which uses a title banner that is consistent on every page of the site and a footer that frames the page. When additional pages are added in Tutorial 2, the site will be navigated by using a simple menu positioned on the left-hand side of the page to allow for easy and quick browsing. The graphical design on this site will be kept simple and uncluttered so that the purpose of the web site, which is to deliver content quickly and easily, is enhanced and not confused.

Please note, the graphics used here are copyright and cannot be freely used. The code, however, can be used and changed as you wish to suit your needs.

Tutorial 1: Creating a basic web site

In Tutorial 1 we will demonstrate basic web site building and coding techniques based on the principles already discussed in the chapter.

All of the tutorial examples shown in this section are provided on the accompanying CD. The examples used in this section are located in the folder named *Tutorial 1*. To view these files, copy the folder from the CD into a folder on your computer's hard drive. The files may be viewed by opening the demonstration HTML files within your web browser (although it is highly recommended to view these files in Dreamweaver Ultradev 4 and the browser of your choice). To view the Tutorial 1 *.html* files in your browser, choose **File** from the top menu and then select **Open** (in Internet Explorer 5). Point to the directory where your files are and open the individual files as required.

Getting started

To start off we will now write some HTML that will instruct the browser how to display the page. The following demonstrations may look daunting at first, but bear with it. By playing around with the given code, you will learn how to create a web site quickly.

Let's begin by opening *Demo.html* from the *Tutorial 1* folder. Clicking on the HTML file will open the file as a web page in your browser. To view the code, use either your chosen web design application (such as Dreamweaver) or a text editor (such as Notepad).

```
<html>
<head>
<title>Dance Music Production</title>
</head>

<body bgcolor="#FFFFFF" text="#000000">
</body>
</html>
```

Figure 20.1 Demo.html.

This is the simplest of web pages because it does not include any textual information, only the title of the page that appears in the very top of the browser window, and the building blocks of every web page.

All web pages must contain an opening <html> tag and a closing </html> tag. All the content of your web site must go between these two tags. In addition, there are two other essential tags in this example. The opening <head> tag defines where the content of the 'head' information starts and the closing </head> tag defines where it ends. Within these tags we've kept things simple for the moment and simply tell the browser what the title of the page is. This title is defined between the <title> and </title> tags and appears at the top of the browser display window. Finally, the <body> and </body> tags are where the body of your web site will go (the displayed content).

Additional information about the characteristics of the body of the page can be defined within these tags using various HTML parameters. In the *Demo.html* file, the bgcolor and text parameters are also defined.

```
bgcolor="#FFFFFF" text="#000000"
```

The bgcolor tells the browser what colour to make the background of the page (#FFFFFF indicates the colour white) and tells it what colour to make the text (#000000 indicates the colour black).

Both #000000 and #FFFFFF are hexadecimal colour codes. To see more colour codes and the colours they produce, open the file *Colourcodes.html* from the *Tutorial 1* folder. The *Colourcodes.html* page demonstrates the 216 web safe colours and their hexadecimal colour codes. These colours are known as 'safe' because both Mac and Windows operating systems display them identically. Although the standard palette contains 256 colours, some of these will be interpreted differently depending on the machine that is used to view them, while the 216 colours demonstrated here appear the same on all platforms. It is worth using a selection of colours from the 216-colour palette to ensure your design looks good wherever it is viewed. A great site for getting ideas about complementary colours for use in web sites is http://www.colormatch.dk/.

Adding text

Now open the *Demo1.html* file. The code used is as shown in Figure 20.2.

```html
<html>
<head>
<title>Dance Music Production</title>
</head>

<body bgcolor="#FFFFFF" text="#000000">
<p>This is the most simple web page you can make!</p>
<p>There are no tables or cells to position the text so this page
will be quite boring.</p>
<p>Using the tags in the HTML, it is possible to tell the browser
to make text
 <b>bold</b>, <u>underlined</u>, or <i>italic</i>.</p>
<p>We could also use tags in the HTML to create a link. For
example, you may wish to <a href="http://www.google.com">click
here</a> to go to www.google.com.</p>
</body>
</html>
```

Figure 20.2 Demo1.html.

We have now added some text to the web site to demonstrate some simple text formatting methods. You will notice quite a few <p> and </p> tags. These tags tell the browser where paragraphs begin and end. You will also notice that text within the <b> and </b> tags becomes bold, between <u> and </u> become underlined, and between <i> and </i> tags text become italicized.

We have also just introduced a way of adding a link to your web site. The format for adding a link is:

```
<a href="[the url of the site you want to link to]">[the link to
appear on screen]</a>
```

If you wanted to add a link to the Yahoo search site (http://www.yahoo.com) and you wanted to make the link from a phrase in your site that says 'here's another search engine', the code would be:

```
<a href="http://www.yahoo.com">here's another search engine</a>
```

Try adding this to *Demo1.html* to see the results. To do this, open *Demo1.html* in a text editor (in Windows go to the **Start** menu, select **Programs**, then **Accessories** and select **Notepad**).

When you have changed the code, save the file under a different name and add a *.html file extension. Now open the file in your browser to see the change.

Creating a layout using tables, rows and data cells

So far we have seen the very basics of creating a web site. We need now to be able to position the text on the page so that we can add our own design and style. To give our web page a specific layout, we need to use tables, rows and table data cells. Tables, rows and table data cells – like most other instructions in HTML – have opening and closing tags defining where tables, rows and table data cells begin and end. Tables always contain rows and table data cells are always contained within these rows. The table data cells contain the actual content, whether that content is Flash animations, text or graphics.

The general layout of a single table is:

```
<table>
 <tr>
  <td>
  [this is the text/images/multimedia to go within the cell]
  </td>
 </tr>
</table>
```

Up to this point you will have noticed how tags always open and close around the text (or graphic) that they are supposed to act upon. In the above case we first tell the browser to start a new table with the <table> tag, then to start a new row within the table by using the <tr> tag. Within this new table row we then start a new table data cell with the <td> tag. Now the table data cell is created we can tell the browser what to put in it.

In *Demo2.html* (Figure 20.3), all the content from *Demo1.html* (Figure 20.2) has been placed within a single table data cell that is contained within a single row, which is contained within a single table. The code that has been added is shown in **bold**.

Demo2.html shows the format of the table tags as previously described with a few added extras.

```
<table width="766" border="0" cellpadding="0" cellspacing="0">
```

The above line is equivalent to beginning a new table with the <table> tag, but with additional attributes to describe its appearance. The width attribute tells the browser to set the table to be 766 pixels wide. The border attribute describes the pixel width of the border that divides table data cells and the table itself. Here it is set to zero. The cellpadding attribute describes the amount of space between the borders of a table and the actual data in the table data cell, and the cellspacing attribute specifies the amount of space to be inserted between the table data cells. The line below describes the opening of a new table data cell. It is the <td> tag with added information.

```
<td width="766">
```

This uses the width attribute to tell the browser to make the table data cell 766 pixels wide.

```
<html>
<head>
<title>Dance Music Production</title>
</head>

<body bgcolor="#FFFFFF" text="#000000">
<table width="766" border="0" cellpadding="0" cellspacing="0">
 <tr>
  <td width="766">
  <p>This is the most simple web page you can make!</p>
  <p>There are no tables or cells to position the text so this
  page will be quite boring.</p>
  <p>Using the tags in the HTML, it is possible to tell the
  browser to make text
  <b>bold</b>, <u>underlined</u>, or <i>italic</i>.</p>
  <p>We could also use tags in the HTML to create a link. For
  example, you may wish to <a href="http://www.google.com">click
  here</a> to go to www.google.com.</p>
  </td>
 </tr>
</table>
</body>
</html>
```

Figure 20.3 Demo2.html.

Demo3.html (Figure 20.4) demonstrates what happens if you set the table border value to 1 and cellpadding value to 8. Now, when you look at the HTML page in your browser, you will now be able to see the border and you will notice that the table data (that is, the content of cell) has moved away from the edges of the border. The code that has been added is shown in **bold**.

Adding images

Now we want to add a picture so that it sits to the left of the text on the page. To do this we need to add another table data cell, as shown in the *Demo4.html* file (Figure 20.5).

A further pair of <td> and </td> tags have been added, creating a new table data cell that contains the text 'Your picture goes here!'. The table is already defined as 766 pixels wide and now we have two table data cells within the same row. You will see that the <tr> and </tr> tags now surround all of the table data (<td> </td>) tags. You can see that the width of the first table data cell that will contain the picture is 155 pixels and the second table data cell (which contains the main text) is now 611 pixels to take into account that the total width of the table is 766 pixels. The cellpadding value has been reset to zero for clarity.

To insert a picture into the table data cell that we have created for it, we will need to tell the browser where to look for the image, as shown in the *Demo5.html* file (Figure 20.6).

```
<html>
<head>
<title>Dance Music Production</title>
</head>

<body bgcolor="#FFFFFF" text="#000000">
<table width="766" border="1" cellpadding="8" cellspacing="0">
 <tr>
  <td width="766">
  <p>This is the most simple web page you can make!</p>
  <p>There are no tables or cells to position the text so this
  page will be quite boring.</p>
  <p>Using the tags in the HTML, it is possible to tell the
  browser to make text
  <b>bold</b>, <u>underlined</u>, or <i>italic</i>.</p>
  <p>We could also use tags in the HTML to create a link. For
  example, you may wish to <a href="http://www.google.com">click
  here</a> to go to www.google.com.</p>
  </td>
 </tr>
</table>
</body>
</html>
```

Figure 20.4 Demo3.html.

```
<html>
<head>
<title>Dance Music Production</title>
</head>

<body bgcolor="#FFFFFF" text="#000000">
<table width="766" border="1" cellpadding="0" cellspacing="0">
 <tr>
  <td width="155">Your picture goes here!</td>
  <td width="611">
  <p>This is the most simple web page you can make!</p>
  <p>There are no tables or cells to position the text so this
  page will be quite boring.</p>
  <p>Using the tags in the HTML, it is possible to tell the
  browser to make text
  <b>bold</b>, <u>underlined</u>, or <i>italic</i>.</p>
  <p>We could also use tags in the HTML to create a link. For
  example, you may wish to <a href="http://www.google.com">click
  here</a> to go to www.google.com.</p>
  </td>
 </tr>
</table>
</body>
</html>
```

Figure 20.5 Demo4.html.

```
<html>
<head>
<title>Dance Music Production</title>
</head>

<body bgcolor="#FFFFFF" text="#000000">
<table width="766" border="1" cellpadding="0" cellspacing="0">
 <tr>
  <td width="155"><img src=" book_cover_section.jpg" width="155"
  height="154" alt="dance image"></td>
  <td width="611">
  <p>This is the most simple web page you can make!</p>
  <p>This text is within single cell, which is inside a single
  table.</p>
  <p>Using the tags in the HTML, it is possible to tell the
  browser to make text
  <b>bold</b>, <u>underlined</u>, or <i>italic</i>.</p>
  <p>We could also use tags in the HTML to
  create a link. For example, you may wish to
  <a href="http://www.google.com">click
  here</a> to go to www.google.com.</p>
  </td>
 </tr>
</table>
</body>
</html>
```

Figure 20.6 Demo5.html.

In the *Demo5.html* example, we have now added a picture instead of the text 'Your picture goes here!'. You can see that where this text was in *Demo4.html* it has been replaced by the code:

```
<img src="book_cover_section.jpg" width="155" height="154"
alt="dance image">
```

The <img> tag tells the browser to look for the image *book_cover_section.jpg*. Because no folder is specified, the browser looks in the same folder as the web page (in this case the **.html* files), so to view this file correctly in your browser window you will have to copy *Demo5.html* and the file *book_cover_section.jpg* into the same folder on your computer. (This should already be the case as they are both contained within the *Tutorial 1* folder.)

Within this tag you will also notice that the width and height of the image are specified using the width and height attributes. In this case, these values reflect the actual size of the image as it was created in Adobe Photoshop.

Take a look at *Demo6.html* (Figure 20.7) in your browser. Visually you will notice nothing at all. The code is as follows, with changes shown in **bold**.

Note: To view this page in your browser, as before ensure the images referred to in the HTML are in the same folder as the **.html* file. In this case copy *Demo6.html* and the file '*book_cover_section_large.jpg*' must be in the same folder.

```
<html>
<head>
<title>Dance Music Production</title>
</head>

<body bgcolor="#FFFFFF" text="#000000">
<table width="766" border="1" cellpadding="0" cellspacing="0">
 <tr>
  <td width="155"><img src="book_cover_section_large.jpg"
  width="155" height="154" alt="dance image"></td>
  <td width="611">
   <p>This is the most simple web page you can make!</p>
   <p>This text is within single cell, which is inside a single
   table.</p>
   <p>Using the tags in the HTML, it is possible to tell the
   browser to make text
   <b>bold</b>, <u>underlined</u>, or <i>italic</i>.</p>
   <p>We could also use tags in the HTML to create a link. For
   example, you may wish to <a href="http://www.google.com"> click
   here </a> to go to www.google.com.</p>
  </td>
 </tr>
</table>
</body>
</html>
```

Figure 20.7 Demo6.html.

The difference between this file and *Demo5.html* is that a different image is loaded into the browser window. It is actually the same image, but is larger both dimensionally and in file size: in *Demo5.html*, the file *book_cover_section.jpg* is 6.67 KB, whereas in *Demo6.html book_cover_section_large.jpg* is 49.0 KB. The point here is that when displayed in the browser over the Internet, *Demo6.html* will take longer to download and display than *Demo5.html*, although the final appearance is the same. The appearance is the same because the width and height attributes in the <img src> tag force the browser to size the image according to the dimensions specified.

Graphical file sizes need to be kept as small as possible to ensure the user of your site gets the quickest download time. As you are viewing this page with all files on your hard drive you will not notice a difference in speed because there is no downloading time to speak of, as there would be over an Internet connection. On the web, however, when someone views your page the images must be downloaded from the hosting company's server, and the bigger the file the longer it will take to reach the browser.

Demo7.html (Figure 20.8) demonstrates how a thumbnail is created from the Dance graphic introduced in *Demo5.html*. By using a thumbnail (a small-scale copy of the image), when this image is clicked on the person viewing the web site will see a full view of the *Dance Music Production and Remixing* book. The code is shown below.

Take a look at the line:

```
<td width="155"><a href="book_cover_full_frontal.jpg">
<img src="book_cover_section.jpg" width="155" height="154"
border="0" alt="dance image thumbnail"></a></td>
```

You will see many similarities between the line just given and the line used to create the link to the Google search engine (http://www.google.com) in the text of the web site. Again, this follows the format shown in *Demo2.html*.

```
<a href="[the url of the site you want to link to]">[the link to
appear on screen]</a>
```

In this case, the <a href="…"> tag contains the location of the graphic that you want to display when the image (*book_cover_section.jpg*) is clicked. As you can see, there is also a closing </a> tag to tell the browser where the actual clickable image ends.

We have almost finished demonstrating many of the key elements to building a web site, but before we build a full site one more important aspect needs to be covered. If you load up *Demo7.html* it should look good on your browser. As I have no idea what browser you are using, I am aware that you will most probably be seeing something a little different to what I see in my browser, the reason being that different browsers will interpret the HTML differently, as mentioned previously. Of particular importance is your monitor resolution. When viewing *Demo7.html*, minimize the display and change the width of the browser window, moving the right side in and out. You will notice that as you do this the text goes off the page and a horizontal scroll bar appears at the bottom of the browser window.

```
<html>
<head>
<title>Dance Music Production</title>
</head>

<body bgcolor="#FFFFFF" text="#000000">
<table width="766" border="1" cellpadding="0" cellspacing="0">
 <tr>
  <td width="155"><a href="book_cover_full_frontal.jpg">
  <img src="book_cover_section.jpg" width="155" height="154"
  border="0" alt="dance image thumbnail"></a></td>
  <td width="611">
   <p>This is the most simple web page you can make!</p>
   <p>This text is within single cell, which is inside a single
   table.</p>
   <p>Using the tags in the HTML, it is possible to tell the
   browser to make text
   <b>bold</b>, <u>underlined</u>, or <i>italic</i>.</p>
   <p>We could also use tags in the HTML to create a link. For
   example, you may wish to <a href="http://www.google.com">click
   here</a> to go to www.google.com.</p>
  </td>
 </tr>
</table>
</body>
</html>
```

Figure 20.8 Demo7.html.

In web sites it is preferable not to have to make your viewers use this bar in the first place to view your page, as scrolling horizontally can be tedious. The grab and drag test that you just did simulates lower screen resolutions.

There are two ways of dealing with the differing viewing resolutions that people will be using to view your web page:

* Make your web pages a specific width to cater for the 'average' user.
* Make you site always fill a percentage of the page width.

If making your pages a specific width, it is recommended that a fixed width of no more than 760 pixels be used. The demonstrations so far are all fixed width pages. You can see that the table containing all the content of the page is 766 pixels wide. Optimally, if you decide to go down the fixed width path, go for a page of around 760. It will allow people with very-low-resolution monitors to have a reasonable viewing experience and also prevents your web page looking too small on high-resolution monitors. The 'pros' of this method are that your page will have a fixed magazine style to it and the layout will appear the same on all browsers, although the viewed size may vary depending on the viewing monitor's screen resolution.

The second method, and one that I prefer, is to make the page a relative width. This has the downside that your page will look slightly different at different resolutions because the width will vary on different screens, but in most cases it allows your viewer to get quick visual access to the information in your page.

Take a look at *Demo8.html* in your browser. If you look at the code for it in Figure 20.9, the main table setting has changed to 100% width from a fixed pixel value. When you grab and drag to change the width of your browser window now, the text scrolls to fit as the page is resized. Changing this value to 90%, for example, means that the page will always fill 90% of the open browser window. You will also notice that the second opening <td> tag now reads <td width=100%>. This is because when you use relative width tables at least one of the table data cells must be a relative percentage.

Now take a look at *Demo9.html* (Figure 20.10).

Here you will see that the two table data cells together fill 100% of the table width, each cell being 50% of the page's width.

Using this knowledge we will now go on to build a more professional-looking site from scratch in Tutorial 2.

Tutorial 2: HTML used for the Dance Music Production web site

The second part of this tutorial can be found in the folder named *Tutorial 2* supplied on the accompanying CD. Again, copy the entire contents of this folder to your hard drive and open the files as required.

We have so far shown how to create a layout that contains some simple content. One of the best ways to learn HTML is to play around with the code. Save it as an **.html* file and see the

```
<html>
<head>
<title>Dance Music Production</title>
</head>

<body bgcolor="#FFFFFF" text="#000000">
<table width="100%" border="1" cellpadding="0" cellspacing="0">
 <tr>
  <td width="155"><a href="book_cover_full_frontal.jpg">
  <img src="book_cover_section.jpg" width="155" height="154"
  border="0" alt="dance image thumbnail"></a></td>
  <td width="100%">
   <p>This is the most simple web page you can make!</p>
   <p>This text is within single cell, which is inside a single
   table.</p>
   <p>Using the tags in the HTML, it is possible to tell the
   browser to make text
   <b>bold</b>, <u>underlined</u>, or <i>italic</i>.</p>
   <p>We could also use tags in the HTML to create a link. For
   example, you may wish to <a href="http://www.google.com">click
   here </a> to go to www.google.com.</p>
  </td>
 </tr>
</table>
</body>
</html>
```

Figure 20.9 Demo8.html.

effect that you have made by altering attributes within the code. Play around and create some layouts of your own. If you understand how tables, rows and table data cells work, you are well on the way to understanding how to build a great web site, as the inputting of visual content is a relatively trivial exercise compared to understanding the layout tables, rows and table data cells.

The aim of this tutorial is to go through the steps involved in creating the demonstration Dance Music Production web site, which will comprise of three main pages and two subpages, the first page of which will look like Figure 20.11 when it is completed. You will notice that the page is composed in three parts – a header, the main content and a footer – and it is in this order that we will be creating this site.

Setting up the basic page

Firstly, we begin with the standard format of all web pages: setting up the opening <html> tag, telling the browser that the description of the page has begun, and the closing </html> tag, which is the corresponding closing tag. Without these the browser cannot know where the page begins and ends. Also included are the <head> and corresponding closing </head> tags. For the moment we won't put anything in here. We have also included the <body> and closing </body> tags. It is between these two latter tags that all of the code describing the layout and

```
<html>
<head>
<title>Dance Music Production</title>
</head>

<body bgcolor="#FFFFFF" text="#000000">
<table width="100%" border="1" cellpadding="0" cellspacing="0">
 <tr>
  <td width="50%"><a href="book_cover_full_frontal.jpg">
  <img src="book_cover_section.jpg" width="155" height="154"
  border="0" alt="dance image thumbnail"></a></td>
  <td width="50%">
   <p>This is the most simple web page you can make!</p>
   <p>This text is within single cell, which is inside a single
   table.</p>
   <p>Using the tags in the HTML, it is possible to tell the
   browser to make text
   <b>bold</b>, <u>underlined</u>, or <i>italic</i>.</p>
   <p>We could also use tags in the HTML to create a link. For
   example,you may wish to <a href="http://www.google.com">click
   here </a> to go to www.google.com.</p>
  </td>
 </tr>
</table>
</body>
</html>
```

Figure 20.10 Demo9.html.

Figure 20.11 The Dance Music Production demonstration web site home page.

content of the page will be written. We have also added the opening <title> and closing </title> tags to give the page a title. The code used in *Part1.html* is as shown in Figure 20.12.

```
<html>
<head>
<title>Dance Music Production and Remixing</title>
</head>
<body>

</body>
</html>
```

Figure 20.12 Part1.html.

Creating the header page

Developing the layout tables

Now we are going to create the header of the web site. This displays an image that will be used throughout all the web pages to give it a consistent feel. The code may look a little scary at first, but let's break it down. The code that has been added to the previous example is shown in **bold** (Figure 20.13).

```
<html>
<head>
<title>Dance Music Production and Remixing</title>
</head>
<body>
<!-- BEGINNING OF CODE TO DISPLAY THE PAGE HEADER -->
<table width="1259" border="0" cellpadding="0" cellspacing="0">
 <tr>
  <td width="291" height="80">Title of the web site image</td>
  <td width="535">
      <!-- BEGINNING OF CODE TO DESCRIBE THE TABLE WITHIN THE
HEADER
  TABLE -->
   <table width="535" border="0" cellpadding="0" cellspacing="0">
    <tr>
     <td width="535" height="80">Adjoining graphical bar</td>
    </tr>
   </table>
          <!-- END OF CODE TO DESCRIBE THE TABLE WITHIN THE
HEADER
   TABLE -->
  </td>
  <td width="433" height="80">Right side image</td>
 </tr>
</table>
<!-- END OF CODE TO DISPLAY THE PAGE HEADER -->
</body>
</html>
```

Figure 20.13 Part2.html.

Firstly, you will notice the line beginning with <!--. This indicates that a comment is inserted into the code. Anything that is written after this and before the --> will not be interpreted by the browser. These comments are purely there to provide information about what the code is doing and to simplify the understanding of how the page works to the developer. The first tag that is actually read by the browser is the line after this comment. We are first setting up the layout of the header of the page, then we'll fill the table data cells with graphics.

The following line defines the outer table of the header:

```
<table width="1259" border="0" cellpadding="0" cellspacing="0">
```

All table data cells must be contained within a row, which must be contained within a table. Table data cells cannot exist alone. We can see that the table is 1259 pixels wide, has no visible border, no cell padding and no cell spacing.

We now tell the browser that a new row starts with the <tr> tag. Remember that table data cells cannot exist unless they are in a table and a table row tag. It is also true that rows cannot exist unless they are in a table. This is why the opening <tr> tag occurs after the opening <table> tag.

Importantly, table data cells – the cells that actually contain visual content for the web site – must exist within a row and these rows must exist within a table.

So, as already demonstrated in Tutorial 1, the format of layout tables is:

```
<table>
<tr>
<td>
[the actual content of the table data cell]
</td>
</tr>
</table>
```

Important to our layout is the possibility of putting tables within tables. This is known as 'nesting'. You can 'nest' tables within tables, within tables, and so on, but be warned, if you nest too many times, some browsers may get upset and 'hang' (stop working). This is certainly true of old Netscape browsers! A nested table has to be within the opening and closing tags of a table data cell. An example of a nested table is as follows, whereby the indented code describes the nested table. The nested table is shown in **bold.**

```
<table>
<tr>
<td>
    <table>
    <tr>
    <td>
    [The content of the nested table]
    </td>
```

```
    </tr>
    </table>
</td>
</tr>
</table>
```

As you are reading this, take a look at *Part2.html* in your browser of choice (preferably a recent Internet Explorer or Netscape browser).

The following lines occur after the table opening tag:

```
<tr>
  <td width="291" height="80">Title of the web site image</td>
  <td width="535">
```

This tells the browser to begin a new row (<tr>) and then a table data cell is created with a width of 291 pixels and a height of 80 pixels. The width and height attributes affect the table data cell being described as they occur within the tag that is closed by the > symbol before the displayed text. Anything that is now put after <td width="291" height="80"> will visually appear within this cell (and therefore on the web site). In this instance it is text that will be displayed in the browser window; as you will see soon, it could be a graphic, an animation or anything you want. To keep things as simple as possible for the moment, we will just display some text. The table data cell's content ends at the </td> tag as in the above code.

You will notice that we are now telling the browser to open another table data cell with the line <td width="535">. You may recall a little time ago that we opened a row with <tr>. We are now telling the browser that we want another table data cell within the same row. The browser will know that this is the case because we haven't yet introduced a </tr> tag to close the row. The next table data cell is 535 pixels wide.

The code after the opening <td width="535"> is:

```
<!--BEGINNING OF CODE TO DESCRIBE THE TABLE WITHIN THE HEADER TABLE-->
    <table width="535" border="0" cellpadding="0" cellspacing="0".
    <tr>
     <td width="535" height="80">Adjoining graphical bar</td>
    </tr>
    </table>
<!--END OF CODE TO DESCRIBE THE TABLE WITHIN THE HEADER TABLE-->
```

Here we are nesting a table within the next table data cell. As you can see, this follows the standard format of opening a table, opening a row, opening a table data cell, inserting the content of the table data cell, closing the table data cell, closing the row, then closing the table. As you can see, this table is 535 pixels wide and the table data cell is 535 pixels wide by 80 pixels high and contains the text 'Adjoining graphical bar', which will be replaced soon with graphical content.

The final bit of code for the header reads:

```
    </td>
    <td width="433" height="80">Right side image</td>
  </tr>
</table>
<!--END OF CODE TO DISPLAY THE PAGE HEADER-->
```

The </td> tag closes the table data cell in which we put the nested table. We now add a third table data cell within, as before, the same row, that is 433 pixels wide by 80 pixels high and contains the text 'Right side image'. Note how, yet again, we tell the browser that the table data cell closes after this text with the </td> tag. We then close the row with the </tr> tag and close the table with the </table> tag.

In summary, we have a table that spans 1259 pixels and contains a single row. This row contains a table data cell that is 291 pixels wide by 80 pixels high; this cell contains the text 'Title of the web site image'. Next to this in the same row is a table that is 535 pixels wide and contains a table data cell that is 535 pixels wide by 80 pixels high. After this table, in the same row, is another table data cell that is 433 pixels wide and contains the text 'Right side image'.

To help you understand how this table actually looks in your browser, open *Part2.html* in a text or HTML editor and change the line that reads:

```
<table width="1259" border="0" cellpadding="0" cellspacing="0">
```

to:

```
<table width="1259" border="1" cellpadding="0" cellspacing="0">
```

This now gives the outermost table a border value of 1. Save this file as *Part2temp.html* and view it in your browser to see how the table layout looks in relation to the text with this new setting.

You may be wondering why a nested table was used in the header bar of the web site described above. This is because we are going to use a relatively complex arrangement of cells within it to create an interesting graphical illusion with the two header graphics that sit either side of this nested table.

Introducing graphics

In *Part3.html* we have made two important changes to *Part2.html* in that we have introduced two graphics into the header of the web site. The changes made are highlighted in **bold**. The code is shown in Figure 20.14.

> Note: If, when you open Part3.html, you do not see two graphics on the page, use the scroll bar in your browser to pan to the right. If your browser is set to low resolution it may currently be off the natural viewing area.

```
<html>
<head>
<title>Dance Music Production and Remixing</title>
</head>
<body>
<!-- BEGINNING OF CODE TO DISPLAY THE PAGE HEADER -->
<table width="1259" border="0" cellpadding="0" cellspacing="0">
 <tr>
  <td width="291" height="80"><img src="HEADER_LEFTSIDE.jpg"
  width="291" height="80" alt="header graphic displaying DANCE
  MUSIC PRODUCTION AND REMIXING"></td>
  <td width="535">
       <!-- BEGINNING OF CODE TO DESCRIBE THE TABLE WITHIN THE
HEADER
  TABLE -->
  <table width="535" border="0" cellpadding="0" cellspacing="0">
   <tr>
    <td width="535" height="80">Adjoining graphical bar</td>
   </tr>
  </table>
           <!-- END OF CODE TO DESCRIBE THE TABLE WITHIN THE
HEADER
    TABLE -->
  </td>
  <td width="433" height="80"><img src="HEADER_RIGHTSIDE.jpg"
  width="433" height="80" alt=""></td>
 </tr>
</table>
<!-- END OF CODE TO DISPLAY THE PAGE HEADER -->
</body>
</html>
```

Figure 20.14 Part3.html.

You can see that the text 'Title of the web site image' has been replaced with the following code:

```
<img src="HEADER_LEFTSIDE.jpg" width="291" height="80" alt="header
graphic displaying DANCE MUSIC PRODUCTION AND REMIXING">
```

You may recall that anything we put within a <td> and corresponding </td> tags will be displayed within the browser window, as this is the visual content of the table data cell and has little to do with the layout (unless you were nesting a table within the <td> and </td> tags).

To place an image in a web page we use the format:

```
<img src="[whatever your filename is]"
```

You can also see that we have made the browser display this image as 291 pixels wide by 80 pixels high. These two attributes affect the image size as they occur within the opening '<' and closing '>' of the <img src> tag. This tag is one of the few tags in HTML that does not need a separate closing tag.

From the code, you can see that the image *HEADER_LEFTSIDE.jpg* should be in the same folder as the corresponding HTML file (in this case *Part3.html*), otherwise the browser will not see the file.

As this point it's worth mentioning how you control where the browser will look for files referred to in you web site code.

If, for example, this image file was in a folder called *Pictures* and this folder was located in the same directory as the **.html* file, in order for the browser to find the image, the line in the code would read as follows:

```
<img src="Pictures/HEADER_LEFTSIDE.jpg" width="291" height="80">
```

This is shown in Figure 20.15.

Figure 20.15

If the file *Part3.html* was in a folder called *WebPages* and this folder were within a folder called *Pictures* that actually contained the file referred to in the html, the line would read as follows:

```
<img src="../HEADER_LEFTSIDE.jpg" width="291" height="80">
```

This tells the browser to look one step in the folder tree backwards (as denoted by the ../) and then look in the graphic, which in this instance is in the folder *Pictures*. This is shown in Figure 20.16.

Figure 20.16

As a final example, let's say you had your directory structure set up as shown in Figure 20.17.

With this folder structure the code in the site would read:

```
<img src="../Pictures/HEADER_LEFTSIDE.jpg" width="291" height="80">
```

Figure 20.17

This tells the browser that *HEADER_LEFTSIDE.jpg* can be located one step backwards in the tree and is in the *Pictures* folder. If the folder containing the picture were two steps backwards in the directory tree, the line would read:

```
<img src="../../Pictures/HEADER_LEFTSIDE.jpg" width="291" height="80">
```

By understanding how to tell browsers to look for aspects of the data in your site, you can create tidy web sites by putting data for it in logical folders. This will make maintaining your site easier in the future.

The dimensions that we have stated the file to be, in this case, are the actual size of the file created in Photoshop.

You can also see that the code <img src="HEADER_RIGHTSIDE.jpg" width="433" height="80"> has replaced the text 'Right side image' in the code *Part2.html* to insert the image *HEADER_LEFTSIDE.jpg* into the web page.

Changing the background colour of the page

At the moment the site does not use a background colour. This can be changed to a specific colour by altering the attributes in the <body> tag. Try changing the line starting with the <body> tag so that it reads <body bgcolor="#FF0000">. Save the file with the *.html* extension and then open it in your browser. You will notice that the background is red. The bgcolor attribute simply means 'background colour' and because bgcolor="#FF0000" has been written within the body tag, this background colour applies to the body of the page, which encompasses the background.

Open *Part4.html*. This is an example where we have simply changed the <body> tag as described above.

The red creates a big clash with the predominantly blue graphics on the page. Ideally, it would be a deep blue to match the graphics already on the page, as shown in *Part5.html*.

In *Part5.html*, the line starting with the <body> tag has simply been replaced with <body bgcolor="#000066">. Take a look at *Colourcodes.html* provided on the CD.

Although there are many colours to choose from here, the colour that will look best within the design is the actual backing colour to the two graphics used on the site. The deep blue that you can see is still not the desired colour, as preferably the background blue would appear exactly the same as the deep blue used in the graphics, so a little trick will be implemented to get things looking right.

So, instead of using the bgcolor attribute we are going to tell the browser to use a graphic for the background colour, as shown in *Part6.html* (Figure 20.18).

```
<html>
<head>
<title>Dance Music Production and Remixing</title>
</head>
<body background="background_colour.gif">
<!-- BEGINNING OF CODE TO DISPLAY THE PAGE HEADER -->
<table width="1259" border="0" cellpadding="0" cellspacing="0">
 <tr>
  <td width="291" height="80"><img src="HEADER_LEFTSIDE.jpg"
  width="291" height="80" alt="header graphic displaying DANCE MUSIC
  PRODUCTION AND REMIXING"></td>
  <td width="535">
        <!-- BEGINNING OF CODE TO DESCRIBE THE TABLE WITHIN THE
HEADER
 TABLE -->
   <table width="535" border="0" cellpadding="0" cellspacing="0">
    <tr>
     <td width="535" height="80">Adjoining graphical bar</td>
    </tr>
   </table>
           <!-- END OF CODE TO DESCRIBE THE TABLE WITHIN THE
HEADER
   TABLE -->
  </td>
  <td width="433" height="80"><img src="HEADER_RIGHTSIDE.jpg"
  width="433" height="80" alt=""></td>
 </tr>
</table>
<!-- END OF CODE TO DISPLAY THE PAGE HEADER -->
</body>
</html>
```

Figure 20.18 Part6.html.

By using <body background="background_colour.gif">, the HTML tells the browser to fill the background of the web site with the *background_colour.gif* image file. It will automatically tile this image to fill the page. So that the web site speed is not compromised, this image is in fact 1 pixel by 1 pixel in size – the smallest image we can use in a web site. It was created by high-lighting an area of 1 pixel by 1 pixel from the original *HEADER_LEFTSIDE.jpg* image created in Adobe Photoshop, then copying and pasting the highlighted area to a new window. We know this is the colour we want, as it was copied directly from the original graphic. This image was then saved as *background_colour.gif*. This approach is useful if you need a background colour with the exact hue of an area of your graphic. You could try another image to see how it looks, but remember to make the file as tiny as possible to avoid slowing down your web site when it's live on the Internet.

Using the 1×1 pixel transparent GIF

Next we need to now fill the nested table to create the illusion that the white lines in the two graphics either side of this table are continuous across the page. To do this we need a series of table data cells, some of which will be filled white and some of which will have no fill colour. In the event of a table data cell having no fill colour, it effectively appears as transparent and the background colour is displayed. You may recall the magic of a 1 pixel by 1 pixel transparent GIF image. We will make extensive use of this here to ensure that this site also works on many older browsers (particularly some older Netscape versions). *Part7.html* (Figure 20.19) shows the changes in the code where the 1×1 transparent GIF has been added.

```
<html>
<head>
<title>Dance Music Production and Remixing</title>
</head>
<body background="background_colour.gif">
<!-- BEGINNING OF CODE TO DISPLAY THE PAGE HEADER -->
<table width="1259" border="0" cellpadding="0" cellspacing="0">
 <tr>
  <td width="291" height="80"><img src="HEADER_LEFTSIDE.jpg"
  width="291" height="80" alt="header graphic displaying DANCE MUSIC
  PRODUCTION AND REMIXING"></td>
  <td width="535">
  <!-- BEGINNING OF CODE TO DESCRIBE THE TABLE WITHIN THE HEADER
  TABLE -->
  <table width="535" border="0" cellpadding="0" cellspacing="0">
   <tr>
    <td><img src="1×1.gif" width="1" height="38" alt=""></td>
   </tr>
   <tr>
    <td bgcolor="#FFFFFF"><img src="1×1.gif" width="1"
    height="2" alt=""></td>
   </tr>
   <tr>
   <td><img src="1×1.gif" width="1" height="10" alt=""></td>
  </tr>
  <tr>
   <td bgcolor="#FFFFFF"><img src="1×1.gif" width="1"
   height="3" alt=""></td>
  </tr>
  <tr>
   <td><img src="1×1.gif" width="1" height="10" alt=""></td>
  </tr>
  <tr>
```

Figure 20.19 *(Continues).*

```
   <td bgcolor="#FFFFFF"><img src="1×1.gif" width="1"
   height="1" alt=""></td>
  </tr>
  <tr>
   <td><img src="1×1.gif" width="1" height="16" alt=""></td>
  </tr>
 </table>
 <!-- END OF CODE TO DESCRIBE THE TABLE WITHIN THE HEADER TABLE -->
  </td>
  <td width="433" height="80"><img src="HEADER_RIGHTSIDE.jpg"
  width="433" height="80" alt=""></td>
 </tr>
</table>
<!-- END OF CODE TO DISPLAY THE PAGE HEADER -->
</body>
</html>
```

Figure 20.19 Part 7.html.

We have now replaced the placeholder 'Adjoining graphical bar' from *Part6.html* with the code shown in **bold** above.

As you can see, within the nested table we have created a series of rows, each containing a table data cell. Within each table data cell, we have inserted a 1 pixel by 1 pixel transparent GIF image (*1×1.gif*) and we have set the height of this image to correspond to where the white lines occur in the *HEADER_LEFTSIDE.jpg* and *HEADER_RIGHTSIDE.jpg* graphics that lie either side of the nested table.

Where a white line occurs the code reads <td bgcolor="#FFFFFF">. This tells the browser that the background colour of the table data cell is white (FFFFFF is the hex code for white). Where we want the background colour of the page to come through we have not given the <td> tag a bgcolor attribute, so this is interpreted as transparent and the colour described within the <body> tag is visible.

If you are viewing this on a 1280 × 1024 resolution browser, you will probably see the entire header in the browser window. If you are viewing this web page on a lower resolution browser you may have to scroll to the right of the page to see it all. As already mentioned, in good web sites horizontal scrolling should to be avoided, as it's cumbersome for your surfers. To sort this out and finish the header of the page, we want to make this header alter its size according to the width of the browser window.

Setting the page width

Open *Part8.html* in your browser. The code that has changed is highlighted in **bold** in Figure 20.20.

As you can see, we have stated that the outermost table will fill 100% of the page width by changing the value of the width attribute from an absolute pixel value to a relative value in the first <table> tag.

```html
<html>
<head>
<title>Dance Music Production and Remixing</title>
</head>
<body background="background_colour.gif">
<!-- BEGINNING OF CODE TO DISPLAY THE PAGE HEADER -->
<table width="100%" border="0" cellpadding="0" cellspacing="0">
 <tr>
  <td width="291" height="80"><img src="HEADER_LEFTSIDE.jpg"
  width="291" height="80" alt="header graphic displaying DANCE MUSIC
  PRODUCTION AND REMIXING"></td>
  <td width="100%">
   <!-- BEGINNING OF CODE TO DESCRIBE THE TABLE WITHIN THE HEADER
   TABLE -->
   <table width="100%" border="0" cellpadding="0" cellspacing="0">
    <tr>
     <td width="100%"><img src="1X1.gif" width="1" height="38"
     alt=""></td>
    </tr>
    <tr>
     <td width="100%" bgcolor="#FFFFFF"><img src="1X1.gif" width="1"
     height="2" alt=""></td>
    </tr>
    <tr>
     <td width="100%"><img src="1X1.gif" width="1" height="10"
     alt=""></td>
    </tr>
    <tr>
     <td width="100%" bgcolor="#FFFFFF"><img src="1X1.gif"
     width="1" height="3" alt=""></td>
    </tr>
    <tr>
     <td width="100%"><img src="1X1.gif" width="1" height="10"
     alt=""></td>
    </tr>
    <tr>
     <td width="100%" bgcolor="#FFFFFF"><img src="1X1.gif"
     width="1" height="1" alt=""></td>
    </tr>
    <tr>
     <td width="100%"><img src="1X1.gif" width="1" height="16"
     alt=""></td>
   </tr>
  </table>
  <!-- END OF CODE TO DESCRIBE THE TABLE WITHIN THE HEADER TABLE -->
  </td>
  <td width="433" height="80"><img src="HEADER_RIGHTSIDE.jpg"
  width="433" height="80" alt=""></td>
  </tr>
```

Figure 20.20 (*Continues*)

```
</table>
<!-- END OF CODE TO DISPLAY THE PAGE HEADER -->
</body>
</html>
```

Figure 20.20 Part8.html.

```
<table width="100%" border="0" cellpadding="0" cellspacing="0">
```

We have also stated that the table data cell that contains the nested tables is also 100% width with the line:

```
<td width="100%">
```

The nested table itself is now also 100% width and all of the table data cells within the nested table are 100% width.

As mentioned before, technically correct HTML will not always work if you are aiming your web site to work on older browsers (which you should, because there are millions of people out there who do not keep their software updated). For example, you could remove many of the 100% width values of the table data cells and this web site will still function completely correctly. If you learn pure HTML and apply this to complex layout structures such as the one above, the web page you create is likely to look very strange on some browsers.

It is important to remember that pure HTML is fundamentally flawed because it must be interpreted by the browser application, which is why it is usually necessary for the HTML language to be adjusted. Other professional web developers may gasp at this statement, but they know all too well the difficulties of getting a complex, structured web site to work well on new and older browsers using pure HTML: different browsers interpret HTML differently. You need to be prepared to write good, clean HTML code and then play with it until it works on all the browsers that you want it to work on. This is why many people prefer to pay someone else to do it, although commonly, many web site designers do not take too much care to get the pages working on many browsers other than the browser used by the person who is paying them.

Some web site designers have given up on HTML altogether and prefer to write web sites purely in Macromedia Flash, creating web sites that, unknown to their clients, are more likely to lose customers than gain them.

Take a look at your browser. This is a nice clean header bar for the web site. As you change your screen resolution or alter the page width by dragging from the right, the page should stay in view, unless you make the screen less than 600 pixels wide.

If you have understood so far, then you're almost a web site builder! The building of the header bar is the most complex layout that we'll use in this site. All the graphics are optimized for size and even on a 56K modem this would load relatively quickly. The nested table contains a total of seven tiny 1×1 pixel transparent images (each 43 bytes in size) and a large effect has been achieved with virtually zero increase in page download time.

Creating the page content

We now need to work on the content of the page. To make this as clear as possible, a new web page will now be built to deal with the content. When this is completed, another page will be built to deal with the footer of the page. The header bar, content and footer will then be combined into one page to create the final web site front page.

Take a look at *Part9.html* (Figure 20.21).

Open this page up in your browser. If you wish to view the actual table layout then, as before, change the value of the border attribute in the description of the outermost table to 1, so that it reads:

```
<table width="1259" border="1" cellpadding="0" cellspacing="0">
```

Save this file as *Part99.html* then view the page in your browser.

You will notice many similarities between the code shown above and the code we used for the header of the site. Hopefully, you will see that even complex web sites are simply a case of planning out your tables, rows and table data cells in pixels then using the code to make the layout of the site. You then fill these cells with images and text as required. When you have finished, open *Part9.html* again.

This page is composed of an outermost table that contains three nested tables. The code corresponding to the three nested tables is highlighted in **bold** in Figure 20.21. You will notice a few new tags. The <br> tag has no closing tag and tells the browser to insert a line break. This means 'line break' and has no closing tag, and is the equivalent to pressing the return key whilst holding the **Shift** key in a word processor such as Microsoft Word. You will also notice the <font> tag, which works with the closing </font> tag.

```
<font color="#FFFFFF">[text that you want to affect with the font
tag]</font>
```

By using the <font> tags like this, you can see that the text of *Part9.html* when viewed in a browser appears white. This is simply used to demonstrate how the <font>tag is used. There are other attributes that can change the actual typeface used, but as you will soon see, we will use CSS later to control how the text looks in the page. For the purpose of this demonstration, we have set the text as white (hex code FFFFFF) so that you can see it clearly on the deep blue page background; the text would otherwise appear as the default colour, black.

The valign attribute within the table data cells' description is set to top and causes the contents of the table data cell to vertically align to the top, hence valign.

If you look at the code you will notice that the second nested table has been set to 100% and the table data cell within has a width set at 100%. This is because this web page always needs to scale to 100% of the width of the page. The menu on the far left of the page and the book picture on the far right need to stay in place, so these two table data cells have fixed values. You will notice that the first and the third nested tables contain a 1 × 1 pixel image that is set to the width of their respective tables. This is to prevent a phenomenon known as 'collapsing'.

```
<html>
<head>
<title>Dance Music Production and Remixing</title>
</head>
<body background="background_colour.gif">
<table width="100%" border="0" cellpadding="0" cellspsacing="0">
 <tr>
  <td width="229"valign="top">
    <!-- Beginning of the 1st nested table - this will hold the left
    menu contents of the Web Page -->
    <table width="229" border="0" cellpadding="0" cellspacing="0">
    <tr>
      <td valign="top" height="609">
       <img src="1×1.gif" width="223" height="1" alt=""><br>
       <font color="#FFFFFF">1st nested table</font><br>
      </td>
     </tr>
    </table>
    <!-- End of the 1st nested table -->
 </td>
 <td width="100%" valign="top">
    <!-- Beginning of 2nd nested table - this will hold the main
    contents of the Web Page -->
    <table width="100%" border="0" cellpadding="0" cellspacing="0">
     <tr>
      <td width="100%" height="610" valign="top">
       <p><font color="#FFFFFF">2nd nested table</font></p>
      </td>
     </tr>
    </table>
    <!-- End of the 2nd nested table -->
 </td>
 <td width="231" valign="top">
    <!-- Beginning of 3rd nested table - this will hold a picture of
    the book cover -->
    <table width="231" border="0" cellpadding="0" cellspacing="0">
     <tr>
      <td height="610" valign="top"><img src="1×1.gif" width="231"
      height="1" alt=""<br>
       <font color="#FFFFFF">Third nested table</font></td>
     </tr>
    </table>
    <!-- End of 3rd nested table -->
 </td>
 </tr>
</table>
</body>
</html>
```

Figure 20.21 Part9.html.

Whenever a web page has a table that is set to 100% width it means literally that. In this case it would make sense to imagine that the second nested table is pushing outwards towards the end of the page, because it's trying to fill 100% of the page. Our code stops it doing this because as it pushes outwards there are two blocks of pixels on either side that are effectively acting as wedges between the second nested table and the edges of the page. These wedges are the 1×1 pixel transparent GIFs. Using the 1×1 pixel transparent GIF images in this way helps your web site work on old browsers and allows people to view your page on low-resolution monitors without having to use the horizontal scroll bar.

Adding a menu to the content page

We will now create a very simple menu to complete the contents of the first nested table.

Take a look at *Part10.html* in your web browser. The code highlighted in **bold** is the code that has been changed from *Part9.html* (Figure 20.22).

```
<html>
<head>
<title>Dance Music Production and Remixing</title>
</head>
<body background="background_colour.gif">
<table width="100%" border="0" cellpadding="0" cellspacing="0">
 <tr>
  <td width="229" valign="top"
  <!-- Beginning of the 1st nested table - this will hold the left
   menu contents of the Web Page -->
  <table width="229" border="0" cellpadding="0" cellspacing="0">
   <tr>
   <td valign="top" height="609">
     <img src="1X1.gif" width="229" height="1" alt=""><br>
     <div align="center"><font
     color="#FFFFFF">INTRODUCTION<br></font></div>
     <div align="center"><font color="#FFFFFF">SOUND
     DEMOS<br></font></div>
     <div align="center"><font
     color="#FFFFFF">EQUIPMENT</font></div>
   </td>
   </tr>
  </table>
  <!-- End of the 1st nested table -->
  </td>
  <td width="100%" valign="top">
  <!-- Beginning of 2nd nested table - this will hold the main
  contents of the Web Page -->
  <table width="100%" border="0" cellpadding="0" cellspacing="0">
   <tr>
    <td width="100%" height="610" valign="top">
     <p><font color="#FFFFFF">2nd nested table</font></p>
```

Figure 20.22 (*Continues*)

```
      </td>
    </tr>
  </table>
  <!-- End of the 2nd nested table -->
  </td>
  <td width="231" valign="top">
  <!-- Beginning of 3rd nested table - this will hold a picture of
  the book cover -->
  <table width="231" border="0" cellpadding="0" cellspacing="0">
    <tr>
      <td height="610" valign="top"><img src="1×1.gif" width="231"
      height="1" alt=""><br>
        <font color="#FFFFFF">Third nested table</font></td>
    </tr>
  </table>
  <!-- End of 3rd nested table -->
  </td>
  </tr>
  </table>
  </body>
  </html>
```

Figure 20.22 Part10.html.

We will go through this code step by step:

```
<table width="229" border="0" cellpadding="0" cellspacing="0">
        <tr>
          <td valign="top" height="609"
```

This first section of code tells the browser that the first nested table is 229 pixels wide and has zero as a value for the other attributes. The <tr> tag opens a new row and the new table data cell (opened with the <td> tag) has a height of 609 pixels and contains the valign attribute. As previously mentioned, the valign attribute sets the vertical alignment of the contents of the table data cell to top, middle or bottom, in this case to the top of the cell.

```
<img src="1×1.gif" width="229" height="1" alt=""><br>
<div Align="center"><font color="#FFFFFF">INTRODUCTION<br></font></div>
<div align="center"><font color="#FFFFFF">SOUND DEMOS<br></font></div>
<div align="center"><font color="#FFFFFF">EQUIPMENT</font></div>
```

The last part of this newly added code inserts a 1×1 pixel transparent GIF image that is 229 pixels wide. This is the 'wedge' described in the explanation of *Part9.html*.

You will notice that we have used <div align="center"> with corresponding closing </div> tags around the displayed text. This sets the alignment of everything within the opening <div align="center"> and closing </div> tags. This is unlike the <valign>tag within the <td> tag above, which affects all of the content of the table data cell. The possible values of <div align> are left, center, right or justify.

You will also see <br> tags, which used in this way make the text of the menu list vertically. At the end of the code you can see the closing </td>, </tr> and </table> tags, with which you are already familiar.

Now open *Part11.html* in your browser. You will now notice that some content has been added to the second nested table. We have simply replaced all of the lines from:

```
<!-- Beginning of 2nd nested table - this will hold the main contents
of the Web Page -->
```

to:

```
<!-- End of the 2nd nested table -->
```

with the code shown below, which describes a complete table:

```
<!-- Beginning of 2nd nested table - this will hold the main contents
of the Web Page -->
  <table width="100%" border="0" cellpadding="0" cellspacing="0">
   <tr>
     <td width="100%" height="610" valign="top">
     <p><b><font color="#FFFFFF">DANCE MUSIC PRODUCTION AND
     REMIXING</font></b><font color="#FFFFFF"><br>
     <br>
     Rick Snoman's no-nonsense guide to writing and producing dance
     music will be all you need to create original tracks of your
     chosen dance genre. He covers everything from the basics of
     sequencers, through to VST instruments, programming music and
     sounds, mixing, remixing and mastering. You will also find
     crucial advice on promoting and distributing your finished
     work.</font></p>
     <p><font color="#FFFFFF"><b>CONTENTS INCLUDE</b><br>
     <br>
     <img src="arrow.gif" width="10" height="10" alt="arrow graphic">
     <b>MIDI Basics</b>
     <i>Introduction to MIDI and sequencing - the binary language and
     bits, programming system exclusive, CC's, NRPN, RPN, MIDI
     limitations, timing, multi MIDI interfaces, MTC and setting up a
     MIDI dance studio.</i>
     <br>
     <br>
     <img src="arrow.gif" width="10" height="10" alt="arrow graphic">
     <b>Basic music theory</b>
     <i>Introduction to music theory in dance - The musical scale, chord
     structures, programming/creating dance grooves, programming
     motifs, riffs, and arrangement principles. Including MIDI and audio
     examples.</i><br>
     <br>
     <img src="arrow.gif" width="10" height="10" alt="arrow graphic">
```

```
<b>Synthesis Basics</b><br>
<i>Introduction to synthesis - Waveforms, Fourier analysis,
B's,acoustic theory, granular, FM and subtractive
synthesis</i><br>
<br>
<img src="arrow.gif" width="10" height="10" alt="arrow graphic">
<b>Sampling and digital audio</b><i> Introduction to sampling -
Key-ranges, sample CD's, the advantage of using more bits in
audio, quantisation problems, noise pollution, studio hum prob-
lems and elimination, problems with clearing samples and copy-
right</i><br>
<br>
<img src="arrow.gif" width="10" height="10" alt="arrow graphic">
<b>Processing and effects</b><i> Compression, valves, gates,
reverb, distortion, chorus, phasers, flangers and their uses in
dance music<br> Inc Audio examples</i><br>
<br>
<img src="arrow.gif" width="10" height="10" alt="arrow graphic">
<b>Mixing desks</b><i>
Mixing desk theory, configuration, busses, auxiliary, inserts,
order of effects and the subsequent results</i><br>
<br>
<img src="arrow.gif" width="10" height="10" alt="arrow graphic">
<b>Sound Design</b><i>
Theory behind programming the sounds for each genre, the
synthesis involved and the further uses of processing and
effects<br> Inc Audio Examples of different stages of sound
design.</i><br>
<br>
<img src="arrow.gif" width="10" height="10" alt="arrow graphic">
.<i>...and more !</i></font></p>

   </td>
  </tr>
 </table>
 <!-- End of the 2nd nested table -->
```

Look at the code above as you are viewing *Part11.html* in your web browser. You will notice an opening <p> and closing </p> tags. These tell the browser where a paragraph begins and ends. We have also inserted the graphic *arrow.gif* to act as bullet points for the text. Also included are the <i> and </i> tags to italicize some text.

Finally, we'll add the image of the book to the third nested table.

Open *Part12.html* in your browser. You will notice that content has been added to the third nested table. Here we have replaced the code between:

```
<!-- Beginning of 3rd nested table - this will hold a picture of the
book cover -->
```

and:

```
<!-- End of 3rd nested table -->
```

with:

```
<!-- Beginning of 3rd nested table - this will hold a picture of the
book cover
-- >
  <table width="231" border="0" cellpadding="0" cellspacing="0">
    <tr>
    <td width="231" valign="top"><img src="1X1.gif" width="231"
height="1"
    alt=""></td>
    </tr>
    <tr>
    <td valign="top" align="center"><img src="book_cover_small.jpg"
width="214"
    height="278" alt="image of book cover"></td>
    </tr>
  </table>
<!-- End of 3rd nested table -->
```

This creates two rows in the third nested table. The top row is filled with the 1×1 pixel transparent GIF that acts as the 'wedge' discussed before. The second row contains an image of the book. You can see that this image is aligned within the table data cell to the centre with the code:

```
<td valign="top" align="center">
```

Now the main content table for the web site is almost completed, apart from making the links work and altering the text format, which we'll do in a moment. The final part of the layout involves creating the footer of the web site.

Creating the page footer

View *Part13.html* in your browser. The code is shown in Figure 20.23.

```
<html>
<head>
<title>Dance Music Production and Remixing</title>
</head>
<body background="background_colour.gif">
<!-- Beginning of code that defines the footer of the web page -->
<table width="100%" border="0" cellpadding="0" cellspacing="0">
 <tr>
  <td width="100%" valign="top"><p><font
color="#FFFFFF">thewebguy@dancemusicproduction.com</font></p>
```

Figure 20.23 (*Continues*)

```
      </td>
    </tr>
    <tr>
      <td valign="top" bgcolor="#FFFFFF"><img src="1X1.gif" width="1"
      height="2" alt=""></td>
    </tr>
    <tr>
      <td valign="top"><img src="1X1.gif" width="1" height="10"
  alt=""></td>
    </tr>
    <tr>
      <td valign="top" bgcolor="#FFFFFF"><img src="1X1.gif" width="1"
      height="3" alt=""></td>
    </tr>
    <tr>
      <td valign="top"><img src="1X1.gif" width="1" height="10" alt=""></td>
    </tr>
    <tr>
      <td valign="top" bgcolor="#FFFFFF"><img src="1X1.gif" width="1"
      height="1" alt=""></td>
    </tr>
    <tr>
      <td valign="top"><img src="1X1.gif" width="1" height="8"
      alt=""></td>
    </tr>
  </table>
  <!-- End of code that defines the footer of the web page -->
  </body>
</html>
```

Figure 20.23 Part13.html.

The footer is a single 100% width table and contains a series of rows, marked by the opening and closing <tr> and </tr> tags respectively. The heights of these rows are effectively 'wedged' open by the 1 × 1 pixel transparent GIFs. This arrangement is very similar to that used for the second nested table in the header of the page.

Combining the header, content and footer into a single page

Having constructed the header, the content and the footer of the home page as individual pages, we now need to combine these into one web page. As you may recall, a web page follows the format:

```
<html>
  <head>
  <title>Dance Music Production and Remixing</title>
  </head>
```

```
  <body>
  </body>
</html>
```

Importantly, a web site will only ever have one pair of opening <html> and closing </html> tags, one pair of opening <body> and </body> tags, and one pair of <head> and </head> tags, in the order shown above.

Recall that the body of the text is where all of the code describing the layout of the page must go. Because HTML is read by the browser from top to bottom, logically we need to paste the sections of code in the order that they will appear on the web site: the header, the main content and the footer. The code that we need to take from *Part8.html* (which is the final code for the header), *Part12.html* (which is the final code for the content of the page) and *Part13.html* (which is the final code for the footer) combined contains all of the code that needs to be placed between the <body> and </body> tags of the final web site – that is, the entire description of the outermost table and all of its contents. We must, however, remember to change the <body> tag in this final composite page so that the blue background image is also included.

To create the final page then, we must create a new page with the contents of the code between the <body> and </body> tags from *Part8.html, Part12.html* and *Part13.html* as shown below.

```
<html>
  <head>
  <title>Dance Music Production and Remixing</title>
  </head>

  <body background="background_colour.gif">
  [insert all the code between the <body> and </body>
  tags of Part8.html here]
  [insert all the code between the <body> and </body>
  tags of Part12.html here]
  [insert all the code between the <body> and </body>
  tags of Part13.html here]
  </body>
</html>
```

Part14.html is home page of the web site, with the header, content and footer sections combined as shown above.

Adding hyperlinks to the menu

With our home page header, content and footer merged to create one page, we now need to add navigation to the menu on the left-hand side of the page by adding active hyperlinks to the text that will enable the user to move around the site.

Take a look at *Part15.html* in your browser. You will notice that the menu text on the left-hand side of the page is now shown as active hyperlinks. They are blue because this is the browser default colour of links (we will change this shortly). If, however, we moved the font tags (<font color="#FFFFFF"> and </font>) so that they were directly either side of the displayed link text, the links would be displayed in white. This is an example of how HTML prioritizes from the innermost content of a table data cell outwards to the outermost table.

We created these links by changing the lines in the code that read:

```
<div align="center">
<font color="#FFFFFF">INTRODUCTION<br></font></div>
<div align="center">
<font color="#FFFFFF">SOUND DEMOS<br></font></div>
<div align="center">
<font color="#FFFFFF">EQUIPMENT</font></div>
```

to:

```
<div align="center">
<font color="#FFFFFF"><ahref="introduction.html">INTRODUCTION</a>
<br></font></div>
<div align="center">
<font color="#FFFFFF"><a href="sounddemos.html">SOUND
DEMOS</a><br></font></div>
<div align="center">
<font color="#FFFFFF"><a href="equipment.html">EQUIPMENT</a>
</font></div>
```

You will notice the use of the opening <a href> and closing </a> to create the links as previously described, and as mentioned before these take the form of:

```
<href="[the web page to be linked to.html]">[the text to appear as a
link]</a>
```

For example, you can see from the code that the word INTRODUCTION links to a web page called *introduction.html*. Similarly, the SOUND DEMOS text links to the *sounddemos.html* page and the EQUIPMENT text links to the *equipment.html* page.

Using Cascading Style Sheets to format the text

We are now going to alter the format of the text using CSS. Using CSS means that the entire look of the text across the web site, or in this case the page, can be managed by altering one file.

When using CSS the browser looks to the CSS file for instructions on how the text on the web page should behave when loading the page. This CSS file is independent of the HTML file and is nothing more than a set of coded instructions (a text document saved with a **.css* extension).

For the Dance Music Production home page, we will create a style sheet that will format the text on the page and allow the text in the menu to change colour when the mouse is rolled over it.

Figure 20.24 shows how the page looks before CSS code is applied.

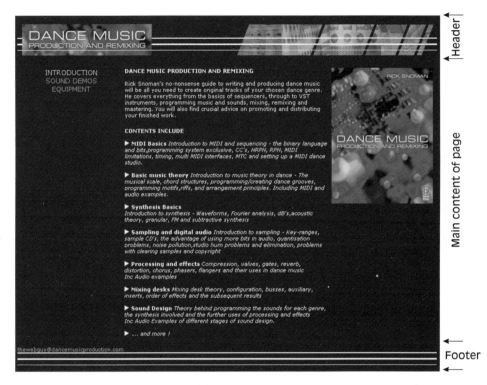

Figure 20.24 Part15.html (no CSS).

Before we create the CSS file, a line of code must be added to the main page within the <head> and </head> tags. This tells the web browser that CSS is being used and where the *.css* file is stored. For this web site we have added the following line to *Part16.html* between the <head> and </head> tags:

```
<link rel="stylesheet" href="music_standard.css" type="text/css">
```

This instructs the browser to look for a file called *music_standard.css* in the same folder that the current page is residing. If, to keep things tidy, you wish to put this file in a folder called *music*, the line would read:

```
<link rel="stylesheet" href="music/music_standard.css"
type="text/css">
```

Now open *Part16.html* and see how the text changes.

You will now notice that the text looks neater and the links are a different colour. The only code that has been added to the file is the line above, to tell the browser where to look for the style sheet, and the lines:

```
<div align="center">
<font color="#FFFFFF"><ahref="introduction.html">INTRODUCTION</a>
<br></font></div>
```

```
<div align="center">
<font color="#FFFFFF"><a href="sounddemos.html">SOUND DEMOS</a>
<br></font></div>
<div align="center">
<font color="#FFFFFF"><a href="equipment.html">EQUIPMENT</a>
</font></div>
```

have been changed to:

```
<div align="center">
<a href="introduction.html"
class="leftmenu">INTRODUCTION</a><br></div>
<div align="center">
<a href="sounddemos.html"
class="leftmenu">SOUND DEMOS</a><br></div>
<div align="center">
<a href="equipment.html"
class="leftmenu">EQUIPMENT</a></div>
```

The only thing that has been added is the code highlighted in **bold**. These 'class' statements tell the browser which style defined in the *.css file is to be applied to the text.

These class statements replace the previously inserted <font> </font> pairs. This is because they are no longer needed to describe text colour or font, as this is now controlled by the CSS information. You will notice that all the <font> and </font> pairs and their attributes have now been removed from *Part16.html*.

So what's in a CSS file? To understand it, you need to look at the style sheet *music_standard.css* (Figure 20.25), which is included in the *Tutorial 2* folder. This can be opened in any standard text editor.

```
table, tr, td {
font-family: Verdana, Arial, Helvetica, sans-serif;
font-size: 11px;
margin: 0px;
color: #FFFFFF
}

a.leftmenu:link {
color: #6699CC;
font-size: 14px;
text-decoration: none;
}

a.leftmenu:link:hover {
color: #FFFFFF;
font-size: 14px;
text-decoration: underline;
}
```

Figure 20.25 (*Continues*)

```
a.leftmenu:visited {
color: #6699CC;
font-size: 14px;
text-decoration: none;
}

a.leftmenu:visited:hover {
color: #FFFFFF;
font-size: 14px;
text-decoration: underline;
}

a.textlink:link {
color: #6699CC;
font-size: 11px;
text-decoration: none;
}

a.textlink:link:hover {
color: #FFFFFF;
font-size: 11px;
text-decoration: underline;
}

a.textlink:visited {
color: #6699CC;
font-size: 11px;
text-decoration: none;
}

a.textlink:visited:hover {
color: #FFFFFF;
font-size: 11px;
text-decoration: underline;
}
```

Figure 20.25 The music_standard.css file.

To get an idea how these work, let's take the first block of code:

```
table, tr, td {
font-family: Verdana, Arial, Helvetica, sans-serif;
font-size: 11px;
margin: 0px;
color: #FFFFFF
}
```

This states that the text in every table data cell in the web page that is reading the file will display with a Verdana typeface and if the browser doesn't have that font it will use Arial, and if
Arial is not available it will use Helvetica, and so on. A selection of the four most popular fonts

is included as possible alternatives: Verdana being the most popular, with sans-serif being the least popular of the four. We need to do this because different browsers and platforms have different fonts available to them. The above code also states that the font will be 11 pixels (px) in size, displayed in white, and without a margin (0 px).

You will notice in the CSS document that the following four blocks of code begin with a 'leftmenu'. This is a label for the CSS and this is why the following code previously seen above has had class="leftmenu" added to it:

```
<div align="center">
<a href="introduction.html"
class="leftmenu">INTRODUCTION</a><br></div>
<div align="center">
<a href="sounddemos.html"
class="leftmenu">SOUND DEMOS</a><br></div>
<div align="center">
<a href="equipment.html"
class="leftmenu">EQUIPMENT</a></div>
```

This is applied as above so that the links on the left of the page behave as described within the CSS file. You can see that *music_standard.css* refers to four versions of a 'leftmenu'. After this file name is the 'state' of the link that will be affected by the CSS. The four states are:

- link, defining the normal state of the links;
- link:hover, defining the state of the link when the mouse is held over it;
- visited, defining the state of the link when it's been visited;
- visited:hover, defining the state when the mouse is put over a visited link.

You can see that the font sizes are constant for the four states of this link. From the code you can see that when the mouse rolls over the link or a visited link, it will turn white (this may not happen in some older browsers). Furthermore, when a link has been visited it will stay white whether the mouse is over the link or not. This will make it easy for visitors of the site to know whether they have previously clicked on a link in the site.

You may wonder why all the other text in the site doesn't have a dedicated class="[class file name]" statement. This is because all the remaining text is within the tables in the web site, which are defined using the first block of code in the CSS file that starts table, tr, td. Only text that does not comply with this formatting needs to be defined with its own class statement.

To get a better idea of how CSS affects the text on the page, play around with values in the CSS file and see the effect it has on the web page displayed in the browser. Please note that to view the change you may have to press **Refresh** in your browser to clear its cache memory or close the browser completely after clearing the cache and open a new instance of it.

You will also notice that a class called textlink is defined in the file *music_standard.css*. This text style will be used to affect links within the page that are not in the left menu and will be used in the pages that follow.

We have only touched upon CSS and any more information here would be beyond the scope of this book. There are many good web sites and books on CSS that you may learn more from if you wish.

The Dance Music Manual

The Dance Music Manual

Adding an email link

The final touch to this page it to make the email link at the bottom of the page active, so that when pressed it opens your email application so that you can contact the site's webmaster. The simplest way to do this is to make the email address webmaster@dancemusicproduction.com an active link. To do this, the line:

```
<td width="100%"
valign="top"><p>thewebguy@dancemusicproduction.com</p>
  </td>
```

simply changes to:

```
<td width="100%" valign="top"><p>
<a href="mailto:thewebguy@dancemusicproduction.com"
class="textlink">thewebguy@dancemusicproduction.com</a></p>
  </td>
```

This makes the text that displays the email address into a link that links to the email address. By placing mailto: in front of the link, the browser knows to open the email client window when this link is clicked. You may also have noticed that a CSS class statement has been added to the code.

```
class="textlink"
```

This tells the browser to use the textlink CSS descriptions that are stored in the *music_ standard.css* file.

The above code has been added to *Part17.html* so that the webmaster's email address link is active and its appearance is controlled by the CSS.

Setting the height of table cells

There is one final thing to do with the home page of the web site. You will notice that there are a few spaces between the main textual content of the page and the footer bar of the page when this page is viewed on a high-resolution browser that is fully expanded. Ideally, for the best look, we want the main content of the page (everything other than the header and the footer of the page) to only be as high as the part of the page with the most content in the vertical axis. In this case you can see that the actual textual content of the page has the most vertical content (it has more vertical content than the menu on the left-hand side or the picture on the right-hand side). To ensure that there are no big gaps between the content of the page and the footer, we need to remove all attributes within the <td> tags that correspond to height.

To demonstrate this, before we apply this to our web site code take a look at *Part18.html*, which is a very simple web site. Then take a look at *Part19.html*. Having removed the height attribute from the table data cell, the cell becomes, for want of a better description, vertically elastic. The table data cell behaves as if it wants to be as small a height as possible. Because there is no height attribute stating how high it should be, it will be as close to 0 pixels high as possible, but because there is content within this table data cell, it prevents the cell collapsing to 0 pixels high. This is in fact very similar in comparison to the 1 pixel by 1 pixel GIF images that we used in previous examples to prevent cells collapsing in a horizontal axis when we made table data cells 100% width.

We shall now apply this technique to the code in *Part17.html* to create *Part20.html*, so in *Part17.html* the code for the first nested table of the page content reads:

```
<!-- Beginning of the 1st nested table - this will hold the left menu
contents of the Web Page -->
  <table width="229" border="0" cellpadding="0" cellspacing="0">
    <tr>
    <td valign="top" height="609">
```

and in *Part20.html* the height attributes have been removed and the code now reads:

```
<!-- Beginning of the 1st nested table - this will hold the left menu
contents of the Web Page -->
  <table width="229" border="0" cellpadding="0" cellspacing="0">
    <tr>
    <td valign="top">
```

Similar changes have also been applied to the height attributes of the second nested table in the page content.

Now if you look at *Part20.html* in your browser you will see that the gap between the page content and footer of the page has gone. The header and footer code has not been changed at all, but all of the height attributes that are contained within the <td> tags that are contained within the outermost table (that contains the nested tables for the left menu, the main content and the picture of the book on the right-hand side of the page) have been removed. As the web browser now has no idea how high these table data cells should be, they all 'collapse' vertically until they can't collapse any further. The effect in this case is desirable and creates a good-looking web site.

If for some reason you didn't want this to happen, you could use a 1 × 1 pixel GIF to 'wedge' the table data cell to a specific height. In this case we don't need to, as the text of the page is acting as such a 'wedge'.

Finishing touches

Take a look at *Part21.html*. The final touches to this web site are the addition of the
 (line break) tag to give a one-character space between the content of the page and the footer of the page. This has been added so that the last line of text in the content does not sit too close to the webmaster's email address.

```
<meta http-equiv="Content-Type" content="text/html; charset=
iso-8859-1">
```

This line has also been added between the <head> and </head> tags; it tells the browser what character set is being used and is an example of a META tag. If, for example, you had written your site using the Chinese character set, this line would be different. A full explanation of this line is beyond the realms of this book, but the above should work for all web sites written in the English language. More will be mentioned on META tags at the end of this chapter.

The front page of this demonstration web site is now completed. By following the steps this far, you will have learnt most of what there is to know about building a basic web site.

Creating additional pages

You will notice that the menu on the Dance Music Production web site we have made links to three web pages:

- *introduction.html*;
- *sounddemos.html*;
- *equipment.html*.

The file we have just finished (*Part21.html*) is the finished *introduction.html* page. This file is in the Tutorial 2 folder and is an exact copy of *Part21.html*.

We now need to create the other two main web pages and two further subpages. This is fairly trivial because we have already created the template for the entire web site when we created the *introduction.html* page. The only content that needs to be changed is the content of the table that contains the main text of the page, which is contained within the tags:

```
<!-- Beginning of 2nd nested table - this will hold the main contents
of the Web Page -->
```

and:

```
<!-- End of the 2nd nested table -->
```

By changing the content of the code between these tags and changing the file name, we can create another page of the web site: nothing else needs to change.

Now take a look at the *equipment.html* page. This is exactly the same as *introduction.html* but with the different content in the table described above. There is no code here that you won't already be familiar with.

Notice how the height of the page has automatically adjusted to the height of the textual content again, all thanks to the removal of the height attributes of the <td> tags within the contents' code.

We will now take a look at how *equipment.html* was created. To do this we simply take the introduction.html page and rename it *equipment.html* using the **Save As** command. To create *equipment.html*, we have created a series of nested tables in which to position the content of the table data cells.

Each of the synth images is a clickable thumbnail image and link to larger versions of the respective images. These images are not just physically larger; they are in fact higher resolution images that are dimensionally larger too. The images that appear on *equipment.html* are set up as thumbnails so that they load quickly and give the user the choice of whether to view the high-resolution images. This technique should always be applied when you wish to display images to prevent your web pages taking a long time to download. The techniques used here are not new and have been previously demonstrated in the tutorials.

Adding audio

There are many different sound formats that can be used to present music on the web. Of these, the most basic (and probably most irritating) is the MIDI format. This is MIDI data that uses the sounds native to your computer's soundcard when played because there are no sounds embedded within the *.mid* file. This type of file contains instructions such as note pitch, timing, velocity, panning and so on. Because it depends entirely on the browsing computer's sound capabilities, different computers may produce different timbres and so the sounds produced are not consistent.

To get a MIDI file playing in a web browser, you use the <EMBED> tag within the HTML code:

```
<EMBED SRC="the_midi_file_you_want_to_play.mid" hidden="true" autostart=
"true" loop="1"><NOEMBED>Your browser cannot play this midi file, if
you want to hear it, <a href="the_midi_file_you_want_to_play.mid">
click here </a><NOEMBED>
```

The above code will play the MIDI file automatically when the page opens. If the browser cannot interpret the <EMBED> tag, the user can click a link to listen to it.

For the purpose of this book, we will not go into any more detail on playing MIDI files via the Internet, as they are not much use for presenting your compositions to interested listeners. They are only useful for giving the listener a vague idea of what your music sounds like without your preferred sounds or any production applied.

As media playing on the web is actually quite in-depth, we will look only at the most common ways in which to play your music on the web. The two major formats for this purpose are the WAV and MP3 file formats.

WAV files are not compressed so are very large compared to MP3 format files. This uncompressed format is the reason that WAV (or AIFF) files are preferred in music production as opposed to MP3. For online purposes, however, MP3 is the choice of professionals, as there is virtually no data loss and file sizes are much smaller than the equivalent WAV file at the same sampling frequency.

With either of these formats, music can be presented either as a file that is downloaded and listened to or a file that 'streams' over the web to the user's computer. There are pros and cons to both these methods. The key difference between non-streaming and streaming audio is how the audio is delivered to your computer.

- *Non-streaming audio.* With non-streaming audio, when a user clicks on the file link from the web site, they must usually wait until the entire file is downloaded before they can listen to the file. Because the file may be around 5000K (5 MB) in size, this may take around 10 minutes over a slow Internet connection. Once downloaded, the user can listen to the music at high quality in its entirety without the sound breaking up. If the user has a broadband connection or similar high-speed connection, this download will happen very quickly so there won't be much waiting around.

- *Streaming audio*. With streaming audio, the user clicks on the file link from the web site and can listen to the music file whilst it downloads. The file is downloaded sequentially from beginning to end, so the data already received can be decoded and listened to. This is great for high-speed Internet users but be cautious, because as mentioned before there are millions of people in the world who are still on 56K connection speeds. With slow Internet connections, streaming media will not work very well and the playback can be jerky and broken.

For either streaming or non-streaming audio, you will need a player and there are several to choose from. A favourite of mine for PC is Winamp, which also has cool visuals that synchronize (although not very well) with your music. Winamp can be downloaded free at www.winamp.com/download/. Another good player for the PC is Windows media player, which is bundled with versions of XP. Be aware that there are effects within Windows Media Player such as SRS WOW Effects that may be enabled in your version once installed. Ensure you switch these off, as they will get in the way of you listening to the real music. The Apple Mac fans amongst you will probably already have QuickTime installed, as this is Apple proprietary technology, whilst PC users will have to download a copy from www.apple.com/quicktime/download/ (correct at the time of writing). Real Player is another alternative that can be downloaded from www.real.com. These three are the most commonly used media players.

The best option for delivering audio to your web page viewers is to provide a variety of ways to listen to it. If you have a great deal of web space available to you, I would recommend giving the option to do a straight download of the audio file and the option to listen to it streaming. This pretty well caters for all types of Internet connection speeds going on in the world at the moment.

As mentioned previously, streaming media is where the file 'streams' to your computer and you can listen to it before all the information of the entire file has been sent. Streaming media to the classical definition will not allow you to obtain the file from the web site and listen to it later; rather you must listen to it whilst you're actually connected to the Internet.

There are key arguments for the use of streaming media on the web:

1 It enables you to listen to files as they are partially sent to your computer so you don't have to wait for the whole file to arrive through the Net before you listen to it. This is all well and good, but for people with slow modems (millions of people), they will probably only be able to listen to the stream in broken sections. Some web sites provide streams for slow modems, but to do this, the actual bit rates and sample rates are commonly lowered, so although you get a constant stream, the sound is terrible!

2 To prevent people from fully downloading the file to the hard drive and sharing it between friends. In fact, at face value this is the case, but it is well known that software exists now to 'capture' such streams.

Streaming media requires that you set up 'streaming servers'. Two of the most common ones are provided by Real Media and QuickTime. Instructions for setting up such servers are beyond the realms of this book, but information on how to do this can be found at www.real.com and www.quicktime.com. QuickTime is generally the preferred method. An exception to setting up a streaming server to stream audio is by using Macromedia Flash (www.macromedia.com). This will allow streaming audio without specific server settings; to learn more about this you are advised to visit the Macromedia web site.

Tutorial 2 will now continue and we will go through how to set up downloadable links to MP3 files. For the purpose of this book, we will not be covering streaming audio, as this requires the setting up of a streaming server, which is beyond the scope of this chapter.

Adding audio to the Dance Music Production web site

You will notice that that the first two 'click here' links on the *sounddemos.html* page in Tutorial 2 actually link to MP3 music files. When you click on a link, your browser will try to open the file with whichever default MP3 player you have installed on your machine. A good one (which is free) is Winamp by Lavasoft, as previously recommended. When you click on the link, you may be asked if you want to save or open the file.

Be aware that this is not streaming media. If this site were live on the web, when selecting **Save**, this file would be downloaded completely onto your computer's hard drive, from which you could open and listen to it. If you clicked on the link and chose **Open**, the file would also download and then play.

The code in *sounddemos.html* that creates the links to the MP3 files is detailed below.

```
<p><b>The Access Virus C Playing a trance riff</b></p>
<p>To download this file as an mp3 file, <a href="trance_riff.mp3"
class="textlink">please click here</a>.</p>
<p><b>A demo of a full mix</b></p>
<p>To download this file as an mp3 file, <a href="full_mix.mp3"
class="textlink">please click here</a>.<br>
</p>
```

You will notice that the formatting of this is very similar to the code used before to insert a hyperlink or a URL.

```
<a href="[the url of the site you want to link to]">[the link to
appear on screen]</a>
```

To allow the MP3 file to be downloadable and allow the user to play it on their computer, we just put the file name of the MP3 file in the <a href> tag.

In the *sounddemos.html* page, you will notice that 'please click here to go to a demonstration of this' links to *sounddemos2.html*. To create this page we simply saved *sounddemos.html* as *soundsdemos2.html* and then changed the code in the content table as before.

If you now open *sounddemos2.html*, you should hear a beat as the page opens. To create this effect, the following line of code has been added just below the opening <body> tag of the web site:

```
<embed src="third_mix_bar_wav.wav" hidden="TRUE" autostart="TRUE"
loop="TRUE"></embed>
```

This code makes the file *third_mix_bar_wav.wav* play in the background when the web page loads.

Although this might seem like a nice touch, playing WAV files in the background is generally considered annoying and is not encouraged for several reasons:

1　Doing this is not encouraged as people will commonly view your page from a place of work and it's annoying for unwanted sound to suddenly jump out of the speakers; it may even lose someone their job.
2　It will slow down the loading of the web site a great deal and the benefit of the sound rarely outweighs the speed at which a web site appears on the browser.
3　This will not work on some browsers, as this type of coding is highly browser dependent.

Next take a look at *sounddemos3.html*. Again, this is the same as *sounddemos2.html* with the code between:

```
<!-- Beginning of 2nd nested table - this will hold the main contents
of the Web Page -->
```

and:

```
<!-- End of the 2nd nested table -->
```

tags changed.

In most browsers, a console should appear that allows you to control when the sound plays. The code that created this effect is:

```
<embed src="third_mix_bar_wav.wav" height="40" width="150"></embed>
```

The height and width of the console are specified using the height and width attributes, as shown above. Please note again that this is likely to not work on some Netscape browsers and it is not advised to use such code heavily within your web sites.

Viewing the finished Dance Music Production web site

Having created all the necessary pages and added the audio files, as required, if you now open up the *introduction.html* home page, you should be able to use the links to surf through the finished web site.

Setting your site up on the web

A final aspect of this chapter is to show you how to get your web site from your computer to be viewed live on the web. This procedure is simple but you will need a good FTP (File Transfer Protocol) program (such as WS_FTP Pro by Ipswitch), some web space and a domain name.

Choosing a basic web space package can be a minefield, as prices and reliability vary a great deal. For a small web site you are unlikely to need more than 25 MB of space, depending on how you build your site and what multimedia content, such as MP3s, you want to include. A small basic site like this should cost no more than about £100, including a .com domain name for one year's hosting.

Many web hosting services offer web space and one domain name as a starter package, and if you're just starting out, this is advisable. Once you have signed up, about 48 hours later your web hosting service should have your domain name live on the web and your web space ready to upload your data to.

Be warned that some hosting services will offer both a domain name and free web space, but there are usually strings attached. In most cases your free web space will have a banner advert on it that will make your site look unprofessional. Furthermore, you will have to pay a premium to remove this, so check first that the web space provided does not contain banner adverts.

Once signed up to a hosting service, they will send you an IP address (this is effectively the web address of your web space), a username and password.

To upload a web site to a hosting service you will need an FTP program. One such recommended piece of software is WS_FTP Pro and Figure 20.26 shows a screenshot of the main page of this software.

As you can see from Figure 20.26, the host name (this could be the IP address), username and password for the web space are filled out and, providing you are online, you will then connect into your web space. Generally, when you connect to your web host, you will be connecting into what is known as the 'root' of your web space. The 'root' is the directory of the site that is reached when the domain name of your site is typed into a web browser.

For example, if your domain name was www.dancemusicproduction.com and you typed this into a browser, you will be viewing the default 'root' page of the web site and it is this folder that you will view when you first access your web space with your hosting service. Generally, servers are set to point to index.html, so make sure that the first page of your site has this name. Alternatively, you could ask your hosting service to change this default. In our example we would ask it to point to *introduction.html* (although it's best that you name your first page as *index.html*). This would mean that when people type in www.dancemusicproduction.com they are automatically taken to www.dancemusicproduction.com/introduction.html.

Once you have done this, you will be able to navigate the files on your hard drive and in the case of WS_FTP Pro you simply highlight the files you wish to transfer and click the green **Upload** arrow. For further information consult the documentation that comes with your chosen FTP program. Your site is now live on the web!

Figure 20.26 Using WS_FTP Pro.

META tags and getting your site viewed on the web

Years ago it was believed that META tags were the key to getting your site indexed at the top of any search engine, but I assure you if there ever was truth in that, there isn't much now. They do help though, so here's an introduction to them. We have already seen one type of META tag in Tutorial 2 defining the character set of the web page.

```
<meta http-equiv="Content-Type" content="text/html; charset=
"iso-8859-1">
```

For our Dance Music Production web site we can use META tags to describe keywords that relate to the site. This will help search engines index your site correctly (as mentioned previously, correct use of ALT statements may also help, depending on the search engine in question). An example set of META tags for our web site in Tutorial 2 would be:

```
<html>
<head>
<title>Dance Music Production and Remixing</title>
<meta http-equiv="Content-Type" content="text/html; charset="iso-
8859-1">
<meta name="description" content="Dance Music Production book by Rick
Snoman">
```

```
<meta name="keywords" content="Dance, music production, midi,
synthesis, vocals, musical genres">
</head>
```

As you can see, there are two new examples of META tags here: one describing the web site and one containing keywords. It is important not to overdo the keywords as this can go against you. Recently, keyword creation in META tags has become somewhat of a science, and software is available to optimize your indexing chances. One such piece of software is advertised at http://www.advancedkeywords.com/ (the author has not tried this software).

Whether you decide to research further into optimizing META tags is up to you, but do ensure that you use them, as they will help indexing of your web site.

To get your site indexed the general method is to go through the various search engines and submit your site to each of them, which is usually free. There are some web sites such as http://www.submitexpress.com/ and http://www.addme.com/ that will do bulk submissions and take the labour out of doing it yourself (these companies have not been verified or used by the author). Some of the most common search engines that you should submit your site to are listed below, but be aware that there are hundreds out there to go for!

www.google.com
www.yahoo.com
www.lycos.com
www.savvysearch.com
www.altavista.com
www.alltheweb.com

Using the techniques shown in this chapter should allow you to go ahead and build any HTML-based web site. With a little further reference into HTML you will be able to learn further techniques and build upon what we have demonstrated here. All the essentials of good web building and design have been covered here: HAVE FUN!

To learn more about web site creation, the following are a few recommended links:

http://www.w3schools.com
http://www.webmonkey.com
http://www.htmlgoodies.com/

21 A DJ's perspective

It's fair to assume that dance music would be nowhere if it wasn't for the innumerable DJs spinning the vinyl and keeping us all up on the floor. With this in mind, it seems that there is no better way to close the book than to interview a DJ and get his perspective on dance music and DJ'ing in general. For this, I decided to interview DJ Cristo, a regular figure on the international circuit, contributor to *DJ* magazine (amongst many others), record reviewer and regular host on a Miami radio show to offer his comments. However, after discovering just how much he had to say on the subject, it seemed more prudent to simply hand over a few topics and leave the entire chapter to him …

I'd like to begin my chapter by saying that it is a great privilege to be asked to contribute my words of wisdom to this phenomenal book by Rick, a dance music stalwart and luminary who is one of the most knowledgeable, astute guys I know on the dance music production scene. When he talks shop, I listen. He has a constant string of quality productions and remixes under his belt – longer than my arm and all under various pseudonyms – even I can't keep up and I'm a reviewer! His dance music and remix seminars are also incredibly insightful and it's great to at last see words written about such a specialist area as dance music by someone who's actually actively involved in it for a change.

Promos and reaction sheets

A promo is a record or CD that record companies release way before it's due to be released to the general public, normally to gauge how well a track is going to do commercially when it eventually goes on general release. For this they use promotional companies who push their tracks and ensure that they land in the right hands – that is, from the top DJs right the way down the scale. These companies have a massive list of nationwide club, bar and radio DJs of all levels, scenes and genres, most of whom are also members of the mailing lists of *Power* in London, which is one of the biggest and most powerful in the industry.

Getting on these lists isn't a walk in the park. All DJs must first apply to an organization such as *Music Power* and pay an annual subscription to them before they can apply to the promotional companies. In order to be accepted on these you don't necessarily have to be Carl Cox, but you do need to be a proficient DJ playing regularly enough to reasonable size crowds. Even then, the selection process can be rigorous because they need to know they can depend on their DJs to support, play and chart the tracks that they want to promote. Notably, most jocks (*a common term for Disc Jockey*) that are on these lists are semi-professional or professional and on the club or radio circuit.

Incidentally, the funding from the DJs/members goes towards keeping the whole ship afloat, as it were. If they didn't pay to subscribe there would be a massive reduction of the number of 'free' records sent out and fewer records being promoted in far smaller quantities. This would have a major impact on the vast majority of DJs, reducing the number of weekly chart returns to hugely influential magazines like *Music Week*. If this ceased, so would one of the central strands of generating interest in music both through club and radio play, and through club chart success. The charts have a fundamental role in gauging records and it would have a devastating effect on promotional companies and smaller labels alike.

Nonetheless, if you are selected to be on these lists you have to maintain your position and abide by certain rules in order to remain on them. It's naive to think that you can just sit back and receive numerous free records, because when you receive a promo copy, you also receive a reaction form that has to be filled in and returned. This is basically a sheet that details the artists, title, label and some history about the track, including where it originated from, who's currently spinning it, which big DJs are supporting it and which radio stations are playing it. At the bottom of the form are a number of boxes for your input, which need to be completed. These are for your opinions on how well the track went down at the club/bar/venue (i.e. your dance floor), along with marks out of 10 for how well it was received by the crowds. It also asks whether the track was enquired about (by Joe Public), which position it occupies in your charts out of 20 other promo records and what clubs you're currently playing at.

When you have gathered all this information and completed the forms, you then need to return them. However, not only do you have to return them to the original promo company, but also to all the relevant magazines, via post, fax and email, because this goes towards compiling charts magazines such as *Record Mirror, DJ, Music Week, Seven* and *DMC/Update* to just a name a few, along all the e-groups and Internet charts such as the influential DMC DJ chart (www.dmcworld.com). What's more, every DJ signed to the promo company lists is closely monitored from their chart performance response, and staying on the list can depend how often and how high you place their records. Of course, you must always stay true to yourself because you are what you play, but you do have to be willing to meet the promo company halfway, otherwise they have the power to remove you if they feel that you're unacceptable. This is a very competitive business and nothing comes easy. That said, one of the many good points (apart from obviously receiving free vinyl!) is being able to see the changes in trends. I'm lucky enough to be sent all the big tracks, along with having the privilege of receiving all the peculiar tracks that are just about to break.

The record companies certainly have it sussed and it's a very finely tuned hype machine, but you must understand that while you'll receive plenty of vinyl way before the general public, the big boys, such as Judge Jules and Pete Tong, will receive more exclusive records than you will. These guys are very highly targeted because of their heavy influence on the radio and club circuits. Average level DJs will receive the same records eventually, but it will be much later than the bigger players. It's a well-oiled machine and it all adds to the allure of the exclusivity of promos.

As many music enthusiasts will already be aware of, it's also easy enough to buy promos from various record shops and web sites, as I used to for many years. By doing so you're onto a winner because you get your hands on a top tune weeks before it's released and if you're especially lucky it might be a TP for a bargain price. This, however, can have a polarity shift effect because when a tune (typically on vinyl) becomes in demand, a buzz will circulate around it and it will increase exponentially in value. This is especially the case if it's on an independent label or a TP/white label and subsequently becomes signed to a big UK record company. With Internet 'shops' such as EBay, previously unsigned material becomes responsible for extreme bidding wars, where DJs and collectors want the promo before the official release date. This happened just recently with a double pack of the artist Jurgen Vries and *The Opera Song* featuring Charlotte Church. When I last looked it was being bid at well over $100 and that's just for a piece of vinyl! It also largely depends on what trends there are, and a simple supply and demand structure is in place which determines the price of the records. For instance, coloured vinyl always seems to be in vogue and despite the fact that some vinyl connoisseurs will argue that the sound quality suffers, it's nevertheless highly sought after and some people will pay ridiculous prices for it.

What 'works' on the dance floor?

I feel that for a track to work on the dance floor all the basic elements have to be there. Firstly, for a DJ to be able to mix in and out of tracks, I find that it's generally best to have a steady build-up of beats and drum loops, along with a solid, crisp pattern. It's a very loud environment in clubs, so it helps if there is a minimum of 16 bars of crisp percussion to begin with. In fact, 32 bars are even better, but this obviously depends on how long the track is and the style of music. Even with shorter tracks of any genre, though, 16 bars of percussion are necessary for the intro and outro, as you can build and structure a set using these, slowly progressing as the night goes on. Also, most jocks will thank you for making tracks 'DJ friendly' by not starting the track with loads of percussion and some melodies too. These are incredibly difficult to mix and they make your life that little bit harder.

I spend a lot of time playing house and trance, and like most genres of music it's true to say that it is somewhat formulated. The drum patterns and rolls are very similar, and many of the melodies share plenty of similarities in the sounds that are used. That said, people who really love dance music understand that it is a vibe as well – a feelgood factor – and this is what really separates the music. Dance music has to have soul and it has to connect with you on the right level. When I buy new tunes (yes, even with promos we still buy lots of vinyl, especially imports), I can hear immediately if a track has the right elements. It has to stand out from the rest with an indefinable groove, that extra special something that makes you want to move your feet. I also often find that sometimes the simplest riffs create the best reaction on the dance floor. Tracks like Layo and Bushwacka's *Love Story*, with the a cappella vocal of *Finally* over the top, is a massive tune in clubs at the moment, reaching anthem status of late. Also, Jam X and De Leon's *Can U Dig It?*, which sampled a famous 1980s riff, in a trance style, is doing incredibly well. Both these use very simple melodies but to a devastating effect that sends the crowds nuts.

I would also say that familiarity in dance records is a good thing. Take Room 5's *Make Luv (And Listen to the Music)*, I, along with a few other DJs, managed to get hold of a copy approximately a year before the UK label Positiva snapped it up. I used to play it out and get an average reaction from the crowd, then, when it became more popular and featured on a certain deodorant advert, the reaction when played out was outstanding. People would cheer and sing along to it as it became the favourite tune of the moment. Familiarity plays a large role in the scene. Ultimately, though, a track has to have a thumping, pounding four-to-the-floor groove that just makes you want to shake your butt. And from the many years I've been spinning vinyl and playing everything from house to progressive to techno to trance, I have to say that the hands in the air, spine-tingling euphoria (trance) always seems to get the best reactions from the crowd.

DJ techniques

This is possibly the most difficult subject to consider, because there are so many genres of dance music and each has a slightly different method of mixing. The basic principles are all the same, but some require more skill than others, along with some dexterity and flair thrown in too. DJ'ing and mixing has come a long way since the early 1980s. It originally started when New York DJs like Cool Herc (one of the founders of hip-hop and DJ'ing) used to use two rickety old turntables with rotary pitch control, a stereophonic mixer (with no cross-fader) to cut, scratch and mix two copies of the same record together. The breaks from old soul, jazz and funk records were mixed so that the music played continuously whilst MCs (acronym for Master of Ceremonies) such as KRS 1, MC Shan, Sugar Hill Gang, Steady B and Afrika Bambaata would rap fresh rhymes over the top. This was the birth of DJ'ing as we know it today, and many hip-hop artists are still cutting up old soul and funk tracks to make fresh beats. Early idols of mine such as the legendary Jazzy Jeff, who invented the transformer cut/scratch, with partner in crime, quirky rapper Will Smith, lead the way for a plethora of DJs to experiment, create and innovate. Here in England back in the late 1970s, a little known fact, believe it or not, is that one of the first non-dance music DJs (as we know it) to 'mix' two records together was none other than Sir Jimmy Saville – how's about that then! In fact, years ago he was one of the first guys on air at the BBC to fashion this technique.

Nowadays, hip-hop DJ'ing aside, this same mixing principle still applies and the typical DJ technique is to mix and 'beat match' two songs together. This is to physically blend and craft the sounds together to make one continuous sound, and to do this, and do it well, is an art form in itself that is dependent on the genre. For example, a techno DJ will mix differently than a drum 'n' bass DJ would, while a house DJ will spin much differently than a hip-hop DJ will. There are lots of different techniques and to describe them all would take me a long, long time, so I'll describe them briefly. Firstly, there's the technique known as the blend. This is used mainly in house style mixing but is also common with trance and consists of merging two tracks together, keeping them in sync so you can fade from one to the other. Personally, I get both tracks in sync and also in key, with room to build both to a crescendo, as they tend to be quite epic.

Hip-hop DJs use 'cutting' techniques where you use the cross-fader in conjunction with the turntable, bringing in parts of the track as you manipulate the faders and 'cut' parts of the track, beat or vocal up. A classic set-up of the decks for hip-hop DJs is to turn them around vertically, so the pitch and tone arm is at the top (as the pitch control is rarely used much with this style), giving more room to manoeuvre for techniques like scratching – which is cueing up a beat, noise or vocal on the up beat, and physically moving the record back and forth quickly to create a 'scratch'.

There's also the rougher cutting and chopping style used in drum 'n' bass and break beat (the music's a lot faster, as is the style of mixing), which consists of using the percussion from one record and mixing it with different elements from another. These are all advance techniques that are difficult to encapsulate in words and the only real way to learn them is to watch the 'masters' at work and pick up on their tactics. What's more, a lot of these techniques can be applied to various styles, it depends how dextrous and creative you are.

Recommended equipment for those wanting to be a DJ

The classic set-up for a DJ is two decks and a mixer (even though you probably want to start mixing on six decks, it's best to start out with the classic set-up for now). To those just starting

Figure 21.1 The hip-hop scratching technique.

out I would recommend investing in a pair of second-hand belt drive decks. However, there are plenty of arguments when it comes to this kind of advice, since some will recommend that you should go straight to the professional decks immediately. This is because when you get to a club they will almost definitely have the 'industry standard' Technics (which they invariably will the world over). I disagree.

I learnt the hard way by using cheap gear and building my way up, and I still believe that this is definitely the way to go. By taking your apprenticeship on less than pro equipment, it stands you in good stead when you get to the pro gear and keeps your skills sharp. If you can mix on wobbly old belt drive decks near perfectly, then you can mix on anything in any condition – once you have learnt to adjust the differences of the two. As a result, I recommend the Gemini PT2000 or the Numark range, but there are literally hundreds of starter decks and packages. Alongside these, you'll also need a cheap two-channel mixer to start mixing and matching beats (if they're not included in the starter pack) and I'd recommend the Kam Made To Fade or the Gemini 626. These can be hooked directly into your hi-fi system through the AUX-IN and it saves you additional expenditure having to purchase an amp and speakers. Conversely, if you want to mix with CD rather than vinyl then, if you're just starting out, get a second-hand pair of Pioneer CDJ 100s or 500s (the old models) and learn the basics on these. When you feel you are ready you can then step up to the intermediate or pro equipment, depending what level you are and what your budget is.

Professional turntables

There was a time when the mighty Technics deck reigned supreme and unchallenged and, although they still are, there are some serious contenders in the turntable ring. There are now three viable and dependable main makes of pro turntables:

- Technics.
- Vestax.
- Stanton.

The Technics 1200 and 1210 Mk2 'wheels of steel' professional turntables have been knocking around for well over 20 years and are still the industry standard turntable in serious clubs the world over. They have heavyweight clout and are the firm favourite of DJs, as well as being the DMC world mixing championships only turntable for well over a decade. That said, they have only released two variations on the theme since initial production of the Mk2. The limited edition 1210 Mk3 was released a few years ago (and is still selling well) and now they are set to release their newest model, the 1210 Mk5 and 5G. These all retain the classic sexy design with slight improvements on performance, yet keeping the old familiar feeling with the torque strength. This has been a problem with other decks from different manufacturers for DJs only used to the Technics design and feel. Any professional relying on a set standard of equipment for a performance wants to know exactly how it works, where everything is and how it reacts. It's dreadful arriving at a gig, especially abroad, only to find a deck you're not familiar with. In fact, because of this I now specify what equipment I will use in my contract, and I only stipulate Technics or Stanton turntables.

Figure 21.2 The requisite Technics SL-1210 Mk5. Photograph printed with permission. © Technics 2003.

Vestax have, over the years, come on leaps and bounds in the turntable market, having been a manufacturer of quality mixers for many years which are a favourite of installations in pro DJ booths worldwide. Of late, they have turned their intricate knowledge to decks with their quality PDX range, including the PDX D3s and PDX 2000/2300, though the features of these decks baffled many a DJ, as they started to include features like ultra pitch, instant reverse, digital pitch display, joystick controls, etc. They proved their worth, though, especially with younger clients and scratch DJs alike, as the straight tone arm skips less when performing in DJ battles. Vestax keep updating their range regularly, with new and innovative models coming out all the time, so it's difficult to keep up and recommend a specific turntable.

Stanton, a name associated and regarded highly for cartridges and styluses (the classic 500 MkII has been an industry standard for many years), have also turned their knowledge to turntables, with groundbreaking effect. A little over a year and a half ago, they released the 'ST' range to critical acclaim. Not only does it come in two different models, the STR8 100 and the ST150 (their latest model), one with straight (STR8) tone arm, the other with the more traditional S-shaped arm, but it also had some very interesting features. Firstly, they are as rock solid as a Technics with a 10% greater motor torque, have an instant reverse (I've never quite figured out the reason for this function yet!), a pitch bend of up to ±25% and a line in directly through the deck for connecting portable media such as MiniDisc, DAT or CD players. This allows you to play them directly though the mixer, so there is no need to plug anything into the sound system halfway through your set! What's more, they also have a master tempo function with key '*aDJust*', which allows you to speed up the track but keep the vocals the same pitch, so as not to wind up with Mickey Mouse vocals! If you have any doubts, Ministry of Sound were so impressed that they have just installed six pairs of them in the main and box room and done away with Technics completely!

CD decks

Over recent years, CD mixing has come to the fore. What was once received as rather shoddy twin-slot CD decks with a basic mixer rack at the bottom and a few buttons and lights have now come into their own. Many DJs used to be totally 100% vinyl purists (some still are), including myself until about 2 years ago, as we love our back-breaking heavy black blobs of plastic with a passion! But with huge advancements in the technology market from the likes of Pioneer and Denon, CD mixing and stand-alone decks have really come up to speed with the twenty-first century. In fact, you'll find that most pro DJ booths will now have quality CD decks or dual players nestling alongside the turntables and mixer. These offer numerous benefits over vinyl since, if you also produce your own music, CD media is incredibly cheap so you can copy your music and road test it at your next gig. What's more, boxes containing 100 CDs will weigh considerably less than 100 vinyl records and, as some professional decks now play the MP3 format (for instance, the Pioneer DMP-555 MP3), it's possible to fit over 30 tracks on one CD that you can spin like vinyl.

Some of the most prominent CD decks are from Pioneer, with CDJ 1000 and CDJ 800. These both use jog-wheel technology that cleverly emulates the practicalities of mixing with vinyl – you can spin the wheel the same as you would spin the vinyl to create scratching effects. They were also the first to introduce smart card slots in their CD players that enable DJs to turn up at any club that is using them, insert their smart card and the decks automatically configure the players for that particular DJ, recalling all the cue points from the DJ's personal collection. These decks also feature a plethora of effects and useful real-time functions, such as instant reverse, internal cue/loop memory, real-time seamless looping, anti-shock and much more. It's a serious deck with heavyweight clout from a well-respected maker of quality DJ gear.

Figure 21.3 The Pioneer CDJ 1000. Photograph printed with permission. © Pioneer 2003.

Slightly more expensive than the 1000 is the Pioneer DMP-555 MP3 CD player, which is the world's first digital media player. It offers all the effects that are available for the CDJ 1000, plus it's also capable of playing MP3 directly from a CD. Again, you can use smart media cards to store all your local settings and it also features full EQ trickery. What's more, you can hook it up to your PC via digital output or USB, and it comes with DJ booth CD software that allows you to control virtual decks and make loops and samples.

Figure 21.4 The Pioneer DMP-555. Photograph printed with permission. © Pioneer 2003.

While these decks are considered to be the best by many DJs, if you require more effects and scratch capabilities for DJ'ing hip-hop, then American Audio's Pro Scratch 2 deck, or the PSX, could be the answer. This offers similar features to the Pioneer decks, along with digital scratching, loads of effects, flash cue, hyper pitch and a host of other facilities that are suited towards 'cutting' beats. Notably, there are also many other fine makes and models of dual CD decks on the market, like Denon, Vestax, Numark, Tascam and Gemini. Some of these have built-in mixer and effects, all of which are good, and they all have ranges to suit different budgets and needs. If you want an option easier on the pocket, then there are many makes out there that will do the job.

Mixers

Having two decks or two DJ CD players isn't much use unless you have a mixer to join the two together, and though many starter kits include these, some of the more professional gear will not. Nevertheless, I'm going to be brief here, as the market is absolutely flooded with DJ mixers, some with fancy flashing lights and knobs, while others have a host of various bells and whistles (that can be confusing if you're just learning). Again, which to choose would depend what you want and how much gear you have to plug into the mixer. For instance, if you have only two decks (CD or vinyl, it doesn't really matter), then you will probably only need a two-channel, four-input mixer, i.e. 2 × Phono and 2 × Line. On the other hand, if you have untold then you will need to invest in more channels.

Some mixers have built-in beat counters and effects, but these won't necessarily help you mix any better. Personally, I've found that the more flashy parameters there are, the more it tends to put you off the job at hand, since they tend to be a hindrance more than anything else. It is worth trying them out, though, and seeing what you think – it's all about what suits you and your budget. If I were to recommend one, however, then it would be the Ecler Smac 32. I had one of these for ages and it was a brilliant mixer with a lovely warm sound and very slimline to fit between the decks. Ecler make some fine stuff, with an expanding range like the impressive HAK 310 and 360 series.

Alternatively, Vestax have a staggering range of quality mixers, like the classic PMC-05 Pro, or the PCV 275. Equally, Stanton also has a good line with the SMX 401 and 501 and SA8, all of which are two-channel, and Pioneer have also made some DJ mixers that are worth checking. The most notable is the DJM 300, 500 and 600 series, with two and four channels respectively. The 500/600s have a built-in FX and sampler that are very user-friendly. All these makes are some of the best I have ever used, but the true *crème de la crème* has to be the Soundcraft range of DJ mixers, but these are very expensive and so are usually limited to the DJs who earn substantial amounts of money.

MP3

I mentioned MP3 earlier when talking about the Pioneer CD DJ decks, but while some say the difference in sound quality is noticeable when compared to CD or MiniDisc, there really isn't a lot in it to your average listener. If it's played on a quality system, it's still going to sound good, but watch it if you're mixing vinyl into it! The sound difference between MP3 and CD, for that matter, is quite substantial because vinyl wins every time for warmth and depth of sound. Nonetheless, the technology around the MP3 format has, as I've mentioned, evolved into the DJ market and programs have started to appear that are specifically for DJ/performance use, whether live at the club/festival or in the studio or at home, fantastic programs like the Native Instruments Traktor DJ studio, and Soundgraph D-Vinyl and PC-DJ, but the real daddy of them all is Stanton's Final Scratch system. This is certainly leading the way for this technology while very cleverly keeping exactly the same techniques DJs use.

Traktor, D-Vinyl and PC-DJ are merely software based in that they allow MP3 and Wave files to be mixed together via a computer, and are extremely good. The look is similar to a dual CD analogue desk and the layout is pretty much the same, with similar buttons, pitch bend/aDJust, etc., with the NI Traktor being one of the leaders. With Stanton's (FS1 and 2) Final Scratch, however, they have gone one step further, and then some!

They have cleverly devised a system that comprises of a central hub, which you hook your turntables up to before connecting to your mixer and then into your PC. By then using two pieces of time-coded 'vinyl' you can 'virtually' play, scratch, mix and blend the Wave and MP3 files together, using a clever piece of software, but exactly the same as mixing two records together on your turntables. There isn't actually any sound on these special 'vinyl' records, they just transmit special signals to the software. Essentially, this means that you can record your music collection into your PC, plus new tracks, and literally have hundreds and thousands of tunes in your 'virtual' record box (hard drive) that you can then mix in the same physical manner as normal vinyl! Plus, this system is truly portable, so all you need to take to a gig is a hi-spec laptop, the FS hub and your two pieces of special vinyl. We are seeing increased popularity at gigs now that include this set-up, alongside the traditional gear, and it seems to be gaining momentum and respect from a quality name in the industry at the forefront of technology.

This obviously makes MP3 as viable a DJ format as any and, although most jocks still have a passion for vinyl (I know I do), this technology has embraced the past as well as the future, so there's no need to give your priceless collection of vinyl to Oxfam just yet! You can indeed use both, to staggering effect. If you want to embrace new technology, yet still use all the essential skills of mixing, then this is certainly a way to go. It's an amazing, forward-thinking system, and is rock solid, as it has been road tested by Ritchie Hawtin for many years.

Tips for aspiring DJs

First, and most important, you need a passion for music. Getting into the scene for just the fame, sex and money will not get you very far; it's a hard slog being a DJ and the only way you'll be able to stick at it is if you love what you do. The obvious place to start is by learning the basic skills of DJ'ing first. This means learning the fundamentals such as cueing up a record (assuming your chosen medium will be vinyl), becoming accustomed to the weight of the vinyl (heavier vinyl has a better sound quality), along with figuring out how to drop a record on the first beat and where the beats are by simply looking at the vinyl. When you are comfortable with this, try the basics of beat mixing by syncing two records together and mixing between them. Of course, at the moment these may all seem incredibly simple and you may be twitching to get to the more advanced techniques, but trying to learn the advanced techniques before you can even beat match will inevitably end in disaster. If you take it easy it will come, and even though you may get the hang of the basics quickly, it doesn't mean that you can DJ.

Once you feel confident enough, start doing small parties, playing on small sound systems and in front of small crowds. It's a whole different ball game than playing at home in your bedroom. You'll be amazed at the difference in sound, plus you have many other distractions in a club – the lights, the smoke, rude people pestering you in the mix! But just do it for a laugh at first and see how you go; if you want to be the next Cutmaster Swift or Judge Jules, you need to learn the basics of mixing first. DJ'ing is in some ways an art, but hip-hop scratching/cutting, etc. is an art form in itself, and like advanced DJ'ing skills it literally takes years to learn and perfect. The only real way is to start from the very bottom and work your way up. This means you will need to do gigs for free until you're at a proficient level and have proved yourself to be worthy of rubbing shoulders with the major players. It certainly won't happen overnight, but be persistent, believe in what you do and try to be individual – it's always a mistake to play all the same tracks as your peers.

When you feel confident enough, it's prudent to start recording your sets and then listen back to them with a critical ear. Ask yourself questions, such as is the mixing tight and are the songs well synced. More importantly, be honest with yourself, because even the DJs at the very top will admit that they do not 'know it all' and are constantly learning new techniques.

If at first you don't succeed, try again and then again, and then again until you do. As boring as this will probably sound, if your 'practised' demo sounds dodgy then you won't get a look in, so keep practising until it is right, and it will gradually become easier. You have to be dedicated, as there are plenty of quality DJs out there and you have to outshine them to stand a chance.

It's also worth putting your best polished demo onto CD or MiniDisc to use as an advertisement for your skills (more on promoting yourself in a moment). The best demo in the world won't get played if it looks pants; a crappy dusty old tape with a name and number scribbled in dodgy biro isn't going to do you any favours. The amount of demos I get that look like a dog's dinner!

If that's the amount of trouble they've gone to on the outside, what's the recording and mixing going to be like? Naturally, I'm not saying you have to get a gold-plated CD and cover, but a little presentation doesn't go amiss. Don't forget that the bigger promoters receive hundreds of demos all the time, so you need to make an effort for them to notice you. Get together a DJ biography, what you've been up to, what clubs and radio gigs you've done, etc., and print it on nice paper. Also, learn how to network, pester promoters and be knowledgeable. The Internet is a good way to promote yourself, so construct a site or get a few pages on various free sites and upload your MP3 mixes to it. I do lots of radio work, both Internet and FM, and that's another angle to get you out to a wider audience; it boosts your confidence and is another string to your musical bow.

How do DJs promote themselves?

There are various ways you can do this, some of which I've just touched upon, but it depends on what level you are at and how much experience you have. As I've mentioned, the Internet is probably one of the best ways, and it's worthwhile building a web site and placing biog and an MP3 mix on there. When you're at a reasonable level, consider joining an agency, as they will profile you, but obviously you have to have a certain amount of experience and achievement behind you.

I still spend many, many, hours tirelessly sending hundreds and hundreds of emails out and spending a lot of time on courtesy and follow-up calls, chasing promoters, making new contacts, following up contacts and networking all the time. You have to try to have something going on all the time and in the future, even several months away. The busier you become, the further down the line you have to plan, and it helps if you have a few fingers in a few musical pies so to speak; be determined, have drive, as it's a very cut-throat business. I review my web site regularly and always make an effort to keep abreast of any new music that's coming out.

More importantly, you need to be adventurous. I've found myself accepting bookings in many weird and wonderful places, resulting in me DJ'ing up mountainsides, at the edge of lakes, in Russia in wintertime, at festivals in Armenia in front of 40 000 people, in Bulgaria in swanky clubs and even in Germany in disused military army bases! If you have an audacious streak, you'll have the chance to play some amazing places, meet some wonderful people and make new friends. What's more, they will value you much, much more than down your local club in whichever major city you live in. This will allow you to build up a massive fan base, many of whom are budding DJs themselves, and will strive to get you more work in these places.

Another good tip to promote yourself is to get a portfolio together – achievements, flyers, discography if that applies, DJ biog, etc. – but make it ostentatious so that you will have something presentable to show promoters and the like. Also, you have to start building a reputation for yourself, and it's hard, you have to take the rough with the smooth. If you're on the lower level you'll often find yourself playing the warm-ups and earlier times, and you'll probably be playing to the floorboards for quite some time. Most of this will be gratis, but it will give you vital experience you need. When I started DJ'ing, it was almost a dirty word and I had to collect glasses in between playing. What's more, when the club shut I had to clear up and sweep the floor before I could get paid. That's what a resident meant back in the early 1980s, there certainly wasn't such a thing as the 'superstar' DJ nonsense that there is today. You learnt the hard way, but gained invaluable experience along the way, playing to tough demanding crowds, learning essential skills of dance floor dynamics. It's the only way to get the experience you need to be a DJ. All too often I see the sorry sight of 'DJs' with 40 records and 12 months experience under their belt with a pair of dodgy Sound Labs who think they're something special. Remember who you are, and be true to yourself, that's what counts, and know where you're at and where you would like to be, because if you disappear up your own backside, you'll quickly lose respect from everyone.

More advice

Possibly the best advice I can offer is to look after yourself. I'm going to go into some aspects quite deeply here, but it's not to put you off; on the contrary, it is an insight into the potential hazards of being a DJ.

Firstly, I'll start with what is the most important aspect of the job – your hearing! It's all to easy to have the opinion that it *'won't bother me'*, but if you spend a lot of time in clubs subjecting yourself to very high levels of sound as DJs, rock musicians, lighting and sound engineers do, then I would fully recommend investing in some professional musicians' ear protectors. These work by using a diaphragm that vibrates at a lower level than your eardrum, therefore reducing the noise by 15–25 dB. In layman's terms, this means that in a loud environment such as a club you can still hear everyone very clearly and the plugs don't spoil any of the enjoyment of the music, it just feels as if someone has turned down the volume control at the back of your head. These are very different from the traditional foam plugs that just block out noise and make everything appear muffled. Without protecting your ears you can suffer from tinnitus.

This is the medical term for a ringing in the ears, which is often the case after a night out working/clubbing. The ringing is our body's natural defence mechanism when we have been subjected to too much noise; for most of us this will pass, but the damage that the ringing has signalled won't. Many old school DJs, including myself, used to play and listen to massive amounts of noise on monstrous sound systems, which admittedly were great at the time, but we were unaware of the damage we were doing to our hearing. We never had this kind of advice back then, so I'm glad to be able to share some of my knowledge with you now.

Tips for you ears

- Wear specialist earplugs if working in or frequently attending loud clubs/concerts, etc.
- Take regular breaks from the dance floor if a clubber or pro dancer.
- Be careful on drink or drugs, as your perception of what is too loud will be impaired.
- Don't get dehydrated, as it can increase the risk of tinnitus.
- Don't push the levels to the max when DJ'ing (as is all to easy to do on a big system).
- Avoid shouting conversations into each other's ears – remember those earplugs!
- Don't stand next to big speaker stacks at big clubs and festivals.

Another hazard for the DJ (especially those that play mostly vinyl) is back pain. I'm a towering 6 ft 2 in and I occasionally get it so chronic that I have to cancel gigs! Carrying heavy record boxes and bags about is fundamental for DJs because you have to get your precious records or CDs to the gig. Over the years of doing this, I now have one shoulder higher than the other, as well as a trapped nerve in my neck (as does Tall Paul), which is worsened with stress, and being cramped in airplane seats and cars for long journeys doesn't help things. This is one of the inherent and unglamorous elements of the job unfortunately.

Apart from these problems, the most damage is done is when you're actually standing, stooped over the decks for hours. This can cause a lot of damage and is hazardous for your spine, especially over a number of years. Mobile jocks are even more susceptible, as they will be carrying even more cumbersome heavy equipment around. Also, another problem is hunching one shoulder, pressing the headphones to one ear, which is not good for you neck. It isn't just me, a lot of DJs, especially tall ones, have the same problem – Judge Jules and Tiesto are two high-profile jocks who are equally affected. Obviously, any stance that stops you bending over the decks unnecessarily is a good thing. As mentioned I'm particularly tall, so sometimes have to part my legs when the decks are a bit low, as this eases the stress on your lumbar region. Try and find a stance that is good for you, as it's in your interest to adopt this pose when you're playing regularly – try and make it a habit.

Tips for your back

- Invest in a pair of decent headphones you can use on one ear without hunching your shoulders too much – it will make a lot of difference.
- Change position and stretch regularly when you're playing.
- Careful on drink, so as not to strain while dancing/throwing shapes!
- Carry equipment/boxes, etc. close to your body, with the heaviest side closest to you. Share heavy loads between friends where possible.
- Get some strong painkillers that will ease the pain if it gets chronic.
- Back pain can become serious – get it checked out and see a specialist if necessary.
- Carry a muscle relief spray or gel with you when you travel.
- Physical exercise plays a big part in keeping your back healthy, so try some stretching exercises, especially after you land from a flight and after your set.
- If playing gigs internationally, have a massage when you get to your destination, at a reputable place, as this helps unwind the stress and tension that causes back and neck pain.

Also tied in with the lifestyle of the DJ is the stress and fatigue factor. On the surface of what appears to be a very glamorous never-ending party image of the busy international DJ, there are gruelling physical and mental demands. With hectic schedules and busy flight patterns, jocks can burn out if not too careful. Many of the top DJs are now coming clean about how demanding this actually is, with people like Sasha admitting that he has suffered years of serious panic attacks, and many other top DJs collapsing from exhaustion. Some not so wise DJs started to abuse heavy amounts of narcotics to deal with the near superhuman aspect of the job and ended up getting in a spectacular mess as a result. Both Brandon Block and Nicky Holloway have been into rehab through this. It's vital that you try and look after yourself; it is still possible to enjoy the highlights of the lifestyle if you take care of yourself and keep your mind and body in good shape.

Tips for fatigue

- Allow yourself to fully recover after a long period awake (sleep and rest well).
- Carry effervescent multivitamin energy tablets with you and take them after a long journey to give you a boost and combat exhaustion.
- Avoid dehydration by staying off booze and caffeine during and after flights; even though it's tempting with complimentary sauce, drink lots of mineral water instead.
- Get a good, undisturbed sleep, using eye mask and earplugs if necessary but not tranquillizers, as they can mess up sleep patterns (if necessary, use herbal remedies).

- Exhaustion affects the immune system, leading to illness, so try and maintain a healthy diet. Eat lots of fruit and take daily vitamin supplements.
- Take Melatonin tablets, a natural remedy thought to help promote sleep and reduce the effects of jet lag (unfortunately not available in the UK, so stock up on them in places like America).

Useful links

- www.energyinternetradio.com (Miami's best radio station).
- www.grooveexponents.co.uk (DJ agency).
- www.elacin.nl or call 0207 323 2076 (musicians' ear protectors).
- www.hearnet.com (useful tips on your hearing gear).
- www.backcare.org.uk (look after that back of yours).
- http://www.pioneer-eur.com/eur (forward thinking DJ gear).
- www.technics.co.uk (home of the world leaders and the mighty Technics deck).
- www.stantonDJ.com (home to the finest cartridges and the world-famous Final Scratch).
- www.vestax.co.uk (purveyors of quality DJ gear).
- www.dmcworld.com.

Appendix A: Binary and hex

All hardware circuits, even computers, are based on a series of interconnected switches that can be in one of two states: on or off. By continually switching these on and off in different configurations, it's possible for the hardware to count and perform complex calculations. This switching is accomplished by sending a series of bytes down an electrical cable, which means that they can only exist in one of two states too. If it is equal to one then the switch is turned on, while if it's a zero then the switch is turned off. Because of this, all computers must count using base 2, or 'binary', rather than our usual method of counting, which is base 10.

Counting in binary

Our normal method of counting was developed because we have 10 figures and using this system any one digit in a number has a value that is based upon the position of the digit in the number. Thus, when we are working to the power of 10, every time a digit moves one to the left, its result is to the power of 10, then by 100, then by 1000 and so forth. Consequently, the number 17 593 could be seen as follows.

10 000 000	1 000 000	100 000	10 000	1000	100	10	1
0	0	0	1	7	5	9	3

Or $(1 \times 10\,000) + (7 \times 1000) + (5 \times 100) + (9 \times 10) + (3 \times 1) = 17\,593$

From the above example we can see that each position in a number essentially adds a 0 to the meaning of a digit. Or, the value of each position is 10 times the value of the previous position (moving from right to left). This method of counting is also implemented in binary, but rather than use a base 10 system, we use a base 2 system. So, rather than saying that 10 to the power of 0 = 1, 10 to the power of 1 = 10, 10 to the power of 2 = 100, 10 to the power of 3 = 1000, it's 2 to the power of 0, 2 to the power of 1, 2 to the power of 2, 2 to the power of 4 and so forth.

128	64	32	16	8	4	2	1
0	0	0	0	0	0	0	0

Looking at the above table we are assuming that there are eight hardware switches grouped together, producing what is essentially an 8-bit byte. Also, as they are all switched off, the sum produced will be zero. However, if we were to introduce some positive bits we could sum them together to produce a decimal number:

128	64	32	16	8	4	2	1
0	1	1	0	1	1	0	1

$(0 \times 128) + (1 \times 64) + (1 \times 32) + (0 \times 16) + (1 \times 8) + (1 \times 4) + (0 \times 2) + (1 \times 1) = 109$

Thus, the decimal equivalent of the binary 01101101 would be 109. Additionally, we could also determine that the maximum number that could be calculated via binary would be 255 (11111111).

As MIDI uses this same 8-bit system when communicating with any devices, it would seem sensible to assume that it should be able to offer this same maximum parameter (of 255), but this isn't the case. Similar to CC messages, two forms of information need to be transmitted; a status bit and a data byte (which is composed of 7 bits). The status bit informs the synth of an incoming message that is arriving, while the following data byte informs it by how much the parameter should be adjusted. Because of this initial status bit, only seven other bits are left to provide the information, resulting in a maximum decimal value of 127; hence, the maximum number of any CC message can only be 127. To transmit numbers larger than this, an 8-bit byte has to be split into two halves and then converted into another numeration format: hexadecimal.

When we split a byte into two halves, both halves are commonly referred to as 'nibbles'. If the previous example were to be broken down into two nibbles it would become 0110 and 1101. These could then be individually summed together as $0110 = 96$ and $1101 = 13$ ($96 + 13 = 109$) to produce the result again. However, by splitting a byte into two and then converting it into a hexadecimal value, it's possible to produce much higher values. The reason behind this is that hexadecimal works to base 16, meaning that it is possible to access up to 16 383 parameters in a synth, much more than the standard 127 offered through Control Change messages.

Counting in hexadecimal

Hex uses a base 16 numbering system, but as there are not enough symbols to represent 16 different digits, as soon as the number 10 is reached it has to be converted into letters. Thus, to represent numbers 10–15 the letters A–F are used.

Recall how we count in decimal. Counting upwards, we count from 1 through to 9 and then we place a 1 to the left of this and go back to 0, making the number 10 (which actually means 1 to the power of 10 plus 0 to the power of 1). With hexadecimal we count beyond 9, but because our entire numeric system is only based on nine digits, with hexadecimal letters are used instead, as below, so that we can count up to 15.

F	E	D	C	B	A	9	8	7	6	5	4	3	2	1	0
15	14	13	12	11	10	9	8	7	6	5	4	3	2	1	0

When we count the zero, just as we do with the decimal counting system, we have used a total of 16 numbers. You can see from the decimal to hexadecimal conversion table that whenever we count up to F, we increase the number to the left by 1. This is also the same as the more common decimal counting system – for example, if you count from 11 through to 19, there are no more digits to use, so you change the 9 to 0 and increase the number to the left by 1, giving us 20.

You can see that hexadecimal counting is exactly the same as decimal, except in hexadecimal we count up to F before increasing the number to the left by a factor of 1.

It follows that in decimal, the number 24 actually means '2 × 10 to the power of $1 + 4 × 10$ to the power of $0 = 24$', which is what you or I already understand it to be because we use this every day, like counting on our fingers.

In hexadecimal, the number 24 actually means '2 × 16 to the power of 1 + 4 × 16 to the power of 0 = 36'. Logically, it follows then that the number CE in hexadecimal means '12 × 16 to the power of 1 + 14 × 16 to the power of 0 = 206' in decimal.

Using this principle, it's easy to convert hexadecimal into decimal.

Using this method, it's possible to count until we reach FF, which essentially means 15 × 16 + 15 units or 255 in decimal. Creating a new position gives 100 or 16 to the power of 2, which is 256 in decimal. Continuing to count using this base system it would be possible to produce two nibbles, both with a value of 7F, resulting in a total of 16 383 (127 × 128 + 127 = 16 383!).

For instance, suppose we wished to convert the decimal number **12 720** to hexadecimal:

- The largest power of 16 that fits into 12 720 is 16^3 = 4096. It fits three times and gives a remainder of 432. This number is derived from the fact that:

  (3 × 4096) = 12 288
  (12 720 − 12 288) = 432

  The first hex number is **3**.
- The largest power of 16 that fits into 432 is 16^2 = 256. It fits one time and gives a remainder of 176. This number is derived from the fact that:

  (1 × 256) = 256
  (432 − 256) = 176

  The second hex number is **1**.

- The largest power of 16 that fits into 176 is 16^1 = 16. It fits 11 times with no remainder (this remainder equates to 1 to the power of nothing which is nothing and will make 0 the last digit in the hex number). The third hex number is therefore **B**.
- The hex equivalent of the decimal number 12 720 is therefore **31B0**.

To clarify, to convert 31B0 hexadecimal into decimal the maths is as follows:

0 × (16 to the power of nothing) = 0
B (which is equivalent to 11 in decimal) × (16 to the power of 1) = 176
1 × (16 to the power of 2) = 256
3 × (16 to the power of 3) = 12 288

0 + 176 + 256 + 12 288 = **12 720** in decimal

As another example, suppose we wanted to convert the decimal number **14 683**.

- The largest power of 16 that fits into 14 683 is 16^3 = 4096. It fits three times and gives a remainder of 2395.

- This remainder is derived from the fact that:

    (3 × 4096) = 12 288
    (14 683 − 12 288) = 2395

    The first hex number is **3**.
- The largest power of 16 that fits into 2395 is 16^2 = 256. It fits nine times and gives a remainder of 91.
- This remainder is derived from the fact that:

    (9 × 256) = 2304
    (2395 − 2304) = 91

    The second hex number is **9**.
- The largest power of 16 that fits into 91 is 16^1 = 1. It fits five times and gives a remainder of 11.
- This remainder is derived from the fact that:

    (5 × 16) = 80
    (91 − 80) = 11

    The third hex number is **5**.
- The remainder of 11 has no power of 16 that will divide into it as a whole number, so the last digit of the hex will be the hex equivalent of 11, which is B. This gives the final hex number as **395B**.

Again, we can check this by converting 395B hexadecimal into decimal. The maths is as follows:

B × (16 to the power of nothing) = 11
5 × (16 to the power of 1) = 80
9 × (16 to the power of 2) = 2304
3 × (16 to the power of 3) = 12 288

11 + 80 + 2304 + 12 288 = **14 683** in decimal

Appendix B: Decimal to hexadecimal conversion table

Dec	Hex	Dec	Hex	Dec	Hex	Dec	Hex	Dec	Hex	Dec	Hex
0	0	44	2C	88	58	132	84	176	B0	220	DC
1	1	45	2D	89	59	133	85	177	B1	221	DD
2	2	46	2E	90	5A	134	86	178	B2	222	DE
3	3	47	2F	91	5B	135	87	179	B3	223	DF
4	4	48	30	92	5C	136	88	180	B4	224	E0
5	5	49	31	93	5D	137	89	181	B5	225	E1
6	6	50	32	94	5E	138	8A	182	B6	226	E2
7	7	51	33	95	5F	139	8B	183	B7	227	E3
8	8	52	34	96	60	140	8C	184	B8	228	E4
9	9	53	35	97	61	141	8D	185	B9	229	E5
10	A	54	36	98	62	142	8E	186	BA	230	E6
11	B	55	37	99	63	143	8F	187	BB	231	E7
12	C	56	38	100	64	144	90	188	BC	232	E8
13	D	57	39	101	65	145	91	189	BD	233	E9
14	E	58	3A	102	66	146	92	190	BE	234	EA
15	F	59	3B	103	67	147	93	191	BF	235	EB
16	10	60	3C	104	68	148	94	192	C0	236	EC
17	11	61	3D	105	69	149	95	193	C1	237	ED
18	12	62	3E	106	6A	150	96	194	C2	238	EE
19	13	63	3F	107	6B	151	97	195	C3	239	EF
20	14	64	40	108	6C	152	98	196	C4	240	F0
21	15	65	41	109	6D	153	99	197	C5	241	F1
22	16	66	42	110	6E	154	9A	198	C6	242	F2
23	17	67	43	111	6F	155	9B	199	C7	243	F3
24	18	68	44	112	70	156	9C	200	C8	244	F4
25	19	69	45	113	71	157	9D	201	C9	245	F5
26	1A	70	46	114	72	158	9E	202	CA	246	F6
27	1B	71	47	115	73	159	9F	203	CB	247	F7
28	1C	72	48	116	74	160	A0	204	CC	248	F8
29	1D	73	49	117	75	161	A1	205	CD	249	F9
30	1E	74	4A	118	76	162	A2	206	CE	250	FA
31	1F	75	4B	119	77	163	A3	207	CF	251	FB
32	20	76	4C	120	78	164	A4	208	D0	252	FC
33	21	77	4D	121	79	165	A5	209	D1	253	FD
34	22	78	4E	122	7A	166	A6	210	D2	254	FE
35	23	79	4F	123	7B	167	A7	211	D3	255	FF
36	24	80	50	124	7C	168	A8	212	D4		
37	25	81	51	125	7D	169	A9	213	D5		
38	26	82	52	126	7E	170	AA	214	D6		
39	27	83	53	127	7F	171	AB	215	D7		
40	28	84	54	128	80	172	AC	216	D8		
41	29	85	55	129	81	173	AD	217	D9		
42	2A	86	56	130	82	174	AE	218	DA		
43	2B	87	57	131	83	175	AF	219	DB		

Appendix C: General MIDI instrument patch maps

Program no.	Instrument set (piano)	Program no.	Instrument (chromatic perc)
1	Acoustic grand	9	Celesta
2	Bright acoustic	10	Glockenspiel
3	Electric grand	11	Music box
4	Honky-tonk	12	Vibraphone
5	Electric piano 1	13	Marimba
6	Electric piano 2	14	Xylophone
7	Harpsichord	15	Tubular bells
8	Clav	16	Dulcimer

Program no.	Instrument (organ)	Program no.	Instrument (guitar)
17	Drawbar organ	25	Nylon acoustic guitar
18	Percussive organ	26	Steel acoustic guitar
19	Rock organ	27	Jazz electric guitar
20	Church organ	28	Clean electric guitar
21	Reed organ	29	Muted electric guitar
22	Accordion	30	Overdrive guitar
23	Harmonica	31	Distortion guitar
24	Tango accordion	32	Guitar harmonics

Program no.	Instrument (bass)	Program no.	Instrument (strings)
33	Acoustic bass	41	Violin
34	Finger bass	42	Viola
35	Pick bass	43	Cello
36	Fretless bass	44	Contrabass
37	Slap bass 1	45	Tremolo strings
38	Slap bass 2	46	Pizzicato
39	Synth bass 1	47	Orchestral
40	Synth bass 2	48	Timpani

Program no.	Instrument (ensemble)	Program no.	Instrument (brass)
49	String ensemble 1	57	Trumpet
50	String ensemble 2	58	Trombone
51	Synth strings 1	59	Tuba
52	Synth strings 2	60	Muted trumpet
53	Choir aahs	61	French horn
54	Choir oohs	62	Brass section
55	Synth voice	63	Synth brass 2
56	Orchestral hit	64	Synth brass 2

Program no.	Instrument (reed)	Program no.	Instrument (pipe)
65	Soprano sax	73	Piccolo
66	Alto sax	74	Flute
67	Tenor sax	75	Recorder
68	Baritone sax	76	Pan flute
69	Oboe	77	Blown bottle
70	English horn	78	Skakuhachi
71	Bassoon	79	Whistle
72	Clarinet	80	Ocarina

Program no.	Instrument (synth lead)	Program no.	Instrument (synth pad)
81	Square lead	89	New age pad
82	Sawtooth lead	90	Warm pad
83	Calliope lead	91	Polysynth pad
84	Chiff lead	92	Choir pad
85	Charang lead	93	Bowed pad
86	Voice lead	94	Metallic pad
87	Fifths lead	95	Halo pad
88	Bass and lead	96	Sweep pad

Program no.	Instrument (synth effects)	Program no.	Instrument (ethnic)
97	Rain FX	105	Sitar
98	Soundtrack FX	106	Banjo
99	Crystal FX	107	Shamisen
100	Atmosphere FX	108	Koto
101	Brightness FX	109	Kalimba
102	Goblins FX	110	Bagpipe
103	Echoes FX	111	Fiddle
104	Sci-Fi FX	112	Shanai

Program no.	Instrument (percussive)	Program no.	Instrument (sound FX)
113	Tinkle bell	121	Guitar fret noise
114	Agogo	122	Breath noise
115	Steel drums	123	Seashore
116	Woodblock	124	Bird tweet
117	Taiko drums	125	Telephone ring
118	Melodic toms	126	Helicopter
119	Synth drum	127	Applause
120	Reverse cymbal	128	Gunshot

General MIDI percussion set

MIDI key	Drum sound	MIDI key	Drum sound
35	Acoustic bass drum	59	Ride cymbal
36	Bass drum	60	Hi bongo
37	Side stick	61	Low bongo
38	Acoustic snare	62	Mute hi bongo
39	Hand clap	63	Open hi bongo
40	Electric snare	64	Low conga
41	Low floor tom	65	High timbale
42	Closed hi-hats	66	Low timbale
43	High floor tom	67	High agogo
44	Pedal hi-hat	68	Low agogo
45	Low tom	69	Cabasa
46	Open hi-hat	70	Maracas
47	Low mid tom	71	Short whistle
48	High mid tom	72	Long whistle
49	Crash cymbal	73	Short guiro
50	High tom	74	Long guiro
51	Ride cymbal	75	Claves
52	Chinese cymbal	76	Hi wood block
53	Ride bell	77	Low wood block
54	Tambourine	78	Mute cuica
55	Splash cymbal	79	Open cuica
56	Cowbell	80	Mute triangle
57	Crash cymbal	81	Open triangle
58	Vibraslap		

Appendix D: General MIDI CC list

CC	Function	CC	Function	CC	Function
0	Bank select	46–47	Undefined	83	General purpose
1	Mod wheel	48	General purpose		controller 5
2	Breath		controller 1	84	Portamento control
	controller	49	General purpose	85–90	Undefined
3	Undefined		controller 2	91	Effects 1 (reverb)
4	Foot	50	General purpose		depth
	controller		controller 3	92	Effects 2 (tremolo)
5	Portamento	51	General purpose		depth
	time		controller 4	93	Effects 3 (chorus)
6	Data entry	52–63	Undefined		depth
7	Channel	64	Damper pedal	94	Effects 4 (detune)
	volume		(on/off)		depth
8	Balance	65	Portamento (on/off)	95	Effects 5 (phaser)
9	Undefined	66	Sustenuto (on/off)		depth
10	Pan	67	Soft pedal (on/off)	96	Data entry+1
11	Expression	68	Legato footswitch	97	Data entry−1
12	Effect control 1	69	Hold 2	98	Non-registered
13	Effect control 2	70	Sound controller		parameter number
14–15	Undefined		1 (sound variation)		LSBa
16	General control 1	71	Sound controller	99	Non-registered
17	General control 2		2 (timbre)		parameter number
18	General control 3	72	Sound controller		MSBb
19	General control 4		3 (release time)	100	Registered parameter
20–31	Undefined	73	Sound controller		number LSB
32	Bank select		4 (attack time)	101	Registered parameter
33	Mod wheel	74	Sound controller		number MSB
34	Breath control		5 (brightness)	102–119	Undefined
35	Undefined	75	Sound controller 6	120	All sound off
36	Foot control	76	Sound controller 7	121	Reset all controllers
37	Portamento time	77	Sound controller 8	122	Local control on/off
38	Data entry	78	Sound controller 9	123	All notes off
39	Channel volume	79	Sound controller 10	124	Omni mode off (+all
40	Balance	80	General purpose		notes off)
41	Undefined		controller 5	125	Omni mode on (+all
42	Pan	81	General purpose		notes off)
43	Expression controller		controller 5	126	Poly mode on/off (+
44	Effect control 1	82	General purpose		all notes off)
45	Effect control 2		controller 5	127	Poly mode on

aLeast Significant Bit.
bMost Significant Bit.

477

Appendix E: Sequencer note divisions

The following charts display the number of clock pulses for each note value for the four most popular PPQN resolutions – 96, 192, 240 and 384.

96 PPQN

Note value	PPQN	Note value	PPQN	Note value	PPQN
Whole	384	Dotted whole	576	Triplet whole	256
Half	192	Dotted half	288	Triplet half	128
Quarter	96	Dotted quarter	144	Triplet quarter	64
Eighth	48	Dotted eighth	72	Triplet eighth	32
16th	24	Dotted 16th	36	Triplet 16th	16
32nd	12	Dotted 32nd	18	Triplet 32nd	8
64th	6	Dotted 64th	9	Triplet 64th	4
128th	3	Dotted 128th	N/A	Triplet 128th	2

192 PPQN

Note value	PPQN	Note value	PPQN	Note value	PPQN
Whole	768	Dotted whole	1152	Triplet whole	512
Half	384	Dotted half	576	Triplet half	256
Quarter	192	Dotted quarter	288	Triplet quarter	128
Eighth	96	Dotted eighth	144	Triplet eighth	64
16th	48	Dotted 16th	73	Triplet 16th	32
32nd	24	Dotted 32nd	36	Triplet 32nd	16
64th	12	Dotted 64th	18	Triplet 64th	8
128th	6	Dotted 128th	9	Triplet 128th	4

240 PPQN

Note value	PPQN	Note value	PPQN	Note value	PPQN
Whole	960	Dotted whole	1440	Triplet whole	640
Half	480	Dotted half	720	Triplet half	320
Quarter	240	Dotted quarter	360	Triplet quarter	160
Eighth	120	Dotted eighth	180	Triplet eighth	80
16th	60	Dotted 16th	90	Triplet 16th	40
32nd	30	Dotted 32nd	45	Triplet 32nd	20
64th	15	Dotted 64th	N/A	Triplet 64th	10
128th	N/A	Dotted 128th	N/A	Triplet 128th	5

384 PPQN

Note value	PPQN	Note value	PPQN	Note value	PPQN
Whole	1536	Dotted whole	2304	Triplet whole	1024
Half	768	Dotted half	1152	Triplet half	512
Quarter	384	Dotted quarter	576	Triplet quarter	256
Eighth	192	Dotted eighth	288	Triplet eighth	128
16th	96	Dotted 16th	144	Triplet 16th	64
32nd	48	Dotted 32nd	72	Triplet 32nd	32
64th	24	Dotted 64th	36	Triplet 64th	16
128th	12	Dotted 128th	18	Triplet 128th	8

Appendix F: Tempo delay time chart

If song tempo is 128 BPM, then set delay time to 469 ms for quarter-note delay, 234 ms for eighth-note delay, 156 ms for eighth triplet delay or 117 ms for 16th-note delay.

Tempo	1/4	1/8	1/8T	1/16	Tempo	1/4	1/8	1/8T	1/16	Tempo	1/4	1/8	1/8T	1/16
80	750	375	250	188	118	508	254	169	127	156	385	192	128	96
81	741	370	247	185	119	504	252	168	126	157	382	191	127	96
82	732	366	244	183	120	500	250	167	125	158	380	190	127	95
83	723	361	241	181	121	496	248	165	124	159	377	189	126	94
84	714	357	238	179	122	492	246	164	123	160	375	188	125	94
85	706	353	235	176	123	488	244	163	122	161	373	186	124	93
86	698	349	233	174	124	484	242	161	121	162	370	185	123	92
87	690	345	230	172	125	480	240	160	120	163	368	184	123	92
88	682	341	227	170	126	476	238	159	119	164	366	183	122	91
89	674	337	225	169	127	472	236	157	118	165	364	182	121	91
90	667	333	222	167	128	469	234	156	117	166	361	181	120	90
91	659	330	220	165	129	465	233	155	116	167	359	180	120	90
92	652	326	217	163	130	462	231	154	115	168	357	179	119	89
93	645	323	215	161	131	458	229	153	115	169	355	178	118	88
94	638	319	213	160	132	455	227	152	114	170	353	176	118	88
95	632	316	211	158	133	451	226	150	113	171	351	175	117	88
96	625	313	208	156	134	448	224	149	112	172	349	174	116	87
97	619	309	206	155	135	444	222	148	111	173	347	173	116	87
98	612	306	204	153	136	441	221	147	110	174	345	172	115	86
99	606	303	204	153	137	438	219	146	109	175	343	171	114	86
100	600	300	200	150	138	435	217	145	109	176	341	170	114	85
101	594	297	198	149	139	432	216	144	108	177	339	169	113	85
102	588	294	196	147	140	429	214	143	107	178	337	169	112	84
103	583	291	194	146	141	426	213	142	106	179	335	168	112	84
104	577	288	192	144	142	423	211	141	106	180	333	168	111	83
105	571	286	190	143	143	420	210	140	105	181	331	167	110	83
106	566	283	189	142	144	417	208	139	104	182	299	166	110	82
107	561	280	187	140	145	414	207	138	103	183	297	165	109	82
108	556	278	185	139	146	411	205	137	103	184	295	164	108	81
109	550	275	183	138	147	408	302	136	102	185	293	164	108	81
110	545	273	182	136	148	405	203	135	101	186	291	163	107	80
111	541	270	180	135	149	403	201	134	101	187	289	162	106	80
112	536	268	179	134	150	400	200	133	100	188	287	161	106	79
113	531	265	177	133	151	397	199	132	99	190	285	161	105	79
114	526	263	175	132	152	395	197	132	99	200	283	160	104	78
115	522	261	174	130	153	392	196	131	98	201	281	159	104	78
116	517	259	172	129	154	390	195	130	97	202	279	158	103	77
117	513	256	171	128	155	387	194	129	97	203	277	157	102	77

1/8 = Eighth-note delay; 1/8T = eighth-note triplet delay; 1/16 = 16th-note delay.

Appendix G: Musical note to MIDI and frequencies

Note	MIDI no.	Frequency	MIDI no.	Frequency
C	0	8.1757989156	12	16.3515978313
D♭	1	8.6619572180	13	17.3239144361
D	2	9.1770239974	14	18.3540479948
E♭	3	9.7227182413	15	19.4454364826
E	4	10.3008611535	16	20.6017223071
F	5	10.9133822323	17	21.8267644646
G♭	6	11.5623257097	18	23.1246514195
G	7	12.2498573744	19	24.4997147489
A♭	8	12.9782717994	20	25.9565435987
A	9	13.7500000000	21	27.5000000000
B♭	10	14.5676175474	22	29.1352350949
B	11	15.4338531643	23	30.8677063285
C	24	32.7031956626	36	65.4063913251
D♭	25	34.6478288721	37	69.2956577442
D	26	36.7080959897	38	73.4161919794
E♭	27	38.8908729653	39	77.7817459305
E	28	41.2034446141	40	82.4068892282
F	29	43.6535289291	41	87.3070578583
G♭	30	46.2493028390	42	92.4986056779
G	31	48.9994294977	43	97.9988589954
A♭	32	51.9130871975	44	103.8261743950
A	33	55.0000000000	45	110.0000000000
B♭	34	58.2704701898	46	116.5409403795
B	35	61.7354126570	47	123.4708253140
C	48	130.8127826503	60	261.6255653006
D♭	49	138.5913154884	61	277.1826309769
D	50	146.8323839587	62	293.6647679174
E♭	51	155.5634918610	63	311.1269837221
E	52	164.8137784564	64	329.6275569129
F	53	174.6141157165	65	349.2282314330
G♭	54	184.9972113558	66	369.9944227116
G	55	195.9977179909	67	391.9954359817
A♭	56	207.6523487900	68	415.3046975799
A	57	220.0000000000	69	440.0000000000
B♭	58	233.0818807590	70	466.1637615181
B	59	246.9416506281	71	493.8833012561

Note	MIDI no.	Frequency	MIDI no.	Frequency
C	72	523.2511306012	84	1046.5022612024
D♭	73	554.3652619537	85	1108.7305239075
D	74	587.3295358348	86	1174.6590716696
E♭	75	622.2539674442	87	1244.5079348883
E	76	659.2551138257	88	1318.5102276515
F	77	698.4564628660	89	1396.9129257320
G♭	78	739.9888454233	90	1479.9776908465
G	79	783.9908719635	91	1567.9817439270
A♭	80	830.6093951599	92	1661.2187903198
A	81	880.0000000000	93	1760.0000000000
B♭	82	932.3275230362	94	1864.6550460724
B	83	987.7666025122	95	1975.5332050245
C	96	2093.0045224048	108	4186.0090448096
D♭	97	2217.4610478150	109	4434.9220956300
D	98	2349.3181433393	110	4698.6362866785
E♭	99	2489.0158697766	111	4978.0317395533
E	100	2637.0204553030	112	5274.0409106059
F	101	2793.8258514640	113	5587.6517029281
G♭	102	2959.9553816931	114	5919.9107633862
G	103	3135.9634878540	115	6271.9269757080
A♭	104	3322.4375806396	116	6644.8751612791
A	105	3520.0000000000	117	7040.0000000000
B♭	106	3729.3100921447	118	7458.6201842894
B	107	3951.0664100490	119	7902.1328200980
C	120	8372.0180896192		
D♭	121	8869.8441912599		
D	122	9397.2725733570		
E♭	123	9956.0634791066		
E	124	10548.0818212118		
F	125	11175.3034058561		
G♭	126	11839.8215267723		
G	127	12543.8539514160		

Glossary

A

AAC (*Advanced Audio Coding*)   An abbreviation for the MPEG-2 Advanced Audio Coding. The term is often used to refer to MPEG-4.

A&R (*Artist and Repertoire*)   A term coined by the record industry to describe the person(s) that work with both the artist and record label. Their job entails finding and signing artists, choosing the songs to place on an album, and deciding which producers, engineers and studios should be used to record the artist.

AC3   The audio coding technique used by all Dolby systems to store five channels of surround sound audio. See also **Dolby Digital**.

Acoustic Feedback   A term used to describe the effect whereby the sound from a loudspeaker is picked up by the microphone. This signal is then amplified again and returned to the speaker, forming a continuous loop, resulting in **howl-round**.

Acoustic Treatment   The process of treating a room to prevent the sound from the loudspeakers from being compromised by furniture and walls. Generally, three methods are employed, consisting of:

1   *Diffusers* – uniformly distribute the sound.
2   *Absorbers* – absorb any rogue frequencies.
3   *Reflectors* – redirect the sound.

ADAT (*Alesis Digital Audio Tape*)   A recording device that uses digital SVHS tapes and allows up to eight tracks at 16-bit, 44.1 kHz.

ADAT Optical   A proprietary system developed by Alesis for use with their ADAT recorders. Using a fibre-optic cable, it permits you to transmit eight channels of digital audio to and from ADAT machines.

ADC (*Analogue-to-Digital Converter*)   An electrical component fitted to all digital devices that are capable of recording analogue signals. They measure the analogue signal at predefined points and convert it into binary language that can be used by the digital device.

AES (*Audio Engineering Society*)   First appeared in 1948 and is the largest professional organization for electronic engineers. It is concerned with standardization.

AES/EBU   A two-way balanced digital connection to interconnect two digital devices together using **XLR** type connectors.

AIFC (*AIFF-C*)   A compressed version of **AIFF**.

AIFF   Default extension for Audio Interchange File Format, an audio file format for PCs and Macs.

Aliasing (*also known as Sample Aliasing*)   Numerous frequency spikes that are introduced into audio when the **ADC** does not sample the audio waveform at the correct rate. See **Nyquist Theorem**.

Ambience   A broad term used to describe the acoustic characteristics of an area with regard to its reverberation. A room with very little reverb is described as being 'dead', while a room that exhibits plenty of reverberation is described as 'lively'.

Amplifier (*Amp*)   A device which increases signal level.

Amplitude   The maximum value of a periodic curve, voltage or waveform, measured along its vertical axis.

Analogue   The way in which all sound was reproduced before digital became available.

Anti-aliasing Filter   A low-pass filter that is employed on digital audio converters to prevent aliasing.

ASCII (*American Standard Code for Information Interchange, pronounced 'askee'*)   A standard data transmission code that consists of 7 bits. These are used to code 128 letters, numbers and special characters. Today it is commonly used artistically by **cracking groups** for **NFO** files.

ASIO (*Audio Stream Input/Output*)   First developed by Steinberg in 1997, this is a transfer protocol that allows audio/MIDI sequencers to communicate directly with soundcards to prevent **latency**.

Attenuate   To reduce a signal in level (most usually frequencies or volume).

Attenuator Pad   Commonly found on mixing desks in the form of a switch that when activated reduces the gain of the incoming audio signal by 12 or 24 dB.

Audiophile   Person(s) with more than an unhealthy interest in sound reproduction and fidelity.

B

Balance Control   A control that allows you to adjust the relative loudness between the left and right channels, making the channel that is not attenuated perceivably (but not physically) louder.

Balanced   In respect to wiring, balanced cables appear in **XLR** or **TRS** form. Three wires are used in all, one of which carries the ground signal and the other two which carry the audio. When the signal is received by a connected device, the two audio-carrying wires are subjected to **phase inversion**, which removes a large percentage of the noise that may have been introduced through the cable.

Bandwidth   The measurement of the total frequency range of a system. Also see **Q**.

Bass Reflex   A loudspeaker that uses a port or duct to expand the low-frequency response.

Baxandall Tone Controls   Developed by British engineer P. J. Baxandall, the circuitry is used on most systems as a bass and treble tone control. It uses the principle of negative feedback to produce low harmonic distortion.

Bell, Alexander Graham   Inventor of the telephone, from whom the term **decibel** is derived.

Bit (*abbreviation for binary digit*)   The smallest possible amount of digital information, it can represent two states, 0 and 1 (*on* or *off*).

Bit Rate   In digital audio, this refers to the rate or frequency that bits appear in a bit stream and defines the overall quality of the results.

Bi-wiring   A description of using two pairs of speaker wire from the same amplifier output, allowing you to power the bass and treble inputs on the speaker separately.

Blumlein, Alan Dower   Responsible for developing the principle stereophonic sound, along with many other designs such as a new disc-cutting system that made modern records possible.

BNC (*Bayonet Neill-Concelman*)   Often wrongly thought to stand for *Baby N Connector, Bayonet Connector, Bayonet Naval Connector* and *British Naval Connector*, the BNC connector is sometimes employed in musical instruments and used to transmit **Word Clock** signals.

Boomy   A term that is used to describe audio that exhibits an excessive bass response.

Boost/Cut EQ   Refers to a graphic equalizer with any number of bands. When the faders are all central there is no effect on the audio, but by raising the fader's boost it is applied at the particular frequency covered by the fader. Similarly, decreasing the fader results in a cut in the frequency.

Bright   A term that is used to describe audio that exhibits an excessive high-frequency response.

Bug   A term often used to denote a problem with a program or computer. Commonly appears at the most inopportune or critical moment or when working to a deadline.

Buss   A term used to refer to the various signal paths that an audio signal can travel through or to.

Byte   The term used to describe a collection of 8 bits (known as a word) operating together.

C

Capacitance   A force that resists the sudden build-up of electric voltage.

Capacitor   A device that introduces capacitance into an electric circuit.

Capacitor Microphone   Refer to **Condenser Microphone**.

Cardioid   A heart-shaped pattern that commonly signifies the response of a cardioid microphone.

CD (*Compact Disc*)   A 12-cm (diameter) plastic disc developed by Sony-Philips. A digital audio optical disc storage system that is capable of storing up to 80 minutes of digital audio or other data.

Checksum   The total sum of a batch of data that is used for error checking. The receiving device calculates the checksum from the data sent, and if it tallies with the one sent in the stream, the device will process the information, otherwise it will ask for the data to be retransmitted.

Chorus   An audio effect that uses multiple delays and pitch-shifting algorithms so it appears as though several instruments are playing simultaneously.

Chromatic Scale   The musical scale, consisting of 12 semitones.

Clipping   A type of distortion that occurs when a signal is recorded too loud in the digital domain. A clipped waveform exhibits a crunchy or harsh sound, as the top of the waveform is removed from the signal.

Codec (*code/decode*)   The name for any electronic device that converts an analogue signal into digital and then uses a proprietary algorithm to compress the data. See also **MPEG** and **MP3**.

Coloration   A term used to describe a sound characteristic that was not present in the original sound.

Compression   To reduce the dynamic range of a signal or the state of air pressure during a sound wave. See also **Rarefaction**.

Compressor   A signal processor that is used to reduce the dynamic range of any signal during recording or mixing.

Concrete   (*French term pronounced 'concratt'*) A French term used to describe music that is constructed entirely of real-world sounds.

Condenser Microphone   (*also known as a Capacitor Microphone or Electrostatic Microphone*) A microphone that captures sound by means of a varying **capacitance**.

Constant-Q EQ   Used to describe the behaviour of an EQ unit where the bandwidth remains the same for all boost/cut levels.

Control Voltage (*CV*)   A DC voltage that is proportional to the amplitude of an incoming audio signal.

Copyright   The legal right granted to an author, composer, publisher or distributor to exclusive publication, production, sale or distribution of a literary, musical, dramatic or artistic work.

Correlation   A mathematical operation that indicates the degree to which two signals are alike.

Crack   Slang term referring to decompiling and removing the copyright protection employed in software.

Cracking Groups   A team of individuals whose pastime involves decompiling the protection methods employed by software manufacturers permitting the software to run without the need for a **dongle** or original CD. This software is then distributed via the Internet as illegal **warez**. There are many cracking groups, the most notable of which are H20, Oxygen, Paradox, Zone and Radium.

Crash   Term used to describe when a piece of software (or hardware) stops working or freezes without any prior warning.

Cross-fade   A term used to describe transition from one audio signal to another.

Crosstalk   A term used to describe the unwanted penetration of one audio channel into another.

Cubase   The most popular and successful audio/MIDI sequencer developed by Steinberg. The latest manifestation is Cubase SX 2.

Cue (*audio industry*)   A term used to describe a mix or sound that is fed to headphones or loud-speakers for monitoring purposes.

Cut-off Frequency   The frequency at which a signal falls off by 3 dB from its maximum value.

D

DA-88   A model number for Tascam's infamous digital multi-track recorder.

DAC (*Digital-to-Audio Converter*)   An electrical component fitted to all digital devices that are capable of playing back analogue signals. They convert the digital bit stream used by the device into an analogue signal.

Damping   A term used to describe the capability of any audio component to stop whatever it was doing when a signal ends.

DASH (*Digital Audio Stationary Head*)   A format developed to ensure compatibility amongst digital multi-track recorders that use stationary heads.

DAT (*Digital Audio Tape*)   A term often used to describe a digital audio recorder that uses rotating heads to record. It may also refer to the magnetic tape that is used by this system. See also **R-DAT**.

DAW (*Digital Audio Workstation*)   A term used to describe a computer that is being used as the basis for producing, editing or storing digital audio.

DCC (*Digital Compact Cassette*)   A digital recorder developed by Philips digital that uses digital cassette tapes (no longer available).

Decibel (dB)   A measurement of the relative loudness of an audio signal. Named after Alexander Graham Bell, it is a logarithmic scale, since our ears become less sensitive to sound as the intensity increases: 0 dB is the threshold of hearing and 130 dB is the threshold of pain.

dBFS   An abbreviation for decibel full scale, a measurement that is used with all **DAC** and **ADC** converters. It specifies the maximum allowable recording or playback level before **clipping** occurs.

dBm   An abbreviation for the power reference point, equal to 1 milliwatt.

dBu (0)   An abbreviation for the voltage reference point, equal to 0.775 V.

dBu (+4)   The standard professional voltage reference point, equal to 1.23 V.

dBV (0)   An abbreviation for the voltage reference point, equal to 1 V.

dBV (−10)   The standard consumer voltage reference level, equal to 0.316 V.

De-esser   A frequency-dependent compressor that operates above 3 kHz to reduce **sibilance**.

Digital Audio   Analogue signals that are stored digitally as numbers.

Disk   A term derived from the Greek word '*diskos*', refers to any magnetic storage media such as computer diskettes or hard disks.

Distortion   A term used to refer to anything that alters a musical signal. Generally refers to undesirable noise introduced by overloading an input or output.

Dither   Refers to noise that is added to a signal prior to quantization to reduce the distortion and noise from quantizing.

Dolby Digital   A five-channel system consisting of left, right, centre, left rear and right rear channels. If AC3 is used, it also utilizes a separate channel to replicate frequencies below 80 Hz.

Dongle   A small hardware device that acts as copyright security and usually connects to the USB port of a computer. When connected it enables a specific copy-protected program to run correctly; if it is not present the program will not run.

Doppler Effect   Named after Christian Johann Doppler, an Austrian physicist who discovered that there is an apparent change in the pitch of a sound when it moves to and from a motion-less listening position.

Dry   In the audio industry, refers to a recorded signal before any effects processing is applied.
Drum Machine   Synthesizer or sample-based instrument that can be programmed to replicate the playing of a drummer. Advantageous towards real drummers, since the rhythm only needs to be punched into a drum machine once.
DSP (*Digital Signal Processing*)   Digital technology that allows for fast calculations, permitting high performance.
DVD   Originally thought to stand for Digital Video Disc or Digital Versatile Disc (yet actually has no meaning), it's a 12-cm compact disc capable of storing 4.7 GB of information. DVDs that hold movies are two discs bound together during the manufacturing process, allowing them to store 8.5 GB.
Dynamic (*Microphone*)   A microphone that captures sound by means of a moving coil suspended in a magnetic field.
Dynamic (*Range*)   The ratio between the loudest signal before clipping and the quietest perceptible signal.

E

E   See **MDMA**.
Early Decay Time   The amount of time measured in seconds that it takes for a signal to decay from 0 to −10 dB.
Early Reflections   Associated with **reverberations**. After the direct sound reaches the listener's ears, further sounds that are reflected from other surfaces take longer to reach the ears; the first few of these are the early reflections.
EBU (*European Broadcasting Union*)   A society that helps to establish audio standards.
Effects Loop   A term used when mixing that refers to an audio signal exiting the mixing desk into an **Effects Unit** and the newly effected signal being returned back into the mixer.
Effects Unit   A hardware (or software) unit that is used to introduce a specific effect onto any audio signal.
Envelope Generator (EG)   An electrical circuit used to control how the properties of a sound will change over a finite period of time. On synthesizers, amplitude always has an envelope generator dedicated to it, thus permitting the user to control how the sound's volume properties will change over time. Some synthesizers also have an EG that can be used to modulate any parameter.
EQ (*Equalizer*)   An electronic filter that is designed to cut or boost specific frequencies in an audio signal.
Euphonic   An audio listening term that refers to any **coloration** that produces sonically pleasing results.
Euphoria   An immense, overwhelming feeling of well-being. See also **MDMA**.
Exciters (*Enhancers*)   A processor that adds either harmonic distortion or adjusts the phase of a stereo signal to make it more pleasing to the ear.

F

Fader   A control with an upward/downward movement that can be used to increase or decrease values.
Feedback   See **Acoustic Feedback**.
FET (*Field Effect Transistor*)   A transistor where the output current is controlled by a variable electric field.
Filter   A device that allows some signals to pass through unaffected and attenuate the rest.
Firewire   See **IEEE 1394**.
Flanging   The process of taking two of the same signals and delaying one in time to the other to produce a series of phase cancellations.

Fletcher–Munson Curve   A measurement that displays how the human ear is more sensitive to some frequencies than others depending on the volume of the source.

FM (*Frequency Modulation*)   A form of digital synthesis invented by Dr John Chowning and first marketed by Yamaha in the DX range of synthesizers. Sounds are generated through interconnecting a series of **operators** in different configurations to produce different timbres. These resulting timbres, while various, are often metallic in character.

Foldback   A term often used to describe the mix that is played out to the performer so they can hear themselves.

Fourier, Joseph   A French physicist who devised a method for analysing periodic waveforms.

Fourier Analysis   Most often the approximation of a function through the application of a Fourier series to periodic data.

Fourier Theorem   The theory that states: every periodic wave can be seen as the sum of sine waves with certain lengths and amplitudes, the wavelengths of which have harmonic relations. In other words, the content of any sound is determined by the relationship between the level of the fundamental frequency and its harmonics, and their evolution over a period of time.

Frequency   The speed at which an object vibrates. See also **Oscillate**.

Fundamental   The lowest frequency of a note in a complex waveform or chord.

FX Unit   Engineering slang for **Effects Unit**.

G

Gain   To increase the amount of amplification of an audio signal, usually expressed in dB (pertaining to the ratio of the output level to the input level).

Gain Riding   A term used to describe the process of constantly monitoring and adjusting the gain of a signal to prevent overloading.

Gang (*Ganged*)   The process of joining two or more controls together either mechanically or electronically, so that moving one control will automatically move others that are ganged with it.

Gate   See **Noise Gate**.

Graphic EQ   A multi-band EQ unit that uses a series of faders to control the amplitude of specific frequencies. So named because, once set, the nature is very graphic.

Groupies (*slang*)   A group of people (commonly young girls) who follow popular artists around, pester for autographs, scream uncontrollably and faint a lot.

Groups (*Subgroup or Submix*)   Similar to **gang**, this is the process where a number of signals are grouped together and controlled as one with just one set of controls (EQ, volume, panning, etc.).

GUI (*Graphical User Interface*)   A generic name for a computer interface that substitutes graphics for characters.

H

Haas Effect (*also known as the Precedence Effect*)   Describes how we can localize any sound even though our heads get in the way, making the sound appear later in one ear than the other (*provided that the subsequent arrivals are within 25–35 milliseconds*).

Hard Disk   A mass storage unit used for storing large amounts of digital data.

Harmonic Distortion   See **Total Harmonic Distortion**.

Harmonic Series   A series of tones consisting of a fundamental and the harmonics produced by it.

Harmonics   Sometimes called overtones, these are vibrations at frequencies that are multiples of the fundamental. Harmonics that are not integers of the fundamental are called partials, which also contribute to the complexity of the sound.

Headroom   The ability of an amp to go beyond its rated power for short durations, permitting it to reproduce peaks without moving into **distortion**.

Hearing Sensitivity   A term used to describe the inaccuracies of the human ear. It is less sensitive at low and high frequencies than in the mid range.

Hertz (*abbreviation Hz*)   A unit of measurement denoting the frequency equal to one cycle per second. See also **kHz**.

Hertz, Heinrich Rudolf   German physicist who was the first to produce radio waves artificially.

Hexadecimal   A numbering system that uses base 16. It is represented by the digits 0–9 and the letters A–F.

High-pass Filter   A circuit that allows high frequencies to pass but rolls off the low frequencies.

Hiss   Random high-frequency noise, most often associated with tape recordings or a badly tuned radio station.

Hot Swappable   A term used to describe the ability to plug and unplug devices into a computer without having to switch it off first.

Howl-round   An uncomfortable (and often speaker destroying) ear-piercing sound resulting from an infinite audio **feedback** loop.

HRTF (*Head Related Transfer Function*)   A term describing how (provided we are facing the centre of the stereo image) some of the higher frequencies emanating from the left speaker will be reduced before they reach the right ear, because sound has to travel around our head.

HTML (*HyperText Markup Language*)   A software language that is used on the Internet to create web sites.

HTTP (*HyperText Transfer Protocol*)   The name for the protocol that is used to transmit documents across the Internet.

Hum   The **harmonics** produced by the AC mains supply.

I

IEC (*International Electrotechnical Commission*)   An organization involved in international standardization within electronics.

IEEE (*Institute of Electrical and Electronic Engineers*)   The largest professional organization for electrical engineers who deal with standardization.

IEEE 1394 (*Firewire*)   A high-speed serial data buss that can move 400 megabits per second and is **hot swappable**.

Impedance   A measurement of electrical resistance that is specified in ohms.

Infrasonic   Sound waves or vibrations with frequencies that are so low they are inaudible.

Inline Mixer   A term used to refer to a mixer that features a narrow vertical strip for each channel.

Inverse Square Law   The calculation that states sound energy is inversely proportional to the square of the distance it has to travel. Since sound waves spread spherically out in all directions from the source, doubling the distance quarters the sound energy.

IP Address   Another term for an Internet address.

J

Jitter   The result of the phase-shifting of digital pulses over a transmission medium; often sounds similar to the wow and flutter experienced with analogue tape cassettes.

K

KaZaa   Peer-to-peer file sharing software.

kHz (*kilohertz*)   One thousand (1000) cycles per second.

Knee   The point on a curve where change begins to occur.

L

Latency   The time taken for a device to respond or the time taken for any signal to pass through a device.

Law of the First Wavefront   A law which states that if two coherent sound waves are separated in time by intervals of less than 30 milliseconds, the first signal to reach our ears will provide the directional information.

LCD (*Liquid Crystal Display*)   A numerical or graphical display that is made of a material whose reflectance changes when subjected to an electric field.

LED (*Light Emitting Diode*)   A self-lighting semiconductor.

LFO (*Low-Frequency Oscillator*)   A **VCO** that produces a frequency so low it is inaudible. Often used to modulate other parameters in a synthesizer.

Limiter   A compressor with a fixed ratio that prevents the audio signal from becoming any larger than the threshold setting.

Line Level   The level of a signal before any power amplification takes place.

Logarithm   A method that uses the powers of 10 (or some other base) to represent the actual number.

Loudness   The perceived volume.

Low-pass Filter   A circuit that allows low frequencies to pass but rolls off the high frequencies.

LSB (*Least Significant Bit*)   The bit within a digital **word** that represents the smallest possible value.

Lyric   A French term literally meaning the words of a song.

M

Mask (*Masking*)   A phenomenon where if two sounds are played simultaneously, only one may be heard.

Mastering   The final step in the recording process, before the record goes off to press.

Matrix   Can be used to describe the routing possibilities of audio equipment. (Also an immensely popular film with a cult following starring Keanu Reeves and Lawrence Fishburne.)

MD (*MiniDisc*)   A digital audio recordable optical storage system developed by Sony that uses data compression to reduce the size of the disc.

MDM (*Modular Digital Multi-track*)   A generic term used to describe any family of digital audio multi-track recorders.

MDMA (*3,4-methylenedioxy-N-methylamphetamine*)   Also referred to as E, disco biscuits, cookies and tabs. A chemical substance developed by Dr Alexander Shulgin and immensely popular with the clubbing generation, it stimulates/increases the production of serotonin in the brain, producing what many describe as a '**euphoric**' feeling.

Mid Range   A speaker used to reproduce the middle range of frequencies.

MIDI (*Musical Instrument Digital Interface*)   Industry standard protocol for interconnection and control of MIDI-equipped musical instruments.

Mixer   An audio device used to combine and adjust the level of multiple inputs into two (or more) outputs.

Modem (*modulator/demodulator*)   A device that permits a computer to connect and communicate to other computers over a standard telephone line.

Modulate   See **Modulation**.

Modulation   The process of one parameter affecting another.

MP3 (*MPEG-1, layer 3*)   An immensely popular digital audio compression technique for encoding music files.

MPEG (*Moving Picture Experts Group*)   A professional body that sets specifications for compression schemes for audio and video transmission.

MSB (*Most Significant Bit*)   The bit within a digital **word** that represents the biggest single bit.
MS-DOS (*Microsoft Disk Operating System*)   Microsoft's registered trademark for their PC operating system.
Muddy   A term used to describe a sound that is poorly defined.
Multi-timbral   A term used to describe an instrument that can play more than one instrument at once. A 16-part multi-timbral instrument can play up to 16 different instruments at once.
Mute   A control found on most mixers that silences a signal path or output.
Muzak   Background music that does not demand (nor deserve) attention, often used in supermarkets and lifts (elevators).

N

NAMM (*National Association of Music Merchants*)   A trade organization for people working in the music business.
Near-field Monitor   A loudspeaker designed to be used at a distance of no more than 4 or 5 feet.
NFO   A text file created by **cracking groups**, describing the software.
Nibble   A group of 4 bits or half a byte (8 bits).
Noise Floor   A measurement of the lowest threshold before there is more noise than signal.
Noise Gate   A device that is used to remove sounds below a specified threshold, commonly used to prevent extraneous sounds from being recorded.
Noise Shaping   A common technique used in **oversampling** to shift the frequency range of **quantizing error**.
Notch Filter   A filter that **attenuates** a narrow band of frequencies.
Nyquist Theorem   A theorem stating that to recreate any waveform accurately in digital form, at least two different points of a waveform's cycle must be sampled.

O

Octave   A doubling or halving of frequency.
Operators (FM)   The term to describe the sound generating oscillators used in FM synthesis. Unlike analogue oscillators, operators are only capable of producing sine waves.
Oscillate   A term used to describe any object that vibrates continually.
Overload   A common cause of distortion, resulting from inputting a signal that is too loud.
Oversampling   A term often used to describe sampling at a rate higher than the **Nyquist Theorem**, but it may also apply to a technique where samples are sampled more than once by an ADC.
Overtones   See **Harmonics**.

P

Pad   See **Attenuator Pad**.
Pan (*abbreviation of panoramic*)   A control implemented on all mixers that permits you to move the position of a single sound to the left or right of the **soundstage**.
Paragraphic EQ   A graphical representation of a parametric EQ.
Parametric Equalizer   A multi-band equalizer offering control over the amplitude, centre frequency and bandwidth.
Partial   See **Harmonics**.
Patch Bay   A panel consisting of two rows of connections that allow you to insert a piece of equipment into a current signal path. The top row consists of sends, allowing you to send the signal path out to an external device, while the lower row consists of inserts, allowing you to insert the signal back into the signal path.

PCI (*Peripheral Component Interconnect*) A format designed by Intel, permitting high-bandwidth communication between the PCI port and the CPU/motherboard.

PCM (*Pulse Code Modulation*) A method used to encode an analogue signal into digital.

PCMCIA (*Personal Computer Memory Card International Association*) A term used to describe small PC soundcards, modems or suchlike for connection to laptop computers.

Peak Program Meter See **PPM**.

Perceptual Coding A digital audio data compression technique based on removing sounds that are deemed to be inaudible.

PFL (*Pre-Fade Listen*) Often found on mixers, the term refers to a signal being monitored before it reaches the main channel fader.

Phantom Power A term used to describe the method of providing power supply to some microphones using the same two lines as the signal path.

Phase Cancellation The product of two signals with some matching frequencies being added together and the matching frequencies being removed from the signal.

Phase Inversion A term used to describe the introduction of a phase difference of 180 degrees between two waveforms. It is commonly produced by the inversion of a symmetrical periodic signal which results in silence.

Phaser The process of taking two versions of the same signal and delaying one in time to the other to produce a hollow, phase-shifted sound. (Also used in the Star Trek universe to disperse aggressive aliens.)

Phat Slang term commonly used by dance musicians to describe something as 'in your face', loud or upfront, but can also mean excellent or first-rate.

Pink Noise Pink noise is random noise that contains differing amounts of energy at different frequencies and produces a heavy, deep hiss.

Pitch-shifting An effect that changes the pitch of musical notes without affecting length or timing.

Pot A diminutive of **potentiometer** (also slang for Cannabis).

Potentiometer Most commonly used as voltage dividers in electrical circuitry, allowing for both negative and positive values.

PPM (*Peak Program Meter*) An audio meter used to measure and display an audio signal's peak.

Pre-amplifier The name for any device that takes a source signal and passes it on to a power amplifier.

Precedence Effect See **Haas Effect**.

Proximity Effect (*microphones*) A term used to describe the increase in low-frequency response that is proportional to how close the vocalist moves towards the microphone.

Psychoacoustics The study of the perception of sound.

Punch-in/Punch-out Term used to describe the sudden engage or disengage of record mode on a previously recorded track, usually for rectifying incorrect pieces of a recording.

PWM (*Pulse Width Modulation*) A conversion method in which the widths of pulses represent the analogue signal.

Q

Q (*or Quality Factor*) Often used to describe the bandwidth.

Quadraphonic Sound A term coined in the 1970s to refer to surround sound.

Quantization The method of digitizing the varying amplitude of an analogue waveform to one of a finite number of distinct levels.

Quantization Distortion See **Quantization Error**.

Quantization Error A term used to describe the errors resulting from quantizing an analogue waveform to a distinct level.

R

Rack Unit   See **U**.

RAM (*Random Access Memory*)   A finite storage medium that can be read from and written to.

Rarefaction   Used to describe a decrease in density and pressure caused by the passage of a sound wave.

RCA Connector   Also known as Phono plugs, they are used as low-level connections between consumer equipment.

R-DAT (*Rotary Head Digital Audio Tape Recorder*)   The true name for DAT recorders – a digital audio recorder that uses a magnetic tape cassette with revolving heads similar to a video recorder.

Resonance (*filters*)   Refers to the amount of the output of the filter that is fed back directly into the input, emphasizing any frequencies that are situated around the cut-off frequency.

Reverb   An electronic device or program that emulates **reverberation**.

Reverberation   The sound (*short echoes*) remaining after the original source has stopped.

Reverberation Time (*RT60*)   The time it takes for reverb to decay away after the original sound has stopped. It's measured by how long it takes the **sound pressure level** to decay to one-millionth of its original value (*one-millionth equals a 60 dB reduction, so reverberation time is often abbreviated as RT60*).

RFI (*Radio Frequency Interference*)   A measurement of radio frequency interference from equipment.

RIAA (*Recording Industry Association of America*)   A trade organization representing the US recording industry.

Ring Modulation   A synthesis technique that generates a signal from the sum and difference compound of two signals (while also removing the original tones).

RMS (*Root Mean Square*)   The square root of the mean of the sum of the squares.

RT60   See **Reverberation Time**.

S

Sample and hold (*S/H*)   A circuit that holds a signal for a short period of time; usually appears as a wave shape on an **LFO**.

Sample Rate Conversion   The method used to convert one sample rate to another.

Sampling   The method of representing the amplitude of a signal at any one particular point in time.

Sampling Frequency   The frequency at which an analogue signal is converted into digital data, expressed in hertz.

Sawtooth Wave   A periodic wave that produces all the even and odd harmonics in a series, creating a bright sound. It is characterized by a 50% duty cycle and a Fourier series consisting of both even- and odd-ordered, equal phase, sinusoidal harmonic components of its fundamental frequency.

SCMS (*Serial Copy Management System*)   A copy protection system built into most consumer digital recorders. It only permits you to make one digital-to-digital copy.

SDIF (*Sony Digital Interface Format*)   A professional digital audio developed by Sony. It is an interface with a **BNC** connector (one for each audio channel) and a separate BNC-type connector for word synchronization.

Semitone   A half tone in the standard diatonic scale.

Sibilance   The result of overemphasized 'sh' sounds, which can create unwanted hissing, most usually associated with vocals.

Side-chain   Often employed in signal processors, allows an inputted signal to control the action of the processor.

Signal-to-Noise Ratio (*SNR*)   The distance between the **noise floor** and the musical signal.

Sine Wave   The simplest wave shape, based on the mathematical sine function. A sine wave consists of the fundamental frequency and does not contain harmonics.

SMPTE (*Society of Motion Picture and Television Engineers*)   An engineering society that establishes standards, including time code.

Snoman   The author of this book.

Snowman   A figure constructed from snow, often encapsulated with a carrot nose and coal for eyes.

Software Piracy   The illegal act of copying and distributing software that is **copyright**. See also **Cracking Groups**, **Cracks** and **Warez**.

Sound Pressure Level (*SPL*)   A measurement of loudness or volume.

Soundcard   A card housing microchips that allows the computer to convert digital signals into analogue sound.

Soundstage   A term used to describe a virtual stage where instruments can be placed during the mixing process.

Spatializer   A name for the technology that modifies an original signal so that the listener perceives the stereo image to go beyond the boundaries of the two speakers.

S/PDIF (*Sony/Philips Digital Interface Format*)   A digital audio connection that uses coaxial cable and RCA connectors, but it may also appear in optical form.

Spectral Balance   Term used to refer to the balance across the entire frequency spectrum of audio.

Spectrum Analyser (*audio test equipment*)   A measurement device or program that displays the frequency components and amplitude of an audio signal.

Square Wave   The simplest waveform for an electrical circuit to generate because it only exists in two states: high and low. This wave produces only odd harmonics.

Stereo (*Stereophonic Sound*)   Term derived from the Greek words 'stereo' (*three-dimensional*) and 'phonics' (*the science of sound*). It provides the illusion of a three-dimensional image between the speakers.

Subgroup   See **Groups**.

Submix   See **Groups**.

Subwoofer   A speaker designed exclusively for low-frequency reproduction.

Sweet Spot   Refers to a spot in between a two-loudspeaker stereo system where the listener receives equal information from both speakers, i.e. he/she is positioned equidistant from each speaker.

T

Talk Over   A term used to describe the use of a ducker on a noise gate to drop the volume of the music while the DJ speaks over the top.

Talkback   A microphone found on more expensive mixing desks. It permits the engineer/producer to speak to any performers in the recording room from his/her soundproofed control room.

Taste Test   A rather dubious technique used by electricians to determine whether a wire has a voltage flowing through it. By placing both ends of the suspected 'live' wire on the tongue, if there is a current there will be some electrolytic decomposition, which produces a taste like no other. This should only be applied on wires that carry a particularly small voltage (less than 1 V); any higher and you may suffer extreme discomfort or even death.

TDIF (*Teac Digital Interface Format*)   An eight-channel digital audio interface developed by Tascam for use with their DA-88.

Theremin   The only musical instrument that can be played without being touched and produces a tone similar to a sine wave. Most theremins have two antennas: the first controls the volume of sound while the second controls the frequency of the sound. Used heavily in old sci-fi movies but also in trip-hop.

Timbre   A term derived from the French meaning colour, it is used to describe the sonic quality of any particular sound.

Time-stretching   A digital signal process that adjusts the length of a sample – changing its BPM – while leaving the pitch unchanged. It is accomplished by adding or removing samples at various intervals during the course of the sample so that it reaches the desired length.

Tone   In music the term refers to a distinct pitch and duration, but can also be used to describe the sound's characteristics. See also **Timbre**.

TOSLINK (*Toshiba Link*)   A fibre-optic interface developed by Toshiba and based around **S/PDIF**.

Total Harmonic Distortion (*THD*)   Often used to describe the amount of harmonics that are added to a signal that was not present in the original.

Transducer   Electrical circuitry that converts one form of energy to another.

Transient Response   The capability of any processor to quickly and accurately respond to **transients**.

Transients   Term used to describe an instantaneous change in dynamics, usually associated with a fast attack speed on the amp EG.

Transparency   A listening term used to describe how 'clear' or 'clean' the audio is.

Triangle Wave   A periodic wave that features two linear slopes that produce a series of odd harmonics.

TRS (*Tip Ring Sleeve*)   A quarter-inch jack connection that can transmit a stereo or balanced signal.

Tuning Frequency   Commonly refers to the resonant frequency of systems.

Turntables   A record deck for playing phonographic records.

Turntablism   A form of music established by **turntablists**.

Turntablist   An artist who uses two or more specially designed **turntable** from which they create original music by mixing various sounds from the records on each turntable.

Tweeter   A speaker designed solely for reproducing high-range frequencies.

U

U (*abbreviation for modular unit*)   A synthesizer, compressor, limiter, pre-amp or other electrical device; may be stacked one above the other in modular rack units. A standard developed by the EIA/ANSI ensures that any racked instrumentation is exactly 19 inches wide and at specific height. For instance 1U is $1\frac{3}{4}$ inches high, 2U is $3\frac{1}{2}$ inches high and 3U is $5\frac{1}{4}$ inches high.

Underground (*music*)   Any ultra-modern or rebellious movement in music.

Unity Gain   A circuit with unity gain will not increase or decrease the volume level.

URL (*Uniform Resource Locator*)   A web site address.

USB (*Universal Serial Buss*)   Sometimes coined 'Useless Serial Buss' due to its slow transfer speeds, it is the standard serial connection on all PCs transmitting data at 12 Mbit/s. It is hot swappable, permitting you to remove or attach any USB devices (**soundcards**, **dongles**, **modems**, printers, etc.) without having to restart the computer.

V

Vacuum Tube   Invented and patented by Ambrose Fleming in 1904, it's a tube that has all the air removed, thus allowing the electrons to move freely. Often employed in **compressors, limiters** and **pre-amplifiers** (amongst others) to produce valve **warmth**.

Valve   See **Vacuum Tube**.

VCA (*Voltage-Controlled Amplifier*)   An electronic circuit controlled by voltages that permits you to shape an audio signal.

VCO (*Voltage-Controlled Oscillator*)   An electrical circuit that creates a basic sound wave.

Vinyl   Term commonly used by DJs to refer to any phonograph record.

Vocoder (*Voice Encoder*)   Invented in 1936 by Homer Dudley and originally developed for the army to permit them to transmit speech over a narrow bandwidth, it permits you to superimpose one **timbre** (usually a voice) onto another (usually a synthesizer) to produce a talking synth effect.

VU Meter (*Volume Unit*)   A unit that is used to measure the perceived loudness of an audio signal.

W

Warez   A slang term used to describe software that has had its copy protection removed and becomes available on the Internet. See also **Crack** and **Cracking Groups**.

Warmth   A term very broadly used to describe the second-order harmonic distortion introduced onto a signal from using valve equipment. It is also sometimes used to describe a system that sounds natural between 100 and 400 Hz.

Wattage   Refers to the unit of power used to rate the output of amplifiers; for it to have any true meaning it should also be accompanied with the distortion level and **impedance**.

WAV   Default extension for a Wave file, the Microsoft audio file format for PCs.

Wavelength   The distance between one apex or base of a sine wave and the next corresponding apex or base.

Web Ring   A group of web sites that all share a common theme and are linked together by **URLs**, permitting the visitor to visit each site in turn.

White Noise   Noise that contains equal amounts of all frequencies and sounds very similar to hiss and radio static.

Windows   Operating system developed by Microsoft for the PC. The most current incarnation is Windows XP.

Woofer   A speaker designed solely for reproducing low-frequency signals.

Word Clock   Sometimes written as WRCLK or WCLK, it's a signal that carries both sample and bit rate, and is used to synchronize two digital devices together.

Word Length   The number of **bits** in a word.

WYSIWYG (*What You See Is What You Get*)   A term that is becoming more popular with audio software manufacturers. It is often used to describe the simplicity of the software.

X

XLR   A three-pin connector that is used for carrying an audio signal. It most commonly appears on microphones, balanced audio components and the AES/EBU digital connection.

Z

Zero Crossing   The point at which a waveform crosses from being positive to negative (or vice versa).

References

Books

Buick, Peter and Lennard Vic (1995) *Music Technology Rpeference Book*, PC Publishing.

Chowning, John and Bristow, David (1986) *Fm Theory and Applications: By Musicians for Musicians*, Yamaha Music Foundation.

Hamer, Lindsay (1998) *House Music: The History, Music and Culture*, PhD Dissertation, King Alfreds University, Winchester.

Katz, Bob (2003) *Mastering Audio: The Art and the Science*, Focal Press.

Roland Corporation (1979) *Practical Synthesis for Electronic Music*, Roland Corporation.

Pierce, John R. (1992) *The Science of Musical Sound*, Scientific American/Freeman.

Watkinson, John (1988) *The Art of Digital Audio*, Focal Press.

Other sources

Access, *Access Virus C Users Manual* – Synthesis.

Computer Music (monthly magazine), Future Publishing, 30 Monmouth Street, Bath BA1 2BW, UK – Computer-related audio.

Future Music (monthly magazine), Future Publishing, 30 Monmouth Street, Bath BA1 2BW, UK – Synthesizers, microphones.

Izotope Software, *Ozone Users Manual* – Mastering.

Native Instruments, *Battery Users Manual* – Sampling.

Native Instruments, *Reaktor Users Manual* – Synthesis.

Novation Corporation, *Supernova II Users Manual* – Synthesis.

www.Bemuso.com – Music publishers.

Yamaha Corporation, *Yamaha 02R Users Manual* – Digital mixing.

Samples

Magix (*DVD Sample Collection*).

Spectrasonics (*Stylus, Atmosphere, Trilogy, Voices of Africa*).

Uberschall (*House Essentials, Disco Deluxe, Houseworx!*).

CD content copyright information

Big Tick Software

Rainbow: © Big Tick Software
Cheeze Machine: © Big Tick Software
EP Station: © Big Tick Software
Rhino Synth: © Big Tick Software

iZotope Software

Ozone: © iZotope, Inc.
Vinyl: © iZotope, Inc.
Trash: © iZotope, Inc.
iZotope, Ozone and Trash are the trademarks of iZotope, Inc.

Kord Kruncher Software

Kord Kruncher: © Andrew Le Couteur Bisson

Native Instruments Software

FM7: © Native Instruments
Pro-53: © Native Instruments
Kontakt: © Native Instruments
Battery: © Native Instruments

Speedsoft Software

VSampler: © Speedsoft All rights reserved

Synapse Audio Software

Junglist: © Synapse Audio
Scorpion: © Synapse Audio

TC Native Software

TC Native Bundle v3: © TC Native

KC Waves Software

Waves Native Bundle: © KS Waves LTD

Square Circle Software

Wavesurgeon: © Square Circle Software. All rights reserved.

PSP Audioware Software

PSP Mixpack: © PSP Audioware
PSP StereoPack: © PSP Audioware
PSP VintageWarmer: © PSP Audioware
PSP 84: © PSP Audioware

Index

+4 dBu, 117
−10 dbV, 117
1210, 459
12-inch, 373–4, 376, 382
130 gram lacquer, 373
180 gram lacquer, 373
56K, 399–400, 402, 429, 448
−60dBu, 182

A cappellas, 385
A Guy Called Gerald, 132, 197, 361
A&R, 363–4
Absynth (Native Instruments), 207–8
Access virus, 135, 162
Acid, 219
Acid house, 156, 165, 271
Acid jazz, 198, 295
Acoustic coupling, 188
AD (attack decay), 145, 155, 158
Adagio, 40
ADAT, 119, 384, 394
ADC (analogue-to-digital converter), 85–7
aDJust, 460, 462
ADSR (attack decay sustain release), 69–71
Advances, 368–9
AES EBU, 119
Aiff, 447
AKAI, 78–80, 200, 239, 242, 272
AKAI MPC, 200, 239, 242, 272
AKG C3000, 184
Algorithms (FM), 73
Aliasing (sample), 85
Allegretto, 40
Allegro, 40
ALT tags (web design), 405
Ambient, 253–67
AMD, 23
Amp EG, 136–7, 147, 149–50, 155, 157–8, 167, 169
Amplification, 57, 121, 123
Analogue FM, 73–4
Andante, 40
Anechoic chamber, 110
Aphex, 355
Apollo 440, 195, 200, 209
Armstrong, Edwin Howard, 101
Arnold, David, 199, 207, 209
Artwork, 368, 370, 374
ATB, 197–8, 200
Atkins, Juan, 115, 285, 293

Attenuate, 64–6
Audio (web design), 447–50
Audio sequencing, 22, 205
AURA, 365–7, 369
Auto-Tune (Antares), 190, 391–2
Automation:
    read, 131
    write, 131
Auxiliary buss, 122, 125–6
Auxiliary returns, 125–6

B Side, 371
B4, 202
Background colour, 424–9, 431–2, 436, 438
Balanced (cables), 115–22
Bambaata, Afrika, 237–8, 457
Band-pass (filters), 65–6, 68, 75
Bandwidth, 319–23, 337
Bars (musical), 41–51
Basic chord groups 34–5
Bass (programming):
    deep heavy, 155
    digital, 158
    kick, 158
    moog, 156
    pop, 159
    sub, 155
    sweeping, 157
Beat juggling, 237
Beat match, 457, 463
Beating (oscillators), 54, 62
Beatless, 254
Beck, 204, 209
Bell, Alexander Graham, 57
Bell Laboratories, 310
Belleville Three, 285
Better Days, 269
Big beat, 18, 100, 136, 147, 152, 156, 165, 167, 175–6
Binary Finery, 236
Binary phrase (musical), 45
Bit rates, 84, 86–7, 448
Björk, 199, 204
Blumlein, Alan, 315
Bootlegs, 384, 386
Boundary effect, 324
BPM (beats per minute), 40
Brand New Heavies, 295
Breakbeat, 270
Brick wall, 357

British hip-hop, 294–5
Broadband, 399–400, 448
Browsers, 398–405, 414–15, 419, 424, 426,
    429, 432, 443, 450
BT, 48, 197, 200, 205, 387
Buss mastering, 20–2
Buss, 123

C major scale, 28–30, 34, 41
Cables, 115–31
Café Del Mar, 254
Cakewalk Project 5, 27
Capacitance (microphones), 181–2
Carrier (vocoders) 192–4
Cartridges (vinyl), 459, 467
Cauty, Jimmy, 253
CC messages, 148, 154, 173, 470
CD, 362–3, 368, 370–1
CDJ, 458–61
Cents, 62
Chattering (noise gate), 107–8
Chemical Brothers, 63, 135, 195, 209, 372
Chicago, 269–71, 285
Chicago house, 271
Chill-out, 38, 42, 136, 141, 156, 167, 253–67, 346
Chords, 33, 35–9
    consonant
    dissonant, 33
    intervals, 33
    inversions, 35, 300
    sequences, 36–7
    triad, 34–6
Chorus, 112
Chowning, John, 73
Clarity, 332, 334–5, 337–8
Clipping, 92, 94–5, 105, 118, 126, 130
Club mix, 387, 394
Club PA systems, 311, 318
Clubs, 310–11, 336
Coil (microphones), 180–1
Coloured vinyl, 456
Comb, 65, 67
Commercial vocals, 190
Complex waveforms, 54, 62
Compression, 92–114
    attack, 95–102
    clipping 92, 94–5, 105
    creative, 99
    hard knee, 95–6, 98
    make-up gain, 96, 101, 103
    mixing, 126–8, 130
    over-compression, 98
    practical, 97
    pumping, 100
    ratio, 95–105,
    release, 95–102
    RMS, 96, 103, 105
    setting up, 98
    side-chaining, 97, 103
    soft knee, 95–6, 98, 102

threshold, 93–9
    vocals 186, 191
Concave, see Envelopes
Condensation (microphone), 185
Consonant, 33, 39, 300, 332
Contracts, 362, 364–5, 369, 373, 393–4, 396–7
Contrary motion, 38–9
Control curve, 310–11
Control voltage, 59, 73, 92, 192–3
Convex see Envelopes
Copyright, 361–2, 366–7, 372
Copyright statement, 361
Corsten, Ferry, 236
Cover preparation, 374
CPU, 22, 126
Craig, Carl, 286, 293
Creeping faders, 317
Cross-fader, 457
Cross-fading, 77, 81, 190, 391
Crossover, 351
Crotchet, 41, 42
CSS (Cascading Style Sheets), 402, 404–5, 430, 439–44
Cubase SX, 14, 83, 388
Culture, 237–8, 245
Cut-off frequency, 64, 68
Cutting, 237
CV, see Control voltage
Cybotron, 285
Cycle, 53, 62–3, 72

Da capo, 45, 225, 250–1
Daft Punk, 113, 138, 144, 160, 164, 195, 204, 209, 271,
    281, 283
Dance2Trance, 219
Darey, Matt, 135, 198
Dark house, 271
DASH, 383
DAT, 120, 183, 302, 358, 363, 371, 384, 394, 460
Data cells, 402, 404, 409–10, 415–16, 419, 426, 429–30,
    445–6
dB, 57, 64–5
dBu, 117
dbV, 117
DDL (digital delay line), 113–14
Decibels, 57
Decks, 457–62, 466
Deep Dish, 382, 387
Deep house, 271, 279
De-esser, 185, 332, 336
Degree (musical), 28–39
Delay, 178
Delerium, 198
Demisemiquaver, 41
Detroit, 285–6
Detuning, 59, 62, 143, 162, 164, 167, 279, 305
Diaphragm, 180–6
Digital clutter, 389
Digital editors, 88
Digital peak meter, 358–9
Digital Signal Processing (DSP), 58

Direct X, 25, 27
Directional Cues, 316
Disco, 268–72, 277, 281–3, 382, 396
Disco house, 271
Disney, Walt, 315
Dissonant, 33, 36, 39
Distortion (harmonic), 345, 355
Distortion, 95, 97, 101, 114
Distributor, 366, 372–3, 375–7
DJ, 285–6, 370–6, 380, 454–67
DJ Dag, 219
DJ International, 270
DJ promotion agency, 375
DJ Sammy, 198, 236
DJ'ing, 237
DMA (Direct Memory Access), 22–3
Domain names, 377–8
Dominant, 29, 32
Dorian mode, 32
Double exposure, 382
Double track, 191
Double tracking, 190
Doubling, 161
Dr Dre, 239, 252
Drive, 44, 272–3
Drums (programming):
  claps, 148–50
  congas, 150, 329
  cowbells, 148, 150, 154
  cymbals, 148–9
  hi hats, 148–53
  kick drums, 145, 175
  tambourines, 150, 152
  toms, 151–2
Drum 'n' bass, 18, 38, 40, 68, 76, 136, 147–8, 156–7,
  165, 387, 457
Drummond, Bill, 253
Dub, 237, 384–5, 392
Ducking, 108
Dust Brothers, 195, 209, 372
D-Vinyl, 462
DX 7 synthesizer, 73, 166
Dynamic equalization, 355
Dynamic microphone, 180, 186–7
Dynamic range, 86, 89, 92–5, 98, 101–5, 128, 130, 349
Dynamic restriction, 349
Dynamics (music) 43–4
Dynamics, 341, 345, 348–56

Ear protectors, 465, 467
Eardrum, 52
Earth system, 118
Editing vocals, 189
Effects, 115–31, 309–10, 315–16, 318, 324–6, 331, 333–5
Effects (order of), 129–30
Electrix EQ Killer, 385
Electromagnetic field, 115, 118
Electrons, 88, 101, 182
Emagic MT8, 20
Eminem, 139, 239, 245, 252

E-MU Drumulator, 144–7, 273–4, 296
E-MU Orbit, 135
E-MU SP12000, 272, 296
E-MU, 78–80, 135, 144–7, 198, 272–4, 296
Engineer, 340–5, 348, 354–5, 358–9
Enharmonic equivalents, 29, 35
Eno, Brian, 253–4
Envelopes, 69, 71, 135, 145, 147
  AD (attack decay), 145, 155, 158
  ADSR (attack decay sustain release), 69–71
  concave, 71, 136, 146–50, 155, 158, 160
  convex, 71, 136–50, 155–60
  linear, 71, 136, 146, 149, 155, 160
  TADSR (time attack decay sustain release), 69, 71
EQ, 120, 122–30, 309, 311, 314–17, 319–33, 336–7,
  342–3, 345–8, 358
EQ, creative, 314, 322
EQ, paragraphic, 323
Euphoric, 66, 160, 162, 195, 219–20, 346
Euro House, 271
Exabyte, 371
Excursion, 104–5, 357

Faders, 122–3, 125, 130
Faithless, 140, 198
Familiarity, 56
Fantasia, 315
Farley, Jackmaster Funk, 269
Fatback Band, 238
FatBoy Slim, 139, 209
Feedback, 188
Feel, 309–10, 325–6
FET, 101–2
File Transfer Protocol, 451
Film charge, 374–5
Filter, 135, 137, 140, 143, 151, 154, 157, 160–3, 167–70,
  174–6, 315, 318–19, 321–2, 326–33, 337
  band-pass, 65–6, 68, 75
  cut-off, 113
  four-pole, 65, 163, 192
  high-pass, 65–6
  low-pass, 64–8
  notch, 65, 67, 113
  parallel, 65
  resonance, 57, 67–8
  series, 65
  transition, 137
  two-pole, 65
Final scratch, 462, 476
Firewire, 23
Flangers, 112–13
Flats (musical), 29, 31–2
Fleming, John Ambrose, 101
Fletcher Munson, 310–11, 315, 336, 358
FM, 166
FM carrier, 73
FM modulator, 73
Foldback, 183
Folders, 406–7, 412, 415, 423–4, 440–1, 446, 451
Footer, 402, 406, 416, 430, 436–40, 444–8

Forced vibration, 56
Fourier theorem, 55
Fourier, Jean, 55
French house, 271
Frequency, 310–14, 317–37
Frequency (acoustic), 53–74
Frequency bands, 192–3
Frequency masking, 328, 333, 335
Frequency modulation, 73–4, 148–51, 166, 169
Frequency ratios, 55
FSOL, 84, 175
FTP, 451–2
Fundamental, 54–7, 59, 63–7, 310, 319, 322, 327–8, 333, 338–9
Future Sound of London, 84, 209, 267

Gain, 57, 314, 317, 319–21, 325, 328–9, 336
Gate *See* Noise gate
Gemini, 458, 461
Gibbons, Beth, 294
Gibbons, Walter, 382
Gif, 403
Glass masters, 371–2
Goa, 219
Graffiti, 237–8
Grandmaster Flash, 237–8
Granular synthesis, 74–5, 114
    density, 75
    grains, 74–5
    length, 75
    pan, 75
    shape, 75
    spacing, 75
Graphic EQ, 323, 330
Gray, Andy, 135, 198, 209
Gritty sounds, 177
GRM tools, 214–15, 392
Groove, 310, 314, 330–2, 336
Groove (musical), 40, 43–50
Groove quantize, 43
Grooves from other records, 138
Ground (electrical), 115–16, 118,
Ground lift switch (mixing), 122
Guidelines for breaking down a timbre, 173
Guitars, 326, 331, 333–4, 338–9

Haas effect, 316
Hammond organ, 166, 202
Hancock, Herbie, 139, 202
Hard house, 271
Hardy, Ron, 269, 382
Harmonic (minor scale), 32
Harmonic exciters, 354
Harmonic mixing, 392
Harmonics, 54–68, 73, 75, 319, 322, 328, 331, 338–9
Harmonization, 39
Harmonizing, 190
Hawtin, Ritchie, 209, 286, 293, 463
Header, 402, 416, 418–21, 427, 429, 437–40, 444–5
Headphones, 183–4, 188–9, 318, 466

Headroom, 359
Heard, Larry, 270
Herrik B, 286
Hertz, 53, 312
Hi-Z, 121
High range, 310, 330, 337, 344, 346, 351, 355
High-shelf, 321, 333
High-pass, 65–6
Hi-high range, 310
Hip-hop, 237–52
Hit sounds, 140
Hocketing, 161
Hook, 385–6, 391, 393
Horn, Trevor, 199
Hours (panning), 317, 324
House, 268–84
Howie B, 294
Head-related transfer function (HRTF), 316
HTML, 398–453
Hum, 115, 118, 122
Human element, 40
Hurley, Steve, 269
Hyperlinks, 438

IEEE 1394, 23
iLabel, 379–80
Illbient, 295
Impedance, 117, 121
Import, American, 374
Imports Etc, 270
Independent releases, 370
Infrasonic, 342
Injecting groove, 43
Insert buss, 122, 125–6
Interference (oscillators), 54
Interference, 115–18
Internet, 362, 377, 379
Internet explorer, 398–9, 402, 405–6, 420
Internet label, 379
Intervals, 33
Intonation, 183–4, 187–92
Intonation problems, 187–8
Inverse law, 313
Inversions, 35, 300
IP addresses, 377
ITD (interaural time delay), 316

Jah Wobble, 204
Jam El Mar, 219
JavaScript, 399
Jazz hop, 295
Jitter, 119–20, 342
Jocks, 454, 456, 463, 466
Johnson, Charles 'Electrifying Mojo', 285
Jones, Quincy, 139, 268
Jones, Rocky, 270
Jpeg, 403

Kazza, 385
Key, 29–39

Key range, 77–81, 132, 134
Key zone, 80–1
Key-follow, 68–9
King Tubby, 237
Knuckles, Frankie, 268–70, 285
Kool Herc, 237–8
Korg Triton, 199–200
Korg Z1, 135
Krome, Thomas, 286

Lacquer, 372–3, 375
Lamb, 295
Larghetto, 40
Largo, 40
Larkin, Kenny, 286, 293
Lavelle, James, 294
Law of first wavefront, 316
Layering, 134, 153, 159–61, 174, 316, 328, 331
Leads (programming):
    hoovers, 165–6
    organs, 166
    pianos, 166
    plucked, 163, 164
    TB303, 164
    theremins, 165
    trance, 160, 162
Lee De Forest, 101
Leftfield, 198, 204, 209
Levan, Larry, 269, 284
Levine, Steve, 199
LFO, 8–12, 72–5, 84, 112–13, 135–8, 141–4,
    148–75, 193–4
LFO fade-in, 72, 261
LFO waveforms, 168, 230
Licensing, 364, 366, 376–7
Limiters, 105
Linear, *see* Envelopes
Locrian mode, 32
Lo-fi, 66, 88, 136, 145, 147, 152, 165, 167, 175–8,
    294–5, 299, 305–6, 382
Logarithmic, 57
Look ahead, 103, 105
Loops, 78, 81–3, 89, 152–3, 176–7
Loudness, 57
Loudness maximizers, 356–7
Low-Z, 121
Low-shelf, 321, 343
Low-pass, 64–8, 321–2
LSD, 269
Lydian mode, 32
Lynx, 399, 405
Lyric sheets, 187–8
LZW, 403

Maas, Timo, 382
Mac AU, 25
Macromedia Dreamweaver, 400
Macromedia Flash, 399–401, 404, 429, 449
Macromedia, 400, 449
Magic Q, 321, 337

Magnetic field, 180–1
Major (Ionian) mode, 33
Make-up gain, 96, 101, 103
Massive Attack, 139, 195, 197, 294–5, 297, 306
Master rights, 396
Mastering, 340–60
Mastering EQ frequencies, 346
May, Derrick, 285, 293
MC'ing, 238
MCPS, 365–7, 369, 372, 376
MDMA, 219–20
Mechanical rights, 396–7
Mechanical royalties, 372
Mediant, 29, 32
Melodic (minor scale), 31
Melodies, 32, 37, 45
Meta tags, 377
Micro house, 271
Microns, 181
Microsoft, 430
MIDI, 3–27
    CC, 6–20
    channels, 19–21
    choke, 18–20, 22
    clock, 82
    controllers, 6
    daisy chain, 18, 120
    GM, 5, 8
    GM 2, 8
    GS, 8
    MTC (MIDI Time Code), 19, 200
    multi interfaces, 20, 23
    NRPN, RPN, 8, 11–12
    PPQN, 15, 19
    resolution, 15
    sequencers, 3, 12–26
    USEM, 8–9
    XG, 8
Mid range, 310–13, 318–19, 323, 328–9, 331, 333–4,
    338, 342–4, 346, 350–1, 356
MiniDisc, 176, 183
Minim, 42, 43
Minor (Aeolian) mode, 33
Minor scale, 31–2
Mixer, 457–62
Mixing, 309–35
Mixing desks, 115, 118, 120–7, 130–1
Mixolydian mode, 32
Modal, 32
Moderato, 40
Modes, 32
Modulation, 62, 71, 73–4, 84, 112–13, 135, 138, 140,
    143–9, 154, 158, 161, 163,
Modulation matrix, 138, 196
Modulation wheels, 73, 154
Modulator, 192–4
Moloko, 198, 204, 306
Monitors, 311, 323–4
Monitoring position, 318
Mono, 115–16, 121, 134, 160, 176, 315–18, 325–6, 336

Montana, Vincent, 268, 284
Morales, David, 135, 138, 198, 382
Moroder, Giorgio, 268, 284, 285
Motifs, 43–51
Motorized faders, 130
Moulton, Tom, 268, 284
MP3.com, 341, 379, 403
MP3-20, 447–9, 451, 460–4
Mr Fingers, 270
MTV, 367–8
Muddy, 312, 315–16, 328, 330, 332–5, 338
Multi band compression, 348, 350–1, 358
Multi sampling, 77–8, 84
Music arpeggiator, 47
Music theory, 28–51
Musical anaylsis, 38
Musical drive, 44
Musical phrase, 44–7
Musicians Union, 397
Mute, 122, 124, 131
Mutronics Mutator, 206–7
Muzak, 253

Native Instruments, 166–7, 201–2, 204–8, 224, 462
    Battery, 204–5, 224
    Kontakt, 205–6
    Reaktor, 207
Natural minor scale, 31–2
Negro, Joey, 195, 204, 209, 382
Nesting, 419–20, 422
Netscape, 398–9, 405, 419–20, 426, 450
New Romantic, 269
Nine Inch Nails, 195, 204, 209
Noise gate, 107
Noise reduction, 341–2, 359
Noise waveform, 61, 72
Non Constant Q, 320–1
Non-streaming audio, 447–8
Notch (filters), 65, 67, 113
Note lengths, 40–3
Novation
    Bass station, 195–6, 225, 304
    ReMote 25, 210
    SuperNova, 135, 162, 196–7, 225, 230, 304
    V station, 162, 196–7, 225, 230
Novelty packaging, 367–8
Nursery rhymes, 45
NWA, 239
Nyquist, 85

Oakenfold, Paul, 135, 198, 209
Oblique motion, 39
Octave, 28–30, 33, 47, 320–2, 337
Off axis, 186, 188
Oliver, Mickey, 269
One shot (samplers), 82
One-pole, 65
Opera, 398–9
Operating systems, 78
Operator, 73, 201

Optical, 101–3
Optimum gain, 88
Orb, The, 198, 200, 204, 253
Orbit, William, 197–200, 209, 254, 267
Orbital, 197–8, 200, 205
Organs, 313
Organs (programming), 166
Oscillators
    pink noise, 61
    pulse wave, 60–1
    sawtooth, 56, 59, 61–2
    sine, 54–60, 68, 72–3
    square, 56, 60
    triangle wave, 57, 59, 61, 72
    white noise, 61
Oscillator sync, 62
Overtones, 63, 65, 338

Packaging, 367–8, 372
Pad (mixing), 121, 124
Pads (programming):
    rise and fall, 143
    swirling, 144
    thin, 144
PAMRA, 365–7, 369
Panning, 315–17, 325–8, 335
Panning law, 317
Panoramic potentiometer, 316
Paragraphic EQ, 323
Parallel motion, 38
Parametric, 319, 321–3
Partials, 55, 61, 68
Patterson, Alex, 253, 309
PC-DJ, 462
PCM, 383
Peak to average ratio, 349, 357
Pentium processors, 23–4
Peoplesound.com, 379–80
Perceptual encoding, 176–7
Percussion guidelines, 151
Perfect fifth, 33, 55
Perfect fourth, 33, 55
Performance rights, 396
Phantom power, 122
Pharmaceuticals, 324
Phase alignment, 355
Phase cancellation, 112–13, 185
Phase, 53–5, 62, 67, 112–14
Phase reverse, 122
Phasers, 112–13
Phrase sampling, 82
Phrygian mode, 32
Pianos, 326, 338–9
Pink noise, 61, 147, 149, 151, 162, 223, 230,
    257–8, 274
Pioneer, 458–62, 467
Pitch arm, 457, 459–62
Pitch bend, 6, 12, 18, 63, 73, 158, 166, 389
Pitch modulation, 7, 145, 148–9, 158, 173, 175
Pitch, 28–47, 183–4, 188–94, 388–9, 391–2

Pitch shifting, 82, 158, 176, 192, 194, 389
Pixel, 402–4, 409–10, 415, 419–21, 425–33, 436–7, 443–5
Planet phatt, 135
Play lists, 380
Plosive popping, 185
PMCD, 371
Polyphony, 8
Pop shield, 185–6
Pops, 341–3
Portishead, 165, 202, 209, 294–5, 297, 306
Positioning, 315, 318, 324, 326
Post fader, 123, 126
PPL, 365–7, 369
Pre-amp, 182, 184, 186–7
Pre-amplifier, 120–2
Precedence effect, 316
Pre-fader, 122–3, 126
Pre-mix, 309
Presence, 340, 346
Press company, 375
Prestissimo, 40
Presto, 40
Principle, Jamie, 270
Pro 53 (VSTi), 201–2
Processing, 310, 334
Programming synthesizers 132–78
Progression (musical), 36–9, 47
Progressive (trance), 219
Progressive house, 271
Promo, 454–6,
Promoters, 464
Promotion, 361–81
Propellerhead, 26, 83, 145, 153, 156, 203
   Reason, 26, 145
   Rebirth RB 338, 203
Proximity effect, 186
PRS, 365–7, 369
PSP Audioware effects, 213
Psychedelic (trance) 219
Psychoacoustic enhancers, 355
Public Enemy, 239, 252
Publishing deal, 364, 369, 372
Pulse wave, 60–1
Pump, 348, 351–2
Pumping (compression), 100
Punch, 340, 343, 346, 348–50
Pythagoras, 54

Q, 319–22, 329–33, 337
Quantization noise, 87
Quantize, 16, 40–3
   Iterative, 16, 297
   Swing, 43
Quaver, 41, 43–4
Quicktime, (Apple), 448–9
Quinones, Lee, 238

Radio Head, 198, 204
RAM, 76–8

Rap, 68, 100, 108, 180, 183–4, 186–7, 271, 294, 301, 382, 457
Rapino Brothers, The, 204
Rarefaction, 52–3, 56
RC circuits, 64–5
Reaction sheet, 454
Read (automation), 131
Real media, 448
Reason software, 26, 145
Rebirth RB-338 software, 203
Record companies, 362–3, 365, 367–8, 379
Recording levels, 78, 84, 89, 93, 342
Recording rights, 396
Recoupables, 368
Recursive (modulation), 146
Recycle (Propellerhead), 79–80, 83, 153, 242
Recycling loops, 83, 153
Red Snapper, 202, 294, 306
Redhead Zync, 286
Relative volumes, 325–6, 336
Remixing, 382–97
Repetition, 386
Reserve royalty account, 369
Residual synthesis, 161–2
Resonance, 57, 67–8
Reverb, 110–11
Rhythm, 38, 40–51
Rhythmic repetition, 46
Ring modulation, 62–3, 148–51, 324
Rivieria, Erasmo, 270
RMS (root mean square), 96, 103, 105
Rodgers, Nile, 199, 268
Roland:
   JP8080, 135, 145
   Juno, 165
   JV1080, 135
   TB303, 145, 156, 164
   TR808, 52, 144–6, 156, 203, 244, 256, 268, 270, 275, 285, 296
   TR909, 52, 144–7, 156, 203, 222–3, 256, 268, 270, 273–5, 285, 296
   XV5080, 199
Roll off, 321–2, 329–31
Romplers, 77–8
Roni Size, 198
Root directory, 451
Root key, 34, 36, 77, 80
Ross, Diana, 139, 268
Rossario, Ralphi, 269
Rotary pitch control, 457
Row (web design), 402, 409–10, 416, 419–21, 427, 430, 433, 436–7
Royalties, 362–72, 376, 379–80
Royalties, 79–80, 91, 362, 364–72, 376, 379–80, 395–6
Royalty-based payments, 394
Rumble, 342
Rushton, Neil, 285

S/PDIF, 119
Safari, 398

Safety buffer, 118
Sales allowance, 368
Sample:
   CDs, 78–81, 84, 134, 145
   clearance, 90, 382–97
   compatibility, 79–80
   licences, 79
   rate, 86, 119–20, 177, 448
   streaming, 77
Sample and hold, 72, 171
Sample packs, 385
Samplers, 76–91, 134, 146, 175, 177, 382, 388
Sampling:
   creative, 84, 175
   key range, 77–81, 132, 134
   key zone, 80–1
   key-follow, 68–9
   multi-sampling, 77–8, 84
   one shot, 82
   phrases, 82
   sustain looping, 77, 84
   synthesizers, 84
   velocity cross-fading, 81
   velocity switching, 81
Sanchez, Roger, 198, 281, 283
Sasha, 46,135, 142, 198, 209, 236, 466
Saturday Night Fever, 268
Saunders, Jesse, 270
Saunderson, Kevin, 195, 285, 293, 382
Sawtooth wave, 56, 59, 61–2
Scale (musical)
   major, 28–41
   minor, 31–2
Scanner, 89,118, 199
Science (acoustic), 52
SCMS, 120
   strippers, 120
Scratching, 237, 457–61, 463
SCSI, 78, 119
SDIF, 119
Self-oscillation, 68, 146, 159, 169, 171, 244
Semi-parametric EQ, 322
Semibreve, 41
Semiquaver, 41
Semitones, 30–6
Serotonin, 219
Seven, 455
Shari Vari, 285
Sharps (musical), 29–32
Shelving, 312, 321–2, 329–31, 333, 337, 339
Sherman, Larry, 270
Sherman Filterbank 2, 209
Shure SM58, 186, 246
Siano, Nicky, 269
Sibilance, 184–5
Side chain, *see* Compression, Noise gate, Vocals
Signal to noise ratio, 115, 121, 327, 359
Simmons SDS 5, 144–6, 273
Sine wave, 54–60, 68, 72–3
Sister Bliss, 198

Slurring gates, 177–8
Snare drums, 102, 105, 147, 329
Sneaker Pimps, 295
SNR, 24, 115
Solid state, 97, 100–1, 103
Song map, 48
Sound design, 132, 137, 140–1, 160, 172, 175
Soundcards (audio quality), 24
Soundcards (drivers), 24
Soundcards (I/O), 23–5, 119
Soundcards, multi-client, 24
Sound effects (programming):
   computer burbles, 171
   Dalek voice, 170
   explosion, 169
   falling rising filter, 170
   ghostly noises, 170–1
   siren, 169
   swoosh, 171
Soundscape, 254
Soundstage, 317
Speakers, 312, 316–18, 323, 326
Spectral anaylser, 132–3
Spectral analysis, 347
Spectrasonics, 210–11, 290, 303
Spill (recording), 184, 188–9
Spin, 457, 460
SPL, 57
Square wave, 56, 60
Staff (musical), 40–1
Stanton, 458–9, 462, 467
Steinberg HALion, 79–80, 146, 205–6
Steinberg Midex, 20
Stereo, 115–16, 119, 121, 123, 134,
   160, 176, 315–18, 326, 329
Stereo widening, 356
Stereophonic, 457
Streaming audio, 447–9
Strips (mixing), 121
Studio, 363
Sub bass, 60, 68, 311, 342
Subdominant, 29, 32
Subgroup (mixing), 120, 123–5
Submediant, 29, 32
Subtonic, 29, 32
Subtractive synthesis, 58, 287
Sugar Hill Gang, 238, 457
Summer, Donna, 89, 138, 283, 285
Supertonic, 29, 32
Suppressors, 118–19
Sustain looping, 77, 84
Swing quantize, 43
Sympathetic vibrations, 54
Sync, 457, 463
Syncopation, 44
Synthesis, 52–75

Tables, 402–3, 408–11, 415–19, 429–30,
   443–6
Tag (graffiti), 238

Tag (Internet), 401, 405, 407–24, 427, 430, 433–40, 444–53
TAKI 183, 238
TC Native Bundle, 211
TDIF, 119
Tech house, 271
Technics, 458–60, 467
Techno, 285–93
Tempo, 40, 44–5
Ten Percent, 382
Ternary, 46
Terry, Todd, 139, 175, 271, 277, 282
Test pressings, 368, 372–3
THD, 191
Thrillseekers, 198
Thumbnails, 404, 447
Timbre, 55–75
Time coded, 462
Time signatures, 40–4
Time stretching, 82–3, 176, 190, 359, 388–91
Tinnitus, 465
Toasts, 238
Tone arm, 457, 459
Tones (music), 29–34
Tonic, 29, 32
TOSLINK, 119
TP, 455–6
TPS, 372, 375
Traktor, 462
Trance, 219–36
Transient, 136–7, 145–51, 154–5, 163, 166, 173–7
    designers, 109
    slicing, 177
    period, 64–5, 68, 319
Transparent gif, 403–5, 426–7, 432–3, 436–7
Trax, 270
Triangle wave, 57, 59, 61, 72
Tricky, 294–5, 306
Trim, 121
Trio minuet, 46
Trip-hop, 294–306
Tritone, 33
TRS, 115–16, 127
True mid range, 312
Turntables, 239
Two-pole, 65, 163, 192, 249
Typeface (web design), 430, 442

UK house, 271
U-Matic, 371
Unbalanced (cables), 115–22
Unison, 33, 55, 161
Unity gain, 121–2, 127
URL (Uniform Resource Locator), 398, 408, 414, 449

US house, 271
USB (Universal Serial Bus), 20–3

Vahl, Steve, 268
Valve, 97, 100–3, 182, 186
Van Helden, Armand, 92, 277, 382, 390
Variable MU, 101
VCA (Voltage Controlled Amplifier), 69, 193
VCF (Voltage Controlled Filter), 63
VCO (Voltage Controlled Oscillator), 59
Velocity, 44–5
Velocity cross-fading, 81
Velocity switching, 81
Vestax, 458–9, 461–2, 476
Videos, 361, 368
Vinyl, 330, 336, 370–1, 373–6
Virtual instruments, 26
Virus (Access), 135, 162
Vocal house, 271
Vocals, recording, 179–94
Vocoders, 192, 194
Volume, 56–72
VST, 25, 27
VU meters, 117, 358

Waldorf Attack, 203–4, 222
Walton, Byron, 270
Warehouse, The, 269, 285
Warmth, 101, 103
Waveforms, 54–5, 59–63, 72, 74–5
Wavelength, 181–2
Waves bundle, 211
Wavesurgeon, 83, 389, 391
WBMX, 269
WCLK (wordclock), 119–20
Weatherall, Andy, 198, 204, 207, 209
Web designer, 377
Webring, 378
Web site design, 400, 407
White label, 362, 374, 377
White noise, 61, 147–51, 162, 224, 258, 274, 297
WJLB-FM, 285
Word clock, *see* WCLK
Write (automation), 131

XLR, 115–16, 119, 121–2

Yamaha, 8, 73, 119, 166, 201

Zap FX, 169
Zero crossing, 53, 77
Zipping FX, 169
Zulu Nation, 237

Focal Press

www.focalpress.com

Join Focal Press online

As a member you will enjoy the following benefits:

- browse our full list of books available
- view sample chapters
- order securely online

Focal eNews

Register for eNews, the regular email service from Focal Press, to receive:

- advance news of our latest publications
- exclusive articles written by our authors
- related event information
- free sample chapters
- information about special offers

Go to www.focalpress.com to register and the eNews bulletin will soon be arriving on your desktop!

If you require any further information about the eNews or www.focalpress.com please contact:

USA
Tricia Geswell
Email: t.geswell@elsevier.com
Tel: +1 781 313 4739

Europe and rest of world
Lucy Lomas-Walker
Email: l.lomas@elsevier.com
Tel: +44 (0) 1865 314438

Catalogue

For information on all Focal Press titles, our full catalogue is available online at www.focalpress.com, alternatively you can contact us for a free printed version:

USA
Email: c.degon@elsevier.com
Tel: +1 781 313 4721

Europe and rest of world
Email: j.blackford@elsevier.com
Tel: +44 (0) 1865 314220

Potential authors

If you have an idea for a book, please get in touch:

USA
editors@focalpress.com

Europe and rest of world
ge.kennedy@elsevier.com

Rick Snoman's Remix Masterclass

- Two day intensive courses face to face with a remixing legend

- Live demonstrations watch listen & learn from a top professional working on real top ten matieral

- The secrets of obtaining & clearing remix material

- Creating killer drum loops & getting that elusive bass sound

Excellent, accessible and good value for money. Aimed at semi-professional musicians who want to take it further, the knowledge I gained will encourage a new way of working and a breath of fresh air. Rick Snoman - an author of interesting stuff!"

DJ Magazine

http://www.remixmasterclass.com

Rick Snoman @ work!